CHEMICAL SENSORS

SIMULATION AND MODELING

VOLUME 4: OPTICAL SENSORS

CHEMICAL SENSORS

SIMULATION AND MODELING

VOLUME 4: OPTICAL SENSORS

EDITED BY

GHENADII KOROTCENKOV

GWANGJU INSTITUTE OF SCIENCE AND TECHNOLOGY
GWANGJU, REPUBLIC OF KOREA

MOMENTUM PRESS

MOMENTUM PRESS, LLC, NEW YORK

Chemical Sensors: Simulation and Modeling Volume 4: Optical Sensors

First published by Momentum Press®, LLC
222 East 46th Street, New York, NY 10017
www.momentumpress.net

ISBN-13: 978-1-60650-318-8 (hard back, case bound)
ISBN-10: 1-60650-318-9 (hard back, case bound)
ISBN-13: 978-1-60650-320-1 (e-book)
ISBN-10: 1-60650-320-0 (e-book)
DOI: 10.5643/9781606503201

Cover design by Jonathan Pennell
Interior design by Derryfield Publishing, LLC

10 9 8 7 6 5 4 3 2 1

Printed in the United States of America

Contents

Preface

This series, *Chemical Sensors: Simulation and Modeling,* is the perfect complement to Momentum Press's six-volume reference series, *Chemical Sensors: Fundamentals of Sensing Materials* and *Chemical Sensors: Comprehensive Sensor Technologies,* which present detailed information about materials, technologies, fabrication, and applications of various devices for chemical sensing. Chemical sensors are integral to the automation of myriad industrial processes and everyday monitoring of such activities as public safety, engine performance, medical therapeutics, and many more.

Despite the large number of chemical sensors already on the market, selection and design of a suitable sensor for a new application is a difficult task for the design engineer. Careful selection of the sensing material, sensor platform, technology of synthesis or deposition of sensitive materials, appropriate coatings and membranes, and the sampling system is very important, because those decisions can determine the specificity, sensitivity, response time, and stability of the final device. Selective functionalization of the sensor is also critical to achieving the required operating parameters. Therefore, in designing a chemical sensor, developers have to answer the enormous questions related to properties of sensing materials and their functioning in various environments. This five-volume comprehensive reference work analyzes approaches used for computer simulation and modeling in various fields of chemical sensing and discusses various phenomena important for chemical sensing, such as surface diffusion, adsorption, surface reactions, sintering, conductivity, mass transport, interphase interactions, etc. In these volumes it is shown that theoretical modeling and simulation of the processes, being a basic for chemical sensor operation, can provide considerable assistance in choosing both optimal materials and optimal configurations of sensing elements for use in chemical sensors. The theoretical simulation and modeling of sensing material behavior during interactions with gases and liquid surroundings can promote understanding of the nature of effects responsible for high effectiveness of chemical sensors operation as well. Nevertheless, we have to understand that only very a few aspects of chemistry can be computed exactly.

However, just as not all spectra are perfectly resolved, often a qualitative or approximate computation can give useful insight into the chemistry of studied phenomena. For example, the modeling of surface-molecule interactions, which can lead to changes in the basic properties of sensing materials, can show how these steps are linked with the macroscopic parameters describing the sensor response. Using quantum mechanics calculations, it is possible to determine parameters of the energetic (electronic) levels of the surface, both inherent ones and those introduced by adsorbed species, adsorption complexes, the precursor state, etc. Statistical thermodynamics and kinetics can allow one to link those calculated surface parameters with surface coverage of adsorbed species corresponding to real experimental conditions (dependent on temperature, pressure, etc.). Finally, phenomenological modeling can tie together theoretically calculated characteristics with real sensor parameters. This modeling may include modeling of hot platforms, modern approaches to the study of sensing effects, modeling of processes responsible for chemical sensing, phenomenological modeling of operating characteristics of chemical sensors, etc.. In addition, it is necessary to recognize that in many cases researchers are in urgent need of theory, since many experimental observations, particularly in such fields as optical and electron spectroscopy, can hardly be interpreted correctly without applying detailed theoretical calculations.

Each modeling and simulation volume in the present series reviews modeling principles and approaches particular to specific groups of materials and devices applied for chemical sensing. *Volume 1: Microstructural Characterization and Modeling of Metal Oxides* covers microstructural characterization using scanning electron microscopy (SEM), transmission electron spectroscopy (TEM), Raman spectroscopy, in-situ high-temperature SEM, and multiscale atomistic simulation and modeling of metal oxides, including surface state, stability, and metal oxide interactions with gas molecules, water, and metals. *Volume 2: Conductometric-Type Sensors* covers phenomenological modeling and computational design of conductometric chemical sensors based on nanostructured materials such as metal oxides, carbon nanotubes, and graphenes. This volume includes an overview of the approaches used to quantitatively evaluate characteristics of sensitive structures in which electric charge transport depends on the interaction between the surfaces of the structures and chemical compounds in the surroundings. *Volume 3: Solid-State Devices* covers phenomenological and molecular modeling of processes which control sensing characteristics and parameters of various solid-state chemical sensors, including surface acoustic wave, metal-insulator-semiconductor (MIS), microcantilever, thermoelectric-based devices, and sensor arrays intended for "electronic nose" design. Modeling of nanomaterials and nanosystems that show promise for solid-state chemical sensor design is analyzed as well. *Volume 4: Optical Sensors* covers approaches used for modeling and simulation of various types of optical sensors such as fiber optic, surface plasmon resonance, Fabry-Pérot interferometers, transmittance in the mid-infrared region, luminescence-based devices, etc. Approaches used for design and optimization of optical systems aimed for both remote gas sensing and gas analysis chambers for the nondispersive infrared (NDIR) spectral range are discussed as well. A description of multiscale atomistic simulation of hierarchical nanostructured materials for optical chemical sensing is also included in this volume. *Volume 5: Electrochemical Sensors* covers modeling and simulation of electrochemical processes in both solid and liquid electrolytes, including charge separation and transport (gas diffusion, ion diffusion) in membranes, proton–electron transfers, electrode reactions, etc. Various models used to describe electrochemical sensors such as potentiometric, amperometric, conductometric, impedimetric, and ion-sensitive FET sensors are discussed as well.

I believe that this series will be of interest of all who work or plan to work in the field of chemical sensor design. The chapters in this series have been prepared by well-known persons with high qualification in their fields and therefore should be a significant and insightful source of valuable information for engineers and researchers who are either entering these fields for the first time, or who are already conducting research in these areas but wish to extend their knowledge in the field of chemical sensors and computational chemistry. This series will also be interesting for university students, post-docs, and professors in material science, analytical chemistry, computational chemistry, physics of semiconductor devices, chemical engineering, etc. I believe that all of them will find useful information in these volumes.

G. Korotcenkov

About the Editor

Ghenadii Korotcenkov received his Ph.D. in Physics and Technology of Semiconductor Materials and Devices in 1976, and his Habilitate Degree (Dr. Sci.) in Physics and Mathematics of Semiconductors and Dielectrics in 1990. For a long time he was a leader of the scientific Gas Sensor Group and manager of various national and international scientific and engineering projects carried out in the Laboratory of Micro- and Optoelectronics, Technical University of Moldova. Currently, Dr. Korotcenkov is a research professor at the Gwangju Institute of Science and Technology, Republic of Korea.

Specialists from the former Soviet Union know Dr. Korotcenkov's research results in the field of study of Schottky barriers, MOS structures, native oxides, and photoreceivers based on Group III–V compounds very well. His current research interests include materials science and surface science, focused on nanostructured metal oxides and solid-state gas sensor design. Dr. Korotcenkov is the author or editor of 11 books and special issues, 11 invited review papers, 17 book chapters, and more than 190 peer-reviewed articles. He holds 18 patents, and he has presented more than 200 reports at national and international conferences.

Dr. Korotcenkov's research activities have been honored by an Award of the Supreme Council of Science and Advanced Technology of the Republic of Moldova (2004), The Prize of the Presidents of the Ukrainian, Belarus, and Moldovan Academies of Sciences (2003), Senior Research Excellence Awards from the Technical University of Moldova (2001, 2003, 2005), a fellowship from the International Research Exchange Board (1998), and the National Youth Prize of the Republic of Moldova (1980), among others.

Contributors

Alexander Bagaturyants (Chapter 1)
Photochemistry Center
Russian Academy of Sciences
Moscow 119421, Russia

Michael Alfimov (Chapter 1)
Photochemistry Center
Russian Academy of Sciences
Moscow 119421, Russia

Sergey Belousov (Chapter 2)
NRC Kurchatov Institute
Moscow 123182, Russia
and
Kintech Lab
Moscow 123182, Russia

Ilya Polishchuk (Chapter 2)
NRC Kurchatov Institute
Moscow 123182, Russia

Boris Potapkin (Chapter 2)
NRC Kurchatov Institute
Moscow 123182, Russia
and
Kintech Lab
Moscow 123182, Russia

Stefano Lettieri (Chapter 3)
Institute for Superconductors, Oxides and Innovative Materials and Devices
National Research Council (SPIN-CNR) U.O.S. Napoli
Complesso Universitario di Monte S. Angelo
Napoli I-80126, Italy

Christophe Caucheteur (Chapter 4)
Faculty of Engineering
University of Mons
Mons, Belgium

Marc Debliquy (Chapter 4)
Faculty of Engineering
University of Mons
Mons, Belgium

Gautier Ravet (Chapter 4)
Faculty of Engineering
University of Mons
Mons, Belgium

Driss Lahem (Chapter 4)
Materia Nova
Mons, Belgium

Patrice Megret (Chapter 4)
Faculty of Engineering
University of Mons
Mons, Belgium

Banshi D. Gupta (Chapter 5)
Physics Department
Indian Institute of Technology Delhi
New Delhi 110016, India

Rajan Jha (Chapter 5)
School of Basic Sciences
Indian Institute of Technology Bhubaneswar
Bhubaneswar 751007, Odisha, India

Guy Louarn (Chapter 6)
Institut des Matériaux Jean Rouxel, UMR 6502
CNRS-Université de Nantes
Nantes 44322, France

Malak Kanso (Chapter 6)
Institut des Matériaux Jean Rouxel, UMR 6502
CNRS-Université de Nantes
Nantes 44322, France

Tahereh Makiabadi (Chapter 6)
Institut des Matériaux Jean Rouxel, UMR 6502
CNRS-Université de Nantes
Nantes 44322, France

Zhengtian Gu (Chapter 7)
Laboratory of Photo-electric Functional Films
College of Science, University of Shanghai for Science and Technology
Shanghai 200093, People's Republic of China

Everardo Vargas-Rodriguez (Chapter 8)
Departamento de Estudios Multidisciplinarios
División de Ingenierías, Campus Irapuato-Salamanca
Universidad de Guanajuato
Yuriria, Gto., México

Harvey N. Rutt (Chapter 8)
Optoelectronics Research Centre
University of Southampton, Highfield Campus
Southampton, United Kingdom

David Claudio-Gonzalez (Chapter 8)
Departamento de Estudios Multidisciplinarios
División de Ingenierías, Campus Irapuato-Salamanca
Universidad de Guanajuato
Yuriria, Gto., México

Roberto Rojas-Laguna (Chapter 8)
Departamento de Electrónica
División de Ingenierías, Campus Irapuato-Salamanca
Universidad de Guanajuato, Palo Blanco
Salamanca, Gto., México

Jose Amparo Andrade-Lucio (Chapter 8)
Departamento de Electrónica
División de Ingenierías, Campus Irapuato-Salamanca
Universidad de Guanajuato, Palo Blanco
Salamanca, Gto., México

Andreas Wilk (Chapters 9 and 10)
Institute of Analytical and Bioanalytical Chemistry
University of Ulm
Ulm 89081, Germany

Boris Mizaikoff (Chapters 9 and 10)
Institute of Analytical and Bioanalytical Chemistry
University of Ulm
Ulm 89081, Germany

Xiaofeng Wang (Chapter 10)
Institute of Analytical and Bioanalytical Chemistry
University of Ulm
Ulm 89081, Germany

Seong-Soo Kim (Chapter 10)
Institute of Analytical and Bioanalytical Chemistry
University of Ulm
Ulm 89081, Germany

Ingo Sieber (Chapter 11)
Karlsruhe Institute of Technology
Institute for Applied Computer Science
Eggenstein-Leopoldshafen 76344, Germany

Ulrich Gengenbach (Chapter 11)
Karlsruhe Institute of Technology
Institute for Applied Computer Science
Eggenstein-Leopoldshafen 76344, Germany

Tatsuo Shiina (Chapter 12)
Graduate School of Advanced Integration Science
Chiba University
Chiba 263-8522, Japan

CHAPTER 1

Atomistic Simulation of Hierarchical Nanostructured Materials for Optical Chemical Sensing

A. Bagaturyants
M. Alfimov

1. INTRODUCTION

Optical chemical sensors are designed to detect various chemical compounds (analytes) by a change in the optical properties of a sensing element (sensor) as a result of its interaction with the analyte. Optical chemical sensors are widely used in various environmental, biomedical, and industrial applications. There is extensive comprehensive literature devoted to optical chemical sensors, their design, principles, implementation, and applications, including books, book chapters, and review articles (see, for example, Baldini 2006; McDonagh et al. 2008).

The first works in which the term "optical chemical sensor" (OCS) was used date back to the mid-1980s (Seitz 1984; Matsubara et al. 1988). Somewhat later, works appeared in which the use of dyes and nanofabrication for optical chemical sensing was published (Blyler 1989; Wolfbeis 1991; Crowther 1995). The use of an array of OCS elements based on fluorescent particles made by immobilizing a fluorescence dye on the surface of a micro- or nanoparticle (e.g., polymer, ceramic, metal) was also being discussed in the literature starting from the

DOI: 10.5643/9781606503201/ch1

end of the 1990s (Dickinson et al. 1996; Choi and Hawkins 1997) and until now (Paolesse et al. 2011). In the last three decades, the number of studies in the field of optical chemical sensors has been growing rather quickly. The availability of low-cost, miniature optoelectronic light sources and detectors and the need for multianalyte array-based sensors (particularly in the area of biosensing) are the main factors determining current interest in this development (Basabe-Desmonts et al. 2007; McDonagh et al. 2008; Borisov and Wolfbeis 2008).

A wide range of sensor materials based on a dye immobilized on a nanoparticle can be created by combining different dyes, particles, and methods of immobilization. The indicator dye and the particle surface can be chemically modified to tailor the functional characteristics of the material for a specific analyte or a group of analytes, while immobilization of an indicator dye on the particle surface significantly reduces the response and relaxation time of materials (Khlebunov et al. 2008). A response of a sensor material to an analyte is the result of complex interactions in a system analyte/indicator/polymer (Plotnikov et al. 2007). A great number of experimental parameters must be selected in order to optimize the sensing properties of such materials (Khlebunov et al. 2008; Sazhnikov and Alfimov 2008).

The diversity of volatile chemical compounds liberated in natural and human-caused processes poses the problem of the design and development of materials and devices for the detection and monitoring of chemical compounds that are comparable with olfactory systems of living organisms. It is necessary to provide a scientific basis for the design and manufacture of monitoring systems for any desired set of volatile chemicals. One of the approaches to the solution of the posed problem can be based on encoding a chemical compound with an optical signal by using materials that change their optical properties upon interaction with chemical compounds. Complex mixtures of volatile chemical compounds can be monitored only with the use of matrix systems composed of a set of different sensing elements.

Materials with a broad spectrum of various prescribed characteristics of optical chemical sensing can be produced using a unified approach to the hierarchical design of nanostructured materials. The design and manufacture of these materials involves several sequential steps. First, a special set of supramolecular receptor centers (RCs) is constructed for a certain set of analytes. These RCs incorporate absorbing or emitting indicator molecules and can interact efficiently with analytes (for example, through host–guest interactions). The interaction of RCs with gas-phase analyte molecules changes their spectral absorption or emission characteristics, which can be used for analyte detection. These receptor centers are either created directly or immobilized on an organic or inorganic nanoparticle. The material is produced by the assembly (self-assembly) of such modified nanoparticles into microstructures (nanoparticle assemblies), which are elements of a matrix chemical sensor (chemochip) (Alfimov et al. 2010).

The predictive modeling of the properties of nanostructured materials is very important for designing sensitive and selective optical chemical sensors. Computer simulation of a device to be constructed can help in finding the necessary parameters, optimizing the structure, and maximizing the sensitivity of the sensing material. In particular, a multiscale approach is well applicable to the simulation of optical chemical sensors. In this approach, the calculations at each level are performed using methods and approximations peculiar to this level, and the results of simulating the structure and properties of the material obtained at a lower level are transferred as input data to the next upper one (Alfimov et al. 2010).

The use of a multiscale approach for the predictive atomistic simulation of materials and, especially, functional materials, began in the late 1990s (see, for example, Phillips 1998 and Nieminen 2002), and these studies have given rise to numerous publications since (Elliott 2011). Actually, some works in which a multiscale atomistic approach was used appeared much earlier. Thus, in our early works, classical mechanics was combined with semiempirical quantum chemistry to calculate the absorption line shapes of some organic chromophores, their dimers, and aggregates (Burshtein et al. 1994, 1995, 1997). An integrated multiscale approach to atomistic simulation of film deposition processes in which a kinetic Monte Carlo method was combined with ab-initio quantum chemical (QC) calculations was described in a series of works (Bagatur'yants et al. 2003, 2004, 2007).

In this chapter we will consider a multiscale atomistic approach to the predictive simulation of hierarchical nanostructured functional materials for optical chemical sensors.

2. HIERARCHICAL NANOMATERIALS: CONSTRUCTION AND ORGANIZATION PRINCIPLES; MATERIALS CONSTRUCTION BY THE BOTTOM-UP PRINCIPLE

2.1. HIERARCHICAL NANOMATERIALS FOR NANOPHOTONICS AND THEIR SENSING POTENTIALITIES

The creation of a hierarchical nanostructured material can improve the sensitivity and selectivity of an optical chemical sensor. Such a material has a very large surface, which provides a very efficient interaction of the analyte with receptor centers, easy penetration of the analyte into the material, and the possibility for the targeted design of all levels of the material architecture. The development and optimization of such complicated devices as optical chemical sensors involves a number of physicochemical problems. In this case, the computer simulation of a device to be constructed can be very important for finding necessary parameters, optimizing the structure, and maximizing the sensitivity. In particular, a multiscale approach is well applicable to the simulation of optical chemical sensors.

In this approach, the calculations at each level are performed using methods and approximations peculiar to this level, and the results of simulating the structure and properties of the material obtained at a lower level are transferred as input data to the next upper one.

A sensing layer of an optical chemical sensor is an ordered system of nanosized particles (from 200 nm to 2 μm in diameter) containing supramolecular receptor centers (dye molecules). Ordered assemblies are formed from thin films or microdrops of a solution by means of a nanoparticle self-assembly process. Self-assembly technologies are based on the self-organization of nanoparticles into an assembly of nanostructures in a dissipative system. The size of the nanoparticles constituting the sensing layer is in the region of visible or near-infrared wavelengths. A multilayer close-packed face-centered cubic structure can form through the self-organization of such particles. This structure has a lattice constant of the order of the size of its constituent nanoparticles. If the particles have a sufficiently regular shape and equal size, this system will operate as a photonic crystal for electromagnetic radiation with a wavelength of the order of the lattice constant. In this case, the range of visible-light wavelengths coincides with the characteristic particle size and represents the working range of optical chemical sensors. Therefore, the effects resulting from the existence of a photonic crystal should be taken into account in designing the geometry of an optical chemical sensor and selecting its parameters (Alfimov et al. 2010).

The hierarchical structure of materials and devices allows one to use nanotechnologies for their production that are based on the "bottom-up" self-assembly of molecules and nanostructures. The material is based on a functional organic molecule, which enters into a supramolecular complex embedded in a nanoparticle. The nanomaterial is formed from such nanoparticles by a self-assembly process. The hierarchical design principle is rather general and can be used as a basis for the design of nanosystems of different types, including inorganic nanosystems. The predictive multiscale simulation of the structure and properties of hierarchical organic nanomaterials is based on the consistent use of atomistic and continual (phenomenological) methods. Such an approach can be used for the development of chemical sensors, microchips, organic light-emitting diodes, quantum dots, etc. (Alfimov and Bagaturyants 2009).

Figure 1.1 presents schematically the structure of a hierarchical nanostructured material for optical chemical sensing. An organic "indicator molecule" (IM) (the Nile Red dye molecule in the figure) is in the basis of the construction of a sensing material and is responsible for the generation of an optical signal upon its interaction with an analyte. The IM imparts the main functional properties to the sensing material. The IM along with its local environment forms a "receptor center" (RC). The local environment can be selected so that the signal can be enhanced and its selectivity can be improved. RCs are arranged on the surface or in the particle bulk (nanoparticles), and the nanoparticles themselves are assembled

Figure 1.1. Illustration of a hierarchically designed nanostructured material for optical chemical sensing: (a) an array of sensing elements; (b) an organized assembly of nanoparticles bearing receptor centers; (c) a nanoparticle with an organic dye molecule adsorbed on its surface and forming a receptor center; (d) an indicator dye (Nile Red) molecule, which imparts functional properties to the sensing material.

in an organized structure (which may, for example, exhibit properties of a photonic crystal) (Alfimov et al. 2009).

Fluorescent optical sensors, in which the output signal is a change in the fluorescence properties of sensing materials (De Silva et al. 1997), are among the most promising ones because of the high sensitivity of the fluorescence signal to various agents and, as a consequence, the possibility of detecting extremely low concentrations of analytes. While designing sensors, one should know the most important properties of the material, such as the energy of analyte interaction with the receptor center and changes in the spectral bands due to this interaction. Such properties of materials can be predicted using computer simulation.

2.2. SPACE-TIME SCALE HIERARCHY AND THE STRUCTURE OF NANOMATERIALS FOR NANOPHOTONICS

An optical chemical sensor using molecular recognition of analyte molecules is based on the effect of a change in the optical properties of an IM upon its interaction with an analyte. An IM is usually a dye molecule (or another substance possessing specific spectral properties). Hence, the material of an optical chemical sensor contains IMs, while a signal from the sensor is a change of its optical properties (positions and intensities of its spectral bands) upon its interaction with an analyte. This change in the optical properties of a molecule or a more complex (supramolecular, nanosized, etc.) system will also be called its optical response. Then, the optical response of a sensor upon its interaction with a gas containing a certain concentration of a given analyte will serve as its analytical signal.

The use of a hierarchically designed, nanostructured material enhances the sensitivity and selectivity of an optical chemical sensor. Such a material exhibits a very large effective surface area, which provides the most efficient interaction of an analyte with receptor centers, the ease of analyte penetration into the material, and also the possibility of designing the material architecture at all levels of scales.

2.3. STRUCTURE OF NANOMATERIALS FOR OPTICAL CHEMICAL SENSORS: FROM A MOLECULE TO A SUPRAMOLECULAR CENTER, NANOPARTICLE, AND NANOMATERIAL

At the lowest, molecular level, an indicator molecule is the main sensing element of such a material. The selection of the indicator molecule is determined by its ability to interact (selectively) with the analyte molecule and its ability to change (selectively) its optical properties (primarily, fluorescence intensity) upon this interaction. It is suggested to make this choice based on the available data on the chemical nature and physical optical properties of the known dyes or other light-absorbing and/or light-emitting compounds.

At the next, supramolecular level, the creation ("design") of a supramolecular receptor center (SRC) is suggested. An SRC includes an IM and its nearest environment. This environment (the architecture of the SRC) is selected (designed) in such a way that the interaction between the analyte molecule and the SRC is optimized, the selectivity of this interaction is maximized, and the optical response of the designed SRC is enhanced as much as possible.

At the higher, nanosized level, a nanoparticle containing SRCs at its surface or in its bulk is created. It is believed that the structures of both the nanoparticle and its surface are known. The construction of a sensing material from nanoparticles provides an increased accessible surface for analytes, easy penetration of their molecules to receptor centers, and a possibility of creating a special structure (architecture) of the material with the best conditions for light scattering (or light absorption), for example, by creating a photonic lattice.

At the next hierarchical, micro level, a (2-D or 3-D) assembly of uniform nanoparticles ordered in a certain way is formed. This assembly contains IMs of only one certain type. This type is characterized by a certain value of the response to a certain analyte molecule. It is this assembly that is the basis for the designed material for optical chemical sensors. An array of assemblies can be constructed from a set of such assemblies. Such an array represents the main sensing element of an optical chemical sensor considered as a device (macro level). Each element (A_i) of such an array might be characterized by its own value (magnitude) of the response (S_i) to interaction with a certain analyte molecule. An analysis of the whole set of responses $\{S_i\}$ from each element of the array upon its interaction with a given multicomponent gas mixture provides information on the mixture composition and on the concentrations of given analytes in the mixture.

The general task of multiscale simulation is the prediction and design of the material structure and response at each scale level. The results of simulation at a certain level are transferred to the next level, and the scheme of multiscale (multilevel) simulation is accomplished in this way. This scheme corresponds to a bottom-up design of the material. Actually, this scheme includes some feedback connections between different levels, when the results of a higher level are used to refine the results of a lower level of modeling (top-down approach). This combined approach (bottom-up/top-down) is widely used for simulations and theoretical predictions of nanomaterials structure and properties for a broad area of applications (optical chemical sensing, light-emitting and photovoltaic materials, materials for photonic crystals, optical memory media, etc.).

The main feature of this multiscale approach is that modeling starts from the lowest molecular level with the use of atomistic ab-initio (first-principles) methods, based on fundamental laws of atomic and molecular interactions. At each next level, the results of the previous level are used as input parameters, while modeling is based on physical models and methods designed specially for the corresponding time and space scales.

In accordance with the above, a multiscale approach to simulations of optical nanostructured materials must include four main levels: a molecular level, a supramolecular level, a level of nanoparticles, and a level of nanoparticle assemblies (see Figure 1.2).

The signal generated by an optical chemical sensor for the detection of an analyte (or a set of analytes) represents a change in the optical characteristics (as a rule, luminescence spectra) of the sensor material upon its interaction with the analyte. This change must be detected by a special detector. The generation

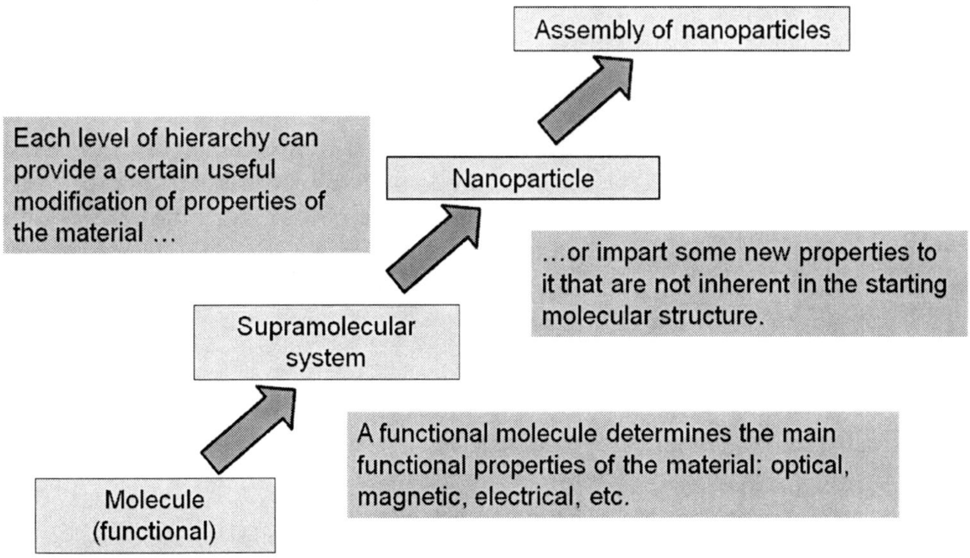

Figure 1.2. Hierarchical levels of a functional material for photonic applications.

characteristics of the signal and, hence, its intensity and selectivity, are determined by physical parameters and processes at each hierarchical level. The corresponding physical characteristics should be modeled at each hierarchical level of the material.

3. HIERARCHY OF ATOMISTIC SIMULATION METHODS CORRESPONDING TO SCALE HIERARCHY

In recent years, multiscale simulation methods have been widely used for the predictions of properties of various nanosized and nanostructured systems, materials, and related processes as reported in numerous reviews and books (see, for example, Guo 2007; Ross 2008; Elliott 2011).

Simulation methods at a molecular or atomistic level are well designed and are implemented in a variety of commercial and noncommercial programs and program packages. Quantum mechanical first-principles (ab-initio) methods and programs differ substantially, depending on whether they are designed for calculations of finite or infinite periodic systems. Conventionally, these methods might be classified, respectively, as molecular (cluster) or solid-state (periodic). It should be noted, though, that many "molecular" programs include the capability of treating periodic systems, while "solid-state" programs can also treat molecular systems.

Semiempirical QC methods can be used for molecular systems of larger sizes. The main deficiency of such semiempirical methods is that their results may depend strongly on parameterization. The possibility of using semiempirical methods is included in most ab-initio packages.

In the cluster approach, the surface of a nanoparticle is modeled by a finite atomic cluster, which is selected so that all the atoms that compose the RC (that is, the IM and its nearest environment) are included. In most typical cases, the number of atoms in the RC does not exceed the limit acceptable for quantum calculations. In the periodic approach, the surface of a nanoparticle is modeled by a slab that is infinite and periodic in two dimensions and with a certain finite thickness in the third direction (the so-called repeated-slab approximation; Schluter 1975). To ensure full 3-D periodicity, the slab is periodically repeated in the third direction in such a way that two neighboring slabs are separated by a vacuum gap of a certain length (for more detail, see, for example, Bechstedt 2003).

Simulation methods that can be used at the next dimensional (supramolecular) level are mostly based on the use of classical interaction potentials. These methods include molecular mechanics (MM), molecular dynamics (MD), and Monte Carlo (MC) techniques with various sets of classical atomistic potentials or force fields, which are implemented in various program packages.

At present, a number of integrated program packages also exist, oriented to the calculation of materials properties based on atomistic approaches. These packages commonly include a variety of applied programs that can be used for modeling crystalline and molecular materials, biological objects, solutions, etc., combined by a well-developed, user-friendly graphical interface. Information on the majority of available specialized and general-purpose programs, both commercial and noncommercial, can be found on the Internet.*

The efficient use of atomistic simulation can reduce substantially the time needed for the design of nanomaterials with desired properties and also reduce their cost. As a rule, various requirements are imposed on the structure and properties of such materials, and atomistic simulation is necessary to obtain the best implementation of these requirements at the microscopic (molecular, supramolecular, and nanosized) levels. For example, in many cases, supramolecular complexes (SMCs), that is, intermolecular complexes formed by two or more molecules or molecular fragments of the nanoparticle, are the main structure elements governing the most practically important properties of nanomaterials. Determining the possible configurations of SMCs for certain molecules and estimating the stability and required properties of the corresponding structures are among the most important tasks of atomistic simulation in the guided design of functional nanomaterials.

One of the main problems arising in determining the possible configurations of SMCs is connected with the necessity of an exhaustive search in the configurational space of the SMC. If the binding sites of the molecules are well defined or their structure is relatively simple, this search can be rather quickly performed manually using molecular visualization and editing tools. Generally, however, the number of possible configurations of an SMC increases drastically (for example, in the case when a dye molecule is bound on the silica or polystyrene surface), and special techniques are necessary to search and select the most stable supramolecular configurations. To do this, various versions of global optimization algorithms are used (Lavor 2003), in which the global potential-energy minimum is searched for an SMC calculated using classical force fields (Halgren 1996). The use of a genetic algorithm for this purposes combined with the use of a classical force field for the calculation of intermolecular interactions between the components of SMCs was also proposed and implemented in a computational program (Grigor'ev et al. 2010).

Because the configurational space volume increases exponentially with the number of degrees of freedom, the variables that are most important for the given problem should be, as a rule, defined for global optimization. For example, for rigid molecules, these variables can be the position vector of the center of inertia of one molecule relative to the other one and three Euler angles characterizing their mutual orientation. However, if the molecules should adopt a certain

*http://en.wikipedia.org/wiki/Molecular_modelling; http://en.wikipedia.org/wiki/List_of_software_for_nanostructures_modeling

conformation to form an SMC, their internal degrees of freedom should also be taken into account. In some cases, the use of force fields to calculate the potential energy during global optimization can result in inaccuracies in both the geometry and relative energy of different configurations of an SMC. In particular, the geometry of conjugated aromatic rings typical in dyes is distorted when calculated with a force field.

To refine the geometry of configurations, one should use QC methods. These methods make it possible to determine the geometry of an SMC and its important properties (absorption spectrum, luminescence spectrum, dipole moment, polarizability, etc.) from the first principles (by solving the Schrödinger equation). Because QC calculations require considerable computational resources, they can be used for the local geometry optimization of an SMC; the structures obtained by global optimization should be used as starting geometries. For SMCs consisting of several hundreds of atoms, various versions of density functional theory can be used for local optimization (Perdew et al. 1996; Adamo and Barone 2002; Becke 1988; Lee et al. 1988; Perdew and Wang 1992). The interaction energy of the most important fragments of an SMC can be calculated using more accurate electron correlation methods, such as perturbation theory (Møller and Plesset 1934; for a modern review, see Cramer 2011) and the coupled cluster method (Paldus and Cizek 1975; Scheiner et al. 1987; Bartlett 1989). After determining the structures of the lowest-energy configurations of an SMC, one should estimate their stability with regard to temperature. The free energy of formation, ΔG_b, which can be calculated using Monte Carlo or molecular dynamics simulation, is a measure of the relative stability of intermolecular complexes. These methods make it possible to average the geometry (and some other properties) of an SMC over microstates with regard to temperature and to calculate ΔG_b from a molecular dynamics or Monte Carlo trajectory using thermodynamic integration (Ytreberg et al. 2006), or harmonic analysis (Brooks et al. 2004). Therefore, in modeling supramolecular structures, it is necessary to use a hierarchical approach, in which atomistic simulation techniques of different levels are employed to solve certain problems arising when different scales of the system are considered.

4. ATOMISTIC MULTISCALE SIMULATION OF HIERARCHICAL NANOMATERIALS FOR OPTICAL CHEMICAL SENSORS: STEP BY STEP

4.1. SUPRAMOLECULAR LEVEL: CALCULATIONS OF MOLECULAR INTERACTIONS BETWEEN GAS-PHASE ANALYTE MOLECULES AND SIMPLE SUBSTRATE MODELS

According to the general strategy for the hierarchical design of nanostructured materials for an optical chemical sensor, an indicator dye molecule is adsorbed on the surface of a nanoparticle (which is used as a substrate). The nanoparticles are assembled into a sensing layer (sensor material), which can be used to construct a sensor array (for more details, see Plotnikov 2007; Sazhnikov and Alfimov 2008; Khlebunov et al. 2008, 2009; Alfimov 2010; Grigoriev 2010). An indicator dye molecule along with its local environment forms a functional supramolecular center, whose properties depend on both the indicator dye itself and its environment. The interaction of an analyte molecule with the supramolecular center generates an optical response of the material. Therefore, a rational approach to the design of optical chemical sensor materials should be based on a consideration of intermolecular interactions among analytes, indicators, and matrices (Sazhnikov and Alfimov 2008).

In a general case, all types of interactions between these components must be taken into account. It may be believed that the possibility of detecting an analyte by an indicator depends on the mode and strength of interaction between the analyte molecule and the indicator. For the successful detection of low concentrations of an analyte, this interaction should be stronger than the interaction of the analyte with the substrate. As a first step of such a large-scale investigation, the interaction of some small molecules (possible analytes) with polystyrene (PS) and amorphous silica substrates and with an indicator dye molecule of the acridine series was considered using simplified molecular models of the substrate and the indicator dye (Safonov et al. 2011). Only the main active surface center (for the substrates) or the fragment (for the dye molecule) responsible for the interaction with analyte molecules was included in these models.

The main goal of this work was to determine the mode of interaction and estimate the interaction energy between the adsorbed molecule, on the one hand, and the substrate and the indicator dye on the other hand. Similar simplified models were used rather long ago in modeling interactions at polyene/graphite and polymer/polymer interfaces (Calderone et al. 1996, 1998) and more recently in modeling interactions of nitrile, aromatic, and olefinic polymers with silica (Perez et al. 2009).

The ethylbenzene molecule (Figure 1.3a) was chosen as the simplest model of PS, which includes its main constituent moieties (phenyl ring and alkyl chain) and represents an elementary link of its polymer chain. The silanol molecule SiH_3OH (Figure 1.3b) was selected as the simplest model of a silanol group on the silica surface. The acridine molecule (Figure 1.3c) was taken as the main fragment of an acridine dye responsible for interaction with analyte molecules.

Formaldehyde, acetaldehyde, ammonia, methylamine, methanol, ethanol, acetone, benzene, acetonitrile, ethyl acetate, chloroform, and tetrahydrofuran were considered as the analyte molecules. QC calculations were made using the DFT-D approach implemented in the ORCA program (Neese et al. 2010). In this approach an empirical correction for dispersion interaction is included (Grimme 2004, 2006, 2011; Grimme et al. 2007). See the original paper (Safonov et al. 2011) for more computational details.

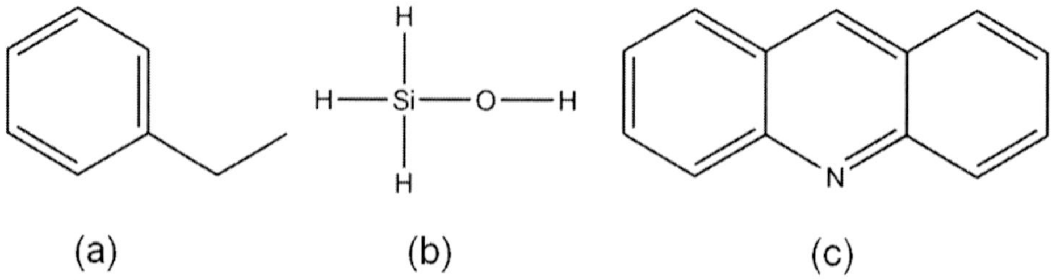

Figure 1.3. Molecules used as simplified substrate models. (Reprinted with permission from Safonov et al. 2011a. Copyright 2011 Springer-Verlag.)

The calculated lengths of shortest contacts in these complexes were used to analyze the corresponding coordination modes, while the interaction energies were used to compare the possibilities of detecting these small molecules, considered as analytes, using an acridine dye adsorbed on the surface of a PS or amorphous silica substrate. It was shown that most complexes of ethylbenzene exhibit a stacking structure in which the analyte molecule is located over the plane of the phenyl ring, with the only exception being the ammonia complex. In the latter, the lateral coordination mode is preferable. All complexes of silanol and most complexes of acridine have a lateral structure formed through an Si–O–H···X hydrogen bond (X is an analyte heteroatom) or an H(analyte) ···N(acridine) hydrogen bond.

The available experimental data on the heats of adsorption on PS and amorphous silica for some of the analytes considered indicate that the adsorbed molecules can interact simultaneously with more than one surface group. The calculated interaction energies indicate that the dyes of the acridine series adsorbed on a PS or silica substrate are not promising indicator molecules for acetone and acetonitrile. For all other analyte molecules, PS can be considered a suitable substrate for an acridine dye indicator. Finally, silica might be considered a suitable substrate for the detection of methanol, ethanol, benzene, ethyl acetate, and chloroform using an acridine dye as a molecular sensor. The properties of the nanoparticles used as chemosensing materials can be improved by increasing the coverage of nanoparticles with indicator dye molecules.

4.2. SUPRAMOLECULAR LEVEL: DFT CALCULATIONS OF THE 9-DIPHENYLAMINOACRIDINE (9-DPAA) FLUORESCENT INDICATOR AND ITS INTERACTIONS WITH ANALYTE MOLECULES

In a more complicated model, QC calculations were performed for a real functional dye molecule to study its interaction with analyte molecules and the effect of this interaction on the absorption and emission spectra of the dye (Safonov et al. 2011,

I and II). A dye of the 9-aminoacridine series, 9-diphenylaminoacridine (9-DPAA) (Sazhnikov et al. 1986), was selected for this study.

This dye exhibits pronounced solvatochromic properties (Sazhnikov et al. 2007) and, therefore, is highly promising to be used in molecular fluorescence sensors for the detection of polar organic molecules (Sazhnikov and Alfimov 2008). A characteristic property of the dye is that its first intense absorption and fluorescence bands lie in the visible spectral region, which simplifies the detection, because a wide variety of detectors, both natural (human eye) and artificial (photosensitive elements), are known for this region. Based on the experimental data, it was assumed that the observed changes in the spectra are due to specific intermolecular interactions in complexes forming between 9-DPAA molecules and one or several analyte molecules. Theoretical calculations were performed for 1:1 complexes between molecules of typical volatile analytes and 9-DPAA, and the resulting changes in the absorption and fluorescence spectra of the dye were analyzed. Polar molecules (methanol, acetonitrile, acetone, tetrahydrofuran, ammonia, formaldehyde, and acetaldehyde) were taken as analytes, while benzene was considered as a simple example of a nonpolar analyte.

The selection of an adequate method for the calculation of electronic transition energies and for geometry optimization is one of the most serious problems in spectral calculations. For rather large molecular systems like the systems under consideration, the most widely accepted choice is time-dependent density functional theory (TDDFT) with different functionals. Neglecting electron correlation effects in the CIS approach leads to an incorrect order of lower excited states (Safonov et al. 2011, I). A reasonable characterization of excited states was attained with the use of TDDFT with hybrid functionals, in particular PBE0, which was recommended for calculations of the electronic spectra of dyes (Jacquemin 2008), and with the 6-31G(d,p) basis set. For systems in the ground electronic states, geometry optimization was performed by the DFT method with the same functional and the same basis set. All calculations were performed by the Firefly program (Granovsky 2009), while GAMESS-US (Schmidt et al. 1993) was used for geometry optimizations in the excited state by the TDDFT method.

The calculated structure of the 9-DPAA molecule in the ground electronic state is shown in Figure 1.4a. The rotation angle of the nearly planar amino group with respect to the acridine fragment is 66°. The phenyl rings of the diphenylamino group are rotated through 33° to the plane of the N atom.

All the analyte molecules under consideration (methanol, acetonitrile, acetone, tetrahydrofuran, benzene, ammonia, formaldehyde, and acetaldehyde) can form 9-DPAA complexes of two types: complexes with a lateral structure and complexes with a stacking structure. The lateral and stacking structures are shown in Figure 1.5 for the 9-DPAA–methanol complex, taken as an example. In the lateral complexes, the analyte molecule is located approximately in the plane of the acridine fragment of 9-DPAA on the side of the N atom and interacts with it through

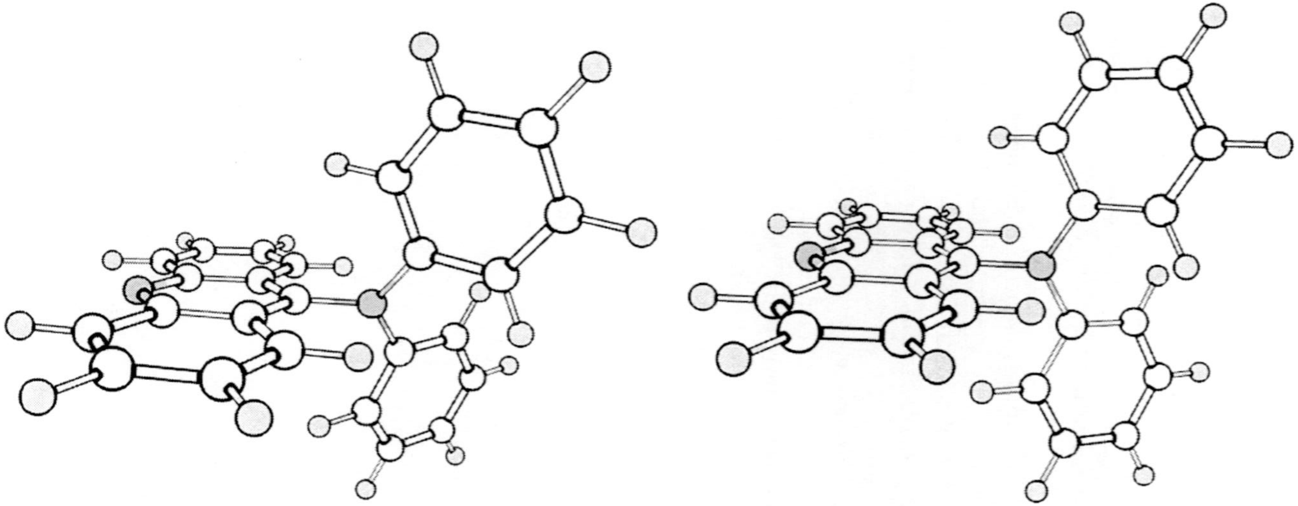

(a) 9-DPAA structure in the ground state

(b) 9-DPAA structure in the S_1 state

Figure 1.4. 9-DPAA structure in the (a) ground and (b) S_1 excited states; atoms C are shown as light circles, and N atoms as dark ones. (Reprinted with permission from Safonov et al. 2011b and 2011c. Copyright 2011 Springer-Verlag.)

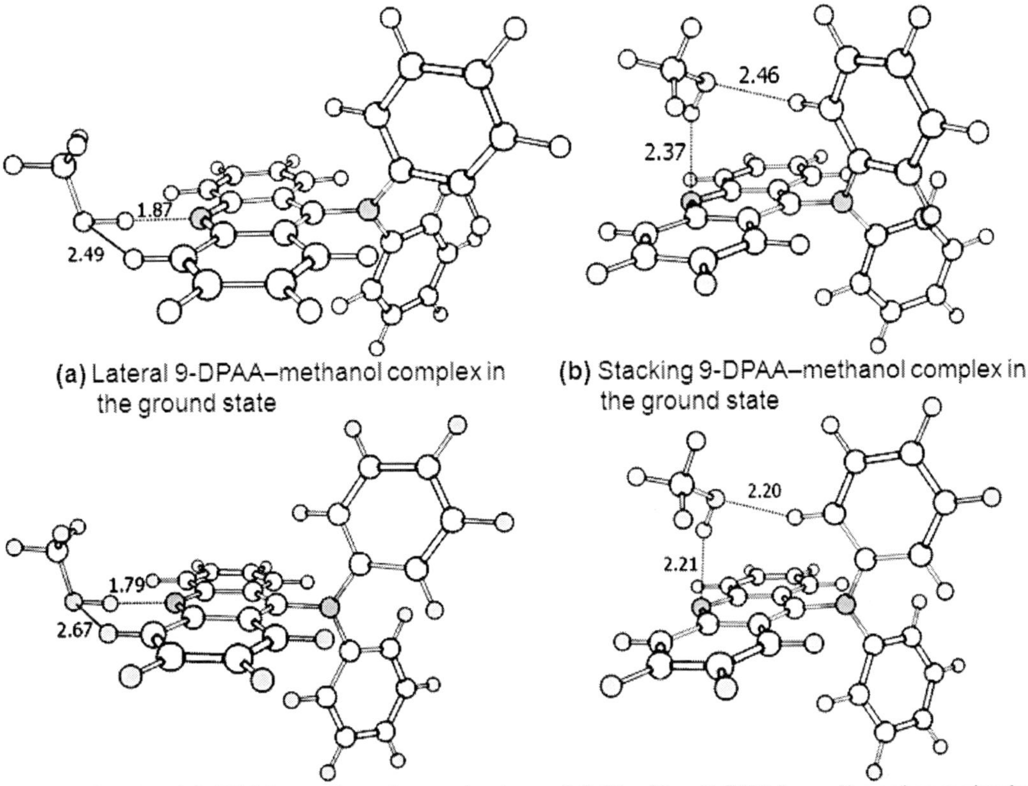

(a) Lateral 9-DPAA–methanol complex in the ground state

(b) Stacking 9-DPAA–methanol complex in the ground state

(c) Lateral 9-DPAA–methanol complex in the excited state

(d) Stacking 9-DPAA–methanol complex in the excited state

Figure 1.5. Lateral and stacking structures of the 9-DPAA–methanol complex in the ground and S_1 excited states. (Reprinted with permission from Safonov et al. 2011b and 2011c. Copyright 2011 Springer-Verlag.)

an X–H···N (X = O, C) hydrogen bond. When the analyte molecule contains an electronegative atom (O or N), an additional contact arises between this atom and one of the positively charged H atoms of the acridine fragment. In stacking complexes, the analyte molecule is located over the plane of the acridine fragment. Similarly to lateral complexes, stacking complexes exhibit a short X–H···N contact with the acridine N atom; however, there is an additional short contact of the electronegative atom of the analyte molecule with the positively charged H atom of the benzene ring of the diphenylamino group.

Calculated vertical transition energies in the free 9-DPAA molecule with large oscillator strengths (at 463, 363, and 292 nm) correspond well to experimental absorption bands with maxima at 450, 360, and 290 nm. The electronic transition between the ground and lowest excited singlet state is accompanied by an electron density transfer from the diphenylamino group to the acridine fragment, so that the dipole moment of the molecule increases from 1.9 D for the ground state to 15.4 D for the excited state. These values are in agreement with experimental estimates (Sazhnikov et al. 2007). Generally, the calculated absorption spectra of 9-DPAA complexes with the analytes considered differ insignificantly from the absorption spectrum of the free 9-DPAA molecule, in agreement with the experimental data for solutions (Sazhnikov et al. 2007).

The calculated structure of the 9-DPAA molecule in the first excited singlet electronic state is presented in Figure 1.4b. The amino group of the molecule is nearly planar. Unlike the case of the ground electronic state, in the excited state the angle between the plane of the amino group and the acridine fragment is close to 90° (86–87°).

Generally the excited-state structures of 9-DPAA complexes are similar to those of the ground-state complexes (Figures 1.5c and 1.5d) The short contacts in the excited states are the same as in the ground states; only their lengths change upon excitation. However, the character of these changes is different for lateral and stacking structures. In lateral structures, X–H···N short contacts becomes shorter upon excitation in nearly all cases (the only exception is acetone), whereas X···H–C contacts become appreciably longer. In stacking structures, both short contacts become appreciably shorter upon excitation. In line with this trend, in stacking structures of nearly all complexes, the 9-DPAA–analyte interaction energy increases significantly (by about 4–5 kcal/mol) in magnitude upon excitation. In the lateral structures, this effect is much less pronounced than in stacking structures. As a result, the formation energies of complexes in excited states for stacking structures become larger in magnitude than for lateral structures in all cases except for the methanol complex.

The HOMO-LUMO excitation makes the dominating contribution to the electronic transition between the ground and lowest excited singlet state at the equilibrium excited-state geometry, which corresponds to the 9-DPAA fluorescence band. The HOMO is localized on the phenyl rings and the nitrogen atom of the diphenylamino group, whereas the LUMO is localized on the acridine fragment. Thus, excitation is accompanied by nearly complete transfer of an electron from the diphenylamino group to the acridine fragment, and the calculated dipole moment of 9-DPAA increases from 1.9 D for the ground state to 17.0 D for the excited state. The corresponding experimental estimates for the methyl-substituted 9-DPAA analog 2,7-dimethyl-9-ditolylaminoacridine are 2.0 and 12 D (Sazhnikov 2007); that is, the TDDFT calculations overestimate charge transfer upon excitation.

For the free 9-DPAA molecule, the calculated position of the fluorescence band at 588 nm is substantially larger than the experimental value of 490 nm for 9-DTAA in a nonpolar hexane solution (Sazhnikov et al. 2007). The calculated Stokes shift was 125 nm, while the experimental value in solution was 40 nm (Sazhnikov et al.

2007). Thus, calculations significantly underestimate the charge-transfer transition energy for the emission band, which is characteristic of TDDFT methods (see, e.g., Baer et al. 2010).

On the other hand, the calculated shifts of the positions of the fluorescence bands due to the formation of the stacking-type 9-DPAA complexes with the analytes correlate well with the experimental solvatochromic shifts of the fluorescence band in the corresponding solvent (R^2 = 0.90). To the contrary, the band shifts upon the formation of lateral complexes are small and do not correlate with the value of solvatochromic effects. Hence, solvatochromic effects in the case of 9-DPAA and its analog 9-DTAA are controlled by specific solvation and the formation of the stacking structures of solvate complexes of the dye.

The structures of 9-DPAA complexes with the analytes have been further refined by DFT-D calculations with the B97-D exchange-correlation functional, in which a correction for dispersion interactions is directly included (Rukin et al. 2011).

The inclusion of the dispersion correction significantly increased the relative stability of stacking complexes (by up to 6.2 kcal/mol). For almost all of the 9-DPAA–analyte complexes, the stacking structures were found to be more favorable than the lateral structures. The only exception was for the methanol complex, for which the lateral structure remained slightly more stable (by 1.6 kcal/mol) than the stacking one. The structures of complexes are characterized by short contracts between the acridine N atom and an H atom of the analyte molecule and between the electronegative X atom (X = O, N, or C) of the analyte molecule and an H atom of either the phenyl ring of the amino group (in the stacking complex) or the acridine moiety (in the lateral complex) (2.2–2.7 Å). The structures remained qualitatively unchanged in comparison with DFT calculations.

4.3. MULTISCALE LEVEL: MD/DFT SLAB MODELING OF THE ADSORPTION OF SIMPLE ORGANIC AND INORGANIC MOLECULES ON AN AMORPHOUS SILICA SURFACE

Silica gel nanoparticles are widely used as a very promising substrate for hierarchical nanostructured materials for optical chemical sensors. Silica gel is known for its good adsorption capacity, porous structure, and large surface area (up to 1000 m^2/g); it also does not absorb light in the visible spectrum region. (see, for example, recent reviews: Bonacchi 2011; Liu 2011; Ramón 2011).

As discussed above, two different approaches are used in modeling nanoparticle surfaces at a quantum mechanical level: periodic (solid-state) and cluster (molecular); see Section 3. The periodic repeated-slab model was applied to studying the structure and electronic properties of an amorphous silica surface and its interaction with small organic and inorganic analyte molecules by Minibaev et al. (2009). In this work, the adsorption of simple molecules (water, ammonia, acetone, and ethanol) on the surface of a silica nanofilm was studied using density functional theory in the generalized gradient approximation for the exchange-correlation potential (PBE; Perdew et al. 1996) and with ultrasoft pseudopotentials (Vanderbilt 1990). The calculations were made using the quantum mechanical program complex Quantum ESPRESSO (Giannozzi et al. 2009).

The amorphous silica surface structure was simulated by classical molecular dynamics (MD) using the interaction potential proposed by Vashishta et al. (1990). A 2 × 2 × 2 supercell of α-quartz consisting of 72 atoms with the lattice period increased by 10% was heated up to 4000 K for 1000 ps. Then, it was held at 2500 K for 500 ps, and the obtained structure was further relaxed by periodic DFT calculations. The atomic structures of the α-quartz 2 × 2 × 2 supercell and the amorphized silica cell are shown in Figure 1.6. The calculated densities of α-SiO_2 and amorphized silica were 2.6 and 2.3 g/cm^3, respectively, in good agreement with the experimental densities.

The amorphized silicon oxide crystal cell was used to construct a periodic slab. All the dangling bonds at the surface oxygen atoms were terminated with hydrogen atoms, thus forming surface silanol groups. Water, ammonia, acetone, and ethanol molecules were adsorbed on the optimized amorphous silica surface. The binding energy of a molecule with the amorphous silica surface was calculated as follows: *E*bind = *E*(slab + mol) − *E*slab − *E*mol, where *E*bind is the binding energy, *E*(slab + mol) is the total energy of an amorphous silica slab with an adsorbed molecule, *E*slab is the total energy of the amorphous silica slab, and *E*mol is the

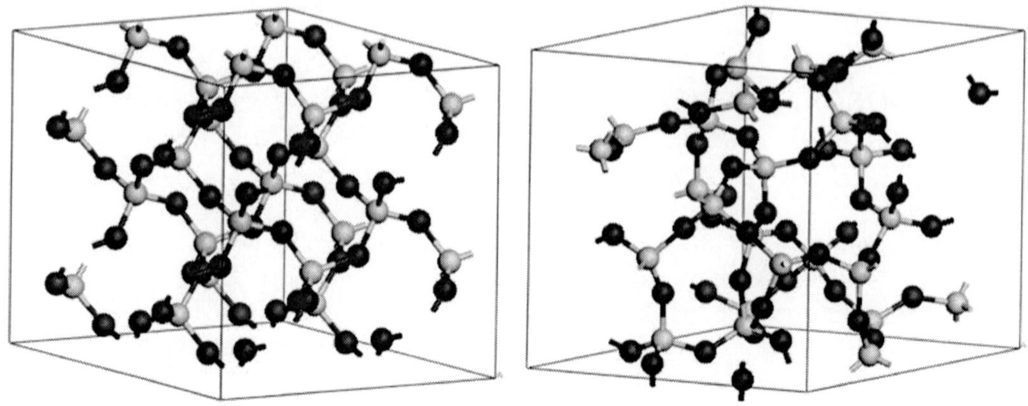

(a) (b)

Figure 1.6. Atomic structures of (a) α-quartz (2 × 2 × 2) supercell and (b) amorphized silica crystal cell; oxygen atoms are shown in black, and silicon atoms are shown in gray. (Reprinted with permission from Minibaev et al. 2009. Copyright 2009 Springer-Verlag.)

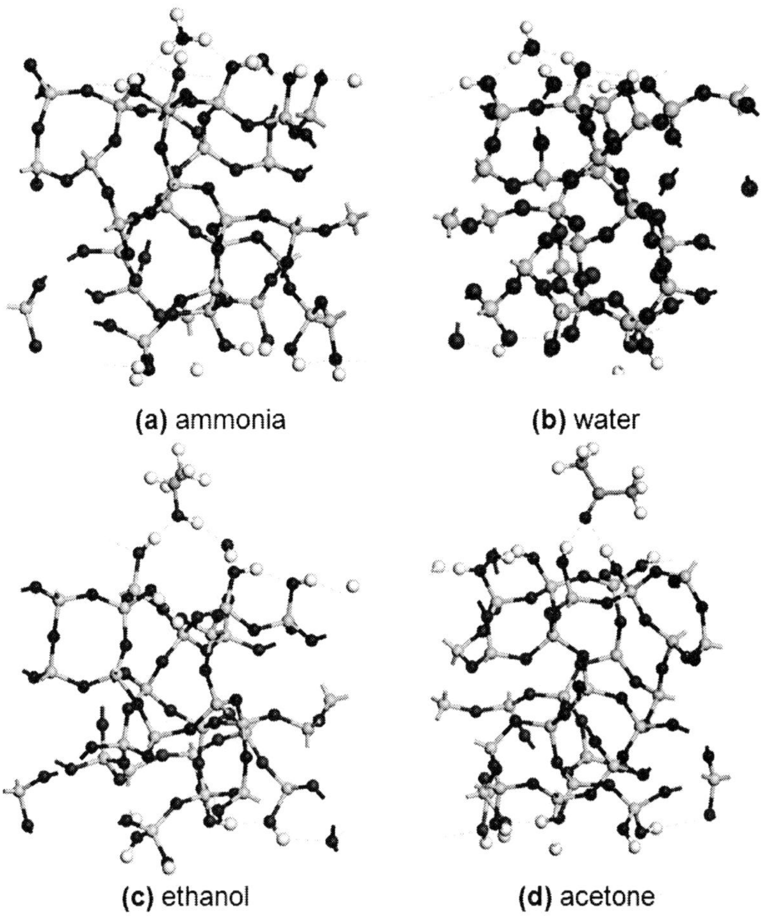

Figure 1.7. Atomic structures of the most stable adsorption complexes for (a) ammonia, (b) water, (c) ethanol, and (d) acetone on the amorphized silica surface. (Reprinted with permission from Minibaev et al. 2009. Copyright 2009 Springer-Verlag.)

total energy of a molecule in vacuum. The energy of a molecule in vacuum was calculated in a vacuum "box" of the same size as that of the silica slab supercell.

It was found that for each molecule, several positions of adsorption exist on the surface. The calculated adsorption energies for the most stable adsorption positions varied from 14 to 18 kcal/mol for water, from 9 to 18 kcal/mol for ammonia, from 11 to 14 kcal/mol for ethanol, and from 7 to 8 kcal/mol for acetone. The atomic structures of the most stable adsorption complexes are shown in Figure 1.7.

4.4. MULTISCALE LEVEL: MD/DFT CLUSTER MODELING OF A 9-DPAA/SILICA RC AND ITS INTERACTION WITH SMALL ANALYTE MOLECULES

The structure of an RC based on an 9-DPAA indicator dye adsorbed on the surface of amorphous silica particles (9-DPAA/silica) and adsorption of small analyte molecules [H_2O, NH_3, C_2H_5OH, and $(CH_3)_2CO$] on the surface of amorphous silica particles and on the RC were studied using the cluster approach at the DFT-D level of theory by Chashchikhin et al. (2011a). Different cluster models were used. The simplest ones were SiH_3OH (cluster **Si1**) and clusters **Si2–Si6** artificially constructed from **Si1** by replacing hydrogen atoms with $Si(OH)_3$ groups followed by full geometry optimization.

The two larger clusters containing 10 and 20 Si atoms (**Si10**, **Si20**) were constructed using classical MD methods. MD simulations provide a most suitable tool for modeling amorphous silica clusters. Using this technique with corresponding force fields, one can amorphize the SiO_2 structure taking the structure of crystalline quartz as the initial approach, changing the unit cell dimensions, and performing simulation in the NVT ensemble. In the cited work, **Si10** and **Si20** clusters were constructed in a similar manner as in the work by Minibaev et al. (2009). The structures of the **Si10** and **Si20** clusters are shown in Figure 1.8.

With increasing silica cluster size from **Si1** to **Si6**, the average energy of the intermolecular interaction between the dye molecule and the silica cluster grows until the number of accessible silanol groups reaches five. Hence, it is sufficient to have approximately five or more silanol groups in contact with the dye for the adequate description of the dye/silica interaction energy. The average energies of analyte interaction with clusters **Si1–Si6** do not increase significantly, while the local binding energies depend strongly on the adsorption geometry.

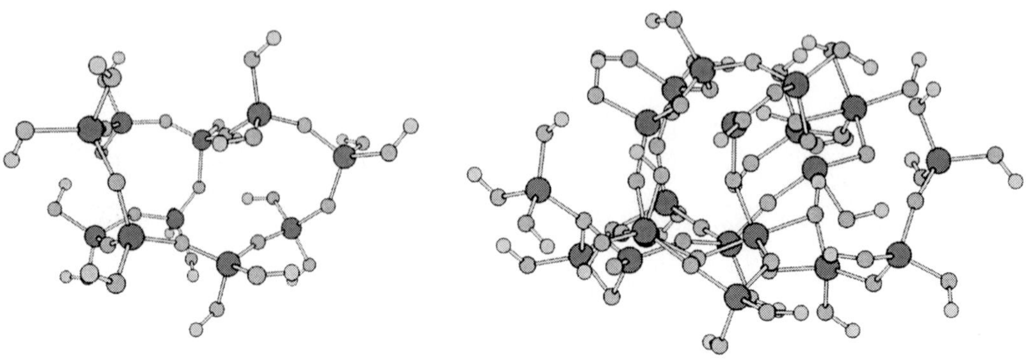

Figure 1.8. Si10 and Si20 cluster models of an amorphous silica surface. (Reprinted with permission from Chashchikhin et al. 2011a. Copyright 2011 the Owner Societies.)

Clusters **Si10** and **Si20** are composed of SiO_4 tetrahedra with Si–O bond lengths varying in a range of 1.60–1.66 Å; alternating Si and O atoms in the inner silica skeleton form six-membered rings with Si–O–Si angles varying in a relatively wide range (130–160°). Along with conventional six-membered rings, the **Si20** cluster also contains one strained four-membered Si–O–Si–O cycle with much smaller Si–O–Si angles (~108° and ~100°). The surfaces of both **Si10** and **Si20** particles are composed of Si–OH groups arranged randomly and linked via Si–O–Si bridges. Their relative concentrations determine the degree of hydroxylation, which is an important parameter controlling the reactivity of the particle. The obtained surface structure is in fairly good agreement with various theoretical models and some experimental data (see, e.g., Jal et al. 2004, Tielens et al. 2008, and Ugliengo et al. 2008, and references therein). The adsorption properties of amorphous silica species are determined by silanol groups, which exhibit acid properties and can interact with Lewis base molecules bearing lone electron pairs. Siloxane bridges, which involve alternating positively (Si) and negatively (O) charged atoms, can participate in electrostatic interactions with polar species. An amorphous silica particle as a whole possesses sufficient polarizability and can also participate in dispersion interactions with aromatic species.

To take into account all these interactions properly, silica clusters should be relatively large (in our case, of the order of 10.5 Å, the size of the dye molecule). Thus, the **Si10** cluster was a little smaller than, while the **Si20** cluster was approximately as large as the 9-DPAA molecule.

It was found that the analyte molecules are attached to the surface by relatively short H-bonds formed by the lone electron pair of their donor N or O atoms with surface silanol groups and also by one or two longer H-bonds between analyte H atoms and O atoms of accessible silanol groups. The analyte adsorption energies depend on the local environment of the adsorbate molecule on the cluster surface. At the best computational level, the calculated adsorption energies of analytes on SiO_2 were found to be approximately 8–9 kcal/mol for H_2O and C_2H_5OH, 11–13.5 kcal/mol for NH_3, and 12–15 kcal/mol for $(CH_3)_2CO$.

The dye forms a relatively strong H-bond with one of the silanol groups by its acridine nitrogen atom (Figure 1.9). The calculated energy of 9-DPAA adsorption on SiO_2 is from two to three times higher (22–29 kcal/mol) than the adsorption energies of the analytes. The presence of the dye molecule on the surface had no significant effect on the energies of analyte adsorption on SiO_2. The simultaneously adsorbed analyte molecule also has no significant effect on the adsorption of the dye. The QC calculations of the 9-DPAA/silica RC will be considered in more detail in the next section.

Thus, the results of this combined MD/QC study provided the geometry of an indicator center and the energies of molecular adsorption on the surface of small silica nanoparticles with a reasonable accuracy at a moderate computational cost. It was shown that the adsorption energy for the large 9-DPAA molecule

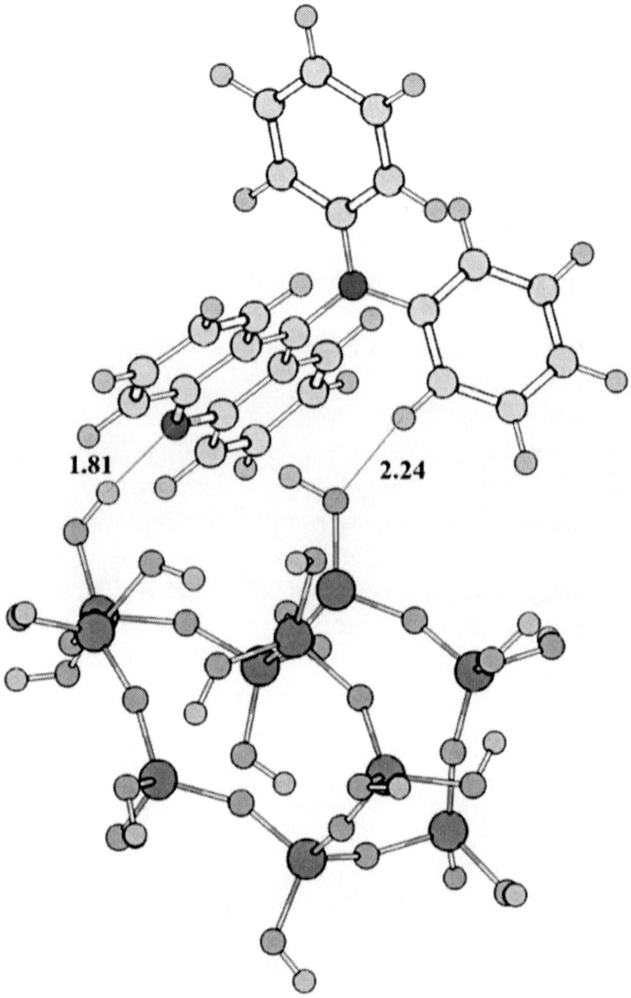

Figure 1.9. 9-DPAA/Si10 model of an RC on the silica surface. (Reprinted with permission from Chashchikhin et al. 2011a. Copyright 2011 the Owner Societies.)

depends on the cluster size and the number of accessible SiOH groups; calculations with small cluster models give a somewhat underestimated binding energy for the 9-DPAA molecule. For small analyte molecules, the energies of adsorption on silica nanoparticles substantially (within several kcal /mol) depend on the adsorption position, and different adsorption sites with extended cluster models should be considered. The dye is much more strongly adsorbed than the small analyte molecules on the surface of amorphous silica nanoparticles. Hence, analytes could not displace the dye from the particle surface, so that sensors based on adsorbed dyes of the 9-DPAA type will possess sufficient chemical stability to analyte vapors.

4.5. MULTISCALE LEVEL: MD/DFT CLUSTER MODELING OF THE EFFECT OF ANALYTE MOLECULES ON THE ABSORPTION AND FLUORESCENCE SPECTRA OF A 9-DPAA/SILICA RC

The investigation of the 9-DPAA/silica RC was further extended by Chashchikhin et al. (2011b) to include more analyte molecules and to study the effect of their interaction with the RC on the RC fluorescence and absorption spectra. Both polar and nonpolar compounds were considered as analytes: acetone, ammonia, methanol, ethanol, water, benzene, naphthalene, toluene, and dinitrotoluene. Their molecules can form complexes with corresponding sensing components of the sensitive layer (receptor center). It has already been discussed (see Sections 2.2 and 2.3) that the operation principle of an optical chemical sensor is based on the change in the intensity or wavelength of the emitted or absorbed light upon the action of a gaseous analyte on the sensing element of the device (McDonagh et al. 2008; Borisov and Wolfbeis 2008). The changes in the fluorescence and absorption spectra of an RC due to its interaction with an analyte represent the output signal (response) indicating the presence of an analyte in the environment. It is expected that the formation of such complexes will change absorption or fluorescence spectra, providing the output signals of the sensor. The electronic fluorescence and absorption spectra of analyte/9-DPAA/silica complexes were calculated within the TDDFT approximation.

The theoretical modeling of the indicator/matrix system included QC calculations of the interaction energies (DFT-D) and the energies and oscillator strengths of electronic transitions (TDDFT) in the dye molecule adsorbed on a model silica particle (as far as possible corresponding to the structure and properties of the chosen substance) and also in the analyte/RC systems. Models used in the cited work were similar to those used by Chashchikhin et al. (2011a); see Section 4.4.

The model of amorphous silica was constructed by classical MD simulations. A hexagonal α-quartz unit cell consisting of nine atoms was built, its dimensions were increased to reproduce the amorphous silica density (the densities of α-quartz and amorphous silica are 2.6 and 2.26 g/cm^3, respectively), and a 7 × 7 × 7 supercell was constructed (about 3000 atoms). The potentials were taken from Feuston and Garofalini (1988, 1990a, 1990b). This supercell was heated for 7 ps at 1-fs steps to the temperature 6000 K and then kept at 4000, 2000, 1000, and 300 K for 7, 7, 17, and 17 ps, respectively. This procedure gave an amorphous bulk SiO_2 structure.

The simplest model of silica was SiH_3OH (cluster **Si1**). The 9-DPAA/**Si1** complex was constructed by forming a H-bond between the nitrogen atom of the acridine fragment of the dye molecule and the silanol group of SiH_3OH (the nitrogen atom of the diphenylamine fragment is sterically inaccessible). Extending the cluster by a sequential substitution of H atoms with H-$Si(OH)_3$ groups showed that the cluster should contain about 10 silanol groups, and four of them should interact directly with the dye, to describe the 9-DPAA–cluster interaction correctly.

For QC calculations, a cluster of the least necessary size (**Si10**, 10 atoms Si, 57 atoms altogether, including the added OH groups) was cut from the obtained bulk structure; the broken bonds of O atoms (so-called dangling bonds) were saturated with H atoms; the positions of thus obtained hydroxyl groups were optimized by DFT-D calculations (PBE0+D, 6-31G**, GAMESS-US). In optimizing the geometry of the cluster, all atoms except for the atoms of newly formed hydroxyl groups were kept frozen at their bulk positions (cluster atoms not interacting with the dye were also kept frozen in the further calculations of complexes with the **Si10** cluster).

For the cluster obtained, the best position was found for the attachment of the dye by maximizing both the number of OH groups interacting with the dye and the projection of the dye onto the cluster (maximum contact area). The latter condition was used in order that the adsorption of the dye on the cluster better described its adsorption on the silica surface. The geometry of the system was finally optimized using the DFT-D approximation (B97-D, 6-31G**); the interaction energy was refined using the extended Dunning cc-pVTZ triple-zeta basis set with the geometries optimized with the double-zeta basis set. The energy of dye adsorption on the silica cluster was as high as 21.6 kcal/mol.

The most probable configurations of analyte complexes with the model RC were found using the INDAM program (Grigoriev et al. 2010) implemented in the NanoModel software complex (NanoModel 2.3, http://www.nanomodel.ru). The INDAM program allows a search for the global energy minimum for a system of two interacting molecules without valence bonds between them. The obtained structures were finally optimized by DFT-D calculations (PBE0+D/6-31G**, GAMESS-US).

To estimate the effect of the extension of the silica cluster on the electronic absorption spectrum of the receptor center and its complexes with small analyte molecules (acetone, ammonia, ethanol), large models based on the **Si10** cluster were also constructed. The structures of complexes were optimized within the DFT-D approximation (PBE0+D/6-31G**). As an example, the dye complex with the ammonia molecule on the **Si10** cluster is shown in Figure 1.10.

To determine the fluorescence maximum of the adsorbed dye, the geometry of the **Si1**/DPAA complex in its first S1 excited state was optimized by TDDFT calculations (PBE0/6-31G**, GAMESS-US). In order to make the calculations of the fluorescence spectra of 9-DPAA feasible, the **Si10** cluster was reduced by removing all –OH and –SiOH groups that do not contact the dye directly; the broken bonds were terminated with hydrogen atoms. Further computational details may be found in the original paper (Chashchikhin et al. 2011b).

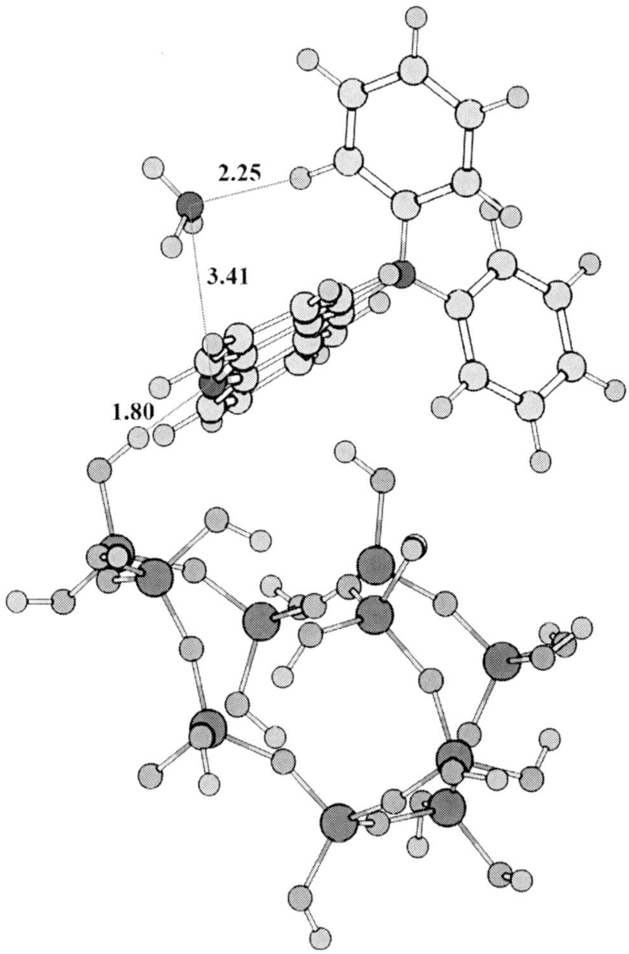

Figure 1.10. Structure of the NH3/9-DPAA/Si10 complex. (Reprinted with permission from Chashchikhin et al. 2011b. Copyright 2011 Springer-Verlag.)

The calculated position of the first absorption bands indicate that, for almost all systems (except for naphthalene), the interaction of the RC with the analytes is accompanied by small (of about 10–30 nm) red shifts of the absorption band. For naphthalene the shift of the absorption band is negligibly small.

The calculated fluorescence spectra were in reasonable agreement with the experimental fluorescence spectra of 2,7-dimethyl-9(ditolylamino)acridine (9-DTAA) adsorbed on spherical silica particles of diameter 10 μm (Khlebunov et al. 2009).

From the calculated positions of low-lying triplet levels (T0 and T1) for the **Si1**/9-DPAA receptor center and **Si1**/9-DPAA/analyte (analyte = acetone, ammonia, water, and ethanol) complexes, it was found that the T0 level is considerably lower than the S1 level. In the **Si1**/9-DPAA receptor center, the T1 level is below S1 and rather close to it. This arrangement is favorable to fluorescence quenching. However, the interaction with the analytes changed the relative positions of S1 and T1 levels in the 9-DPAA/SiH_3OH complexes so that the T1 level becomes markedly higher than S1. Thus, interactions with analytes in the 9-DPAA/silica RC should prevent fluorescence quenching and lead to rise in fluorescence intensity, which was actually observed in the experiments (Khlebunov et al. 2009).

Thus, it was shown that the 9-DPAA/silica receptor center can be used to detect vaporous acetone, ammonia, methanol, ethanol, benzene, and toluene by changes in fluorescence and absorption spectra.

4.6. MULTISCALE LEVEL: MODELING THE STRUCTURE AND SPECTRA OF AN RC BASED ON THE NILE RED DYE ADSORBED ON THE SURFACE OF POLYSTYRENE

Models of the structures of complexes of the Nile Red dye (NR) molecule on the surface of various types of polystyrene (PS) were constructed using molecular dynamics simulations by Tikhomirov et al. (2011). The structure of the NR dye is shown in Figure 1.11. The molecule of the dye consists of a rigid aromatic fragment and a relatively mobile diethylamino group, which can rotate around the C–N bond, adopting various positions.

The aim of this work was to model the structures of the NR/PS receptor center and to analyze the accessibility of the dye to the analyte in the composition of such a center. The molecule of the NR dye was suggested as a fluorescence indicator molecule by Dutta et al. (1996) and used by Khlebunov et al. (2009b) and quite recently by Behnke et al. (2011) as a sensing element. In the latter work, PS was used directly as a matrix.

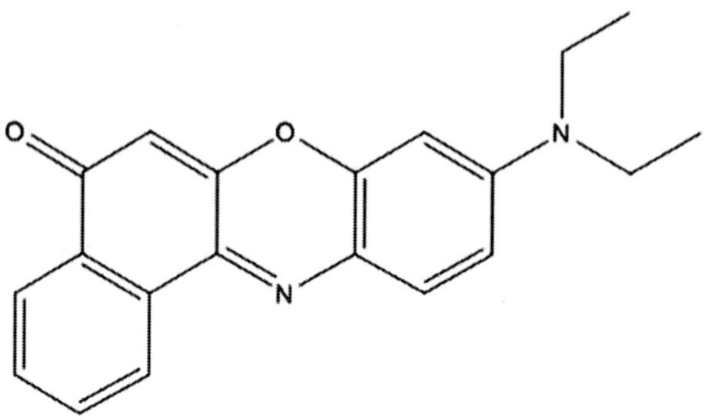

Figure 1.11. Structure of the NR dye. (Reprinted with permission from Tikhomirov et al. 2011. Copyright 2011 Springer-Verlag.)

Tikhomirov et al. (2011) used an all-atom approach in order to fully retain the chemical structure for both the matrix and the dye molecule during the modeling of their complex. All the MD simulations were made using the GROMACS package (Hess et al. 2008; Van der Spoel et al. 2010) and the OPLS-aa force field (Jorgensen et al. 1996) for the PS and NR inter- and intramolecular interaction potentials. The initial geometry and the charges of the NR dye molecule were obtained from TDDFT QC calculations (B3LYP/6-31G*).

The surface area for NR adsorption is small and approximately equal to 1.4 nm^2. It was assumed that the surface structure of a polymer with such a small area has enough time to relax during the modeling time, so it was not necessary to attain equilibrium in the entire volume of the polymer.

In order to construct a PS film, a system of 10 linear PS chains containing 50 monomer links each was used (500 monomer links altogether). This system was first thermalized at 500 K in order to remove the initial anisotropy. The PS density was deliberately lower than the experimental value, which facilitated the relaxation and mixing of the chains. After thermalization, the system was subjected to long-time shrinkage in the NPT ensemble to reach the experimental density of 1060 kg/m^3.

When a cubic PS cell was obtained, the periodic boundary conditions along the z axis were removed, and the size of the cell in this direction was restricted by the two repulsing walls. These walls were moved slowly together, so that the system in the xy plane existed in an NTP ensemble, while its size was variable and depended on the pressure. The cubic volume of the polymer became flattened and adopted the form of a planar structure. At the end, the walls moved away so that one of them restricted the polymer from below while the other was 10 nm above it, leaving a vacuum of sufficient size above the polymer layer. This method of film modeling was described by Doruker and Mattice (1998).

Harmonic confining potentials with a force constant of 1000 $kJ/mol{\cdot}nm^2$ acting along the z axis were applied to the heavy atoms of the main chain of the PS in the central part of the film. As a result, the PS bulk as a whole remained in place during the subsequent calculations, while the motion of separate sections of the chains was not restricted. After the final thermalization, the monomers at the PS surface on the vacuum side relaxed, and the film adopted the final form (see Figure 1.12a).

A globule was constructed in the cited work using an isolated polymer chain as the initial structure. The chain contained 500 monomer links. The system was heated to 500 K, thermalized for a few nanoseconds, and cooled to 298 K. As a result, the chains rolled up into a globule of an imperfect spherical form (see Figure 1.12b). The NR molecule was initially located close to the PS surface, and then the NR/PS system was relaxed at T = 298 K to a constant average NR/PS interaction energy.

The structure of the PS surface layer in both models (film and globule) differed significantly from the bulk structure. The density of the polymer along the normal

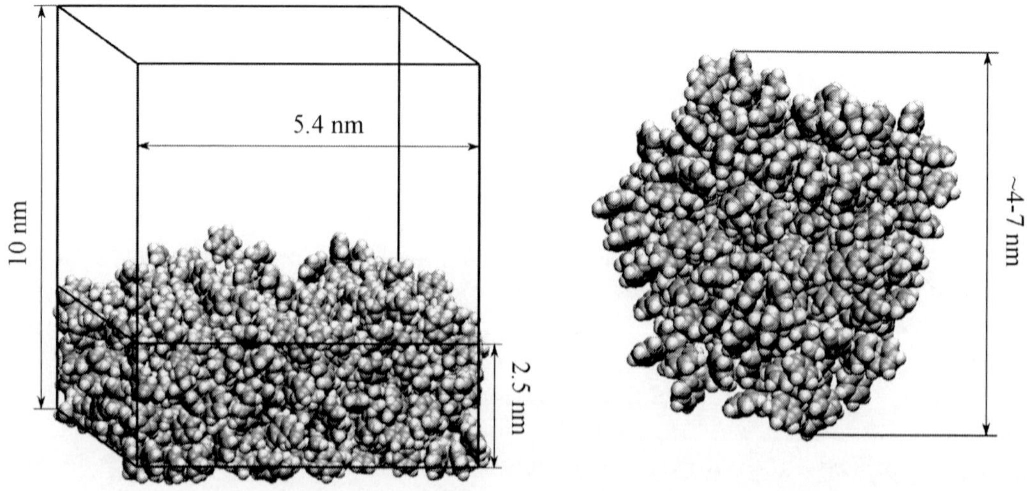

Figure 1.12. Polystyrene structures used as a substrate for Nile Red adsorption: (a) film, the integration cell is shown; (b) globule, the integration cell in this case has no boundaries. (Reprinted with permission from Tikhomirov et al. 2011. Copyright 2011 Springer-Verlag.)

to the surface had the form of a plateau with a value close to the experimental value, which gradually decreased to zero in the surface area. The transverse length of the transitional surface layer in the film model was about 5 Å, which corresponds approximately to the size of the phenyl group. The adsorbed NR molecule penetrated partly into the space between PS links. This position corresponds to a local energy minimum. Thus, a quasi-equilibrium configuration was obtained for the dye in the environment of the PS links. The amine fragment remained above the surface and had enough space for the rotation of the amine group. This last circumstance is important because the amine group in the excited state can rotate out of the plane of the aromatic system (Tuck et al. 2009).

An analysis of the solvent-accessible surface (SAS; Richards 1977; Varshney et al. 1994) indicated that in the film model an NR molecule adsorbed on the PS surface and buried in the space between PS links is accessible for a typical analyte molecule.

The PS density inside the globule was appreciably less than inside the film, while the thickness of the transitional layer was greater. The NR molecule penetrated into the globule surface by the aromatic fragment deeper than in the film model and formed a strong complex with PS. However, as in the case of the film, the NR amine group remained in the free space accessible for interaction with analyte molecules and allowed the rotation of the ethyl groups around the C–N bond.

Thus, the MD simulations of the PS surface using both the periodic film model and the globule model showed that the dye molecule should partly penetrate into the PS surface and form a stable complex with surrounding segments of polymer chains. The morphology of the polymer surface had an effect on the sensor characteristics of the material, and this effect can be investigated by molecular dynamics. The resulting sensitivity of the sensor material depends on the distribution of the investigated substrate/chromophore surface complexes over the degree of chromophore accessibility to a certain analyte. In principle, the obtained sensitivity may differ for analytes of different size and form, which provides the basis for predicting the selectivity of a potential sensor material.

Based on the results obtained by Tikhomirov et al. (2011), the structure and absorption spectra of the NR dye adsorbed on the surface of a PS nanoparticle was studied by DFT by Freidzon et al. (2012a, 2012b). A PS nanoparticle with an adsorbed NR molecule was simulated by a large cluster consisting of PS chains using molecular dynamics in the OPLS-aa force field in the NVT ensemble at 298 K. Different surface models were tried: a surface of a single-chain coil (up to 1000 monomeric units) and of a periodic box. Next, the surface was truncated so that only the nearest neighboring phenyl rings surrounding the chromophore were treated either explicitly by DFT or as effective fragment potentials (EFPs) as proposed by Jensen et al. (1994); a recent review of EFPs and their applications was done by Gordon et al. (2012). The structure of the dye obtained by Tikhomirov et al. (2011) from MD calculation was reoptimized by DFT in the frozen environment, and its absorption and emission spectra were calculated by TDDFT in this environment and compared with those in vacuo and in toluene (simulated through a polarizable continuum model).

The structures of clusters used for quantum chemical calculations are shown in Figure 1.13. It was shown that the transition energies calculated in an explicit

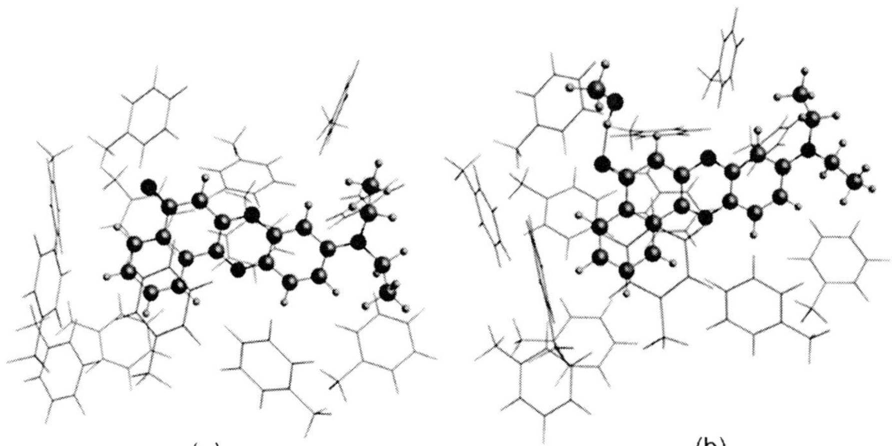

(a) (b)

Figure 1.13. Examples of clusters used in DFT calculations of the absorption spectra of the NR dye adsorbed on the surface of a PS nanoparticle: (a) Nile Red and (b) Nile Red–methanol complex surrounded by EFPs that describe the neighboring polystyrene links surrounding the NR molecule. (Reprinted with permission from Freidzon et al. 2012b. Copyright 2012 IOP Science.)

environment and in EFPs are almost the same, but the use of EFPs significantly reduces the calculation time.

5. PROSPECTS AND OUTLOOK

The general considerations and various examples presented above demonstrate that the multiscale atomistic modeling of nanomaterials for optical chemical sensors allows one to predict the structure of an RC, its interactions with analyte molecules, and spectral response of the RC to these interactions. These results provide a basis for a directional search for new, most promising materials for optical chemical gas-phase sensing possessing best sensitivity and selectivity.

An important new step toward a more comprehensive description of the optical response of a nanomaterial was made quite recently by Yurenev et al. (2010). In this work, a simple and convenient approach was proposed to evaluating band shapes in the electronic spectra of complex molecular systems from first principles based on the classical multiphonon model (Pekar 1953). In this model, each electronic transition is broadened into a structureless band of approximately Gaussian shape due to the linear electron–vibration coupling. The required parameters for these model calculations can be calculated by QC methods, such as DFT and TDDFT, as was described in detail by Yurenev et al. (2010), so that the bandwidths and shapes of interest can easily be calculated from the results of QC calculations.

This rather promising approach was quite recently applied to modeling the band shapes in the electronic absorption spectra of a silica/DPAA RC and its complexes with analytes by Chashchikhin et al. (2011c, 2012). A set of both polar and nonpolar analytes included acetone, ammonia, methanol, ethanol, water, benzene, naphthalene, toluene, and dinitrotoluene. The results of calculations of the bands were in good agreement with the experimental data for DPAA in solutions, which led to the conclusion that the proposed approach can be used for the estimation of the shapes of spectral bands in the absorption spectra of organic dyes adsorbed on silica particles and their complexes with analytes and that the DPAA/silica receptor center can be used to detect compounds in a gas phase by changes in fluorescence and absorption spectra.

ACKNOWLEDGMENTS

This work was supported by the Russian Foundation for Basic Research, project no. 12-03-01103, and the Russian Ministry of Education and Science, state contracts nos. 02.523.11.3014 and 16.523.11.3004, and project no. 8031. The facilities of the Joint Supercomputer Center of the Russian Academy of Sciences were used in some calculations.

REFERENCES

Adamo C. and Barone V. (2002) Physically motivated density functionals with improved performances: The modified Perdew–Burke–Ernzerhof model. *J. Chem. Phys.* **116**, 5933–5940. DOI: 10.1063/1.1458927

Alfimov M.V., Bagaturyants A.A., and Potapkin B.V. (2009) Multiscale computer design of materials for optical chemosensors based on photonic crystals. In: *Abstract Book of the Second International Nanotechnology Forum, Moscow, October 6–8, 2009,* 145. http://rusnanotech09.rusnanoforum.ru/Public/LargeDocs/theses/eng/pdf/Rusnanotech09_abstracts_en.pdf

Alfimov M.V. and Bagaturyants A.A. (2009) Multiscale computer design of hierarchical nanomaterials. In: *Abstract Book of the Second International Nanotechnology Forum, Moscow, October 6–8, 2009,* 151. http://rusnanotech09.rusnanoforum.ru/Public/LargeDocs/theses/eng/pdf/Rusnanotech09_abstracts_en.pdf

Alfimov M.V. and Sazhnikov V.A. (2009) Hierarchical chemosensor materials. In: *Abstract Book of the Second International Nanotechnology Forum, Moscow October 6–8, 2009,* 151–152. http://rusnanotech09.rusnanoforum.ru/Public/LargeDocs/theses/eng/pdf/Rusnanotech09_abstracts_en.pdf

Alfimov M.V., Bagatur'yants A.A., Safonov A.A., Scherbinin A.V., Vladimirova K.G., Belousov S.A., Bogdanova M.V., Valuev I.A., Deinega A.V., Lozovik Yu.E., and Potapkin B.V. (2010), Multiscale computer design of photonic crystal based materials for optical chemosensors. *Nanotechnol. Russia* **5**, 250–258. DOI: 10.1134/S1995078010030134

Baer R., Livshits E., and Salzner U. (2010) Tuned range-separated hybrids in density functional theory. *Annu. Rev. Phys. Chem.* **61**, 85–109. DOI: 10.1146/annurev.physchem.012809.103321

Bagatur'yants A.A., Korkin A.A., Novoselov K.P., Savchenko L.L., Umanskii S.Ya. (2003) Integrated approach to atomistic simulation of film deposition processes. In: Catlow C.R.A. and Kotomin E. (eds.), *Computational Materials Science.* IOS Press, Amsterdam, 388–418.

Bagatur'yants A.A., Minushev A.Kh., Novoselov K.P., Safonov A.A., Umanskii S.Ya., Vladimirov A.S., and Korkin A. (2004) Atomistic simulation of silicon nitride CVD from dichlorosilane and ammonia. In: Dabrowski J. and Weber E.R. (eds.), *Predictive Simulation of Semiconductor Processing Status and Challenges,* Springer Series in Materials Science, 72. Springer-Verlag, Berlin/Heidelberg, 295–356.

Bagatur'yants A.A., Deminskii M.A., Knizhnik A.A., Potapkin B.V., and Umanskii S.Ya. (2007) Integrated approach to dielectric film growth modeling: Growth mechanisms and kinetics. In: Trakhtenberg L.I., Lin S.H., and Ilegbusi O.J. (eds.), *Thin Films and Nanostructures: Physico-Chemical Phenomena in Thin Films and at Solid Surfaces.* Elsevier, Amsterdam, 468–522. DOI: 10.1016/S1079-4050(06)34009-4

Baldini F., Chester A.N., Homola J., and Martellucci S. (eds.) (2006) *Optical Chemical Sensors.* Springer-Verlag, Dordrecht, The Netherlands. DOI: 10.1007/1-4020-4611-1

Bartlett R.J. (1989) Coupled-cluster approach to molecular structure and spectra: A step toward predictive quantum chemistry. *J. Phys. Chem.* **93**, 1697–1708. DOI: 10.1021/j100342a008

Basabe-Desmonts L., Reinhoudt D.N., and Crego-Calama M. (2007) Design of fluorescent materials for chemical sensing. *Chem. Soc. Rev.* **36**, 993–1017. DOI: 10.1039/b609548h

Becke A.D. (1988) Density functional exchange energy approximation with correct asymptotic behavior. *Phys. Rev. A: At., Mol., Opt. Phys.* **38**, 3098–5940. DOI: 10.1103/PhysRevA.38.3098

Bechstedt F. (2003) *Principles of Surface Physics.* Springer-Verlag, Berlin. DOI: 10.1007/978-3-642-55466-7

Behnke T., Wurth C., Hoffmann K., Hubner M., Panne U., and Resch-Genger U. (2011) Encapsulation of hydrophobic dyes in polystyrene micro and nanoparticles via swelling procedures. *J. Fluoresc.* **21**, 937–944. DOI: 10.1007/s10895-010-0632-2

Blyler L.L. Jr., Lieberman R.A., Cohen L.G., Ferrara J.A., and MacChesney J.B. (1989) Optical fiber chemical sensors utilizing dye-doped silicone polymer claddings. *Polym. Eng. Sci.* **29**, 1215–1218. DOI: 10.1002/pen.760291715

Bonacchi S., Genovese D., Juris R., Montalti M., Prodi L., Rampazzo E., Sgarzi M., and Zaccheroni N. (2011) Luminescent chemosensors based on silica nanoparticles. *Top. Curr. Chem.* **300**, 93–138. DOI: 10.1007/128_2010_104

Borisov S.M. and Wolfbeis O.S (2008) Optical biosensors. *Chem. Rev.* **108**, 423–461. DOI: 10.1021/cr068105t

Bromley S.T., Moreira I.P.R., Neyman K.M., and Illas F. (2009) Approaching nanoscale oxides: Models and theoretical methods. *Chem. Soc. Rev.* **38**, 2657–2670. DOI: 10.1039/b806400h

Brooks B.R., Jane I.D., and Karplus M. (2004) Harmonic analysis of large systems: Methodology. *J. Comput. Chem.* **16**, 1522–1542. DOI: 10.1002/jcc.540161209

Burgess L.W. (1995) Absorption-based sensors. *Sens. Actuators B* **29**, 10–15. DOI: 10.1016/0925-4005(95)01657-0

Burshtein K.Ya., Bagatur'yants A.A., and Alfimov M.V. (1994) Shapes of absorption bands in solutions: Organic chromophores. *J. Mol. Struct., THEOCHEM* **314**, 311–320. DOI: 10.1016/0166-1280(94)03731-Y

Burshtein K.Ya., Bagatur'yants A.A., and Alfimov M.V. (1995) MO calculations on the absorption spectra of organic dimers. The interaction energy between dipole moments of electronic transitions in monomers and the shape of absorption bands. *Chem. Phys. Letters* **239**, 195–200. DOI: 10.1016/0009-2614(95)00443-8

Burshtein K.Ya., Bagatur'yants A.A., and Alfimov M.V. (1997) Computer simulation of the shape of absorption bands in electronic spectra of J-aggregates. *Russ. Chem. Bull.* **46**, 62–64. DOI: 10.1007/BF02495348

Calderone A., Lazzaroni R., and Bredas J.L. (1996) Geometric and electronic structure of the butadiene/benzene complex as a model for the polyene/graphite interface. *Synth. Met.* **82**, 225–229. DOI: 10.1016/S0379-6779(96)03795-2

Calderone A., Lazzaroni R., and Bredas J.L. (1998) Geometric and electronic structure of alkane/benzene, ethylbenzene/benzene, and alkane/ethylbenzene complexes: Towards the characterization of polymer alloy composites. *Synth. Met.* **95**, 1–15. DOI: 10.1016/S0379-6779(98)00023-X

Chashchikhin V., Rykova E., and Bagaturyants A. (2011a) Density functional theory modeling of the adsorption of small analyte and indicator dye 9-(diphenylamino)acridine molecules on the surface of amorphous silica nanoparticles. *Phys. Chem. Chem. Phys.* **13**, 1440–1447. DOI: 10.1039/c0cp00821d

Chashchikhin V., Rykova E., and Bagatur'yants A. (2011b) Effect of analyte molecules on the electronic absorption and fluorescence spectra of a receptor center based on the 9-(diphenylamino)acridine dye adsorbed on silica clusters. *Nanotechnol. Russia* **6**, 579–586. DOI: 10.1134/S1995078011050065

Chashchikhin V., Rykova E., and Bagaturyants A. (2011c) Molecular design of a sensor for small analyte molecules based on a dye adsorbed on silica nanoclusters and nanopores. In: *Poster Abstract Book of the TNT2011 Trends in Nanotechnologies Conference, November 21–25, 2011, Tenerife, Canary Islands (Spain).* http://www.tntconf.org/2011/abstracts_TNT2011/TNT2011_Chashchikhin.pdf

Chashchikhin V., Rykova E., Scherbinin A., and Bagaturyants A. (2012) DFT modeling of band shifts and widths in the absorption spectrum of a 9-(diphenylamino)acridine/silica receptor center upon its interaction with gas-phase NH_3, C_2H_5OH, and $(CH_3)_2CO$ molecules. *Int. J. Quant. Chem.* **112**(18), 3110–3118.

Chipot Ch. and Pohorille A (eds.) (2007) *Free Energy Calculations: Theory and Applications in Chemistry and Biology.* Springer-Verlag, Berlin/Heidelberg. DOI: 10.1007/978-3-540-38448-9

Choi M.F. and Hawkins P. (1997) The development of optical chemical sensors for the detection of volatile compounds from spoiled hams. *Sens. Actuators B* **38–39**, 390–394. DOI: 10.1016/S0925-4005(97)80240-7

Cramer D. (2011) Møller–Plesset perturbation theory: From small molecule methods to methods for thousands of atoms. *WIREs Comput. Mol. Sci.* **1**, 509–530. DOI: 10.1002/wcms.58

Crowther D. and Liu X. (1995) Covalent immobilisation of solvatochromic dyes. *J. Chem. Soc. Chem. Commun.* **1995**, 2445. DOI: 10.1039/c39950002445

De Ruiter A. and Oostenbrink C. (2011) Free energy calculations of protein–ligand interactions. *Curr. Opin. Chem. Biol.* **15**, 547–552. DOI: 10.1016/j.cbpa.2011.05.021

De Silva A.P., Gunaratne H.Q.N., Gunnlaugsson T., Huxley A.J.M., McCoy C.P., Rademacher J.P., and Rice T.E. (1997) Signaling recognition events with fluorescent sensors and switches. *Chem. Rev.* **97**, 1515–1566. DOI: 10.1021/cr960386p

Dickinson, T.A., White J., Kauer J.S., and Walt D.R. (1996) A chemical-detecting system based on a cross-reactive optical sensor array. *Nature,* **382**, 697–700. DOI: 10.1038/382697a0

Doruker P. and Mattice W.L. (1998) Simulation of polyethylene thin films on a high coordination lattice. *Macromolecules* **31**, 1418–1426. DOI: 10.1021/ma971322z

Dutta A.K., Kamada K., and Ohta K. (1996) Spectroscopic studies of nile red in organic solvents and polymers. *J. Photochem. Photobiol. A* **93**, 57–64. DOI: 10.1016/1010-6030(95)04140-0

Elliott J.A. (2011) Novel approaches to multiscale modelling in materials science. *Int. Mater. Rev.* **56**, 207–225. DOI: 10.1179/1743280410Y.0000000002

Feuston B.P. and Garofalini S.H. (1988) Empirical 3-body potential for vitreous silica. *J. Chem. Phys.* **89**, 5818–5824. DOI: 10.1063/1.455531

Feuston B.P. and Garofalini S.H. (1990a) Oligomerization in silica sols. *J. Phys. Chem.* **94**, 5351–5359. DOI: 10.1021/j100376a035

Feuston B.P. and Garofalini S.H. (1990b) Onset of polymerization in silica sols. *Chem. Phys. Lett.* **170**, 264–270. DOI: 10.1016/0009-2614(90)87126-C

Freidzon A., Tikhomirov V., Odinokov A., and Bagaturyants A. (2012a) Multiscale approach to the structure and spectra of Nile Red adsorbed on polystyrene nanoparticles. In: *The Book of Abstracts of the International Conference "Functional Materials and Nanotechnologies," FM&NT-2012, April 17–20, 2012, Riga (Latvia)*, 198. http://www.fmnt.lu.lv/fileadmin/user_upload/lu_portal/projekti/konference_fmnt/2012/FM_NT2012_FINAL_for_internet_2.pdf

Freidzon A., Tikhomirov V., Odinokov A., and Bagaturyants A. (2012b) Multiscale approach to the structure and spectra of Nile Red adsorbed on polystyrene nanoparticles. *IOP Conf. Ser.: Mater. Sci. Eng.* **38**, 012036. DOI: 10.1088/1757-899X/38/1/012036

Giannozzi P., Baroni S., Bonini N., Calandra M., Car R., Cavazzoni C., Ceresoli D., Chiarotti G.L., Cococcioni M., Dabo I., Dal Corso A., Fabris S., Fratesi G., de Gironcoli S., Gebauer R., Gerstmann U., Gougoussis C., Kokalj A., Lazzeri M., Martin-Samos L., Marzari N., Mauri F., Mazzarello R., Paolini S., Pasquarello A., Paulatto L., Sbraccia C., Scandolo S., Sclauzero G., Seitsonen A.P., Smogunov A., Umari P., and Wentzcovitch R.M. (2009) Quantum ESPRESSO: A modular and open-source software project for quantum simulations of materials. *J. Phys. Condens. Matter* **21**, 395502. DOI: 10.1088/0953-8984/21/39/395502

Gordon M.S. and Schmidt M.W. (2005) Advances in electronic structure theory: GAMESS a decade later. In: Dykstra C.E., Frenking G., Kim K.S., and Scuseria G.E. (eds.), *Theory and Applications of Computational Chemistry: The First Forty Years*. Elsevier, Amsterdam, 1167–1189. DOI: 10.1016/B978-044451719-7/50084-6

Gordon M.S., Fedorov D.G., Pruitt S.R., and Slipchenko L.V. (2012) Fragmentation methods: A route to accurate calculations on large systems. *Chem. Rev.* **112**, 632–672.

Granovsky A.A. (2009) Firefly version 7.1.G. www http://classic.chem.msu.su/gran/firefly/index.html.

Grigor'ev F.V., Romanov A.N., Laikov D.N., Zhabin S.N., Golovacheva A.Yu., Oferkin I.V., Sulimov A.V., Bazilevskii M.V., Bagatur'yants A.A., Sulimov V.B., and Alfimov M.V. (2010) Molecular modeling methods for supramolecular complexes: A hierarchical approach. *Nanotechnol. Russia* **5**, 290–298. DOI: 10.1134/S1995078010050022

Grimme S. (2004) Accurate description of van der Waals complexes by density functional theory including empirical corrections. *J. Comput. Chem.* **25**, 1463–1476. DOI: 10.1002/jcc.20078

Grimme S. (2006) Semiempirical GGA-type density functional constructed with a long-range dispersion correction. *J. Comput. Chem.* **27**, 1787–1799. DOI: 10.1002/jcc.20495

Grimme S, Antony J, Schwabe T, and Muck-Lichtenfeld C (2007) Density functional theory with dispersion corrections for supramolecular structures, aggregates, and complexes of (bio)organic molecules. *Org. Biomol. Chem.* **5**, 741–758. DOI: 10.1039/b615319b

Grimme S. (2011) Density functional theory with London dispersion corrections. WIREs *Comput. Mol. Sci.* **1**, 211–228. DOI: 10.1002/wcms.30

Guo Z.X. (ed.) (2007) *Multiscale Materials Modelling: Fundamentals and Applications*. CRS Press, Boca Raton, Boston, New York, Washington, DC.

Haus J. (2010) *Optical Sensors: Basics and Applications*. Wiley-VCH Verlag, Weinheim, Germany. DOI: 10.1002/9783527629435

Halgren T.A. (1996) Merck molecular force field: I. Basis, form, scope, parameterization, and performance of MMFF94. *J. Comput. Chem.* **17**, 490. DOI: 10.1002/(SICI)1096-987X(199604)17:5/6<490::AID-JCC1>3.0.CO;2-P

Hess B., Kutzner C., van der Spoel D., and Lindahl E. (2008), GROMACS 4: Algorithms for highly efficient, load-balanced, and scalable molecular simulation. *J. Chem. Theory Comput.*, **4**, 435–447. DOI: 10.1021/ct700301q

Jacquemin, D., Perpete, E.A., Scuseria, G.E., Ciofini, I., and Adamo, C. (2008) TD-DFT performance for the visible absorption spectra of organic dyes: Conventional versus long-range hybrids. *J. Chem. Theory Comput.* **4**, 123–135. DOI: 10.1021/ct700187z

Jal P.K., Sudarshan M., Saha A., Patel S., and Mishra B.K. (2004) Synthesis and characterization of nanosilica prepared by precipitation method. *Colloids Surf. A Physicochem. Eng. Asp.* **240**, 173–178. DOI: 10.1016/j.colsurfa.2004.03.021

Jensen J.H., Day P.N., Gordon M.S., Basch H., Cohen D., Garmer D.R., Krauss M., and Stevens W.J. (1994) Effective fragment method for modeling intermolecular hydrogen bonding effects on quantum mechanical calculations. In: Smith D.A. (ed.), *Modeling the Hydrogen Bond.* ACS Symp. Series **569**, 139–151.

Jorgensen W.L., Maxwell D.S., and Tirado-Rives J. (1996) Development and testing of the opls all-atom force field on conformational energetics and properties of organic liquids. *J. Am. Chem. Soc.* **118**, 11225–11236. DOI: 10.1021/ja9621760

Khlebunov A., Ionov D., Aristarkhov V., Sazhnikov V., and Alfimov M. (2008) A study of optical chemosensor materials. *Chem. Eng. Trans.* **15**, 293–300.

Khlebunov A.A., Komarov P.V., Ionov D.S., Koshkin A.V., Sazhnikov V.A., and Alfimov M.V. (2009a) Spectral and sensing properties of 9-DTAA dye in polymer matrices. In: *Abstract Book of the ICON-RUSSIA Conference 2009 June 21–28, 2009, St. Petersburg, Russia,* 61.

Khlebunov A.A., Ionov D.S., Komarov P.V., Aristarkhov V.M., Sazhnikov V.A., Petrov A.N., and Alfimov M.V. (2009b) An experimental system for investigating the characteristics of optical sensor materials. *Instrum. Exp. Technol.* **52**, 132–136. DOI: 10.1134/S0020441209010229

Knizhnik A.A., Bagaturyants A.A., Belov I.V., Potapkin B.V., and Korkin A.A. (2002) An integrated kinetic Monte Carlo molecular dynamics approach for film growth modeling and simulation: ZrO_2 deposition on Si(100) surface. *Comput. Mater. Sci.* **24**(1–2), 128–132. DOI: 10.1016/S0927-0256(02)00174-X

Korotcenkov G., Cho B.K., Narayanaswamy R., and Sevilla F. III (2011) Optical and fiber optic chemical sensors. In: Korotcenkov G. (ed.), *Chemical Sensors: Comprehensive Sensor Technologies, Vol. 5: Electrochemical and Optical Sensors.* Momentum Press, New York.

Krishnan R., Binkley J.S., Seeger R., and Pople J.A. (1980) Self-consistent molecular orbital methods. XX. A basis set for correlated wave functions. *J. Chem. Phys.* **72**, 650–654. DOI: 10.1063/1.438955

Lavor C. (2003) A deterministic approach for global minimization of molecular potential energy functions. *Int. J. Quant. Chem.* **95**, 336–343. DOI: 10.1002/qua.10701

Levitsky I., Krivoshlykov S.G., and Grate J.W. (2001) Rational design of a Nile Red/polymer composite film for fluorescence sensing of organophosphonate vapors using hydrogen bond acidic polymers. *Anal. Chem.* **73**, 3441–3448. DOI: 10.1021/ac0014966

Lee C.T., Yang W.T., and Parr R.G. (1988) Development of the Colle-Salvetti correlation-energy formula into a functional of the electron density. *Phys. Rev. B* **37**, 785–789. DOI: 10.1103/PhysRevB.37.785

Liu J., Yang X., He X., Wang K., Wang Q., Guo Q., Shi H., Huang J., and Huo X. (2011) Fluorescent nanoparticles for chemical and biological sensing. *Sci. China—Chem.* **54**, 1157–1176. DOI: 10.1007/s11426-011-4350-7

Lobnik A. (2006) Absorption-based sensors. In: Baldini F., Homola J., Martellucci S., and Chester A. (eds.), *Optical Chemical Sensors.* Springer-Verlag, Dordrecht, The Netherlands, 77–98. DOI: 10.1007/1-4020-4611-1_5

Lobnik A. and Urek S.K. (2011) Nano-based optical chemical sensors. *J. Nano. Res.* **13**, 99–110. DOI: 10.4028/www.scientific.net/JNanoR.13.99

Matsubara K., Kawata S., and Minami S. (1988) Optical chemical sensor based on surface plasmon measurement. *Appl. Optics,* **27**, 1160–1163. DOI: 10.1364/AO.27.001160

McDonagh C., Burke C.S., and MacCraith B.D. (2008) Optical chemical sensors. *Chem. Rev.* **108**, 400–422. DOI: 10.1021/cr068102g

Møller C. and Plesset M.S. (1934) Note on an approximation treatment for many-electron systems. *Phys. Rev.* **46**, 618–622. DOI: 10.1103/PhysRev.46.618

Nieminen R.M. (2002) From atomistic simulation towards multiscale modelling of materials. *J. Phys.: Condens. Matter* **14**, 2859–2876. DOI: 10.1088/0953-8984/14/11/306

Neese F., with contributions from Becker U., Ganiouchine D., Koßmann S., Petrenko T., Riplinger C., and Wennmohs F. (2010) ORCA, an ab initio, density functional and semiempirical program package. *University of Bonn, Germany.* http://www.thchuni-bonnde/tc/orca

Orellana G. (2006) Fluorescence-based sensors. In: Baldini F., Chester A.N., Homola J., and Martellucci S. (eds.), *Optical Chemical Sensors.* Springer-Verlag, Dordrecht, The Netherlands, 99–116. DOI: 10.1007/1-4020-4611-1_6

Paldus J. and Cizek J. (1975) Time-independent diagrammatic approach to perturbation theory of fermion systems. *Adv. Quantum Chem.* **9**, 105–197. DOI: 10.1016/S0065-3276(08)60040-4

Paolesse R., Monti D., Dini F., and Di Natale C. (2011) Fluorescence based sensor arrays. *Top. Curr. Chem.* **300**, 139–174. DOI: 10.1007/128_2010_97

Pekar S.I. (1953) On the effect of lattice distortion by electrons on the optical and electrical properties of crystals. *Usp. Fiz. Nauk* **50**, 197–252.

Perdew J.P., Burke K., and Ernzerhof M. (1996) Generalized gradient approximation made simple. *Phys. Rev. Lett.* **77**, 3865–3868. DOI: 10.1103/PhysRevLett.77.3865

Perdew J.P. and Wang Y. (1992) Accurate and simple analytic representation of the electron-gas correlation energy. *Phys. Rev. B* **45**, 13244–13249.

Perez L.D., Florez E., Mark J.E., and Lopez B.L. (2009) Estimation of the interaction energy between small molecules and a silica model as an approach for predicting the interaction order between elastomers and silica. *Polym. Int.* **58**, 811–816. DOI: 10.1002/pi.2596

Phillips R. (1998) Multiscale modeling in the mechanics of materials. *Curr. Opin. Solid State Mater. Sci.* **3**, 526–532. DOI: 10.1016/S1359-0286(98)80020-X

Plotnikov V.G., Sazhnikov V.A., and Alfimov M.V. (2007) Intermolecular interactions and spectral and luminescent properties of optical molecular sensors. *High Energy Chem.* **41**, 299–311. DOI: 10.1134/S0018143907050013

Ramón M.-M., Félix S., Mandy H., Mustafa B., and Knut R. (2011) Nanoscopic optical sensors based on functional supramolecular hybrid materials. *Anal. Bioanal. Chem.* **399**, 55–74. DOI: 10.1007/s00216-010-4198-2

Richards F.M. (1977) Areas, volumes, packing, and protein structure. *Annu. Rev. Biophys. Bioeng.* **6**, 151–176. DOI: 10.1146/annurev.bb.06.060177.001055

Ross R.B. and Mohanty S. (eds.) (2008) *Multiscale Simulation Methods for Nanomaterials.* John Wiley, Hoboken, NJ. DOI: 10.1002/9780470191675

Rukin P.S., Safonov A.A., and Bagaturyants A.A. (2011) *Nanotechnol. Russia* **6**(5–6), 298–302. DOI: 10.1134/S1995078011030141

Safonov A.A., Rykova E.A., Bagaturyants A.A., Sazhnikov V.A., and Alfimov M.V. (2011a) Atomistic simulations of materials for optical chemical sensors: DFT-D calculations of molecular interactions between gas-phase analyte molecules and simple substrate models. *J. Mol. Mod.* **17**, 1855–1862. DOI: 10.1007/s00894-010-0882-6

Safonov A.A., Bagaturyants A.A., Sazhnikov V.A., and Alfimov M.V. (2011b) Density functional calculations of 9-diphenylaminoacridine fluorescent indicator and its interactions with analyte molecules: I. Structures of complexes in the ground electronic states and absorption spectra. *High Energy Chem.* **45**, 229–236. DOI: 10.1134/S0018143911030131

Safonov A.A., Bagaturyants A.A., Sazhnikov V.A., and Alfimov M.V. (2011c) Density functional calculations of 9-diphenylaminoacridine fluorescent indicator and its interactions with analyte molecules: II. Structures of complexes in the excited electronic states and emission spectra. *High Energy Chem.* 2011, **45**, 265–272. DOI: 10.1134/S001814391104014X

Sazhnikov V.A., Strukov A.G., Stunzhas M.G., Efimov S.P., Andreev O.M., and Alfimov M.V. (1986) Photoinduced formation of 9-substituted acridine from diphenylamine and tetrabromomethane. *Dokl. Akad. Nauk SSSR* **288**, 172–176.

Sazhnikov V.A., Khlebunov A.A., and Alfimov M.V. (2007) Solvatofluorochromic properties of 2,7-dimethyl-9-(ditolylamino)acridine, *High Energy Chem.* **41**, 25–28. DOI: 10.1134/S0018143907010055

Sazhnikov V. and Alfimov M. (2008) Rational design of optical chemosensor materials. *Chem. Eng. Trans.* **15**, 267–274.

Scheiner A.C., Scuseria G.E., Rice J.E., Lee T.J., and Schaefer H.F. III (1987), Analytic evaluation of energy gradients for the coupled cluster wave function: Theory and applications. *J. Chem. Phys.* **87**, 5361–5373. DOI: 10.1063/1.453655

Schluter M., Chelikowsky J.R., Louie S.G., and Cohen M.L. (1975) Self-consistent pseudopotential calculations for Si (111) surfaces: Unreconstructed (1×1) and reconstructed (2×1) model structures. *Phys. Rev. B,* **12**, 4200–4214. DOI: 10.1103/PhysRevB.12.4200

Schmidt M.W., Baldridge K.K., Boatz J.A., Elbert S.T., Gordon M.S., Jensen J.J., Koseki S., Matsunaga N., Nguyen K.A., Su S., Windus T.L., Dupuis M., and Montgomery J.A. (1993) General atomic and molecular electronic structure system. *J. Comput. Chem.* **14**, 1347–1363. DOI: 10.1002/jcc.540141112

Seitz W.R. (1984) Chemical sensors based on fiber optics. *Anal. Chem.* **56**. 16A–34A. DOI: 10.1021/ac00265a001

Tam J.M., Song L., and Walt D.R. (2005) Fabrication and optical characterization of imaging fiber-based nanoarrays. *Talanta* **67**, 498–502. DOI: 10.1016/j.talanta.2005.06.021

Tielens F., Gervais C., Lambert J.-F., Mauri F., and Costa D. (2008) Ab initio study of the hydroxylated surface of amorphous silica: A representative model. *Chem. Mater.* **20**, 3336–3344. DOI: 10.1021/cm8001173

Tikhomirov V.A., Odinokov A.V., Bagatur'yants A.A., and Alfimov M.V. (2011) Modeling the surface of polystyrene and the adsorption of dye molecules on this surface. *Theor. Exp. Chem.* **46**, 342–349. DOI: 10.1007/s11237-011-9164-3

Tuck P.O., Mawhinney R.C., and Rappon M. (2009) An ab initio and TD-DFT study of solvent effect contributions to the electronic spectrum of Nile Red. *Phys. Chem. Chem. Phys.* **11**, 4471–4480. DOI: 10.1039/b902528f

Ugliengo P., Sodupe M., Musson F., Bush I.J., Orlando R., and Dovesi R. (2008) Realistic models of hydroxylated amorphous silica surfaces and MCM-41 mesoporous material simulated by large-scale periodic B3LYP calculations. *Adv. Mater.* **20**, 4579–4583. DOI: 10.1002/adma.200801489

Van der Spoel D., Lindahl E., Hess B., van Buuren A.R., Apol E., Meulenhoff P.J., Tieleman D.P., Sijbers A.L.T.M., Feenstra K.A., van Drunen R., and Berendsen H.J.C. (2010) Gromacs User Manual, version 4.5.4. www.gromacs.org.

Vanderbilt D. (1990) Soft self-consistent pseudopotentials in a generalized eigenvalue formalism. *Phys. Rev. B* **41**, 7892–7895. DOI: 10.1103/PhysRevB.41.7892

Varshney A., Brooks F.P., and Wright W.V. (1994) Computing smooth molecular-surfaces. *IEEE Comput. Graph. Appl.* **14**, 19–25. DOI: 10.1109/38.310720

Vashishta P., Kalia R.K., Rino J.P., and Ebbsjo I. (1990) Interaction potential for SiO_2: A molecular-dynamics study of structural correlations. *Phys. Rev. B* **41**, 12197–12209. DOI: 10.1103/PhysRevB.41.12197

Tan W., Shi Z.Y., Smith S., Birnbaum D., and Kopelman R. (1992) Submicrometer intracellular chemical optical fiber sensors. *Science* **258(**5083), 778–781. DOI: 10.1126/science.1439785

Wu W.-B., Wang M.-L., Sun Y.-M., Huang W., Cui Y.P., and Xu C.-X. (2008) Fluorescent polystyrene microspheres with large Stokes shift by fluorescence resonance energy transfer. *J. Phys. Chem. Solids* **69**, 76–82. DOI: 10.1016/j.jpcs.2007.07.132

Wolfbeis O.S. (1991) Optical sensing based on analyte recognition by enzymes, carriers and molecular interactions. *Anal. Chim. Acta* **250**, 181–201. DOI: 10.1016/0003-2670(91)85071-Y

Wolfbeis O.S. (2005) Materials for fluorescence-based optical chemical sensors. *J. Mater. Chem.* **15**, 2657–2669. DOI: 10.1039/b501536g

Ytreberg F.M., Swendsen R.H., and Zuckerman D.M. (2006) Comparison of free energy methods for molecular systems. *J. Chem. Phys.* **125**, 184114. DOI: 10.1063/1.2378907

Yurenev P.V., Kretov M.K., Scherbinin A.V., and Stepanov N.F. (2010) Environmental broadening of the CTTS bands: The hexaammineruthenium(II) complex in aqueous solution, *J. Phys. Chem. A*, **114**, 12804–12812. DOI: 10.1021/jp1031477

CHAPTER 2

Self-Assembling and Modeling of Sensing Layers

Photonic Crystals

S. Belousov

I. Polishchuk

B. Potapkin

1. INTRODUCTION

A sensing layer of an optical chemical sensor is an ordered system of nanosized particles (from 200 nm to 2 μm in diameter), containing supramolecular receptor centers (dye molecules). Ordered assemblies are formed from thin films or micro-drops of a solution by means of a nanoparticle self-assembly process (Nicolis and Prigogine 1977; Park et al. 2006; Andreeva et al. 2007). Self-assembly technologies are based on the self-organization of nanoparticles into an assembly of nanostructures in a dissipative system. The size of the nanoparticles constituting the sensing layer is in the region of visible or near-infrared wavelengths. A multilayer close-packed face-centered cubic structure can form through the self-organization of such particles. This structure has a lattice constant of the order of the size of its constituent nanoparticles. If the particles have sufficiently regular shape and equal size, this system will operate as a photonic crystal (PC) for electromagnetic

DOI: 10.5643/9781606503201/ch2

radiation with a wavelength of the order of the lattice constant. In this case, the range of visible-light wavelengths coincides with the characteristic particle size and represents the working range of optical chemical sensors. The effect of a photonic crystal on the embedded dye molecules consists of modification of their spontaneous emission (Yakimanski et al. 2009; Alfimov et al. 2010; Megens et al. 1999; Nikolaev et al. 2005).

Control over the spontaneous emission is very important in many applications, such as chemical and biosensing, lasing, etc. (Lakowicz 2003; Shinar and Shinar 2008). At the heart of the control lies the modification of spontaneous emission of luminescent species such as atoms, molecules, laser dyes, and colloidal quantum dots placed in close proximity to nano-objects, or positioned near or inside a nanostructured dielectric environment, such as nanopatterned surfaces and photonic crystals. Examples of surface-enhanced fluorescence include sources near single metallic particles (Lakowicz 2003), metallic 1-D or 2-D gratings (Fort 2008; Wedge and Barnes 2004), or arrays of plasmonic nanoantennas (so-called plasmonic crystals) (Vecchi et al. 2009). Sufficient modification of spontaneous emission (over 100 times) has been also achieved with dielectric 2-D nanohole arrays (Ganesh et al. 2007). The nanostructure scatters the propagating surface waves, opening up a radiative decay channel for these waves, so that the energy of the source coupled to the surface waves is outcoupled and reradiated back in space.

In 1-D and 2-D structures (metallic gratings, nanohole arrays, etc. (Wedge and Barnes 2004; Ganesh et al. 2007; Fort 2008; Vecchi et al. 2009), strong enhancement is achieved by matching the emission frequency to the leaky surface modes existing in such structures. The mechanism behind this enhancement is simplest to understand in the case of the metallic interface, where the surface plasmons can propagate. When the surface is smooth, an emitter positioned close to the metal will be subject to strong quenching, as its radiation is coupled to the plasmons through the near-field and in this way the energy is lost, as the plasmons do not radiate. The plasmon wave propagating along the smooth surface decays as its energy is absorbed by the metal. However, the energy transferred to the surface plasmons can be outcoupled and reradiated into the far field by adding a grating on the metal surface. The grating acts as a Bragg scatterer for the surface waves. If the wave vector of the plasmon wave is matched to the in-plane component of the radiating wave, scattering can occur and the energy is radiated back in space (Wedge and Barnes 2004; Yates et al. 2006). This scattering opens a radiative channel for the plasmon wave decay. A similar picture holds for nanohole arrays (Ganesh et al. 2007), where leaky modes are also present. It has been shown that such leaky modes can lead to hundred-fold enhancements of fluorescence.

Strong modification (enhancement or suppression) of spontaneous emission can also be achieved with microcavities of different geometries (Englund et al. 2005; Chen et al. 2010), by creating localized photonic states similar to defect states in semiconductor crystals.

Fluorescence enhancement can be achieved with 1-D and 2-D structures, but 3-D photonic crystals provide a much more flexible way of controlling the spontaneous emission, as both enhancement and inhibition of the luminescence can be achieved. The strongest modification of spontaneous emission is possible for a PC with a photonic band gap (PBG). In this case, emission is totally inhibited for the frequencies inside the band gap. However, PBG structures possess very complicated geometry and require high dielectric contrast and therefore are very hard to fabricate. However, sufficient modification of spontaneous emission can be achieved in PCs without PBGs. Examples of such structures include titania inverse opals (Koenderink et al. 2003; Nikolaev et al. 2005), silica, and polysterene opals.

In chemical sensing systems, both luminescent dyes and analytes can be distributed over a wide region inside the sensor (Alfimov et al. 2010). Using photonic crystals as a platform for optical chemical sensing makes it possible to achieve modification of the spontaneous emission of all emitting sources at the same time.

2. PHOTONIC CRYSTALS

Photonic crystals are generally artificial structures in which the refractive index varies periodically along one, two, or three dimensions on length scales comparable to the wavelength of light, forming one-, two-, or three-dimensional PCs correspondingly. The propagation of electromagnetic waves within a PC is completely different from the plane waves in homogeneous media. The strong difference is due to the fact that the optical properties of a PC are very complicated and do not in the least resemble those of its constituent elements. Owing to the refractive index periodicity, the propagation of light in a PC resembles the propagation of electrons in the periodic potential of crystalline solids. The periodicity leads to Bragg diffractions associated with the frequency ranges in which the propagation of light is forbidden for a certain direction corresponding to a specific set of lattice planes. Such frequency ranges are called stop gaps (Soukoulis 2001; Sakoda 2004). Stop gaps even arise in dielectric mirrors, which represent a special case of a one-dimensional PC. In certain conditions a photonic band gap may arise as a result of overlapping of stop gaps. In the PBG the propagation of light is forbidden for all directions (Yablonovitch 1987; John 1987).

The presence of a PBG means that no optical modes can exist within a PC over a certain frequency range. The tremendous significance of this effect for manipulating spontaneous emission was first noted by Yablonovitch (1987) and John (1987). For example, an atom placed in a PC with a PBG tuned so that it overlaps its emission band will not be able emit photons. Inhibition of spontaneous emission in periodic structures was first predicted by Bykov (1972). More generally, any interaction mediated by vacuum fluctuations, such as van der Waals or

Casimir forces, black-body radiation, and resonant dipole–dipole interactions, is affected by their suppression in a PBG (Koenderink 2003; Moroz 2005).

Contrary to the localized inhibition of spontaneous emission, which can be achieved in microcavities, in PCs the emission is inhibited anywhere and is only limited by the extent of the crystal. The very important point here is that inhibition in the PC occurs not because light cannot penetrate through its surface, as would be the case for, e.g., a metal box. A source in a metal box would still emit light, though it could not be seen from outside the box, while in a PC inside a PBG frequency range there are no electromagnetic states available for the source, so it simply cannot emit at all.

However, the spontaneous emission can be strongly modified even in the absence of PBGs. The rate of spontaneous emission, that is, the radiative decay rate, is, according to Fermi's golden rule, proportional to the local radiative density of optical states (LRDOS), which counts the number of electromagnetic modes wherein photons can be emitted at a specific location of the source. Strong variations of the LRDOS happen even when no photonic band gap is present in 3-D PCs and manifest in variations of total emission. These effects have been confirmed experimentally using both continuous-wave (cw) total emission power (Koenderink 2003, Nikolaev et al. 2005; Yakimanski et al. 2009) and time-resolved lifetime techniques (Lodahl et al. 2004; Nikolaev et al. 2007; Vallée et al. 2007; Julsgaard et al. 2008). The latter allows direct measurement of the decay rates of emitters inside the structure, while cw experiments are based on a comparison of the total emission spectra with a homogeneous medium. In the latter case, a complete understanding of all angle-dependent effects such as Bragg diffraction on the light propagation is very important. Furthermore, the angular distribution of spontaneously emitted light is modified in the presence of the stop gaps (Bogomolov et al. 1997; Blanco et al. 1998; Vlasov et al. 2000; Yakimanski et al. 2009). Apart from the Bragg diffraction, the interference effect linked to the PC periodicity, which leads to the suppression of light propagation in certain directions, the disorder in the PC also plays a role. This disorder is inherent to the PC, fabricated by the self-assembling technique (such as opals and inverse opals) and affects the Bragg diffraction in PCs. It has been shown that the disorder determines the depth of stop bands (Megens et al. 1999). The redistribution of emitted light due to the disorder leads to the enhancement of emission at the stop-band edges (Koenderink 2003, Nikolaev et al. 2005). The method for modeling the influence of the disorder (Koenderink 2003; Nikolaev et al. 2005) will be described later in this chapter.

3. METHODS OF MODELING SPONTANEOUS EMISSION MODIFICATION

The spontaneous emission in atoms, molecules, and other types of emitters such as, for example, colloidal quantum dots, occurs as a result of a dipole transition of an electron from an excited state to the ground state. The decay of the excited state is determined by both the chemical and dielectric environment of the emitter (see, e.g., Fort 2008), and is a sum of two processes, namely, radiative and nonradiative decay. The dielectric environment affects the radiative decay rate, which is determined by the local density of radiative states (modes) available to the emitter (Sprik et al. 1996), and a part of the nonradiative decay rate associated with the absorption of the emitted energy in the material (Ohmic losses). This latter component describes the well-known metallic quenching of spontaneous emission, which is of primary importance in the area of spontaneous emission studies (Chance et al. 1978; Fort 2008). The other part of the nonradiative decay rate is determined by the intrinsic emitter properties as well as by its chemical environment (intrinsic quenching). The intrinsic quenching is modeled on the atomistic scale, so below we leave it out and deal with the radiative decay and Ohmic losses.

3.1. CORRESPONDENCE PRINCIPLE

The correspondence principle plays a key role in the classical electromagnetic approach to the calculation of decay rates. According to this principle, a power radiated by the spontaneous emission of an excited state of an emitter in an electric dipole transition is calculated as a power radiated by a classical electric dipole with the dipole moment magnitude equal to the dipole transition matrix element. The decay rate is then given by the following expression:

$$\gamma = \frac{P}{\hbar\omega} \tag{2.1}$$

where P is the radiated power and ω is the radiation frequency. The same holds for the magnetic dipole transition. It is assumed here that neither the frequency nor the dipole transition matrix element are substantially affected by the dielectric environment of the emitter (Moroz 2005; Novotny and Hecht 2006). The radiative decay rate can be obtained by integrating the far field of the emitter over a surface enclosing the emitter. The nonradiative decay rate can be obtained by integrating the energy of the radiation field over the absorbing regions.

3.2. DIPOLE NEAR A SURFACE

The correspondence principle was first applied for the modeling of spontaneous emission of Eu^{3+} complexes placed in close proximity to metallic surfaces (Chance et al. 1974, 1975, 1978). In contrast with previous works on the subject, in which a mechanical model had been utilized (Kuhn 1970), the method described in these studies provided a rigorous way of separating radiative and nonradiative losses of a source and thus information about both radiative and nonradiative decay rates as well as the apparent quantum yield of the source in the complex environment could be obtained. The apparent quantum yield is the amount of real photons associated with the radiation field of the dipole per created luminescent state, in contrast to the quantum yield of the free dipole, which equals the amount of emitted photons per luminescent state. In the case of a dipole near the metallic surface, only a part of the photons emitted by the dipole will actually radiate, while the other part will be absorbed by the surface. Modifications of luminescence intensity were investigated for a source with respect to its distance from the metallic surface. Other examples included a source near an absorbing slab of finite thickness, as well as one between two partially reflecting surfaces. The method developed by Chance et al. was further expanded to account for nonlocal effects on the dielectric permittivity, which play a crucial role at very small distances (typically, less than 10 nm) between the molecule and the metal interface (Ford and Weber 1984).

The method predicts with high accuracy the behavior of luminescent molecules near absorbing surfaces. In particular, it enables correct prediction of the nonradiative decay rate of a molecule at small distances from the absorbing surface.

3.2.1. Dipole Near or Inside a Sphere

Understanding of spontaneous emission of excited atoms near curved surfaces is very important in the field of luminescence chemical and biosensing. It concerns such topics as amplification of luminescence near metallic nanoparticles (Lakowicz 2003) for single molecule probing, etc. Pioneering work by Chew (1987, 1988) and Ruppin (1982) provided the full classical description of the problem for the case of a dipole near or inside a homogeneous dielectric or metal spherical particle. The spherical case involves some interesting features not present in the planar case, such as possible resonant interactions between the dipole and the particle when the emission frequency is close to the resonant frequency of the particle. The method itself is similar to the Mie scattering formalism (Bohren and Huffman 1998). Both the incident field (dipole field) and the scattered field are expanded in terms of vector spherical harmonics, and the boundary conditions are applied on the sphere surface. The expansion coefficients are the well-known Mie coefficients. The obtained fields are integrated in the far zone over the spherical surface S_∞ enclosing both the emitter and the sphere.

$$P^{\text{rad}} = \frac{c}{8\pi} \operatorname{Re} \int_{S_\infty} \left(\mathbf{E} \times \mathbf{H}^*\right) \mathbf{n}\, dS \qquad (2.2)$$

The results give information about the radiative decay rates of dipole sources near or inside the sphere.

3.2.2. Dipole in a Layered Spherical Particle

The fluorescent properties of atoms in layered spherical particles are of special interest (Moroz 2005). For example, dielectric-core metal-shell particles, also called metal nanoshells, consisting of a dielectric core with a metallic shell of nanometer thickness (50 nm), made it possible to vary the optical resonances of such nanoparticles over hundreds of nanometers in wavelength by varying the relative dimensions of the core and shell (Oldenburg 1998, 1999), which is much greater than the tunability of the single-sphere Mie resonances in the case of metal-core particles. An extra dielectric coating can be also added over the metallic shell (Graf et al. 2002), and fluorescent organic groups can be placed inside the dielectric core or dielectric shell, which can be done with a very high accuracy (on the scale of nanometers) (Blaaderen et al. 1992). With this structure, great freedom in engineering inelastic light-scattering properties such as surface enhanced Raman scattering (SERS), as well as atomic, molecular, and nanoprobe fluorescence properties such as spontaneous and stimulated decay rates, can be achieved, depending on the desired applications.

To treat the case of a spherical particle with several concentric shells, the method developed by Chew (1987, 1988) and Ruppin (1982) was enhanced by Moroz (2005). This case requires accounting for the multiple shell boundaries. This is achieved by utilizing the transfer-matrix formalism for the multilayered sphere shown in Figure 2.1.

The procedure consists of sequentially applying the boundary conditions at each shell interface, taking into account which layer (shell) contains the dipole source. Ultimately, the scattering properties of the whole multilayered sphere are obtained and the fields can be calculated. According to the correspondence principle, the radiative decay rate is then calculated by integrating the fields in the far zone around the structure. The nonradiative decay rate (due to Ohmic losses in the absorbing shells) can also be calculated as

$$P^{\mathrm{nrad}} = \frac{ck_0}{8\pi} \int_A \left(\mathrm{Im}\,\varepsilon \left|\mathbf{E}\right|^2 + \mathrm{Im}\,\mu \left|\mathbf{H}\right|^2 \right) d\mathbf{r} \tag{2.3}$$

We note at this point, that for the textbook cases discussed above such as a dipole near a partially reflecting surface, or a sphere, it can be explicitly shown that the results obtained by using the corresponding principle are equivalent to the rates calculated quantum mechanically in the linear-response approximation (Chance et al. 1978; Chew 1987; Moroz 2005; Novotny and Hecht 2006).

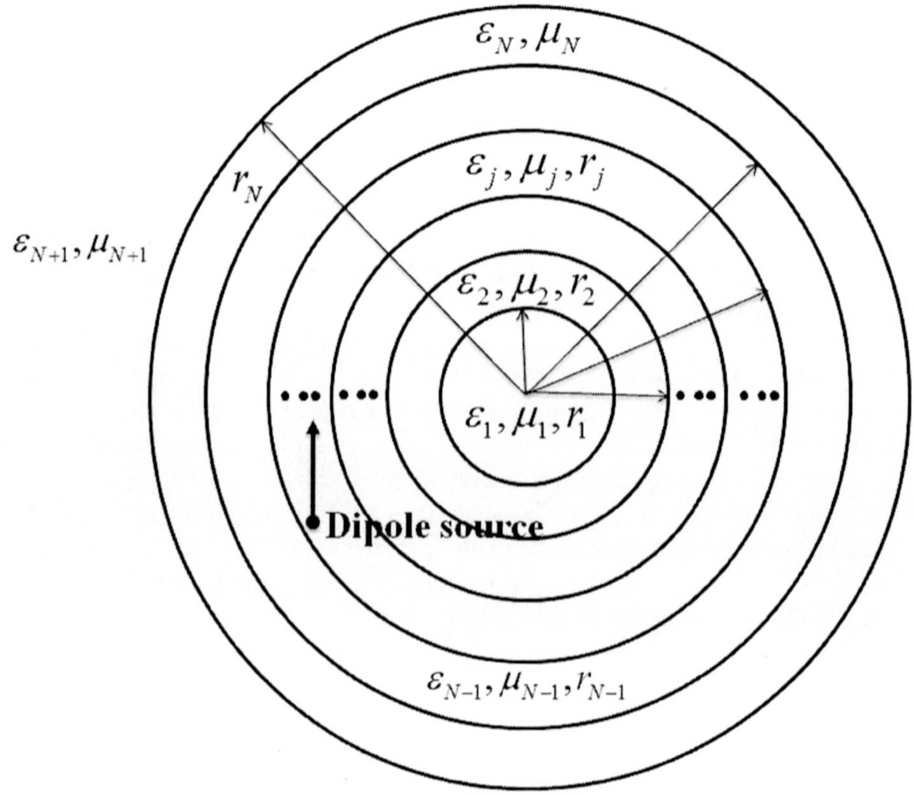

Figure 2.1. Schematic diagram of a multilayered sphere with a dipole source inside the *j*th shell.

3.2.3. Green's Function Formalism

The influence of the dielectric environment on the properties of an emitting source can be described by the local radiative density of optical states with the help of the Fermi's "golden rule." The LRDOS quantifies the amount of optical modes available for the decay of the excited state of the emitter. The LRDOS $N(\mathbf{r}, \omega, \mathbf{e}_d)$ depends on the frequency ω, position $\mathbf{r}$, and orientation $\mathbf{e}_d$ of the emitter. It is related to the dyadic Green's tensor as

$$N(\mathbf{r}, \omega, \mathbf{e}_d) = \frac{6\omega}{\pi c^2}\left\{\mathbf{e}_d^T \cdot \operatorname{Im}\mathbf{G}(\mathbf{r}, \mathbf{r}, \omega) \cdot \mathbf{e}_d\right\} \tag{2.4}$$

where $\mathbf{G}(\mathbf{r}, \mathbf{r}, \omega)$ is the dyadic Green's function which is a solution of an inhomogeneous wave equation

$$\left[\frac{\omega^2}{c^2}\varepsilon(\mathbf{r}) - \nabla\times\nabla\times\right]\ddot{\mathbf{G}}(\mathbf{r},\mathbf{r}',\omega) = -\ddot{\delta}_{\varepsilon_\perp}(\mathbf{r}-\mathbf{r}') \tag{2.5}$$

where $\varepsilon(\mathbf{r})$ is the dielectric constant varying in space, and $\ddot{\delta}_{\varepsilon_\perp}(\mathbf{r}-\mathbf{r}')$ is an ε-transverse dyadic delta function (Glauber 1991; Dowling 1992; Chigrin 2009).

In a lossless medium, the imaginary part of the Green's tensor describes the radiative decay rate of the emitter, while in a lossy medium it also accounts for the nonradiative quenching of the emitter by the environment.

The decay rate of a dipolar emitter in the weak-coupling approximation (Sprik et al. 1996; Busch et al. 1998, 2000; Vats et al. 2002; Novotny and Hecht 2006) reads

$$\Gamma(\mathbf{r},\omega,\mathbf{e}_d) = \frac{\pi d^2\omega}{\hbar\varepsilon_0}N(\mathbf{r},\omega,\mathbf{e}_d) \tag{2.6}$$

where d is the modulus of the matrix element of the transition dipole moment.

The Green's function and the LRDOS can be calculated analytically for the case of a planar interface and a sphere (Tai 1993). For complex structures, such as PCs, the plane-wave formalism (Nikolaev et al. 2009) can be used to compute the Green's function numerically. The finite-difference time method can also be used for this purpose, however, the accuracy of the calculation of the off-diagonal components of the Green's tensor is hampered by the fact that the electric and magnetic field values are calculated at different points in space.

3.2.4. Spontaneous Emission Rate in Three-Dimensional PCs via LDOS. Orientation-Dependent Rates

As has been shown above, the decay rate of the emitter depends on its frequency and position, as well as its orientation. So control over the orientation is of crucial importance for the purpose of manipulating spontaneous emission. It turns out that one does not need to sample the decay rate for a large set of emitter orientations $\mathbf{e}_d$ in order to adequately describe the orientational dependence of the rate. Vos et al. (2009) describe a method which allows modeling of the spontaneous emission rate surfaces (that is, the decay rate as a function of the emitter orientation) for a given emitter position, frequency-based calculating the rates at the fewest possible orientations (less than 10). The method is based on the Green's function formalism and makes special use of symmetry properties of the Green's function in a given environment. The approach provides a clear physical picture of emission rate dependencies in an arbitrary nanostructured environment. The authors also show how control over the emitter orientation opens up new possibilities for switching emission from inhibited to enhanced and vice versa.

The dependence on the orientation of the emitter can be clearly understood for the case of a dipole in front of a perfect mirror. Straightforward analysis in the framework of image theory shows that, for a dipole parallel to the mirror, the emission rate vanishes as the emitter approaches the mirror surface, because of destructive interference with the oppositely oriented image. On the other hand, for a perpendicularly oriented dipole, one has constructive interference and therefore the emission rate equals twice the value of the free dipole at zero distance from the mirror. We can see that the symmetry of the problem implies that there are two principal axes for the dipole in front of the mirror, corresponding to the minimum and maximum decay rates. The decay rate for any other direction equals the weighted average of the principal values.

It turns out that the concept of the principal axes holds for a dipole in an arbitrary dielectric environment. Below we follow the derivation in the work by Vos et al (2009). It can be inferred from the properties of the Green's tensor with the use of the reciprocity principle,

$$\mathrm{Im}\left[\mathbf{G}(\mathbf{r},\mathbf{r}',\omega)\right]^T = \mathrm{Im}\left[\mathbf{G}(\mathbf{r}',\mathbf{r},\omega)\right] \tag{2.7}$$

that the imaginary part of the Green's tensor is a real and symmetric 3×3 matrix, and therefore can always be diagonalized. It has three eigenvalues (g_1, g_2, g_3) corresponding to the three eigenvectors. Reshuffling the eigenvalues in ascending order, we can rewrite them as $(g_{\mathrm{min}}, g_{\mathrm{med}}, g_{\mathrm{max}})$, whereas the corresponding eigenvectors are then $(\mathbf{v}_{\mathrm{min}}, \mathbf{v}_{\mathrm{med}}, \mathbf{v}_{\mathrm{max}})$. They form a basis comprising the three principal axes of the dipole emitter, which vary with its position and frequency. The dipole orientation can be decomposed over this basis as

$$\mathbf{e}_d = \beta_1 \mathbf{v}_{\mathrm{min}} + \beta_2 \mathbf{v}_{\mathrm{med}} + \beta_3 \mathbf{v}_{\mathrm{max}} \tag{2.8}$$

where $\beta_1(\mathbf{e}_d) = \mathbf{e}_d \cdot \mathbf{v}_{\mathrm{min}}$, etc., and $\sum_i \beta_i^2 = 1$.

Ultimately, for the decay rate we arrive at

$$\Gamma(\mathbf{e}_d) = \beta_1(\mathbf{e}_d)\Gamma_{\mathrm{min}} + \beta_2(\mathbf{e}_d)\Gamma_{\mathrm{med}} + \beta_3(\mathbf{e}_d)\Gamma_{\mathrm{max}} \tag{2.9}$$

where the principal values of the decay rate are given by [see Eqs. (2.4) and (2.6)]

$$\Gamma_{\mathrm{min}}(\mathbf{r},\omega) = \frac{6d\omega^2}{\hbar\varepsilon_0 c^2}\left\{\mathbf{v}_{\mathrm{min}}^T \cdot \mathrm{Im}\,\mathbf{G}(\mathbf{r},\mathbf{r},\omega)\cdot \mathbf{v}_{\mathrm{min}}\right\} \equiv \frac{6d\omega^2}{\hbar\varepsilon_0 c^2} g_{\mathrm{min}} \tag{2.10}$$

etc.

Thus, we arrive at the main result of the analysis, which states that the dependence of the decay rate on the emitter orientation is a quadratic form with three principal axes on the unit sphere. As the decay rate is physically constrained to be positive for all orientations, polar plots of $\Gamma(\mathbf{e}_d)$ can only assume specific shapes determined by the ratios between Γ_{min}, Γ_{med}, and Γ_{max}.

This result greatly simplifies the task of finding $\Gamma(\mathbf{e}_d)$ in many cases of special interest, such as photonic crystals, where the Green's tensor, the principal axes, and the principal values of the decay rate are not known a priori. It seems that one needs to perform a full mapping of $\Gamma(\mathbf{e}_d)$, which amounts to a 100–1000 costly calculations of the rate at the unit sphere, making the task computationally prohibitive. However, it can be shown that knowledge of the quadratic form of $\Gamma(\mathbf{e}_d)$ allows full reconstruction of the orientational dependence $\Gamma(\mathbf{e}_d)$ by calculating it in only for just nine key orientations, which do not depend on the geometry of the physical system (Mohlenkamp 1999; Vos et al. 2009), thus greatly reducing the computation cost by avoiding a full mapping of $\Gamma(\mathbf{e}_d)$.

The above method is directly applicable to controlling emission properties of fluorescent molecules such as laser dyes (Schäfer 1977; Novotny and Hecht 2006), which belong to a class of quantum emitters with a single transition dipole moment. The orientation of such emitters can be controlled, e.g., by attaching them to liquid crystal molecules that are oriented in external fields (Gottardo et al. 2006). This opens a novel opportunity to "switch" spontaneous emission from inhibited to enhanced and vice versa. The emission rate surfaces reveal that optimal switching always requires a dipole rotation by 90°, since minimal and maximal emission rates always occur along the mutually perpendicular main axes (Vos et al. 2009).

Knowledge of the principal rates and axes is a key to proper interpretation of the dynamics of orientational dipole ensembles (Gérard et al. 1998; Lodahl et al. 2004; Muskens et al. 2007). Decay of such ensembles is a sum of single exponentials with a rate distribution given by the emission rate surface. Therefore, any observable derived from time-resolved decay beyond the orientation-averaged rate $\mathrm{Tr}\,\mathrm{Im}\,\mathbf{G}$ requires knowledge of the principal rates, which is thus relevant to many physical situations in nanophotonics.

3.2.5. Angular Distribution of Spontaneous Emission from PCs. Influence of Disorder

While the decay rates of colloidal quantum dots and laser dyes in periodic nanostructures, such as PCs, can be conveniently described within the Green's function formalism with help of either plane-wave or other multi-scattering formalisms, the real case of partially disordered systems requires accounting for the chaotic scattering of photons in the environment.

The emission spectra of sources embedded in PCs are strongly directional (Bogomolov et al. 1997; Blanco et al. 1998; Vlasov et al. 2000; Nikolaev et al. 2005; Yakimanski et al. 2009). There are two mechanisms that lead to strong directionality of the spectra. The first is the Bragg diffraction, which is the interference effect connected with the periodicity of the PC, and is manifested through stop bands (that is, suppression of the light propagation in certain directions), with center frequencies and widths described by the photonic band structure. The other important mechanism is the disorder of the PC, such as polydispersity, roughness, dislocations, etc. This disorder is inherent to PCs fabricated by self-assembling techniques (such as opals and inverse opals) and affects the Bragg diffraction in the PC. It has been shown that the disorder determines the depth of stop bands (Megens et al. 1999). The redistribution of emitted light due to the disorder leads to enhancement of emission at the stop band edges (Koenderink 2003). A model has been proposed that provides a quantitative description of this effect (Nikolaev et al. 2005). The model treats both interference and random scattering effects simultaneously, based on a combination of diffusion theory for the light transport and the PC band model.

Defects are always present in experimentally realized PCs, leading to random multiple scattering of light. Therefore all light emitted in such PCs becomes diffuse on length scales of the order of the transport mean free path l, which is often much smaller than the thickness of the sample L. For example, opals and inverse opals have $l = 15$ μm, while $L = 200$ μm (Megens et al. 1999; Koenderink et al. 2003; Nikolaev et al. 2009). This means that even though photons generated inside a PC are diffracted by the crystal structure, this effect is washed out by the random multiple scattering as the photons propagate toward the PC surface. Only at distances $z < l$ from the surface, where the photons propagate ballistically toward the surface after a last scattering event, is the effect of Bragg diffraction not destroyed by the scattering. So the diffuse emission becomes directionally dependent only when it exits the crystal.

The transport mean free path can be extracted from experiments by measuring the attenuation in the stop band of a PC. It is equal to $1 - L_B/l$, where L_B is the Bragg attenuation length, which can be estimated from band structure calculations (Koenderink 2003; Nikolaev et al. 2005).

The method for angular-dependent intensity calculation is based on the diffusion theory (Lagendijk et al. 1989). According to the diffusion theory, the angular-dependent intensity with frequency ω exiting the sample at external angles between θ_e and $\theta_e + d\theta_e$ relative to the surface normal (denoting $\mu_e = \cos\theta_e$) equals

$$I(\omega, \mu_e)\, d\mu_e = I_{\text{tot}}(\omega) P(\omega, \mu_e)\, d\mu_e \tag{2.11}$$

where $I_{\text{tot}}(\omega)$ is the total spontaneous emission spectrum of the light sources integrated over the exit angles. The distribution $P(\omega, \mu_e)$ is called the escape function and is defined as (Zhu et al. 1991; Durian 1994)

$$P(\omega,\mu_e) = \mu_e \frac{n_e^2}{n_i^2}\left[\frac{1+\bar{R}(\omega)}{1-\bar{R}(\omega)} + \frac{3}{2}\mu_i\right]\left[1 - R(\omega,\mu_i)\right] \tag{2.12}$$

where n_e and n_i are the average external and internal refractive indices, while μ_e and μ_i are related according to Snell's law. $R(\omega, \mu_i)$ is an angle-dependent internal reflection coefficient through which an angle-averaged internal reflection coefficient $\bar{R}(\omega)$ is defined as

$$\bar{R}(\omega) = \frac{3C_2(\omega) + 2C_1(\omega)}{3C_2(\omega) - 2C_1(\omega) + 2}$$

$$C_n(\omega) = \int_0^1 R(\omega,\mu_i)\mu_i^n \, d\mu_i \qquad n = 1,2 \tag{2.13}$$

The escape function describes the distribution of emission intensity over the escape angles. It has the form of the Lambertian distribution of diffuse surfaces in the absence of reflection effects.

The key feature of the method lies in the calculation of the internal-reflection coefficient. $R(\omega, \mu_i)$ is almost frequency-independent in random media (powders, macroporous sponges) (Schuurmans 1999), where the light propagation through the interface can be described with the use of the Fresnel coefficients, assuming an average refractive index. However, this approximation becomes incorrect if we consider a highly dispersive PC. In this case, the light escaping from $z < l$ undergoes a strong interference due to the PC periodicity and is Bragg-reflected for angles and frequencies inside a stop band, which leads to a reduced escape function inside the stop band [c.f. Eq. (2.12)]. Interestingly, from the analysis of the same equation, it follows that the increased averaged reflection coefficient $\bar{R}(\omega)$ leads to an increase of the escape function outside the stop band.

The internal-reflection coefficient is modeled with a modified Gaussian,

$$R(\omega,\mu_i) = A(\mu_i)\exp\left\{-\frac{\left[\omega - \omega_C(\mu_i)\right]^4}{2\left[\Delta\omega_C(\mu_i)\right]^4}\right\} \tag{2.14}$$

where the peak reflection $A(\mu_i)$ is extracted from the experiment. $\omega_C(\mu_i)$ is the central frequency of the stop band, $\Delta\omega_C(\mu_i)$ its width, as estimated from band structure calculations.

The results (Nikolaev et al. 2005, see Sections 3.3.1. and 3.3.2. below) show clearly the enhancements and suppressions in the escape function. A notable fact is the enhancements at the stop band edges, which come from redistribution of light being unable to exit at a specific angle in the stop-band frequency range to the angles outside the stop band.

3.3. MODELING THE MODIFICATION OF SPONTANEOUS EMISSION BASED ON THE FINITE-DIFFERENCE TIME-DOMAIN METHOD

Many problems concerned with the interaction of electromagnetic waves with nanostructured media deal with systems of very complex geometry. In such cases it appears more appropriate to solve the Maxwell equations numerically rather than exploit special symmetries of the problem. The examples include both LEDs and OLEDs, where the efficiency and directional properties of luminescence are optimized by adding nanostructured features to the layers and interfaces of the device (Lee et al. 2003, 2005; Chen et al. 2010), as well as luminescence-based bio- and chemical sensors, where the luminescence of specific molecules is enhanced by using arrays of metallic and dielectric nanoparticles (Lakowicz 2003).

The finite-difference time-domain (FDTD) method is one of the most powerful methods in the field of numerical modeling of 3-D nanostructured media (Taflove and Hagness 2005). It was developed first in 1960s as a tool for modeling radiation of antennas and was used heavily in military applications. Much later, in the 1990s, as the technology developed, it acquired tremendous popularity in the field of modeling optical phenomena in nanostructured systems. With the ever-growing efficiency of computer simulation it became possible to model quite complicated systems with the help of supercomputers. Nowadays, the FDTD method is one of the key tools in the modeling of optics in various technological applications, ranging from chemical and biosensors (Alfimov et al. 2010) to OLEDs (Lee et al. 2003, 2005; Chen et al. 2010) and to the modeling of magnetic resonance tomography (MRT) devices (Wang et al. 2007) and ultralow-frequency geophysics phenomena.

The FDTD method is a technique for direct numerical solution of time-dependent Maxwell equations (Taflove and Hagness 2005; Valuev et al. 2007, 2008; Deinega et al. 2007, 2009). The Maxwell equations in the absence of charge density (with c, ε_0, and 4π omitted for brevity) are

$$\begin{aligned} &\nabla \times \mathbf{E} = -\frac{\partial \mathbf{B}}{\partial t} \\ &\nabla \times \mathbf{H} = \mathbf{J} + \frac{\partial \mathbf{D}}{\partial t} \\ &\nabla \mathbf{D} = 0 \\ &\nabla \mathbf{B} = 0 \end{aligned} \tag{2.15}$$

where **E** and **H** are the electric and magnetic field vectors, respectively, while **D** and **B** are the electric and magnetic induction vectors, respectively. $\mathbf{J} = \sigma \mathbf{E}$ is the electric current density. Material equations relate fields to the respective inductions. In linear and isotropic materials these equations have the form

$$\mathbf{D} = \varepsilon\mathbf{E}$$
$$\mathbf{B} = \mu\mathbf{H} \tag{2.16}$$

where ε and μ are dielectric permittivity and magnetic permeability, respectively.

In the following, for simplicity, we assume the absence of currents and non-magnetic medium ($\mu = 1$). The discretization on the Cartesian grid proceeds in the following way. The first pair of Maxwell equations is decomposed into six equations for each field component as (only the equations for the x components are shown below)

$$\frac{\partial E_x}{\partial t} = \frac{1}{\varepsilon}\left(\frac{\partial H_z}{\partial y} - \frac{\partial H_y}{\partial z}\right)$$
$$\frac{\partial H_x}{\partial t} = \frac{1}{\varepsilon}\left(\frac{\partial E_y}{\partial z} - \frac{\partial E_z}{\partial y}\right) \tag{2.17}$$

These equations are further discretized on the Yee grid (Yee 1966) (see Figure 2.2).

The size of the cell depicted in Figure 2.1 is $\Delta x, \Delta y, \Delta z$. According to the stability criterion, the time step should be

$$dt \le \frac{\sqrt{(\Delta x)^2 + (\Delta y)^2 + (\Delta z)^2}}{c} \tag{2.18}$$

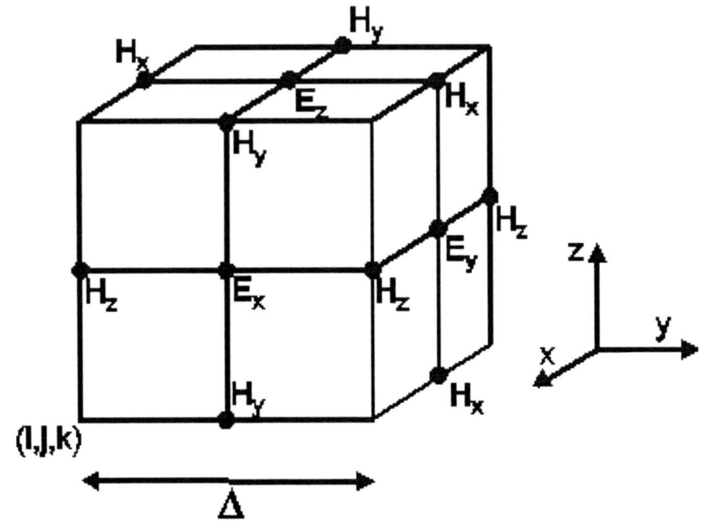

Figure 2.2. Yee grid for discretization of the Maxwell equations.

The fields $\mathbf{E}(x,y,z,t),\mathbf{H}(x,y,z,t)$ are approximated by the grid functions $\mathbf{E}^{n}_{i,j,k},\mathbf{H}^{n+1/2}_{i+1/2,j+1/2,k+1/2}$, where n corresponds to the moment $t = n\,\Delta t$, and i,j,k to the spatial coordinates $i\,\Delta x, j\,\Delta y, k\,\Delta z$, respectively. The field components at (i,j,k) are taken as (only x components are shown)

$$\begin{aligned}(E_x)_{i+1/2,j,k} &= E_x\left[(i+1/2)\,\Delta x, j\Delta y, k\Delta z\right]\\ (H_x)_{i,j+1/2,k+1/2} &= E_x\left[i\Delta x,(j+1/2)\,\Delta y,(k+1/2)\,\Delta z\right]\end{aligned} \tag{2.19}$$

The y and z components are written in a similar way.

The advantage of the Yee approximation consists in automatically fulfilled boundary conditions at a surface. Finally, using the second-order approximation over both the space and time coordinates, we arrive at the system of grid equations:

$$\begin{aligned}&(E_x)^{n+1}_{i+1/2,j,k} = (E_x)^{n}_{i+1/2,j,k}\\ &+\frac{\Delta t}{\varepsilon}\left[\frac{(H_z)^{n+1/2}_{i+1/2,j+1/2,k}-(H_z)^{n+1/2}_{i+1/2,j-1/2,k}}{\Delta y}-\frac{(H_y)^{n+1/2}_{i+1/2,j,k+1/2}-(H_y)^{n+1/2}_{i+1/2,j,k-1/2}}{\Delta z}\right]\\ &(H_x)^{n+1/2}_{i,j+1/2,k+1/2} = (H_x)^{n+1/2}_{i,j+1/2,k+1/2}\\ &+\frac{\Delta t}{\mu}\left[\frac{(E_y)^{n}_{i,j+1/2,k+1}-(E_y)^{n}_{i,j+1/2,k}}{\Delta z}-\frac{(E_z)^{n}_{i,j+1,k+1/2}-(E_z)^{n}_{i,j,k+1/2}}{\Delta y}\right]\end{aligned} \tag{2.20}$$

Here, in order to achieve second-order precision over the time coordinate, the electric and magnetic fields are calculated in a leapfrog fashion with the time shift $\Delta t/2$.

The material equation for the electric field is rewritten as (Lozovik et al. 2006)

$$\mathbf{E}^{n+1} = f\left(\mathbf{D}^{n+1},\ldots,\mathbf{D}^{n-M+1};\mathbf{E}^{n+1},\ldots,\mathbf{E}^{n-M+1}\right) \tag{2.21}$$

For the M-order decomposition, at a given time step this equation also involves fields and inductions from $M-1$ previous steps and must be solved inside the volume of the material. The dielectric properties of the material are given in the Drude–Lorentz form as

$$\varepsilon(\omega) = \varepsilon_1 - \sum_i \frac{\left(\omega^{(i)}_{pl}\right)^2}{\omega\left(\omega + i\gamma^{(i)}\right)} - \sum_j \frac{\varepsilon^{(j)}\left(\omega^{(j)}_0\right)^2}{\omega^2 + 2i\omega\delta^{(j)} - \left(\omega^{(j)}_0\right)^2} \tag{2.22}$$

One of the key features of the FDTD method consists of acquiring the wide-band spectral properties of the investigated structure in a single calculation. For this purpose, the structure is irradiated either by a plane-wave, or by a point-source impulse, which has a simple Gaussian form and is called a Berenger impulse:

$$E(t) \sim (t - t_0)\exp\left[-\left(\frac{t - t_0}{t_{\mathrm{decay}}}\right)^2\right] \tag{2.23}$$

After the impulse wave has interacted with the structure and has fully decayed, the resulting fields are Fourier-transformed and the required spectra can be obtained. The parameters of the impulse are chosen so it matches the frequency range of interest upon Fourier transform.

The radiation sources are modeled within the FDTD framework as dipole sources. According to the correspondence principle, the radiative decay rate in a given environment is calculated as a ratio of the total emitted power to the total emitted power of the dipole in the free space.

$$\Gamma_{\mathrm{rad}} = \frac{P_{\mathrm{tot}}}{P_0} \tag{2.24}$$

The nonradiative decay rate due to the absorption in the environment can be obtained by calculating the total inflow of emitted radiation in the absorbing material:

$$P_{\mathrm{abs}} = \frac{c}{8\pi}\mathrm{Re}\int_A \left(\mathbf{E} \times \mathbf{H}^*\right)\mathbf{n}\, dA \tag{2.25}$$

where the surface A encloses the absorbing material and excludes the emitter itself (see Figure 2.3).

3.3.1. Averaging over Orientation and Position

The dye molecules in PCs can be adsorbed either on the surface of the nanoparticles or embedded in porous nanoparticles and are randomly oriented (Nikolaev et al. 2005; Yakimanski et al. 2009; Alfimov et al. 2010). This means that the experimentally observed emission spectra are averaged over both the position and orientation of emitters. From the modeling point of view, the spectra produced by the emitting dipoles should also be averaged.

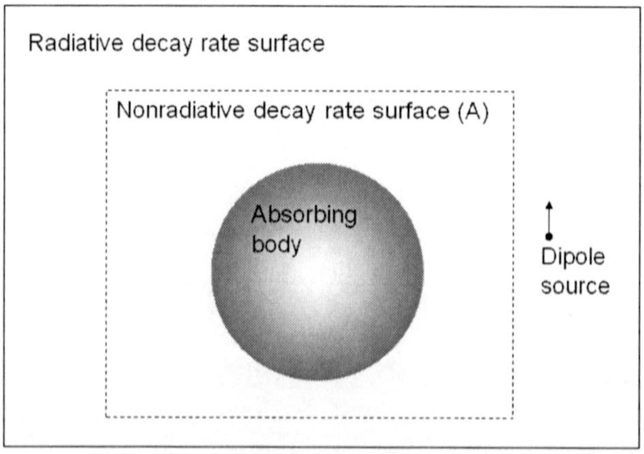

Figure 2.3. Calculation of decay rates with FDTD.

There are two approaches to averaging the dipole emissions in the FDTD method. The first approach makes use of the self-averaging of emitters which are distributed randomly over the computational volume and are ignited at random time steps during the computation (Lee et al. 2005). A sufficient number of sources is required in order to smear the interference effects (of the order of $N \sim 10^4$ and more) in order to ensure that the integral intensity grows as $I \sim N$. However, in our experience, even such a number of sources is insufficient to obtain correct far-field distributions.

The second approach (Greiner et al. 2008) relies on a simple consideration, that an orientational average of dipole emission intensity is a mean value of the intensities of three mutually transverse dipoles:

$$I_{\text{mean}} = \frac{1}{3}\left(I_x + I_y + I_z\right) \tag{2.26}$$

Thus, in order to obtain the orientational average for a source at a given position, three separate calculations are performed. The space averaging is performed by calculating the emission intensity of a dipole in different positions in a unit cell in the PC and subsequent averaging.

3.3.2. *Examples of Using FDTD for Calculating Modification of Spontaneous Emission*

3.3.2.1. DIPOLE NEAR A METALLIC SPHERE

As an example of modification and angular redistribution of a dipole emission in a complex dielectric environment, we calculate the emission spectrum and angular

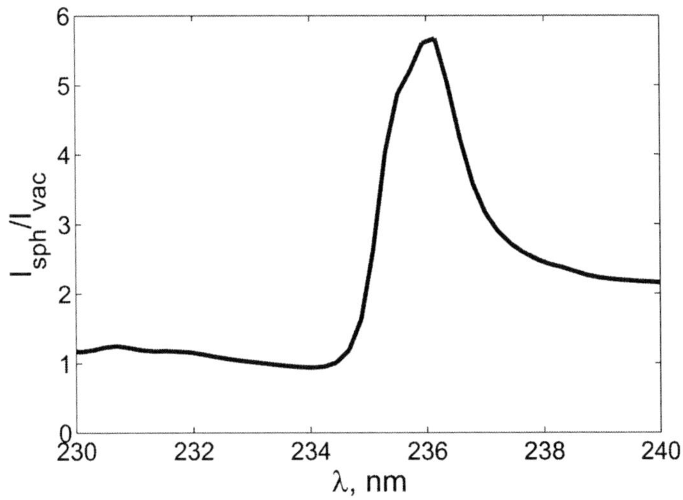

Figure 2.4. Emission enhancement factor for a dipole near an Ag sphere (see text).

distribution of a dipole near a metallic sphere. The sphere is made of silver [Ag; the dielectric constant is taken from Palik (1985)] and has radius $R = 80$ nm, with the dipole positioned at $d = 20$ nm from the sphere surface and oriented parallel to the surface. The spectrum has a strong maximum at the local surface plasmon resonance $\omega = \omega_p/\sqrt{3}$ (Figure 2.4). The angular distribution of emission below the resonance has a notable five-lobed structure (Figure 2.5) (Novotny 2006).

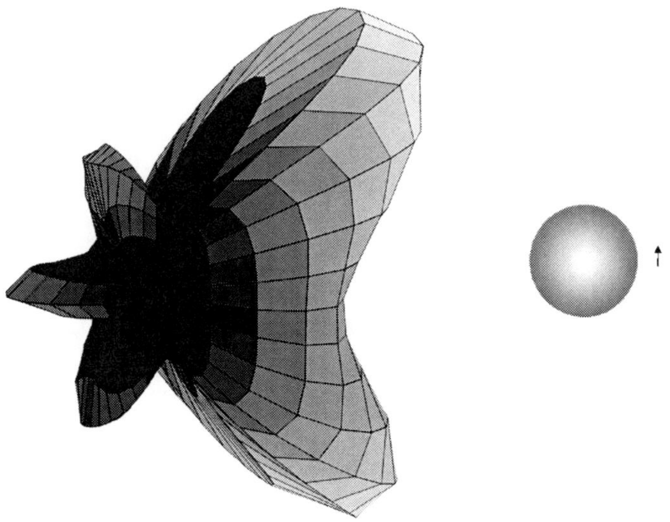

Figure 2.5. (Left) Angular pattern of the dipole emission near the Ag sphere below the plasmon resonance. (Right) Sketch of the modeled system.

3.3.2.2. DIPOLE INSIDE A LOG-PILE PC

A PC with a PBG totally inhibits the emission of an embedded source. We illustrate this by calculating the emission spectra of a dipole source embedded in a log-pile PC (Li et al. 2003; Chen et al. 2010). The PC is composed of dielectric rods stacked in a lattice, with each layer consisting of parallel rods with spacing a between the centers of the rods. Each rod has a rectangular cross section with width w and height h. The rods are stacked in layers along the z axis. The rods in each successive layer are arranged by rotating by 90° with respect to the previous layer. The second neighboring layer is shifted by $a/2$. Thus, 4 layers comprise the period in the z direction. The structure has face-centered tetragonal symmetry. We consider a log-pile PC with $w / a = h / a = 0.25$. The material of the rods is GaAs, with $n = 3.5$. Such a PC has a wide PBG in the frequency range $a/\lambda = 0.43 - 0.50$ (Chen et al. 2008, 2010).

To model the modification of spontaneous emission in a PC, we consider a sufficiently large finite fragment of the PC so as to minimize the effects of finiteness on the properties of radiation from inside the bulk of the crystal. The log-pile PC

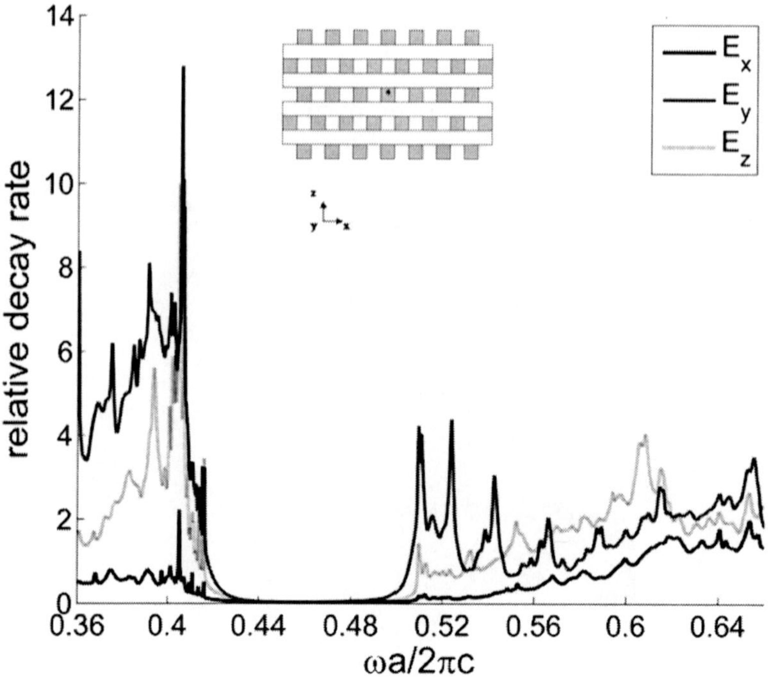

Figure 2.6. Relative radiative decay rates of a dipole in the PC for the different dipole orientations. Inset shows the sketch of the log-pile PC and the position of the source.

is modeled by a finite structure consisting of 14 periods in the transverse directions and 9 layers. The dipole is positioned at the center of the rod as shown in the inset in Figure 2.6. The three mutually transverse orientations are considered. The relative radiative decay rate is computed as the ratio of the emission intensity of the dipole embedded in the PC to the emission intensity of the dipole in free space. The results are shown in Figure 2.6.

The emission is strongly suppressed inside the PBG for all dipole orientations. The strong enhancement on the lower-frequency (red) side of the gap for the *z*- and *y*-oriented sources is due to the band edge resonance. Such resonances occur whenever the wave vector of the wave propagating inside the PC is close to the edge of the Brillouin zone, where the standing wave forms and the group velocity become low. The low group velocity corresponds to the large density of states, which reveals itself through the emission enhancement (see, e.g., Sakoda 2001; Chigrin 2009). In other words, the enhancement is caused by longer field–emitter interaction times due to the slowing of the waves near the edge of the gap.

The above results show that a strong modification of spontaneous emission can be obtained with a PC composed of semiconducting material (GaAs), which is the essential component of LED-based devices, such as screens, biosensors, and low-threshold lasers. Using nanostructured semiconducting components in designing such devices paves the way to significant improvement of their efficiency (Fan et al. 1997; Veronis et al. 2005; David et al. 2006; Bermel et al. 2006; Chen et al. 2010).

The above results also serve as a verification of our method versus the modeling results in literature (Chen et al. 2010).

3.3.3. Spontaneous Emission of Dye Molecules from a PC: Modeling Examples and Discussion

Let us consider the emission from a polystyrene opal doped with Rhodamine 6G dye molecules in air (cf. Nikolaev et al. 2005). The PC consists of close-packed polystyrene spheres with refractive index $n_{PC} = 1.59$ and filling factor $f = 0.74$. The conventional cubic lattice constant of the PC $a = 365$ nm. The dye molecules are located at the surface of the polystyrene spheres. In the FDTD simulation, the PC is modeled by a parallelepiped consisting of 20 layers of spheres arranged into an FCC lattice (the total height is 4.262 μm), stacked along the (111) crystallographic axis, with a total transverse size of 5.1619 × 8.9752 μm. The emitting dye is modeled as a dipole source and is placed at the surface of a sphere located at the geometric center of the structure.

The calculation is organized as follows. First, the radiative decay rate corresponding to a relative emission intensity is computed as described in the previous section by normalizing the result for the dipole in the structure to the result for the dipole in free space. The emission spectra of the dye is then accounted for by multiplying the relative emission intensity by the publicly available emission data for Rhodamine 6G (Du et al. 1998).

The polystyrene opal is only weakly photonic due to the low dielectric contrast between the polystyrene spheres and the air voids. Thus the relative emission intensity spectra obtained for this type of structure will show the PC features [first-order stop gap associated with the (111) direction in the FCC opal] overlapping with the unwanted Fabry-Pérot–like fringes resulting from reflections from the edges of the parallelepiped. The period of the fringes for the modeled structure is less than the width of the stop gap. Therefore, these fringes can be effectively removed by applying a running average to the spectra without losing the essential features of the spectra arising due to the PC periodicity. The spectra are calculated for the three mutually transverse orientations (Figure 2.7). All the spectra possess a shallow dip corresponding to the stop gap in the (111) direction. Note that the results for different orientations are different in the long-wavelength regime. This anisotropy seems surprising, as one would expect some sort of effective media behavior for the long wavelengths with

$$n_{\text{eff}} = n_{\text{PC}} f + (1 - f) \approx 1.43 \tag{2.27}$$

and the intensity of the dipole source emitting inside the medium with refractive index n_{eff} :

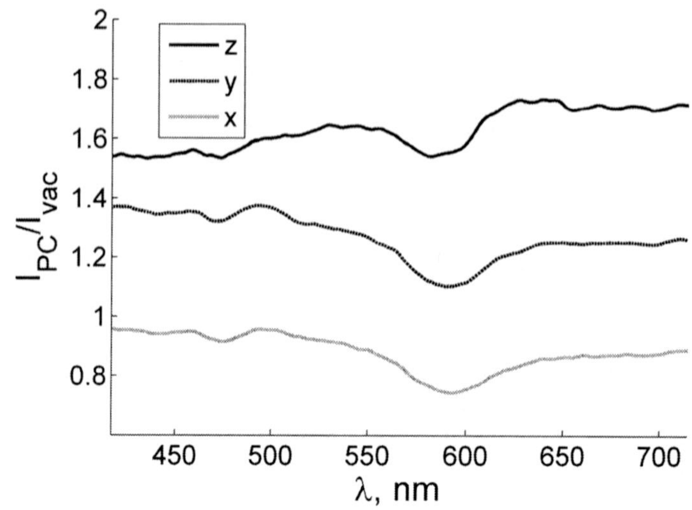

Figure 2.7. Relative radiative decay rate of a dipole source embedded in the polystyrene opal PC. Results for the three mutually transverse orientations are shown. The Fabry-Pérot fringes are removed by applying the running average.

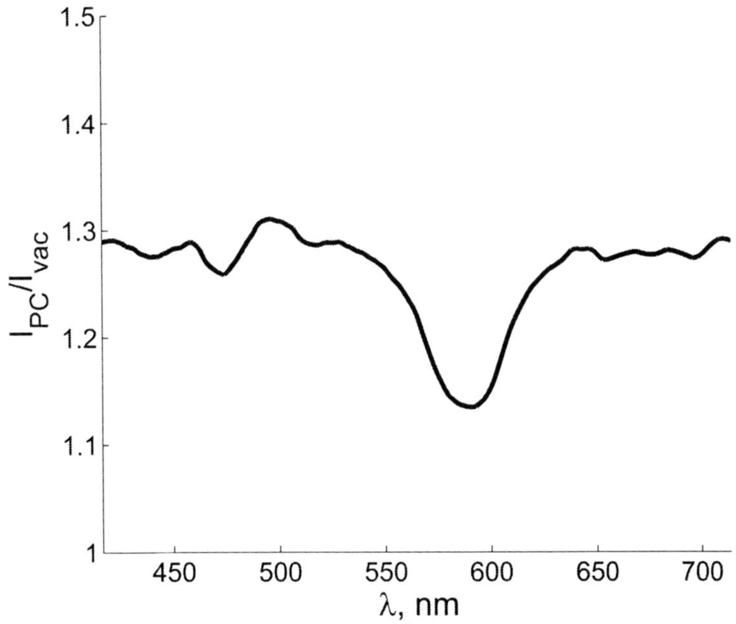

Figure 2.8. Relative radiative decay rate of a dipole source embedded in the polystyrene opal PC. Averaging over orientations is performed. The Fabry-Pérot fringes are removed by applying the running average.

$$I_{\text{eff}} = I_0 n_{\text{eff}} \tag{2.28}$$

where I_0 is the dipole emission intensity in vacuum.

However, the anisotropy is explained by the electrostatic depolarization effects, which have been discussed in the literature (Miyazaki and Ohtaka 1998; Rogobete et al. 2003), which lead to different results for every orientation for a dipole positioned at the surface of a sphere inside the opal.

To obtain a result for a randomly oriented source, averaging over the orientations is performed.

In Figure 2.8 the angle-integrated relative emission spectrum is shown. The figure clearly demonstrates the dip corresponding to the first-order stop gap in the (111) direction, in good agreement with the experimental data (Nikolaev et al. 2005).

In Figure 2.9, the spectra $P(\omega,\theta)$ for different observation angles with respect to the (111) direction (surface normal) are presented. All spectra are integrated over the axial axis and then normalized to the angle-integrated emission spectra, so that the emission intensity in an infinitesimal element of the solid angle around the given direction is given by

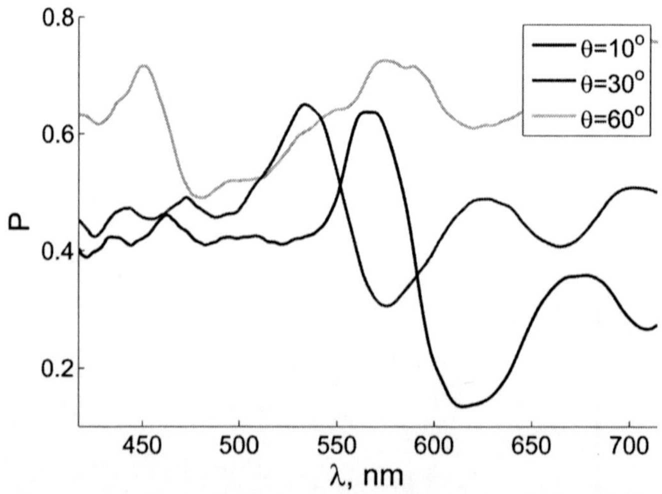

Figure 2.9. Escape function $P(\omega, \theta)$ for different observation angles for a source embedded in the polystyrene opal PC. Averaging over orientations is performed. The Fabry-Pérot fringes are removed by applying the running average.

$$I(\omega,\theta)\sin\theta\, d\theta = I_{\text{tot}} P(\omega,\theta)\sin\theta\, d\theta \tag{2.29}$$

In Figure 2.10, we plot the angle-integrated emission spectra of the dye in the PC as well as the spectra for the specified observation angles, according to:

$$I_{\text{R6G}}(\omega,\theta) = I_{\text{R6G}}^{(0)} P(\omega,\theta) \tag{2.30}$$

$$I_{\text{R6G,tot}} = I_{\text{R6G}}^{(0)} I_{\text{tot}}(\omega,\theta) \tag{2.31}$$

A running average is performed to remove the unwanted fringes as in the case of the total emission spectra. For each observation angle, the result clearly demonstrates the stop gaps associated with the (111) direction in the FCC opal PC. The shift of the stop gap to the lower wavelengths (higher frequencies) with increasing observation angle is in full accordance with Bragg's law (Sakoda 2001; Bogdanova et al. 2010),

$$\lambda_B = 2d_{111}\sqrt{n_{\text{eff}}^2 - \sin^2\theta} \tag{2.32}$$

where λ_B is the stop-gap position, $d_{111} = a/\sqrt{3}$ is the interlayer spacing in the (111) direction, and the refraction of the light coming from inside the PC on its surface is taken into account through Snell's law.

The normalization we use here is identical to that used in the literature (Nikolaev et al. 2005). The function $P(\omega,\theta)$ is the escape function. However, our results are different from those described in the cited work (Nikolaev et al. 2005) in two respects: (1) we model a perfectly periodic structure, so random scattering is not present in our simulation; (2) the modeled structure is only 20 layers thick compared to several hundred layers in the work of Nikolaev et al. (2005).

By changing the dielectric contrast, material filling fraction, and the lattice constant of a PC, the regions of both suppression and enhancement can be either broadened or narrowed and shifted. Fitting the PC parameters so that dye emission lines fall within a suppression or enhancement region can lead to enhancement of the optical response of the dye in presence of an analyte, improving the efficiency of the chemical sensing system.

The dye molecules in real dye-doped PC samples can be distributed either in voids between the spheres, or at the surface of the spheres, or inside the porous spheres, depending on both the fabrication technique and the materials used (Alfimov et al. 2010). In modeling the spontaneous emission modification of dyes in PCs, we assume that the dyes do not interact with each other, that is, the processes of reabsorption and energy transfer between the neighboring molecules are irrelevant (see, e.g., Moroz 2005 and citations therein). Such an assumption is fairly justified, as a typical density of dye molecules in a dye-doped PC is of order of tens per the PC unit cell, which is two orders of magnitude lower than the typical molecular density at which these processes play a significant role (Lakowicz 1999).

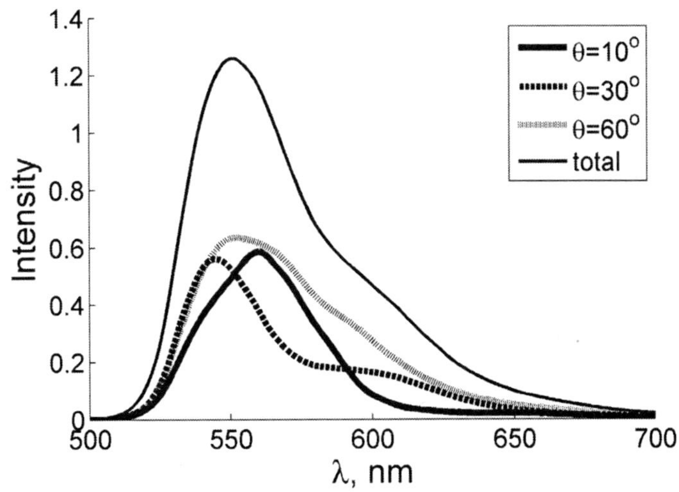

Figure 2.10. Emission intensity $I_{R6G}(\omega,\theta)$ for different observation angles for a source embedded in the polystyrene opal PC, compared to the total emission intensity $I_{R6G,tot}(\omega,\theta)$. Averaging over orientations is performed. The Fabry-Pérot fringes are removed by applying the running average. The Rhodamine 6G spectra (Du et al. 1998) is normalized to its maximum intensity.

In order to fully characterize the emission properties of dyes in PCs, it is necessary to average over both the orientations and the positions of the sources. Special care should be taken regarding sources positioned close to the sample borders (less than the Bragg attenuation length L_B), for which the influence of the PC is weaker than for sources inside the bulk of the sample. Thus, in the case of relatively thin samples (with thickness L corresponding to 20–50 layers), taking account of the emission of the sources near the surface is relevant, as the ratio L_B/L is not small. In the case of thick samples, with the number of layers greater than 100, however, it appears necessary to average only over the inner area of the PC, away from the sample borders, as the influence of the sources near the surface is small. This statement is confirmed by experimental studies (Nikolaev et al. 2005), in which the photobleaching technique was used to deactivate sources near the sample surface, in order to compare the fluorescence spectra of an unbleached sample to that of a bleached one.

There are two approaches to the averaging, as we have discussed above. The first is based on the self-averaging of a dipole ensemble with random position, orientation, and moment of ignition within a single FDTD run. In order to achieve self-averaging, a sufficiently large ensemble of dipoles N should be considered. From our experience, the onset of the self-averaging regime occurs when $N > 10^3$, and reveals itself as the linear dependence of the total intensity on the dipole number $I \sim N$. However, this approach appears to fail due to persisting correlations in the case of angle-resolved pattern calculation. Thus, the second approach should be used instead, where a single source per run is simulated. The results of such averaging for the case of the 3-D PC will be presented elsewhere.

4. CONCLUSION

Modification of the spontaneous emission in nanostructured media and approaches to the modeling of this effect have been reviewed in this chapter. Two types of methods have been discussed. The first is based on the correspondence principle, which relates the decay rate of a quantum dipole emitter to the power radiated by the classical dipole with the dipole moment and orientation equal to those of the original quantum emitter. The second approach relies on calculating the dyadic Green's function of the electromagnetic problem, which provides information on the local density of optical states (LDOS). The latter is directly related to the decay rate of an emitter through Fermi's "golden rule." The approach based on the FDTD method (Taflove and Hagness 2005) has been discussed in particular related to modeling of the spontaneous emission in photonic crystals. The decay rates are calculated with the help of the correspondence principle. Examples of applying FDTD to the modeling of emitters in a complex photonic environment have been presented. The results were obtained with the program

FDTD-II (Kintech 2011), realizing the FDTD algorithm with a very high parallel efficiency, developed by the Kintech Lab.

REFERENCES

Alfimov M.V., Bagatur'yants A., Safonov A., Scherbinin A., Vladimirova K., Belousov S., Bogdanova M., Valuev I., Deinega A., Lozovik Yu., and Potapkin B. (2010) Multiscale computer design of photonic crystal based materials for optical chemosensors. *Nanotech. Russ.* **5**(3–4), 250–258. DOI: 10.1134/S1995078010030134

Andreeva L.V., Koshkin A.V., Lebedev-Stepanov P.V., Petrov A.N., and Alfimov M.V. (2007) Driving forces of the solute self-organization in an evaporating liquid microdroplet. *Colloids Surf. A* **300**, 300–306. DOI: 10.1016/j.colsurfa.2007.02.001

Bermel P., Lidorikis E., Fink Y., and Joannopoulos J.D. (2006) Active materials embedded in photonic crystals and coupled to electromagnetic radiation. *Phys. Rev. B* **73**, 165125-1–162125-8. DOI: 10.1103/PhysRevB.73.165125

Blanco A., López C., Mayoral R., Míguez H., Meseguer F., Mifsud A., and Herrero J. (1998) CdS photoluminescence inhibition by a photonic structure. *Appl. Phys. Lett.* **73**, 1781–1783. DOI: 10.1063/1.122280

Bogdanova M.V., Lozovik Yu.E., and Eiderman S.L. (2010) An optical analog of the Borrmann effect in photonic crystals. *JETP* **110**(4), 604. DOI: 10.1134/S1063776110040072

Bogomolov V.N., Gaponenko S.V., Germanenko I.N., Kapitonov A. M., Petrov E. P., Gaponenko N.V., Prokofiev A.V., Ponyavina A.N., Silvanovich N.I., and Samoilovich S.M. (1997) Photonic band gap phenomenon and optical properties of artificial opals. *Phys. Rev. E* **55**, 7619–7625. DOI: 10.1103/PhysRevE.55.7619

Bohren C.F. and Huffman D.R. (1998) *Absorption and Scattering of Light by Small Particles.* John Wiley, New York. DOI: 10.1002/9783527618156

Busch K. and John S. (1998) Photonic band gap formation in certain self-organizing systems. *Phys. Rev. E* **58**, 3896–3908. DOI: 10.1103/PhysRevE.58.3896

Busch K., Vats N, John S., and Sanders B.C. (2000) Radiating dipoles in photonic crystals. *Phys. Rev. E* **62**, 4251–4260. DOI: 10.1103/PhysRevE.62.4251

Bykov V.P. (1972) Spontaneous emission in a periodic structure. *Sov. Phys. JETP* **35**, 269–273.

Bykov V.P. (1975) Spontaneous emission from a medium with a band spectrum. *Sov. J. Quantum Electron.* **4**, 861–871. DOI: 10.1070/QE1975v004n07ABEH009654

Chance R.R., Prock A., and Silbey R. (1974) Lifetime of an emitting molecule near a partially reflecting surface. *J. Chem. Phys.* **60**, 2744–2748. DOI: 10.1063/1.1681437

Chance R.R., Prock A., and Silbey R. (1975) Decay of an emitting dipole between two parallel mirrors. *J. Chem. Phys.* **62**, 771–772. DOI: 10.1063/1.430523

Chance R.R., Prock A., and Silbey R. (1975) Comments on the classical theory of energy transfer. *J. Chem. Phys.* **62**, 2245–2253. DOI: 10.1063/1.430748

Chance R.R., Prock A., and Silbey R. (1978) Molecular fluorescence and energy transfer near interfaces. *Adv. Chem. Phys.* **37**, 1–65. DOI: 10.1002/9780470142561.ch1

Chen J.F., Hong R.-T., and Yang J.-Y. (2010) Modified spontaneous emission rate in three-dimensional layer-by-layer photonic crystals with planar defects. *J. Appl. Phys.* **107**, 023110-1–023110-9. DOI: 10.1063/1.3279659

Chew H. (1987) Transition rates of atoms near spherical surfaces. *J. Chem. Phys.* **87**, 1355–1360. DOI: 10.1063/1.453317

Chew H. (1988) Radiation and lifetimes of atoms inside dielectric particles. *Phys. Rev. A* **38**, 3410–3416. DOI: 10.1103/PhysRevA.38.3410

Chigrin D.R. (2009) Spatial distribution of the emission intensity in a photonic crystal: Self-interference of Bloch eigenwaves. *Phys. Rev. A* **79**, 033829-1–033829-9. DOI: 10.1103/PhysRevA.79.033829

David A., Fujii T., Moran B., Nakamura S., DenBaars S.P., Weisbuch C., and Benisty H. (2006) Photonic crystal laser lift-off GaN light-emitting diodes. *Appl. Phys. Lett.* **88**, 133514-1–133514-3. DOI: 10.1063/1.2189159

Deinega A., Belousov S., and Valuev I. (2009) Hybrid transfer-matrix FDTD method for layered periodic structures. *Opt. Lett.* **34**, 860–862. DOI: 10.1364/OL.34.000860

Deinega A. and Valuev I. (2007) Subpixel smoothing for conductive and dispersive media in the finite-difference time-domain method. *Opt. Lett.* **32**, 3429–3431. DOI: 10.1364/OL.32.003429

Dowling J.P. and Bowden C.M. (1992) Atomic emission rates in inhomogeneous media with applications to photonic band structures. *Phys. Rev. A* **46**, 612–622. DOI: 10.1103/PhysRevA.46.612

Du, H., Fuh R.-C.A., Li J., Corkan, L.A., and Lindsey J.S. (1998) PhotochemCAD: A computer-aided design and research tool in photochemistry. *Photochem. Photobiol.* **68**, 141–142. DOI: 10.1111/j.1751-1097.1998.tb02480.x

Durian D.J. (1994) Influence of boundary reflection and refraction on diffusive photon transport. *Phys. Rev. E* **50**, 857–866. DOI: 10.1103/PhysRevE.50.857

Englund D., Fattal D., Waks E., Solomon G., Zhang B., Nakaoka T., Arakawa Y., Yamamoto Y., and Vučkovic J. (2005) Controlling the spontaneous emission rate of single quantum dots in a two-dimensional photonic crystal. *Phys. Rev. Lett.* **95**, 013904-1–013904-4. DOI: 10.1103/PhysRevLett.95.013904

Fan S., Villeneuve P.R., and Joannopoulos J.D. (1997) High extraction efficiency of spontaneous emission from slabs of photonic crystals. *Phys. Rev. Lett.* **78**, 3294-3297. DOI: 10.1103/PhysRevLett.78.3294

Ford G.W. and Weber W.H. (1984) Electromagnetic interactions of molecules with metal surfaces. *Phys. Rep.* **113**(4), 195–287. DOI: 10.1016/0370-1573(84)90098-X

Fort E. and Grésillon S. (2008) Surface enhanced fluorescence. *J. Phys. D: Appl. Phys.* **41**, 013001-1–013001-31. DOI: 10.1088/0022-3727/41/1/013001

Ganesh N. Zhang W., Mathias P.C., Chow E., Soares J.A.N.T., Malyarchuk V., Smith A.D., and Cunningham B.T. (2007) Enhanced fluorescence emission from quantum dots on a photonic crystal surface. *Nat. Nanotech.* **2**, 515–520. DOI: 10.1038/nnano.2007.216

Gérard J.-M., Sermage B., Gayral B., Legrand B., Costard E., and Thierry-Mieg V. (1998) Enhanced spontaneous emission by quantum boxes in a monolithic optical microcavity. *Phys. Rev. Lett.* **81**, 1110–1113. DOI: 10.1103/PhysRevLett.81.1110

Giannini V. Sánchez-Gil J.A., Muskens O.L., and Gomez-Rivas J. (2009) Electrodynamic calculations of spontaneous emission coupled to metal nanostructures of arbitrary shape: Nanoantenna-enhanced fluorescence. *J. Opt. Soc. Am. B* **26**, 1569–1577. DOI: 10.1364/JOSAB.26.001569

Glauber R.J. and Lewenstein M. (1991) Quantum optics of dielectric media. *Phys. Rev. A* **43**, 467–491. DOI: 10.1103/PhysRevA.43.467

Gottardo S., Burresi M., Geobaldo F., Pallavidino L., Giorgis F., and Wiersma D.S. (2006) Self-alignment of liquid crystals in three-dimensional photonic crystals. *Phys. Rev. E* **74**, 040702R-1–040702R-4. DOI: 10.1103/PhysRevE.74.040702

Graf C. and van Blaaderen A. (2002) Metallodielectric colloidal core–shell particles for photonic applications. *Langmuir* **18**, 524–534. DOI: 10.1021/la011093g

Greiner H.J.R., Bienstman P., Pond J., and Vandersteegen P. (2008) FDTD and RCWA simulations of OLED light extraction structures. *SPIE Photonics Europe Conference, France*, 174.

John S. (1987) Strong localization of photons in certain disordered dielectric superlattices. *Phys. Rev. Lett.* **58**, 2486–2487. DOI: 10.1103/PhysRevLett.58.2486

Julsgaard B., Johansen J., Stobbe S., Stolberg-Rohr T., Sünner T., Kamp M., Forchel A., and Lodahl P. (2008) Decay dynamics of quantum dots influenced by the local density of optical states of two-dimensional photonic crystal membranes. *Appl. Phys. Lett.* **93**, 094102-1–094102-3. DOI: 10.1063/1.2977605

Kintech Lab. http://fdtd.kintechlab.com.

Koenderink A.F. (2003) Emission and transport of light in photonic crystals. Ph.D. thesis, Universiteit van Amsterdam, Amsterdam, The Netherlands.

Kuhn H. (1970) Classical aspects of energy transfer in molecular systems. *J. Chem. Phys.* **53**, 101–108. DOI: 10.1063/1.1673749

Lagendijk A., Vreeker R., and de Vries P. (1989) Influence of internal reflection on diffusive transport in strongly scattering media. *Phys. Lett. A* **136**, 81–88. DOI: 10.1016/0375-9601(89)90683-X

Lakowicz J.R. (1999) *Principles of Fluorescence Spectroscopy*, 2nd ed. Kluwer, New York.

Lakowicz J.R., Malicka J., Gryczynski I., Zygmunt Gryczynski Z., and Geddes C.D. (2003) Radiative decay engineering: The role of photonic mode density in biotechnology. *J. Phys. D: Appl. Phys.* **36**, R240–R249. DOI: 10.1088/0022-3727/36/14/203

Lee Y.-J., Kim S.-H., Huh J., Kim G.-H., Lee Y.-H., Cho S.-H., Kim Y.-C., and Do Y. R. (2003) A high-extraction-efficiency nanopatterned organic light-emitting diode. *Appl. Phys. Lett.* **82**, 3779–3781. DOI: 10.1063/1.1577823

Lee Y.-J., Kim S.-H., Kim G.-H., Lee Y.-H., Cho S.-H., Song Y.-W., Kim Y.-C., and Do Y.R. (2005) Far-field radiation of photonic crystal organic light-emitting diode. *Opt. Express* **15**, 5864–5870. DOI: 10.1364/OPEX.13.005864

Li Z.-Y., El-Kady I., Ho K.-M., Lin S.Y., and Fleming J.G. (2003) Photonic band gap effect in layer-by-layer metallic photonic crystals. *J. Appl. Phys.* **93**, 38–42. DOI: 10.1063/1.1529294

Lodahl P., van Driel A. F., Nikolaev I.S., Irman A., Overgaag K., Vanmaekelbergh D., and Vos W.L. (2004) Controlling the dynamics of spontaneous emission from quantum dots by photonic crystals. *Nature* **430**, 654–657. DOI: 10.1038/nature02772

Lozovik Yu.E. and Eiderman S.L. Calculation of the band structure of three-dimensional metallic photonic crystals with FDTD method. *Mat. Model.* **18**(12), 35–42 (in Russian).

Megens M., Wijnhoven J.E.G.J., Lagendijk A., and Vos W.L. (1999) Light sources inside photonic crystals. *J. Opt. Soc. Am. B* **16**, 1403–1408. DOI: 10.1364/JOSAB.16.001403

Mohlenkamp M.J. (1999) A fast transform for spherical harmonics. *J. Fourier Anal. Appl.* **5**, 159–184. DOI: 10.1007/BF01261607

Moroz A. (2005) A recursive transfer-matrix solution for a dipole radiating inside and outside a stratified sphere. *Ann. Phys.* **315**, 352–418. DOI: 10.1016/j.aop.2004.07.002

Muskens O.L., Giannini V., Sánchez-Gil J.A., and Gómez Rivas J. (2007) Strong enhancement of the radiative decay rate of emitters by single plasmonic nanoantennas. *Nano Lett.* **7**, 2871–2875. DOI: 10.1021/nl0715847

Miyazaki H. and Ohtaka K. (1998) Near-field images of a monolayer of periodically arrayed dielectric spheres. *Phys. Rev. B* **58**, 6920–6937. http://dx.doi.org /10.1103/PhysRevB.58.6920

Nicolis G. and Prigogine I. (1977) *Self-organization in Non-equilibrium Systems: From Dissipative Structures to Order Through Fluctuations.* John Wiley, New York.

Nikolaev I.S., Lodahl P., and Vos W.L. (2005) Quantitative analysis of directional spontaneous emission spectra from light sources in photonic crystals. *Phys. Rev. A* **71**, 053813-1–053813-10. DOI: 10.1103/PhysRevA.71.053813

Nikolaev I.S., Vos W.L., and Koenderink A. F. (2009) Accurate calculation of the local density of optical states in inverse-opal photonic crystals. *J. Opt. Soc. Am. B* **26**, 987–997. DOI: 10.1364/JOSAB.26.000987

Nikolaev I.S., Lodahl P., van Driel A.F., Koenderink A.F., and Vos W.L. (2007) Strongly nonexponential time-resolved fluorescence of quantum-dot ensembles in three-dimensional photonic crystals. *Phys. Rev. B* **75**, 115302-1–115302-5. DOI: 10.1103/PhysRevB.75.115302

Novotny L. and Hecht B. (2006) *Principles of Nano-Optics.* Cambridge University Press, Cambridge, UK.

Oldenburg S.J. (1998) Nanoengineering of optical resonances. *Chem. Phys. Lett.* **288**, 243–247. DOI: 10.1016/S0009-2614(98)00277-2

Oldenburg S.J. (1999) Infrared extinction properties of gold nanoshells. *Appl. Phys. Lett.* **75**, 2897–2899. DOI: 10.1063/1.125183

Palik E.D. (ed.) (1985) *Handbook of Optical Constants of Solids.* Academic Press, Orlando, FL.

Park J., Moon J., Shin H., Wang D., and Park M. (2006) Direct write fabrication of colloidal photonic crystal microarrays by ink jet printing. *J. Colloid Interface Sci.* **298**, 713–719. DOI: 10.1016/j.jcis.2006.01.031

Rogobete L., Schniepp H., Sandoghdar V., and Henkel C. (2003) Spontaneous emission in nanoscopic dielectric particles. *Opt. Lett.* **28**, 1736–1738. DOI: 10.1364/OL.28.001736

Ruppin R. (1982) Decay of an excited molecule near a small metal sphere. *J. Chem. Phys.* **76**, 1681–1684. DOI: 10.1063/1.443196

Sakoda K. (2004) *Optical Properties of Photonic Crystals,* 2nd ed. Springer-Verlag, Berlin.

Schäfer F.P. (ed.) (1977) *Dye Lasers.* Springer-Verlag, Berlin.

Schuurmans F.J.P., Vanmaekelbergh D., van de Lagemaat J., and Lagendijk A. (1999) Strongly photonic macroporous gallium phosphide networks. *Science* **284**, 141–143. DOI: 10.1126/science.284.5411.141

Shinar J. and Shinar R. (2008) Organic light-emitting devices (OLEDs) and OLED-based chemical and biological sensors: An overview. *J. Phys. D: Appl. Phys.* **41**, 133001-1–133001-26. DOI: 10.1088/0022-3727/41/13/133001

Soukoulis C.M. (ed.) (2001). *Photonic Crystals and Light Localization in the 21st Century.* Kluwer, Dordrecht, The Netherlands.

Sprik R., van Tiggelen B.A., and Lagendijk A. (1996) Optical emission in periodic dielectrics. *Europhys. Lett.* **35**, 265–270. DOI: 10.1209/epl/i1996-00564-y

Taflove A. and Hagness S.C. (2005) *Computational Electrodynamics: The Finite-Difference Time-Domain Method,* 3rd ed. Artech House, New York.

Tai C.-T. (1993) *Dyadic Green Functions in Electromagnetic Theory,* 2nd ed. IEEE, New York.

Vallée R.A.L., Baert K., Kolaric B., Van der Auweraer M., and Clays K. (2007) Nonexponential decay of spontaneous emission from an ensemble of molecules in photonic crystals. *Phys. Rev. B* **76**, 045113-1–045113-9. DOI: 10.1103/PhysRevB.76.045113

Valuev I., Deinega A., and Belousov S. (2008) Iterative technique for analysis of periodic structures at oblique incidence in the finite-difference time-domain method. *Opt. Lett.* **33**, 1491–1493. DOI: 10.1364/OL.33.001491

Valuev I., Deinega A., Knizhnik A., and Potapkin B. (2007) Creating numerically efficient FDTD simulations using generic C++ programming. *Lecture Notes Comput. Sci. (LNCS)* **4707**, 213–226. DOI: 10.1007/978-3-540-74484-9_19

van Blaaderen A. and Vrij A. (1992) Synthesis and characterization of colloidal dispersions of fluorescent, monodisperse silica spheres. *Langmuir* **8**, 2921–2931. DOI: 10.1021/la00048a013

Vats N., John S., and Busch K. (2002) Theory of fluorescence in photonic crystals. *Phys. Rev. A* **65**, 043808-1–043808-13. DOI: 10.1103/PhysRevA.65.043808

Vecchi G., Giannini V., and Gomes-Rivas J. (2009) Shaping the fluorescent emission by lattice resonances in plasmonic crystals of nanoantennas. *Phys. Rev. Lett.* **102**, 146807-1–146807-4. DOI: 10.1103/PhysRevLett.102.146807

Veronis G., Suh W., Liu Y., Han M., Wang Z., Dutton R.W., and Fan S. (2005) Coupled optical and electronic simulations of electrically pumped photonic-crystal-based light-emitting diodes. *J. Appl. Phys.* **97**, 044503-1–044503-6. DOI: 10.1063/1.1848194

Vlasov Y.A., Deutsch M., and Norris D.J. (2000) Single-domain spectroscopy of self-assembled photonic crystals. *Appl. Phys. Lett.* **76**, 1627–1629. DOI: 10.1063/1.126117

Vos W.L., Koenderink A.F., and Nikolaev I.S. (2009) Orientation-dependent spontaneous emission rates of a two-level quantum emitter in any nanophotonic environment. *Phys. Rev. A* **80**, 053802-1–053802-7. DOI: 10.1103/PhysRevA.80.053802

Wang H., Trakic A., Xia L., Liu F., and Crozier S. (2007) An MRI-dedicated parallel FDTD scheme. *Con. Magn. Reson. B (Magn. Reson. Eng.)* **31B**(3), 147–161. DOI: 10.1002/cmr.b.20092

Wedge S. and Barnes S.L. (2004) Surface plasmon-polariton mediated light emission through thin metal films. *Opt. Express* **16**, 3673–3685. DOI: 10.1364/OPEX.12.003673

Yablonovitch E. (1987) Inhibited spontaneous emission in solid-state physics and electronics. *Phys. Rev. Lett.* **58**, 2059–2062. DOI: 10.1103/PhysRevLett.58.2059

Yakimanski A.V., Menshikova A.Yu., Shevchenko N.N., Shabsels B.M., Bazhenova A.G., Sel'kin A.V., Sazonov S.K., Vedernikov A.I., Gromov S.P., Sazhnikov V.A., and Alfimov M.V. (2009) From polymeric nanoparticles to dye-containing photonic crystals: Synthesis, self-assembling, optical features, and possible applications. *Polym. Adv. Technol.* **20**, 581–588. DOI: 10.1002/pat.1400

Yates C.J., Samuel I.D. W., Burn P.L., Wedge S., and Barnes W.L. (2006) Surface plasmon-polariton mediated emission from phosphorescent dendrimer light-emitting diodes. *Appl. Phys. Lett.* **88**, 161105-1–161105-3. DOI: 10.1063/1.2193795

Yee K. (1966) Numerical solution of initial boundary value problems involving Maxwell's equations in isotropic media. *IEEE Trans. Antennas Propagation* **14**(3), 302–307. DOI: 10.1109/TAP.1966.1138693

Zhu J.X., Pine D.J., and Weitz D.A. (1991) Internal reflection of diffusive light in random media. *Phys. Rev. A* **44**, 3948–3959. DOI: 10.1103/PhysRevA.44.3948

CHAPTER 3

Optical Sensing by Metal Oxide Nanostructures

Phenomenology and Basic Properties

S. Lettieri

1. INTRODUCTION

One of the most common and useful approaches to chemical sensing relies on the chemoresistive effect, which is the phenomenon where the electrical conductivity of a material is modified by adsorption (and/or desorption) of molecules on the surface of the material itself. Since the first demonstration of the chemoresistive effect in 1962 using zinc oxide (ZnO) thin-film layers (Seiyama et al. 1962), it has been known that many metal oxides exhibit a relevant chemoresistive response. This property represents the basic feature that allows the use of such materials for realization of conductometric gas sensors, where the presence and/or concentration of adsorbed gas species is monitored through measurable variations in electrical current flow or in voltage drop across an electrode. As a consequence, it can be easily seen that materials exhibiting too large or too low resistance are not very suitable for realizing efficient devices; this rules out pure insulators and pure metals as good functional materials for realization of sensing devices. In practice, materials representing good choices as sensing element should exhibit semiconducting behavior. Many different metal oxide (MOX) materials typically

DOI: 10.5643/9781606503201/ch3

exhibit both semiconductor behavior and chemoresistive effects, so that they can be successfully used as sensing elements in practical devices.

Among the MOX materials that are most used for gas sensing, one can cite some transparent conductive oxides (TCOs) characterized by a large band gap, such as zinc oxide (ZnO), tin oxide (SnO_2), titanium oxide (TiO_2), indium oxide (In_2O_3), or other semiconductor oxides such as tungsten oxide (WO_3) or cupric oxide (CuO). It is worth noting that TCO materials are of great interest also for applications other than gas sensing (one can cite, as representative examples, the use of TiO_2 as a distributed photo-anode in hybrid photovoltaic cells and the use of ZnO as a light emitter). Due to their intrinsic versatility and multiple functionalities, research on these materials has proceeded constantly and at a fast pace in recent years.

The morphology of the active layers plays a key role in sensing devices. In fact, chemoresistive effect is clearly enhanced when the sensing material has a very large surface area so that it adsorbs as much of the target analyte as possible. For this reason, standard choices for the active components of gas-sensing devices are highly porous and particulate thick films. Enhanced porosity and granularity—exhibited, for example, by sintered metal oxide powders and micro- or nanograins—lead to very high specific surface area and to end devices whose sensitivity to a gas can be as low as parts per billion (ppb) (Gurlo et al. 1998; Guo et al 2006; Elmi et al. 2008; Kida et al. 2009). Because grain sizes and surface-to-volume ratios are crucial parameters affecting the sensitivity of the chemoresistive response, the development of novel efficient and reliable sensing devices is thus strongly related to improvement in MOX synthesis and deposition techniques. Progresses in synthesis of MOX nanocrystals represent a major milestone in the field of gas-sensing applications; in fact, MOX nanocrystals typically exhibit well-defined single-crystal surfaces, thermal stability, and large surface area, all features favoring the realization of reliable and stable devices. Moreover, due to their reduced transverse dimensions (a few tens of nanometers is typical), the depletion layer induced by gas adsorption may in principle extend even over the whole nanostructure, thus enhancing the chemoresistive sensitivity.

Apart from conductivity changes in the gas-sensing material, the detection of external species can also be achieved by other means, such as monitoring changes of material capacitance, work function, or mass (Korotcenkov 2007). Such approaches share the use of electrical devices for electrical transduction of the signal and the need to realize electrical contacts on the material.

Different routes to gas sensing, alternative to approaches based on conductometric response (or, more generally, on electrical properties) are based on modification of the optical characteristics induced by chemical interactions with external species. This latter kind of approach can be classified as "optochemical sensing" (or, more shortly, optical sensing). Optical sensing may present a number of advantages over the traditional approach. First, it can be easily implemented to give multiparametric response and information. Because different gas species may lead to distinct responses in various available detection parameters, the optochemical approach has the potential to detect multiple gases from differences in the intensity, wavelength, phase, and polarization of the output light signals (Nishizawa 1988; Eguchi 1992; Wolfbeis 2002, 2004; Shi et al. 2004). Moreover, optical sensors do not need electrical contacts, do not suffer from electromagnetic noise, and may allow optoelectronic integration.

Optical chemical sensors may be divided into several subgroups depending on the working principles applied. By "working principles" we mean here the physical quantity that is modified by chemical interaction(s) that one wants to monitor. Such physical quantities have of course to be determined, in the case of optical sensing, through a suitable spectroscopy technique or method. In this sense, practical methods may include reflectometry, absorptometry, infrared or Raman spectroscopy, photoluminescence, surface plasmon resonances, etc., in bench-top setups or in combination with optical fibers for remote sensing. The range of possible setups and technical approaches in optical sensing is very broad and flexible, and this can be regarded as an additional advantage of this kind of approach.

The largest single group of chemical sensors based on an optical working principle is probably represented by fluorescent organic molecules acting as probes or labels. However, we will not discuss these sensors, as the aim of this chapter is to deal with solid-state inorganic (oxide) nanomaterials and/or nanoporous materials, mainly obtained from MOX materials that are commonly used for electrical sensors. In particular, we will focus on optochemical sensing based on modification of photoluminescence induced in oxide nanostructures by chemical interaction with gas molecules. The two main sections of this chapter are devoted entirely to this topic, discussing different possible models for photoluminescence quenching phenomena and reviewing some specific literature on this topic.

The chapter is organized as follows. In Section 2, applications of oxide optical sensors that are not based on photoluminescence modification are reviewed, considering in particular those based on absorbance and on refractive response.

Sections 3 and 4 are devoted to the core topic of this contribution, namely, basic theory and and practical application of photoluminescence properties of oxide nanostructures for gas sensing. In particular, In Section 3, basic elements of photoluminescence (PL) are first introduced, then cases of "static" versus "dynamic" PL quenching due to gas adsorption are discussed through a phenomenological approach. As part of the discussion we show why studying the PL intensity modifications as a function of gas concentration is, in a very general case, not enough to understand the fundamental chemical-physical features that characterize the observed phenomena. In this regard, we will underline the role of time-resolved PL analysis to extract more precise information on effective gas–nanoparticle interaction.

Section 4 is devoted to a review of some experimental results regarding gas-sensitive photoluminescence properties of nanostructured oxide materials, discussing (when possible) possible interpretation of the basic physical processes that lie behind the observed phenomena. It is worth noting that most of the literature results considered in Section 4 refer to oxide nanostructures having a high degree of crystal order (in short, "nanocrystals"). This is not a coincidence: Indeed, as previously mentioned, such materials promise specific advantages over traditional sensors based on particulate films. Therefore, before reviewing specific literature, we will briefly review (in Section 4.1) why a specific interest exists in studying nanometer-size single-crystal structures. Next (in Sections 4.2–4.4), practical cases of gas-sensitive luminescence observed in oxide nanocrystals and nanoporous structures will be reviewed, considering each different nanostructured material in a separate paragraph. We will consider the cases of zinc, silicon, and tin oxides (ZnO, SiO_2, and SnO_2), due to their importance in sensing applications and to the slightly more developed literature on these materials.

2. OPTOCHEMICAL SENSING BY OXIDE MATERIALS: METHODS NOT BASED ON PHOTOLUMINESCENCE

2.1. APPROACHES TO OPTICAL SENSING

As discussed in Section 1, optical characteristics and/or optical responses of a material can undergo modifications induced by chemical interaction with other species such as gas molecules. This class of effects can represent an effective basis for optochemical sensing (or, more simply, optical sensing).

One useful feature that characterizes the optical approach to chemical sensing is in allowing, in principle, multiparametric response and selective information. Electromagnetic waves are described by a certain number of degrees of freedom (intensity, wavelength, polarization), which may all represent suitable detection parameters. This is not possible in electrical sensing, where only a single scalar quantity (electrical conductivity) is involved. An increase in the number of detection parameters available also means an increase in the possibility of detecting multiple gaseous species, as different molecules may lead to distinct responses in the different parameters available for detection. Optical sensing devices can also exhibit other advantages, such as low fabrication costs, high sensitivity, and no requirements for electrical contacts.

Optical sensors reported to date can be divided into different categories according to the detection method used. A first classification may distinguish sensors based on modification of (1) absorbance and/or reflectance, (2) light propagation properties (i.e., refractive index modifications), (3) Photo-thermal and photo-acoustic response, (4) surface plasmon resonance, and (5) photoluminescence.

Approaches based on photoluminescence (PL) modifications will be discussed in more detail LAter. Here we will just anticipate that such approaches appear to be particularly suited to the use of MOX nanostructures and nanocrystals. As a matter of fact, the MOX materials most commonly used in traditional sensing applications, such as ZnO, TiO_2, and SnO_2, typically exhibit efficient visible (VIS) or even ultraviolet (UV) PL emission once synthesized in nanosized structures. To cite a relevant example, one may consider ZnO: Apart from being a gas-sensitive material, thanks to its chemoresistive properties, it is also known as an important optical material for applications as UV emitters, due to a number of properties (wide band gap, large exciton binding energy, efficient near-band-edge radiative recombination). Indeed, a wide research literature exists dealing with just the light emission properties of single-crystal and nanostructured ZnO, apart from other eventual possible applications of such materials in chemical sensing.

As this chapter focuses on metal oxide materials, we will not discuss the huge range of (inorganic) semiconductor materials and, mostly, organic molecules and polymers that can be used as sensors using one of the many possible optical-based working principles listed above. Instead, here we will deal only with materials (and devices) based on oxide semiconducting materials in various forms (thin films as well as nanostructures). However, it is also worth noting that useful devices can be prepared using a "hybrid" approach, in which functionalities of different materials (for example, the PL response of organic dyes and the conductometric response of inorganic oxides) can be integrated in a single device for multiparametric sensing.

In the following we proved a brief review of approaches based on modification of optical absorption ("absorbance response") and optical refraction ("refractive response") in oxide-based materials and devices. A detailed discussion of sensing approaches based on photoluminescence modifications is the main topic of this chapter and will be the subject of the next sections, along with a review of related literature and experimental evidence on this topic.

2.2. OXIDE-BASED OPTOCHEMICAL SENSING USING ABSORBANCE RESPONSES

In the case of chemo-optical sensing based on optical absorbance (absorptometry), it is useful to distinguish between the case where a material—acting as sensing element—shows a change in its optical absorption properties due to exposure to given species (e.g., gas molecules) from the case where the optical absorption of the molecules to be detected is probed directly at a specific wavelength (by means of a suitable optical spectroscopy setup). This latter approach is particularly useful in the case of gas detection, because gas molecules exhibit specific absorption resonances (related, for instance, to vibrational modes of the molecule) to which one can tune the spectroscopic setup in order to achieve very high sensitivity and selectivity. It is also worth noting, however, that methods based on direct spectroscopy also have a severe drawback in being expensive and in requiring specific instrumentation.

As this chapter is devoted to optical sensing developed using specific (nano) materials as sensing elements, here we will consider only the first approach, limiting our discussion to oxide semiconducting materials. Some specific literature cases will be reviewed here. In order to limit the discussion to a reasonable length, we will just review some basic evidence of practical interest.

As already underlined previously, the most relevant characteristic of optical approaches to sensing is probably the possibility to obtain selective responses to different gas species, thus allowing detection of multiple gases. In this regard, some successful examples have been demonstrated in the literature using an optical absorption-based response. In this case, a key element to achieve selective responses turns out to be the double functionality of metal nanoparticles dispersed into an inorganic sensing oxide semiconductor. As a matter of fact, metal nanoparticles can act as both a catalytic enhancer of redox chemical reactions between the gas and the semiconductor surface, and as a "resonant optical enhancer" for radiation–matter interaction due to plasmon resonance. While the first feature is commonly used in conductometric sensors, the second is not available unless optical excitation is somehow involved. Once the presence of optical resonant interaction with light at a wavelength close to the plasmon resonance of the metal nanoparticles is found, some additional specific interaction may take place that does not occur when the material interacts with off-resonance light. In some literature cases, such additional plasmon-related interactions have been shown to be selective toward specific gas species. Some significant examples of selective gas detection through absorbance-based effects on metal oxide films were published in the 1990s by Kobayashi et al. (Ando et al. 1997a, 1997b). They reported a reversible absorbance change in nickel and cobalt oxides (NiO, Co_3O_4) upon exposure to reducing species such as CO and H_2, occurring with response times ranging from several to tens of minutes at 250–350°C. The authors found that using light with wavelength detuned with respect to the typical plasmon absorption band of Au (about 600 nm in the works cited), exposure to both reducing gas species lead to a nonselective decrease of optical absorption. Such modification was attributed to changes in hole density in the film (which behave as *p*-type semiconductors) due to catalytic oxidation of the gases. The reaction leading to oxidation of gas species indeed consumes holes, according to reactions such as $CO + 2O^{2-} + 2V_O^{2+} \rightarrow CO_3^{2-} + 2V_O^{+}$ (where V_O^{2+} and V_O^{+} are divalent and monovalent oxygen vacancy (Ando et al. 1997b). This effect, leading to reversible decrease of optical absorption, was accompanied by a selective and specific increase of absorption to H_2 at the Au plasmon resonance peak (near 600 nm). Because the value of the absorption coefficient at plasmonic resonance for a metal particle depends on many factors (such as particle size, shape, and the local value of the dielectric constant of the surrounding medium), the origin of this second phenomenon was tentatively attributed to modifications of local dielectric constant near the Au nanoparticle surface or to modification of the morphology at the Au–Co_3O_4 interface due to the fact that smaller H_2 molecules may penetrate into the inner part of the Au–Co_3O_4 interface region more deeply than the larger CO molecules (Ando et al. 1997b).

Similar approaches for H_2 and CO detection through absorbance modification have been reported using composite films of tungsten oxide doped with palladium and gold (Pd:WO_3, Au:WO_3) (Ando et al. 2001) and of gold-doped cupric oxide (Au:CuO) (Ando et al. 2003). Ando et al. (2001) obtained a 1% detection of H_2 in air by monitoring the increase in absorbance of Pd:WO_3 in the visible and near-infrared optical region (centered at $\lambda = 1000$ nm). The effect was attributed to formation of tungsten bronze (H_xWO_3), which is enhanced by the dissociation of H_2 on Pd. In the case of Au:WO_3 films, the material was found to be sensitive mainly at the Au plasmon absorption region in the visible range (increasing change of absorbance at 550 nm occurring upon exposure to H_2), attributed to a change in the dielectric constant of the surrounding tungsten oxide induced by H_2 adsorption. In the case of Au:CuO composite film, a CO detection range of 50–10,000 ppm at 300°C was achieved by employing reversible absorbance change near the plasmon-absorption band of the Au particles, with a response time of several minutes at 300°C (see Figure 3.1).

Chemochromic detection of H_2 was reported by Liu et al. (2003) using the absorption modification of a Pd-doped mesoporous vanadium oxide (V_2O_5). An abrupt increase of the film transmittance was observed by exposing the material to 4% H_2/N_2 gas, resulting from the formation of palladium hydride, followed by a slower transmittance decrease due to coloration of the vanadium oxide layer. Improved kinetic performance was exhibited by the mesoporous oxide with respect to a thermally evaporated V_2O_5 film, demonstrating the advantage of employing highly porous structures.

Optical and electrical gas-sensing properties can be simultaneously monitored by preparation of suitable hybrid metal/oxide films or systems. One example is given by Ando et al. (2004), in which both optical absorption and electrical conductivity responses of porous $In_xO_yN_z$ films covered by a thin overlayer of Au nanoparticles (20–30 nm in diameter) were investigated. These exhibited a reversible increase of absorbance upon exposure to H_2 in the optical region of Au-plasmon absorption (typically near 550–50 nm), again triggered by dielectric constant modification of the oxide layer surrounding the metal nanoparticles. The presence of the oxide film also allowed monitoring optical and conductometric transduction simultaneously. The advantage of this possibility was outlined as the authors found selective response to H_2 and NO_2, the first species acting on optical response (as discussed before), while the second was evidenced by electric

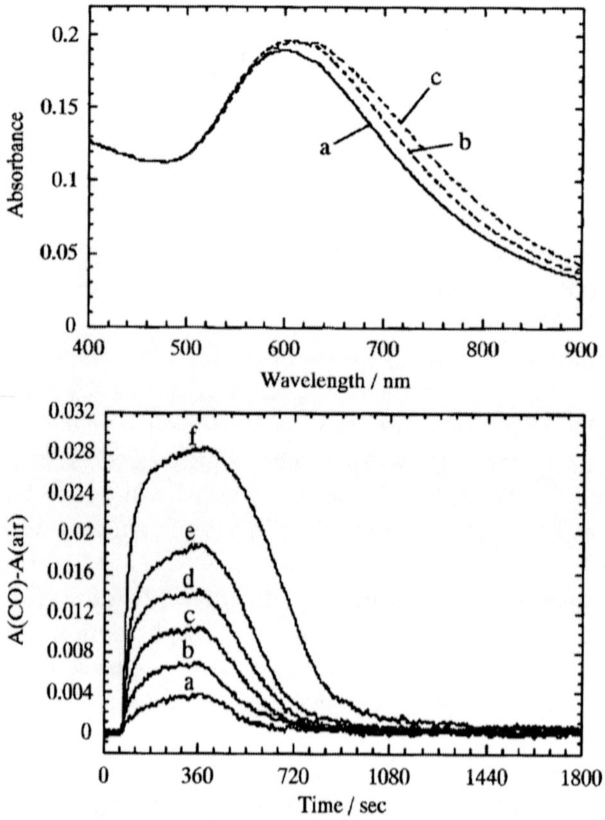

Figure 3.1. (A) Absorption spectrum of the Au:CuO film in air and change caused by CO (a) in air and (b) in air containing 1000 ppm CO. (B) Absorbance change of the Au:CuO film as a function of time under atmosphere periodically changing from air (t = 0–70 and 370–1800 s) to air containing CO (t = 70–370 s): (a) by 50 ppm CO; (b) by 200 ppm CO; (c) by 500 ppm CO; (d) by 1000 ppm CO; (e) by 2000 ppm CO; (f) by 1 vol% CO. Wavelength 700 nm. Temperature 300°C. (Reprinted with permission from Ando et al. 2003. Copyright 2003 Elsevier.)

resistance changes. In the latter case, weak response to H_2 was observed, while strong variations (2–3 orders of magnitude) in dc electrical resistance were observed upon NO_2 exposures.

More recently, optical detection with improved selectivity toward H_2 and CO has been shown for the case of SiO_2 sol-gel films containing NiO and Au nanoparticles (Buso et al. 2007). Infiltration of Au nanoparticles (at controlled concentration) and crystalline NiO aggregates inside the SiO_2 network allowed exploitation of the surface plasmon resonance (peaking at $\lambda = 550$ nm), as noticeable differences in the detected transmittance of the films were observed after exposure to CO and H_2, with major differences observed in the spectral region near the surface plasmon excitation band (500–650 nm). In particular, CO induced an increase in absorption but with a depression of absorption change values near the plasmon resonance. On the other hand, H_2 induced a large decrease in absorption in the same spectral region. The observation was accompanied by a shift in surface plasmon resonance peak which was not found for the case of CO exposure, thus indicating a different detection mechanism for the two species. The wavelength dependence of transmittance changes could thus be exploited to tailor film selectivity toward detection of the two species with reduced interference, as can be seen in Figures 3.2 and 3.3. The optical transmittance changes ratio (OCTR), defined as

$$\mathrm{OCTR} = \frac{A_{\mathrm{gas}}(\lambda) - A_{\mathrm{air}}(\lambda)}{A_{\mathrm{air}}(\lambda)} \tag{3.1}$$

are reported in Figure 3.2 for the case of Au:NiO film with 6% Au content. The spectral dependence of the OCTR allows selective detection by properly choosing the detection wavelength of optical transmittance as shown in Figure 3.3, where opposite changes induced by exposure to H_2 and CO can be observed by monitoring the optical transmittance at 470 nm wavelength.

Even if it is quite likely that metal nanoparticles acted actively as plasmonic elements and not merely as enhancers of chemical reactions, a clear interpretation of the findings could not be given easily, as the H_2 detection mechanism

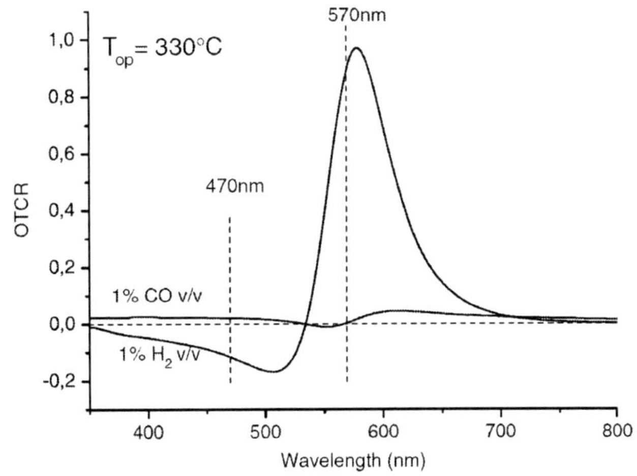

Figure 3.2. (a) Comparison between optical transmittance changes ratio (OTCR) values induced by 1% CO and 1% H_2 measured at temperature T = 330°C for Au/NiO films with 6% Au content. Vertical dotted lines indicate wavelengths at which dynamic selective response can be obtained. Near surface plasmon resonance at 570 nm response to CO is almost negligible, while at 470 nm opposite transmittance variations are induced by the two species. (Reprinted with permission from Buso et al. 2007. Copyright 2007 IOP.)

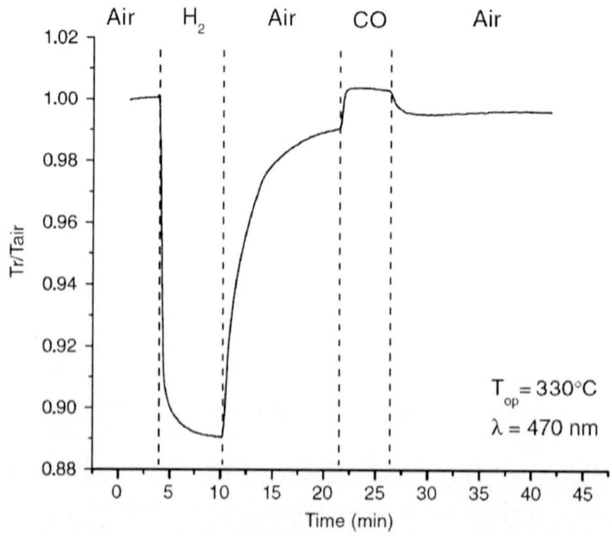

Figure 3.3. Dynamic optical transmittance response of film containing 6% Au exposed to air/1% H_2 and air/1% CO cycles. The test was performed at 330°C and optical transmittance recorded at 470 nm. (Reprinted with permission from Buso et al. 2007. Copyright 2007 IOP.)

appears to be specific to the Au:NiO system. Indeed, studies to determine the effects of H_2 on the optical properties of other Au–metal oxide systems, such as the cited Au:Co_3O_4, Au:WO_3, and Au:$In_xO_yN_z$ systems, report that H_2 exposure causes an increase of optical absorption in surface plasmon resonance frequencies, together with no substantial change in plasmon absorption peak wavelength. These findings are in contrast to those described by Buso and co-workers (Buso et al. 2007) for Au:NiO. Further studies seem to be needed to elucidate the key processes leading to Au:NiO optical absorption response.

2.3. OXIDE-BASED OPTOCHEMICAL SENSING USING REFRACTIVE RESPONSES

Another class of optical sensors is based on the use of a sensing material as core or cladding of a light-guiding structure, such as optical fibers or optical waveguides. The light propagation properties of a guiding structure (planar waveguide or optical fibers) depend strongly on the refractive index values of the media surrounding the one that guides the light. Chemical interactions between gas and a gas-sensitive materials leads to a variation in local refractive index of the gas-sensitive material itself. Therefore, launching a light beam into a waveguide prepared with gas-sensitive materials allowing good waveguiding (i.e., transparent and homogeneous at the micrometer scale) can lead to gas detection by monitoring changes in the transmitted optical power at the waveguide output (Mazingue et al. 2005, 2006).

Various TCO materials such as ZnO, SnO_2, In_2O_3, TiO_2, and SnO_2/NiO can in principle be used for realization of optical waveguide sensors, given their optical transparency in the visible range. A vast literature exists on this topic, and no exhaustive report is possible here. We thus only cite a few examples, without attempting to give a complete review.

Since early reports on electro-optic modulators in multilayered ZnO waveguides for optical sensing applications (Horsthuis 1986), ZnO has been considered a promising candidate for such applications, due to its high transparency in the visible wavelength range, its chemical reactivity to oxidizing and reducing gas species (Dikovska et al. 2007; Dimitrov et al. 2008), and the eventual possibility of employing its electro-optical properties to realize more sophisticated devices (Horsthuis and Pannekoek 2005).

Dikowska and co-workers (2007) prepared ZnO films by pulsed laser deposition on periodically structured substrates, prepared by applying a combination of a modified lithography with a typical holographic exposure. The periodically structured substrates enhanced the light coupling into the ZnO film so that the latter could act as waveguide and ensure optical detection by exposure to gas species. Using such a device, the authors authors proved sensitivity to 1000-ppm butane with a response time of about 1 min, as shown in Figure 3.4.

More recently, the idea of using individual ZnO nanowires as an optical waveguide has been explored. In particular, Borner and co-workers (2007) investigated the use of single ZnO nanorods as evanescent-wave optical sensors. The basic idea here is to use the resonant coupling between the evanescent field leaking toward the border of the waveguide (the rod itself) and gas molecules adsorbed on the guide surface. To this aim, the near-infrared region is of particular interest, since vibrational or vibrational-rotational transitions of species such as O_2, CO, and CO_2 can be stimulated—and therefore detected—in this spectral region. Borner and co-workers computed optical waveguide properties of both single ZnO nanorods and ordered nanorods arrays, calculating evanescent field distributions and penetration depths into the surrounding media as function of wavelength and rod diameter. Some of the computed results are shown in Figure 3.5. The analysis performed by Borner and co-workers represents a useful suggestion for future experimental work on nanoscale optical sensors based on evanescent field interactions.

Using the same principle already discussed for optical waveguides, optical sensing based on refractive response can also be achieved by depositing active sensing materials on external cladding of optical fibers. One example is represented by work by Shukla et al. (2004), who reported an optical-fiber humidity sensor in which a porous MgO film was deposited on a U-shaped bent glass rod acting as light guide. The porous MgO film, consisting if small particles (50–60 nm

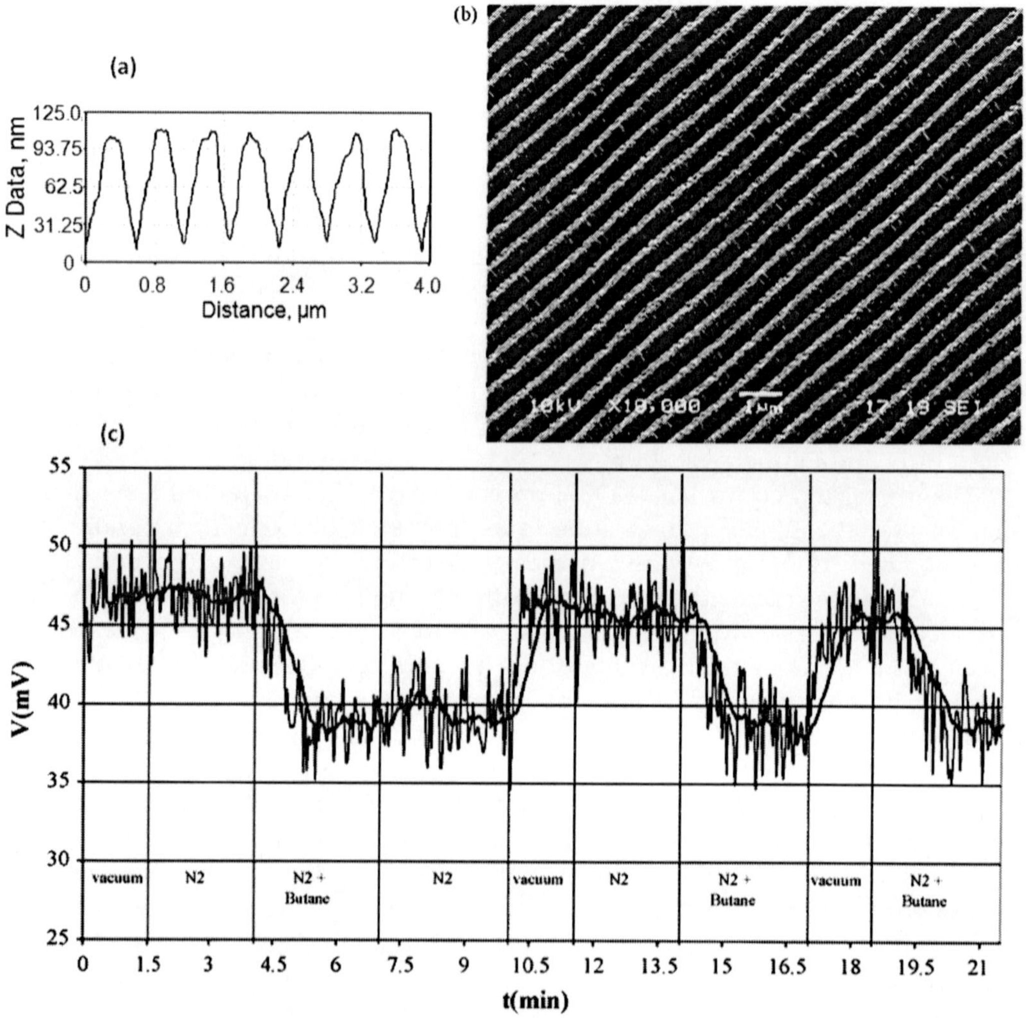

Figure 3.4. (a) Height profile of the periodically structured ZnO waveguide. (b) Scanning electron microscopy (SEM) image of ZnO waveguide surface (magnification × 10,000). (c). Sensitivity measurements of the ZnO periodically structured waveguide upon exposure at 1000 ppm butane diluted in nitrogen. (Reprinted with permission from Dikovska et al. 2007. Copyright 2007 Elsevier.)

in diameter) prepared by a sol-gel technique, acted as a water vapor adsorber due to the presence of unsaturated, partially coordinated sites over its surface. Injecting the light of a He–Ne laser into the glass rod, authors observed that output power from the glass fiber increased with increasing relative humidity in the range of 5–80%, as adsorption of water (refractive index = 1.3) on the surface/pores of MgO (refractive index = 1.7) led to a continuous decrease of effective refractive index of the MgO coating that clad the glass-rod light guide. Similar

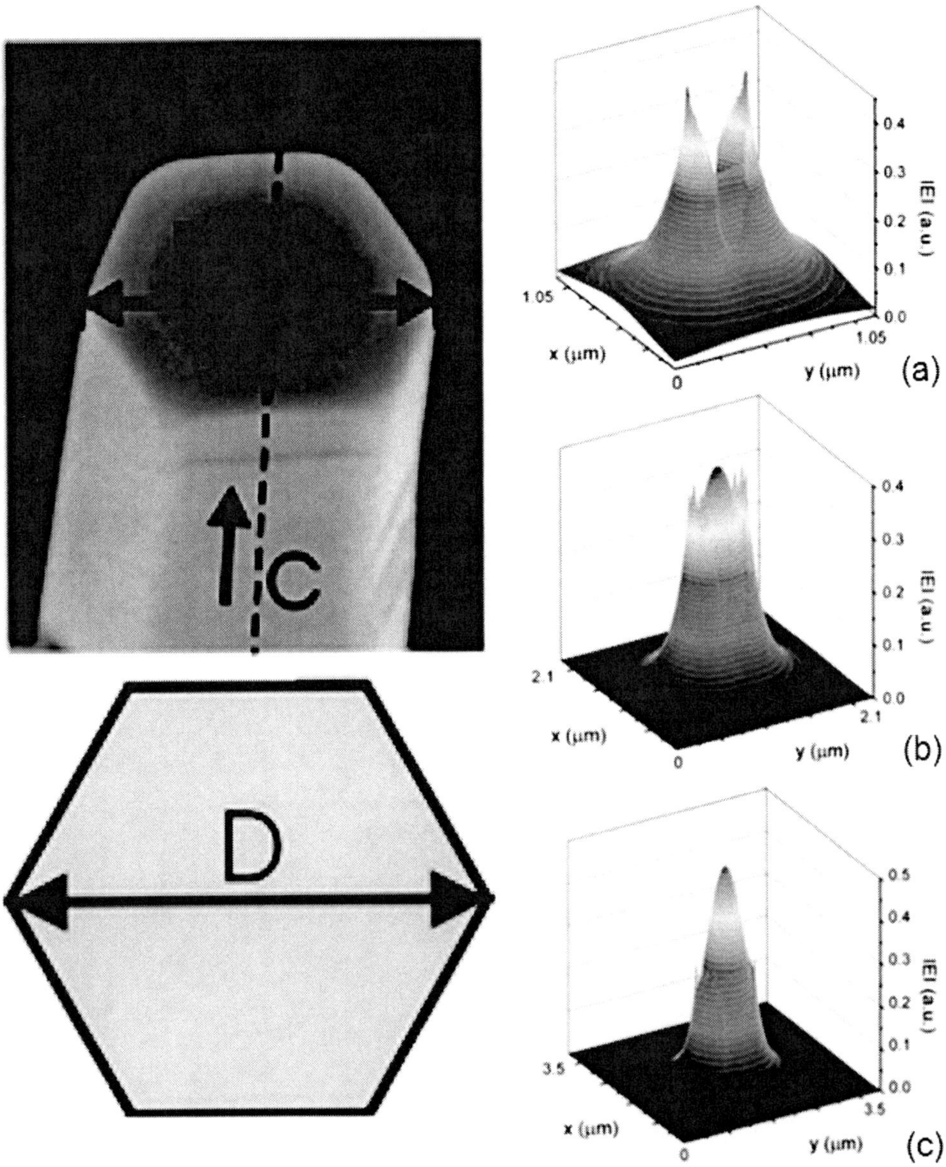

Figure 3.5. (Left) SEM image (top) and scheme (bottom) of a hexagonal-shaped ZnO nanorod. Diameter is indicated by *D* and crystalline *c* axis which coincides with the propagation direction *z*. (Right) Three-dimensional model of the absolute value of the electric field $|E|$ of nanorods with diameters (a) D = 300 nm, (b) D = 600 nm, and (c) D = 1000 nm at λ = 761 nm. (Reprinted with permission from Borner et al. 2007. Copyright 2007 Wiley-VCH.)

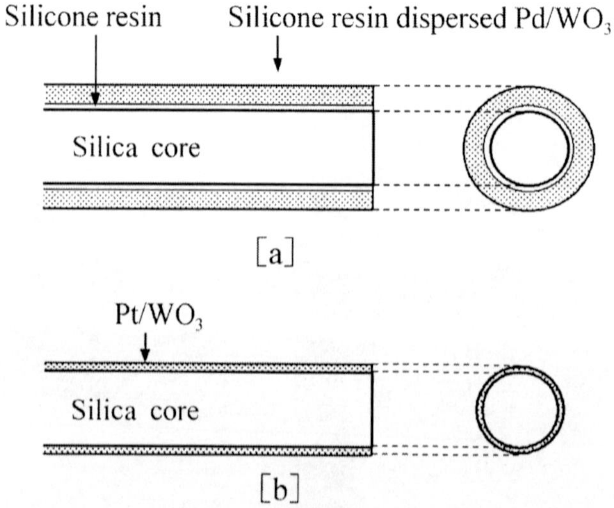

Figure 3.6. Schematic diagram of the optical-fiber H_2 sensor devices reported by Sekimoto et al. (2000). In (a), the fiber uses as cladding coating a silicone resin in which palladium-supported tungsten oxide powder is dispersed. In (b), fiber cladding is coated by Pt/WO_3 film prepared by sol-gel process. (Reprinted with permission from Sekimoto et al. 2000. Copyright 2000 Elsevier.)

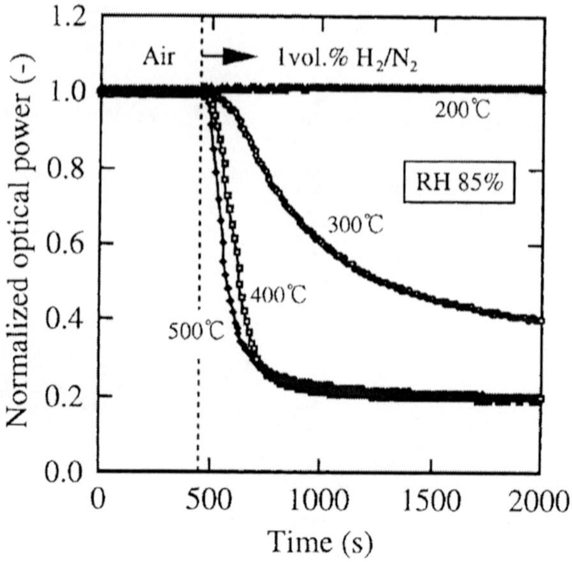

Figure 3.7. Response of optical fiber coated with Pt/WO_3 film toward repeated exposure to 1 vol% H_2/N_2 and air. The first arrow indicates the exposure to 1 vol% H_2/N_2, and the second indicates the exposure to air. (Reprinted with permission from Okazaki et al. 2003. Copyright 2003 Elsevier.)

examples are reported by Okazaki and co-workers (Sekimoto et al. 2000; Okazaki et al. 2003), who realized a H_2 gas sensor consisting of a catalyst-supported WO_3 thin film coated around the silica core of an optical fiber (a schematic diagram of the sensor device is shown in Figure 3.6). The dielectric-constant modification induced by formation of tungsten bronze led to a reduction of more than 75% in optical transmission upon exposing the fiber to 1% H_2 dissolved in N_2 at room temperature, as shown in Figure 3.7.

3. PHOTOLUMINESCENCE-BASED OPTOCHEMICAL SENSING BY SEMICONDUCTING MATERIALS: MODELS

The term "luminescence" is used in a general manner to indicate a range of phenomena in which a material emits light when excited by some external stimulus (during as well as after excitation). The wavelength spectrum of emitted light is material-specific, depending on the energy distribution of its quantum mechanical electronic states.

From a microscopic point of view, the luminescence process can be simply explained as follows. Excited states created by an external energy source ("excitation") are unstable and, once the external source is removed, such states will decay to the ground state. Due to the energy conservation principle, the decay processes are accompanied by release of energy; "radiative" processes are defined as those in which the energy is released via emission of photons (i.e., emission of light). The spectrum of emitted light is representative of the energy distribution of the ground- and excited-state levels. In fact, the fraction of excited electrons that decays via radiative transition (spontaneous emission) from an excited level to a ground-state level whose energy separation is ΔE will emit photons with wavelength λ given by $\Delta E = hc/\lambda$ (where h is Planck's constant). If the external source is switched off abruptly, the material will recover its initial equilibrium in a finite time; and once equilibrium is restored, it of course will no longer emit light. Thus, for the case of pulsed or modulated excitation, the intensity of the emitted light will exhibit material-specific time dynamics, reflecting the lifetimes and, more generally, the decay kinetics of the excited states. In this case, one speaks about time-resolved photoluminescence (TRPL). On the other hand, if the excitation source is stationary, a dynamic equilibrium is created between the excitation and recombination processes, so the intensity of luminescence emission will also be stationary. In this latter case, one talks about continuous-wave photoluminescence (CWPL).

The excitation can be obtained in several ways and, depending on how energy to excite the material is supplied, one refers to a different kind of luminescence. Schematically, one can distinguish, for example, photoluminescence (excitation provided by absorbed optical radiation), chemiluminescence (excitation provided by chemical reactions), or thermoluminescence (excitation provided by a heat source), etc. Here we will consider only the process of photoluminescence (PL), to introduce the topic of PL-based optical sensing. In this approach, one monitors modification of PL emission (meaning changes in emission intensity, in spectral shape, or even in both) induced by interaction with chemical species (such as gas molecules). As will be discussed next, such an approach is attractive in that it can provide good room-temperature sensitivity (as opposed to conductometric sensors, which usually require working temperature of about 300°C or more) with stable and reversible responses. A possible drawback to this approach may be difficulty in realizing sensing devices inexpensively and in compact structures, since they require, apart from the sensing material, a light source to excite the material, a PL detector, and processing and output circuitry. However, these problems can be overcome, and indeed, compact and simple PL-based sensors have been recently realized (Savvate'ev et al. 2002; Choudhury et al. 2004).

In this section we briefly review the fundamental physical processes that give rise to PL, namely, optical excitation, electron thermalization, and radiative recombination. Next, we will focus on description of photoluminescence quenching phenomena that, as we will see in the next section, can represent a useful route for optochemical sensing of gas.

3.1. BASIC PRINCIPLES OF PHOTOLUMINESCENCE

As mentioned above, photoluminescence is a process in which a material absorbs and then re-emits photons by radiative transitions. In a semiconductor this process corresponds to the recombination of electron–hole pairs generated by

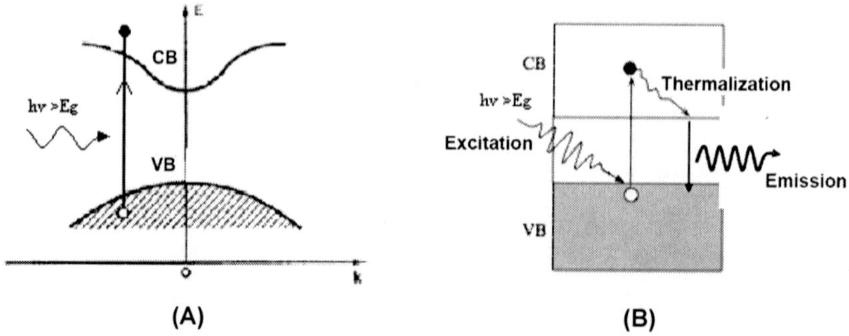

Figure 3.8. (a) Optical interband excitation of an electron–hole pair. Excitation photons having energy greater than the optical band gap of the material are required to excite the material. For the case of transparent conductive oxides (TCOs) this corresponds, in practical cases, to excitation wavelengths in the blue or near-ultraviolet range. (b) Schematic representation of the basic steps (excitation, thermalization, and emission) that constitute the photoluminescence phenomenon. For simplicity, the contribution to PL emission at photon energies lower than the optical band gap (such as exciton emission and deep-defect emission) are not shown in the figure.

optical absorption (i.e., "photoexcited"). In direct gap semiconductors, the room-temperature photoluminescence spectrum usually peaks at a photon energy close to the optical band-gap value of the semiconductor. Such maximum light emission may correspond to band-edge emission or, when exciton formation is favored, to excitonic emission. The former results from interband radiative recombination of electrons and holes positioned near the "edges" or respective bands (the top of the conduction band and the bottom of the valence band, respectively). Exciton formation results in usually sharp emission lines occurring at photon energies slightly less than the band-gap energy. Other peaks or broader features can also be observed at energies significantly lower than the band-gap energy, due to radiative transitions of charge carriers from deep defect levels, or to tightly bound exciton–defect complexes.

The photoluminescence process is usually divided into three temporal stages, shown schematically in Figure 3.8.

1. **Optical excitation:** After absorption of electromagnetic radiation, electron–hole pairs are created inside the semiconductor.
2. **Thermalization:** Due to their mutual interactions and to interactions with lattice vibrations (electron–electron and electron–phonon scattering), photo-excited free carriers reach a near-thermodynamic equilibrium. Typically, the time required to a reach complete thermalization of excited carriers may vary from tenths of femtoseconds (fs) to few picoseconds (ps) after optical excitation.
3. **Recombination:** Thermalized electrons and holes lying near the bottom of the conduction band and near the top of the valence band (respectively) recombine through different routes. One of these is represented by radiative recombination, which gives rise to photon creation (i.e., light emission or PL). This process typically begins after complete thermalization of charge carriers.

Some important considerations concern the characteristics times for each of the above-listed processes. For practical purposes, creation of photo-excited electron–hole pairs in a semiconductor can be considered as occurring "instantaneously." Next, electron–hole pairs lose their excess energy through scattering processes. By this means, electrons and holes separately reach thermal equilibrium within the semiconductor, moving toward the extremes of the bands (see Figure 3.8b) before recombining. In this situation, electron and hole populations can be described by means of quasi-equilibrium Fermi distributions. The characteristic time duration of the thermalization stage can vary between a few tenths of femtoseconds to a few picoseconds, where the time of the interaction depends on the density of charge carriers that are present.

The photoluminescence spectrum of a given material thus reflects its electronic spectrum, i.e., the distribution of excited and occupied states of the material itself. Therefore, in order to give a clearer picture, the most important contributions to a generic PL spectrum are briefly discussed next.

3.2. MAIN PROCESSES CONTRIBUTING TO PHOTOLUMINESCENCE

The simplest way to describe a semiconductor is to consider it as an infinite crystal in which charge carriers (electrons and holes) move, with negligible mutual interactions, in a fixed periodic potential whose periodicity is dictated by the symmetries of the crystalline lattice. These assumptions give rise to the well-known band model for energy states. Therefore, the most intuitive recombination path for excited charge carriers (electrons and holes) is interband recombination, i.e., their mutual recombination from states lying in different energy bands (conduction and valence, respectively).

On the other hand, effects that characterize a real crystal produce additional contributions to PL spectra. As first point, "free" charges are not really free and independent particles, due to reciprocal Coulomb interactions. The mutual electron–hole electrostatic interactions lead to formation of excitons that make a distinct contribution to PL spectra. Also, the periodic potential cannot be strictly considered as fixed, due to lattice vibration. The quasi-particles that quantize the vibrational energy of the crystal lattice, called phonons, may scatter with electrons and/or holes. In this process, charge carriers may lose the energy $\hbar\omega_{LO}$ needed to create a longitudinal optical phonon before recombining, giving rise to "phonon replicas" of emission lines. These are simply replicas of emission lines ordinarily exhibited at energy E by the material, but characterized by a photon energy $E - n\hbar\omega_{LO}$ (n integer), corresponding to the fact that n optical phonons have been created before the photon emission. Finally, real crystals also contain various kinds of defects (shallow as well as deep), whose corresponding energy levels may be positioned well below the conduction band edge, thus giving rise to photoluminescence lines at photon energies less than the band-gap energy.

We briefly describe below the three main contributions to PL emission in a semiconductor (interband, excitonic, defect-related).

3.2.1. *Interband Transitions*

Considering that typical recombination lifetimes for holes and electrons are usually longer than thermalization times, it is reasonable to assume that the charge carriers are distributed along the edges of their respective bands (as predicted by the Fermi distribution) before recombining. As a consequence, most radiative transitions in an ideally perfect semiconductor will take place near the band-gap energy. On the other hand, the presence of defect states in the "forbidden" band-gap region will instead introduce additional states that are available for excited carriers during the thermalization stage.

Even if real systems are far from being flawless, the PL contribution from band-edge states indeed represents a major contribution to PL emission in real cases. It has to be noted, however, that the efficiency of the radiative interband recombination process depends strongly on the electronic structure of the energy bands. In particular, very strong radiative emission is usually observed in direct gap semiconductors, while much smaller light emission efficiency is observed in indirect gap semiconductors. The difference stems from the momentum conservation principle, which leads to the fact that only "vertical" interband optical transitions (i.e., those that preserve the crystalline momentum) are allowed due to electron–photon interactions (Bassani and Pastori Parravicini 1975). In indirect gap semiconductors, thermalized electrons and holes have different crystalline momenta (k and k') and thus their recombination is allowed only if assisted by a third particle, namely, a phonon, providing the momentum $\Delta k = k - k'$ needed for the radiative electron–hole recombination to take place. As a process requiring assistance by a third particle is much less probable than a process in which only two particles interact (as in direct band-gap recombination), this explains why much stronger radiative emission is usually observed in direct gap semiconductors compared to indirect gap semiconductors.

Some simple considerations allow one to extract an approximate expression of PL line shape related to interband recombination in a direct gap semiconductor. In general, the intensity of PL emission at a given photon energy $\hbar\omega$ will be proportional to the product of two factors: (1) the number of states per unit volume whose energy distance is equal to $\hbar\omega$ times and (2) a temperature-dependent occupation factor that expresses the number of electrons and holes that occupy such states. The first quantity (1) represents the joint density of states (DOS) that can be calculated by standard perturbation theory. Using a simple Boltzmann factor to describe the occupancy factor of states (term *b*), one can obtain a simplified expression for the band-edge profile emission (Yu and Cardona 1995):

$$I_{PL}(\hbar\omega) \propto \begin{cases} \left(\hbar\omega - E_g\right)^{1/2} \exp\left[-\left(\hbar\omega - E_g\right)/k_B T\right] & \text{for } \hbar\omega > E_g \\ 0 & \text{otherwise} \end{cases} \tag{3.2}$$

It is worth noting that the temperature T explicitly appearing in the above relation describes the statistical thermal distribution of electrons and holes. In practice, it represents the electron temperature, which is in principle different from the lattice temperature that describes the vibrational state of the crystal (i.e., phonon population). In the case of very-high-intensity excitation, electron–electron and hole–hole interactions can become stronger than carrier–phonon interactions. In such cases, it may be common for charge carriers to attain a temperature higher than that of the lattice. Photoluminescence is indeed one of the few methods allowing a direct determination of carrier temperature (Yu et al. 1978).

3.2.2. Exciton Transitions

In real crystals, Coulomb interaction occurs between free charges. As a consequence, in photoluminescence experiments one expects the photo-excited electrons and holes to be attracted to each other by Coulomb interactions. This leads to formation of excitons, i.e., "bound" states formed by oppositely charged pairs. Excitons can next recombine radiatively, leading to emission of light exactly as in the case of interband recombination. However, the photons produced by exciton recombination do not have the same photon energy as those produced by band-edge recombination. In fact, once exciton formation takes place, the minimum energy distance between a conduction electron and a valence hole no longer corresponds to the interband distance, because the "binding energy," i.e., the energy that must be supplied to the bound pair to decouple the individual charges, has to be subtracted from the interband distance.

It is to be noted that if the system is at a finite temperature T, electron and holes are also subject to thermal motion. If the exciton binding energy is less or much less than the average thermal energy k_BT (where k_B is Boltzmann's constant), excitons easily dissociate and no "energy shift effect" is observed in the PL emissions. Practical observation of excitonic emission thus depends strongly on the relative strength of the exciton binding energy and of typical thermal energy. At low temperatures one expects that emission of light should be dominated by exciton emission; however, this is of course not of practical importance for materials acting as gas sensors, since one usually has to heat the sample in order to speed up the response time of the device.

Literature findings suggest that in cases such as TiO_2 and SnO_2 nanostructures, one cannot easily obtain evidences of excitonic emission at room temperature. The situation is strikingly different in the case of ZnO, as its exciton emission is clearly observable even at room temperature, due to its very large exciton binding energy (Ozgur et al. 2005).

3.2.3. Transitions Involving Deep Defects

Real crystals are far from being ideal, and various kind of defects occur in them. As a general rule, crystal defects are associated with electron (or hole) states that would not be present in an ideal crystal. The energy levels of such states do not suffer from constraints such as a forbidden energy region (band-gap region), which applies only to electrons traveling in a perfectly periodic potential. Indeed, the loss of periodicity induced by any kind of defect (missing atoms, dislocations, interstitials, atomic substitutions, extrinsic doping) leads to more or less localized states whose energy levels may lie well inside the band-gap region.

For many TCO nanostructures (Lei et al. 2001; Kundu and Biswas 2005; Lettieri et al. 2008a) and even some insulators as SiO_2 (Wehrspohn et al. 1999; Anedda et al. 2005), their PL emission band lies at a photon energy well below the band-gap edge. Thus, these emissions cannot be attributed either to interband recombination or to excitonic emission (or excitonic phonon replica, etc.). For example, a well-known green line emission that peaks at about 3.4 eV is ubiquitously found in ZnO (whether in the form of single crystals, thin films, nanostructures, etc.), even if its band edge is at about 3.3 eV. In such cases, it is customary to attribute the emission of light to transitions involving some deep defect states.

It is to be noted that observation of sub-band-gap PL alone does not allow attributing the observed transition to a specific chemical species with which the crystal may be doped. A specialized literature exists on this topic, and we will not discuss it here.

As a point worth remembering, it is often noted that in many cases evidence suggests that deep states leading to sub-band-gap transitions are likely to be associated to surface sites (Zhu and Ding 1999; Shalish et al. 2004; Liao et al. 2008). This is very interesting in terms of PL-based gas sensing, because the surface nature of visible light–emitting centers in many nanostructured TCOs means enhanced sensitivity to adsorbed chemical species.

3.3. MODELS FOR GAS-INDUCED PHOTOLUMINESCENCE QUENCHING

In this section, experimental results relative to PL efficiency modifications induced by adsorption of gas molecules will be reviewed. In this kind of analysis, one typically detects the spectral PL emission of a material in an inert environment (e.g., vacuum or nitrogen) and monitors it during the flow of controlled concentrations of some specific gas of interest. It appears that specific behavior of the PL efficiency curve versus gas concentration occurs very often. In past years, we investigated this specific point (Lettieri et al. 2008a, 2008b; Setaro et al. 2008) and developed a phenomenological modeling of static PL (CWPL) intensity quenching induced by adsorption of gas molecules without the need to specify the physical/chemical processes involved. The modeling is reported here; it assumes some very general propositions and considers two different kinds of quenching mechanisms that can be induced by the presence of molecules acting as photoluminescence quenchers.

Insight into the processes occurring at a more fundamental level cannot be obtained by simple PL quenching analysis, but useful information can be obtained by coupling the CWPL investigations to other analyses. In particular, we will outline the importance of also investigating recombination lifetimes by time-resolved PL (TRPL).

As a first point, we describe the "light-emitting centers" (LECs) by using a sort of "two-level approximation"; that is, photoluminescence emission is described by optical transitions of electrons from an excited state (indicated as X^*) to the fundamental state (indicated by X) of the LEC. No other states besides X and X^* are considered in order to simplify the discussion and allow us to obtain an explicit expression of the PL intensity. It can be noted, again, that by using this approach we are not making any statement about the real chemical nature of the LEC (this topic, however, will be discussed in the next section).

The PL intensity is obviously determined by the total number of radiative events that take place in the material after optical excitation. This total number depends on two parameters: the probability that an excited state (X^*) can decay radiatively to the fundamental state (X), and the number (or density) of radiative states (indicated in the following as $[X]$ and $[X^*]$). As both quantities can be affected by the presence of a quencher, this indicates that two kind of mechanisms can lie behind a PL quenching phenomenon. We will indicate them by "dynamic" and "static" quenching, referring respectively to mechanisms in which the quencher induces modifications in recombination probability or in the density of emitting centers, respectively.

3.3.1. Dynamic Quenching

We assume the following basic points. When the material is isolated, optical excitation of the state X^* can release excess energy only through radiative (i.e., photoluminescence) or nonradiative recombination. No other recombination paths are considered, since the material is isolated and no interactions take place with outer (external) physical entities. When the material is instead interacting with gas molecules (indicated as Q) adsorbed at the material surface and acting as PL quenchers, a third de-excitation path can be introduced as exchange of energy between the excited states of the material (X^*) and the molecules. In practice, the presence of environmental quenching molecules interacting with LECs can introduce an additional nonradiative recombination channel, reducing the probability that excited emission centers give rise to photon emission. Thus the following four channels can be recognized.

Excitation: $$X + \hbar\omega \xrightarrow{\alpha I_\omega} X^* \tag{3.3a}$$

Radiative recombination: $$X^* \xrightarrow{k_r} X + \hbar\omega' \tag{3.3b}$$

Nonradiative recombination: $$X^* \xrightarrow{k_{\mathrm{nr}}} X + \delta E_{\mathrm{nr}} \tag{3.3c}$$

Quenching: $$X^* + Q_{\mathrm{ads}} \xrightarrow{k_q} X + Q^* \tag{3.3d}$$

Each of the four processes is characterized by a recombination (reaction) rate and a coupling constant (indicated here as k_r, k_{nr}, k_q). The recombination rates (indicated by W below) are defined as the number of processes taking place per unit volume and time, and their expression is reported in Eqs. (3.4a)–(3.4d). The coupling constant relates the recombination rates to the density of states that are involved in the specific elementary process. The expression of reaction rates for the above-listed four elementary processes can be expressed as follows.

Excitation rate: $$W_{exc} = \alpha_\omega I_\omega [X] \quad (3.4a)$$

Radiative recombination rate: $$W_{PL} = k_r [X^*] \quad (3.4b)$$

Nonradiative recombination rate: $$W_{nr} = k_{nr} [X^*] \quad (3.4c)$$

Quenching rate: $$W_Q = k_q [X^*][Q_{ads}] \quad (3.4d)$$

We note that the rate of the excitation process (3.4a) is of course ruled by the intensity of the external radiation (e.g., laser) source. We indicate here with I_ω and aω the intensity of the laser source and the absorption coefficient at laser photon energy $\hbar\omega$. The quantities $[X]$ and $[X^*]$ represent the density of light-emitting centers in the fundamental and exited states, respectively, while $[Q_{ads}]$ indicates the density of molecules that are exerting the PL quenching activity on the material. Thus, in the case of a gas, this quantity is in principle not equal to the environmental density of the gas itself. For this reason we adopt the label "ads" to specify that only the "adsorbed" molecules are important to describe the process.

Assuming a simple two-level system, i.e., that no state other than X and X^* plays a role in recombination processes, the quantity $[X] + [X^*]$ is time-independent:

$$\frac{d}{dt}([X]+[X^*]) = 0 \quad (3.5)$$

or, equivalently, the number of transitions from X to X^* per unit time is equal to that characterizing inverse transition.

Using the above expressions (3.4a)–(3.4d) for the recombination rates and working in two-level approximation and using Eq. (3.5), one can thus write the rate equation for $[X]$ and $[X^*]$ as

$$\frac{d[X]}{dt} = W_{PL} + W_{NR} + W_Q - W_{exc} = (k_r + k_{nr} + k_q[Q_{ads}])[X^*] - aI_\omega[X] \quad (3.6)$$

$$\frac{d[X^*]}{dt} = -\frac{d[X]}{dt} = -(W_{PL} + W_{NR} + W_Q) + W_{exc} \quad (3.7)$$

An even more stringent equality is defined by the steady-state condition, which characterizes a situation where excitation is provided by a stationary source (for example, a continuous-wave laser in CWPL). In this case, one gets

$$\frac{d}{dt}[X]_{ss} = \frac{d}{dt}[X^*]_{ss} = 0 \tag{3.8}$$

which simply states that the density of the fundamental and excited states at steady state (indicated by "ss") are stationary quantities. Of course, this situation applies only to the case where excitation energy is supplied by a stationary source (i.e., I_ω = constant).

Our goal is to obtain a useful expression for the CWPL intensity. Apart from an unimportant proportionality constant, the total CWPL intensity can be expressed as the number of radiative transitions per unit time, and thus by the value of the radiative recombination rate [Eq. (3.4b)] at steady state. Therefore, the expression for $[X^*]_{ss}$ first has to be determined.

We begin by considering the case of absence of quencher ($[Q_{ads}] = 0$). Using Eqs. (3.6), (3.7), and (3.8), we obtain

$$\frac{d[X^*]_{ss}}{dt} = 0 \Leftrightarrow -(k_r + k_{nr})[X^*]_{ss} + \alpha I_\omega [X]_{ss} = 0 \Leftrightarrow [X^*]_{ss} = \frac{\alpha I_\omega \cdot [X]_{ss}}{k_D} \tag{3.9}$$

where $k_D = k_r + k_{nr}$. As already mentioned, photoluminescence intensity (here indicated by Φ) can be expressed as proportional to radiative recombination rate W_r as in Eq. (3.4b). Using (3.9), one gets (apart from a noninfluential proportionality constant)

$$\Phi^0_{PL} = k_r [X]_{ss} = \frac{\alpha I_\omega \cdot [X]_{ss}}{k_D} k_r \tag{3.10}$$

This result is easy to understand: The total yield of photons emitted by photoluminescence in the absence of a quencher (Φ^0) is proportional to the excitation density provided by the external (laser) source (we notice that the absorption coefficient α takes into account the fraction of external energy that is really transferred from the external electromagnetic beam to the material), to the density of emitting centers available, and to the quantity $k_r/(k_r + k_{nr})$, representing the internal quantum efficiency of the radiative process, i.e., the probability that an excited state will recombine radiatively. The label "0" indicates that the PL yield shown here is for the absence of external gas molecules.

Once the presence of a quencher is also considered, the steady-state equation reads

$$\frac{d[X^*]_{ss}}{dt} = 0 \Leftrightarrow -k_D[X^*]_{ss} + \alpha I_\omega [X]_{ss} - k_q[X^*]_{ss}[Q_{ads}] = 0 \tag{3.11}$$

and thus

$$[X^*]_{ss} = \frac{\alpha I_\omega [X]_{ss}}{k_D + k_q[Q_{ads}]} \tag{3.12}$$

As a consequence, photoluminescence intensity can be expressed as

$$\Phi_{PL} = k_r[X]_{ss} = \alpha I_\omega \cdot [X]_{ss} \frac{k_r}{k_D + k_q[Q_{ads}]} \tag{3.13}$$

Here it is important to underline the following point: To obtain the above expression, we assumed that the steady-state density of emitting centers $[X]$ is the same for both expressions [Eq. (3.10) and Eq. (3.13)] and thus that the presence of the quencher is not altering the concentration of LECs. This point represents a major difference with respect to the case of static quenching, as will be more clear later.

The expression (3.13) shows that, under the assumptions here, the measurable PL intensity depends explicitly on the concentration of quenching molecules adsorbed on the material through a coupling term k_q. The ratio between the PL intensity in the case of isolated material (no environmental gas) and in the presence of quenching molecules can thus be expressed as

$$\frac{\Phi_0}{\Phi} = 1 + \frac{k_q}{k_D}[Q_{ads}] \tag{3.14}$$

More insight can be gained by focusing attention on the term $[Q_{ads}]$. A quite general description can be obtained by introducing some assumptions about adsorption kinetics and, consequently, by using some related quantities. A general way to see an adsorption process is to consider the gas–surface interaction at specific surface sites that can be "occupied" by chemisorbed molecules, and where gas adsorption is self-limited, i.e., each site can be occupied only by a given number of molecules. Such assumptions describe, for example, Langmuir adsorption kinetics (discussed next). Accordingly, we thus assume the existence of a density of quenchable surface sites, indicated by Q_0. This represents the maximum density of gas molecules that may be accommodated on the material surface and given an observable PL quenching effect. Introducing also the "coverage" θ, which expresses the fraction of occupied quenchable sites, one can write $[Q_{ads}] = \theta \cdot Q_0$ and hence Eq. (3.14) can be rewritten as

$$\frac{\Phi_0}{\Phi} = 1 + \frac{k_Q}{k_D}\theta \qquad (3.15)$$

where $k_Q = Q_0 \cdot k_q$. We notice that k_Q as defined has the physical dimension of a rate coefficient (units: 1/s) and is therefore dimensionally homogeneous with the decay coefficient k_D. Determination of the PL quenching ratio is therefore equivalent to determination of the surface coverage of the quencher species. This point will be examined later, in the framework of the Langmuir adsorption model.

3.3.2. *Static Quenching*

Here we consider another mechanism by which the presence of a chemisorbed gas species leads to quenching of photoluminescence intensity. In this case, which we call "static quenching," the effect of the quencher is somehow more drastic than simply introducing an alternative recombination channel for excited radiative states. Indeed, chemical interaction between the material and chemisorbed species can lead to substantial modification of the physical properties of electron states and/or of the electronic spectrum of the material (some evidence will be discussed later for the case of SnO_2). In this case, PL intensity modification can be predicted even without assuming introduction of the additional recombination channels expressed by Eq. (3.3d) and, thus, without changes in radiative and nonradiative recombination rates. This can be modeled by using the expression of steady-state LEC concentrations obtained previously in the absence of gas species, which led to the equalities

$$[X^*]_{ss} = \frac{\alpha I_\omega [X]_{ss}}{k_D} \qquad \text{and} \qquad \Phi^0_{PL} = \alpha I_\omega \cdot [X]_{ss} \cdot \frac{k_r}{k_D} \qquad (3.16)$$

and explicitly introducing a dependence between $[X]$ss and the density of adsorbed quencher molecules $[Q_{ads}]$ ($= \theta \cdot Q_0$).

As we will show, even assuming the simplest dependence can lead to realistic and interesting results. We assume here a linear dependence such as

$$[X]_{ss}(Q) = [X]^0_{ss}(1 - \alpha'[Q_{ads}]) = [X]_{ss}{}^0(1 - \alpha\theta) \qquad (3.17)$$

where $[X]$ss represents the steady-state density of recombination centers in the absence of quencher and $\alpha = \alpha_\omega Q_0$ (not to be confused with the optical absorption coefficient α_ω of the material at the laser excitation frequency). Expression (3.17), with coverage θ and density of quenchable surface sites Q_0 as already defined, is likely to be the simplest that one can assume. Using Eqs. (3.16) and (3.17) for the PL intensity, we thus obtain

$$\frac{\Phi_0}{\Phi} = \frac{1}{1-\alpha\theta} \tag{3.18}$$

This expression is different from the one previously obtained for the case of purely dynamic quenching. However, for low values of surface coverage (i.e., $\alpha\theta << 1$), the expression reduces to one equivalent to that previously obtained and characterizing the regime that we will indicate next as "proportional quenching." In fact, one can notice that, in order to retain physical meaning for Eq. (3.17), the quantity $\alpha\theta$ has to be less than unity. As a consequence, the above function can be expressed as the sum of a geometric series,

$$\frac{\Phi_0}{\Phi} = \sum_{n=1}^{\infty} (\alpha\theta)^n = 1 + \alpha\theta + (\alpha\theta)^2 + \cdots + (\alpha\theta)^n + \ldots \tag{3.19}$$

and for low values of surface coverage ($\alpha\theta << 1$) we get the an expression such as $\Phi_0/\Phi \cong 1+\alpha\theta$. Comparing this expression to Eq. (3.15), we see that both predict a quenching ratio Φ_0/Φ proportional to the surface coverage θ. In the case of Eq. (3.15), the proportionality constant is given by the ratio k_Q/k_D, which measures the "strength" of the additional nonradiative recombination channel introduced by the quencher. In the case of Eq. (3.18), the proportionality constant is represented by a parameter that can be expressed as

$$\alpha = \frac{1}{\theta}\frac{\Delta[X]}{[X]_0} \tag{3.20}$$

where $\Delta[X]/[X]_0$ represents the relative change in the number of radiative states induced by the presence of a quencher at coverage θ.

3.4. PRACTICAL ISSUES IN ANALYSIS AND INTERPRETATION OF PL QUENCHING DATA

After exposing the above modeling, some practical considerations can be underlined. First, relating the CWPL yield to surface coverage is not practical unless one relates the coverage itself to a measurable quantity. This point will be discussed here in the framework of Langmuir adsorption kinetics. Second, we obtained that for low coverage values the two mechanisms previously shown ("static" and "dynamic") could be indistinguishable on the basis of CWPL measurements only. In such cases, we here discuss why also performing time-resolved analysis of PL emission becomes important and useful.

3.4.1. Analysis of CWPL Quenching: Langmuir Adsorption Kinetics

We have seen how the CWPL yield can be related to surface coverage, which is a quantity that measures the density of adsorbed molecules. Such a quantity cannot be determined directly in a simple manner. Instead, in practical gas-sensing experiments, one usually controls the composition of the environment by flowing known concentrations (or partial pressures) of an analyte species diluted in an inert gas carrier, or by flowing mixtures of different gas species having known partial pressures. Thus, a fundamental practical need is a relationship between the local surface density of adsorbed molecules and the parameter that one controls experimentally, such as the partial pressure of the gas species of interest.

To this aim, we will follow the Langmuir model for adsorption processes. In this model, one assumes that adatoms are held at specific sites on the surface, that the value of desorption enthalpy is the same for all sites, that there is no motion of atoms over the surface, and that the only effect that the presence of one adatom has on any other atom is that atoms striking a surface site that is already occupied do not adsorb. This implies that the total adlayer coverage can never exceed the total number of sites, and coverage is limited to a single monoatomic layer.

Under the above assumptions, it can be demonstrated (De Boer 1953) that the relation between the surface coverage (θ) and the partial pressure (P) of adsorbing gas species is given by the following classical equation:

$$\theta(p) = \frac{bP}{1+bP} \tag{3.21}$$

where the quantity b is given by

$$b = \frac{\tau_0}{Q_0} \frac{1}{(2\pi m k_B T)^{1/2}} \exp\left(\frac{\Delta H_{\text{des}}}{k_B T}\right) \tag{3.22}$$

τ_0 represents the mean staying time for adsorption in the limit of very high temperature, and ΔH_{des} is the desorption enthalpy (Hudson 1998).

Using (3.21), one can get an explicit expression of PL yield versus the external partial pressure (P) of the quencher. For the case of dynamic quenching, Eq. (3.15) becomes

$$\frac{\Phi_0}{\Phi} = 1 + \frac{k_Q \cdot bP}{k_D(1+bP)} \tag{3.23}$$

For the case of static quenching, using (3.21) and (3.18), one gets

$$\frac{\Phi_0}{\Phi} = \frac{1}{1-\left[\alpha bP/(1+bP)\right]} = \frac{1+bP}{1+(1-\alpha)bP} \tag{3.24}$$

As noticed before, in the low-pressure range, all expressions assume a similar form where the PL yield decreases linearly with quencher pressure. It is interesting to underline the pressure limits that define such behavior.

For the case of dynamic quenching, Eq. (3.15) for $bP << 1$ (i.e., $\theta \approx bP$) leads to the well-known Stern-Volmer equation:

$$\frac{\Phi_0}{\Phi} = 1 + \left(\frac{k_Q \cdot b}{k_D}\right)P \tag{3.25}$$

where we see that the quantity $P_{\text{eff}} = k_D/bk_Q$ acts as an "equivalent pressure" parameter quantifying the quenching effect of a gas at a given pressure P.

By writing the Stern-Volmer equation as $\Phi/\Phi_0 = \left[1+\left(P/P_{\text{eff}}\right)\right]^{-1}$, we see that for $P << P_{\text{eff}}$, one obtains the simple expression $\Phi = \Phi_0\left[1-(P/P_{\text{eff}})\right]$, i.e., a PL signal that scales linearly as a function of the quencher pressure. Such a result is obtained, for the static case, in the limit $bP << 1$, as from (3.18) and (3.21) one obtains $\Phi = \Phi_0(1-\alpha\theta) \cong \Phi_0\left(1-\alpha bP\right)$.

We can thus summarize the results we obtained for the two cases, as listed in Table 3.1:

Table 3.1. Summary of the expressions obtained for the PL intensity ratio (Φ_0/Φ) and for the PL intensity as a function of external gas pressure $\Phi(P)$—according to the Langmuir adsorption model—for the two cases of "dynamic" and "static" quenching mechanisms

Type	"Dynamic"	"Static"
Quenching mechanism	Quencher introduces additional nonradiative recombination channels	Quencher modifies the number of available emitting centers
PL intensity ratio versus quencher coverage	$\frac{\Phi_0}{\Phi} = 1 + \frac{k_Q}{k_D}\theta$	$\frac{\Phi_0}{\Phi} = \frac{1}{1-\alpha\theta}$
PL intensity using Langmuir model	$\Phi(P) = \Phi_0 \cdot \frac{1+bP}{1+(a+b)P}$	$\Phi(P) = \Phi_0\left(1-\frac{\alpha bP}{1+bP}\right)$

3.4.2. "Proportional Quenching" Regime: The Importance of Time-Resolved PL Analysis

The model we used for the "static" mechanism assumed that the number of emitting centers that are "switched off" due to interaction with gas is linearly proportional to the density of adsorbed molecules. This leads to a PL intensity that is proportional to surface coverage. However, it is worth noting that also for the case of "dynamic" mechanism we obtained an expression of PL intensity that is linearly proportional to surface coverage when quencher coverage is not very high (i.e., in the limit $bP << 1$). Thus, in conditions of low coverage, both mechanisms lead to an expression for PL quenching that can be expressed as

$$\Phi = \Phi_0(1 - \alpha\theta) = \Phi_0\left(1 - \frac{\alpha bP}{1 + bP}\right) \tag{3.26}$$

This expression indicates a PL intensity that scales (decreases) linearly proportion with surface coverage. For this reason, we can refer to it as a "proportional quenching" regime.

As a result, experimental analyses showing a PL intensity scaling linearly versus quencher coverage do not by themselves allow us to recognize one or the other of the mechanisms discussed here.

Deeper understanding can be obtained by studying the time dynamics of recombination processes, allowing some distinguishing between the two kinds of quenching processes. To show why, we recall the main difference between the "dynamic" and "static" processes. In the first process, the quenching species produces additional nonradiative recombination channels, thus modifying the recombination probabilities (rates). This fact reflects the direct dependence of PL lifetimes on the quencher concentration. We can see it by writing the total rate equation $-d[X^*]/dt = W_{\mathrm{PL}} + W_{\mathrm{nr}} + W_Q$ using the expression for recombination rates given by Eqs. (3.4a)–(3.4d) as

$$-\frac{d[X^*]}{dt} = \left(k_D + k_q[Q_{\mathrm{ads}}]\right)[X^*] \tag{3.27}$$

Equation (3.27) indicates that the rate coefficient in the presence of a quencher is given by $k_T = k_D + k_q[Q_{\mathrm{ads}}]$, or, equivalently, introducing the effective lifetime for recombination process $\tau (= 1/k_T)$, we get $(1/\tau) = (1/\tau_0) + k_q[Q_{\mathrm{ads}}]$ (where $1/\tau_0 = k_D$). The PL yield (Φ) is proportional to the radiative rate integrated over time, so, from (3.23), one obtains

$$\Phi \propto \int k_r[X^*]_{t=0} \exp\left(-\frac{t}{\tau}\right)dt = k_r\tau[X^*]_{t=0} \tag{3.28}$$

and thus, $\tau_0/\tau = \Phi_0/\Phi$ (where Φ_0, τ_0, and Φ,τ are the PL yield and effective recombination lifetime in the absence and presence of a quencher, respectively). Taking into account (3.11), one finally has the following Stern-Volmer expression:

$$\frac{\tau_0}{\tau} = 1 + k_q \tau_0 [Q_{ads}] \tag{3.29}$$

In conclusion, when a quencher introduces competitive nonradiative recombination rates, a definite change in recombination lifetime can be expected.

The static quenching mechanism, on the other hand, is based on modification of the density of available emitting centers, and thus in principle no modification of recombination rates and lifetimes should be expected; that is, $\tau_0/\tau = 1$ regardless the value of $[Q_{ads}]$. These considerations sketch out that performing time-resolved PL analysis under flowing gas represents a possible way to distinguish between the two mechanisms. Of course, in principle it cannot be expected that only one of the two is active, and some superposition of both mechanisms may occur in practical cases; however, dominance of one over the other can be indeed evidenced by time-resolved analysis.

4. PHOTOLUMINESCENCE-BASED OPTOCHEMICAL SENSING BY OXIDE NANOCRYSTALS AND NANOSTRUCTURES: RESULTS AND INTERPRETATIONS

The goal of this section is to review recent literature on application of metal oxide semiconductors in photoluminescence-based chemical sensing, discussing the results in the light of the modeling described in Section 3. The literature that we will consider deals with oxide materials in the form of nanocrystals as well as of nanoporous structures.

It is quite evident that the huge specific surfaces associated to nanoporous materials and to individual nano-sized structures (in particular to quasi-one-dimensional ones) is a parameter of crucial importance in tailoring the performances of a sensing material. However, the specific surface is not the only reason for the current interest in nanocrystalline MOX semiconductors. To further explain this interest, we should underline the roles of both crystal size and crystal order.

Next, we will finally consider some practical examples from the literature, discussing each material while also sketching a picture for the physical interpretation of the observed phenomena. It is worth underlining that the literature on the subject is, in this author's opinion, still not completely established, so only incomplete conclusions can be drawn in some cases. A somewhat more established picture has emerged for tin oxide (SnO_2) nanostructures, due to a number of theoretical works that have been developed in the last years to explain the experimental findings.

4.1. WHY NANOSTRUCTURES? THE ROLES OF CRYSTAL ORDER AND SIZE

Although nanocrystals made of several MOX materials can today be synthesized with low-cost techniques and at a reasonable production rate, the production of ordered (and even single-crystal) structures defined at the nanometer scale (for example, one-dimensional structures) remains a technological task. High specific surface alone would probably not be enough to justify the practical use of nanomaterials, as thick particulate films of many oxides can exhibit very high porosities, allowing extensive access to surface adsorption and high sensitivities.

A peculiar feature of MOX nanocrystals resides in their crystal order, a feature that is not shared with granular MOX films, which mostly exhibit a polycrystalline morphology resulting from aggregation of crystallites separated by grain boundaries. To explain the present interest in nanocrystals, we should underline the role played by this latter difference—in other words, the role played by material morphology (including both crystal size and crystal order) in determining the performance of standard (conductometric) sensor devices.

As discussed in Section 1, the basic principle of operation of conductometric solid-state gas sensors consists of the modulation of electrical conductivity of a gas-sensing material induced by changes in composition of the gas atmosphere surrounding it. The role played by the morphological structure in the chemoresistive effect is of course crucial. In fact, it is well known that grain conductivity is basically controlled by the eventual presence (or the eventual modulation) of a space charge region (SCR) extending around the grain borders. Generally speaking, SCR formation in a semiconductor is a consequence of accumulation (or depletion) of localized charge in specific regions of the material and is a natural and ubiquitous phenomenon occurring at interfaces between two semiconductor materials or at semiconductor/environment interfaces (i.e., at surfaces).

Variations of charge density and of potential energy are "required" to screen out the localized charge accompanying the formation of a semiconductor interface or the adsorption of a molecular species that can give rise to redox chemical reactions. Such variations are by no means "local": Redistribution of charge may extend up to hundreds of atomic layers, leading to dramatic changes in the conductivity of the material (a typical example is the electrical current rectification observed in *p-n* diodes, due to formation of charge-depleted regions at the junction interface).

This effect can be even more dramatic as the size of the semiconductor is reduced. We consider, for example, the case of a granular *n*-type semiconductor in which localization of net negative charge at surfaces is associated to electron depletion in the subsurface internal volume of the crystallite. In this case, a decrease in the grain size will be equivalent to an increase of the semiconductor volume fraction in which the charge flow is inhibited by formation of a electron-depleted region. It is thus not surprising that, for instance, Yamazoe (1991) showed that reduction of crystallite (grain) size led to considerable improvement in sensor performance. Therefore, in tailoring sensor performance, it is important to prepare materials with small crystallize size which are stable when operated at high temperature for long periods.

First-generation devices were prepared by thick-film technology starting from powders. This route, however, had some drawbacks. First, in such thick films, macroscopic charge flow is assured by the presence of percolating paths connecting injection and collection electrodes through intergrain regions. Therefore, even small variations in the preparation process (and thus in the resulting morphology) may lead to sensor devices that differ slightly in their initial characteristics. Moreover, electrical properties of granular media are strongly related to overall morphology and thus drift due to "ageing" effects such as grain coalescence, porosity modification, and grain-boundary alteration. Such effects may in general be accelerated by prolonged heating of the material, which is required in metal oxide sensors to guarantee the reversibility of chemical reactions (i.e., to increase the desorption kinetics of analytes) at the surface.

A relevant example is represented by the problem of incomplete stoichiometry: In the case of metal oxide particulate films, the metal-to-oxygen stoichiometry ratio in grains is often not exact, because of oxygen vacancies. Incomplete oxidation of metals can be detrimental to the stability of the sensor and may lead to drift in its response (Dieguez et al. 1999) once the working temperature of the device is raised to increase its response time. Moreover, in powderlike thick films, particulates may tend to sinter and form larger aggregates due to prolonged heating in ambient air, again leading to long-term drift in device performance.

In more recent years, it emerged that these major limitations could be avoided successful preparation of single-crystal nanostructures. In particular, for many materials of practical interest for the realization of sensing devices (SnO_2, TiO_2, ZnO, CuO), the possibility to grow ordered and "quasi-one-dimensional" (quasi-1-D) structures has been demonstrated (Pan et al. 2001; Dai et al. 2002; Yuan et al. 2002; Wen et al. 2003; Vayssieres 2003) by simple preparation routes starting from commercial metal oxide powders. The crystal order exhibited by many of those structures [such as, for example, nanobelts (Pan et al. 2001), nanowires (Huang et al. 2001; Vayssieres 2003), or nanoribbons (Dai et al. 2002; Yuan et al. 2002)] assures improved stability, while at the same time assuring high sensitivity due their reduced lateral dimensions, which may range from 100 nm to even smaller. Such peculiarities make MOX nanocrystals interesting for both fundamental studies and potential applications as novel sensor devices.

As discussed previously, the possibility of growing gas-sensing semiconducting oxide materials in single-crystalline form is an important step toward improving the performance of possible devices. Single-crystal order, by definition, means optimal stoichiometry, thus reducing the above-mentioned problems related to

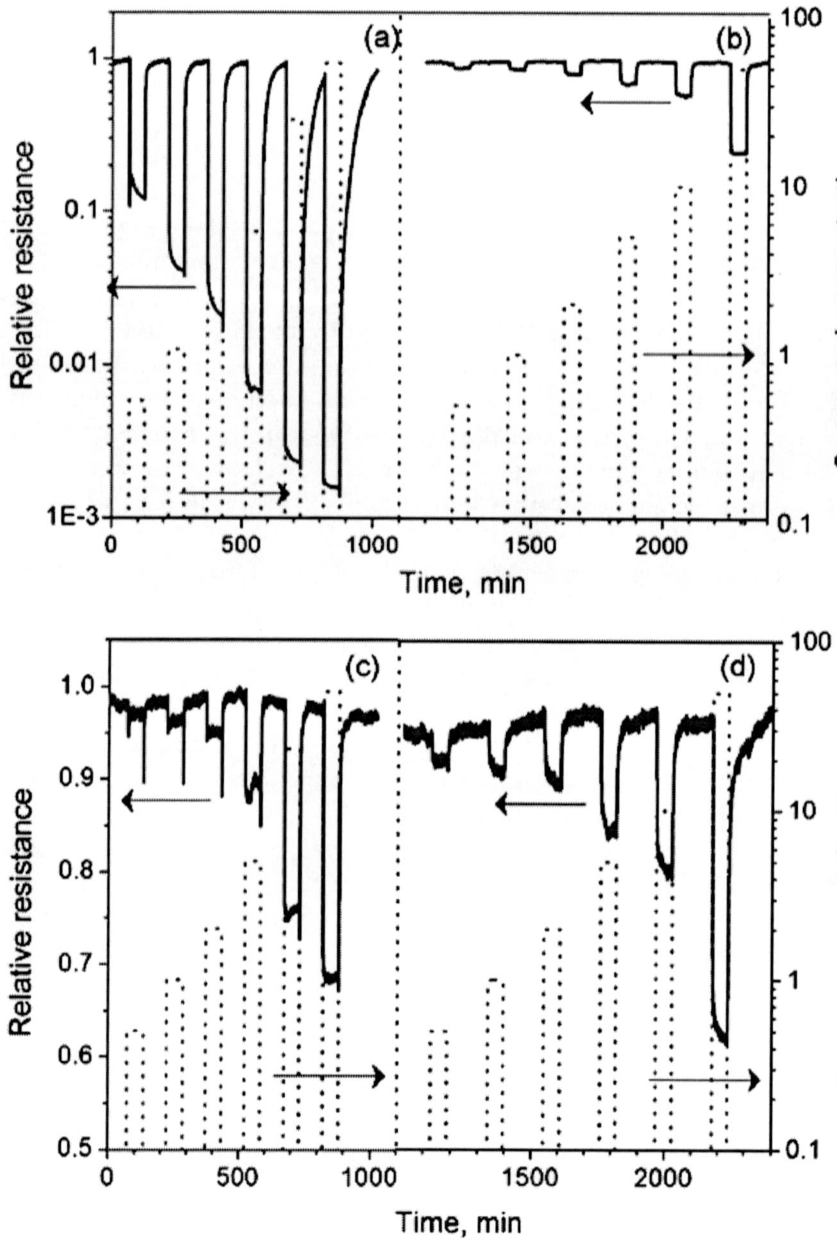

Figure 3.9. Change in SnO_2 nanostructures' resistance under exposure to 2-propanol vapors of step-changed concentration on 1st day (a and c) and 46th day (b and d). (a and b) Nanoparticulate SnO_2. (b–d) crystalline nanowire-based SnO_2. (Reprinted with permission from Sysoev et al. 2009. Copyright 2009 Elsevier.)

incomplete stoichiometry and to excessive defect densities. This leads to improved thermal stability, reducing device ageing problems related to grain re-oxidation and long-term sintering.

An example of comparative study of long-term stability was carried out by Sysoev et al. (2009), who compared the sensing performance of chemoresistors made of randomly oriented single-crystal SnO_2 nanowires and of SnO_2 nanoparticle thin layers. Sensors based on nanowires exhibited improved long-term stability toward traces of 2-propanol in air versus those based on SnO_2 nanoparticles. Figures 3.9a and 3.9c show the gas response measured for as-prepared nanoparticulate-based (a) and nanowires-based (c) SnO_2 chips to 2-propanol vapors of step-changed concentrations (0.5–50 ppm, in air). Figures 3.9a and 3.9b show results for the first day of device operation. The response of the same samples as measured after 46 days of continuous operation is plotted in Figures 3.9c and 3.9d. As can be seen, the device based on nanowire nanocrystals has stable gas-sensing performance over the whole time period.

Better stability of nanowires was explained by Chaim et al. (2008) as an increased propensity of nanoparticles to sinter and form larger aggregates atelevated operating temperature conditions (300°C). As a result of morphology changes due to such nanoparticle aggregation, the diffusion of gas molecules through the pores of the sensing material becomes hampered (Sakai et al. 2007). On the other hand, the absence of such processes in single-crystal SnO_2 nanowires explains the improved stability of devices that use them as sensing elements.

Once reliable techniques for synthesis of reproducible and stable nanostructures had been established, pivotal works on highly sensitive gas sensors based on quasi-1-D nanostructures began to be published, employing as active materials semiconducting metal oxides (Comini et al. 2002; Wan et al. 2004; Chen et al. 2005) or even other materials such as carbon nanotubes (Li et al. 2003; Modi et al. 2003). Since then, a vast amount of research has been devoted to possible applications of single-crystal MOX nanostructures as gas sensors based on conductometric and, more recently, optical transduction. Size control is of course a major issue, which can be achieved only by careful characterization of the specific deposition apparatus. For further details on growth procedures, one should refer to specialized literature (see, for example, Comini et al. 2006).

Literature results on PL-based optical sensing and on basic physical mechanisms of PL quenching and gas–nanostructure interaction are discussed here for some specific oxide nanomaterials.

4.2. ZINC OXIDE NANOSTRUCTURES

Zinc oxide nanostructures have been extensively investigated as active materials for gas sensors and for applications in biosensing, as witnessed by the great number of tested gases, vapors, and organic components, including hydrogen, ammonia, carbon monoxide, ethanol, methanol, acetone, butane, glucose, cholesterol, etc. (Mazingue et al. 2005; Wang et al. 2006; Bismuto et al. 2006; Li et al. 2007; Sberveglieri et al. 2007; Singh et al. 2007; De Julián Fernández et al. 2008).

As a semiconductor, ZnO exhibits n-type behavior due to its native defects (e.g., oxygen vacancies and zinc interstitials). Therefore, oxidizing molecules acting

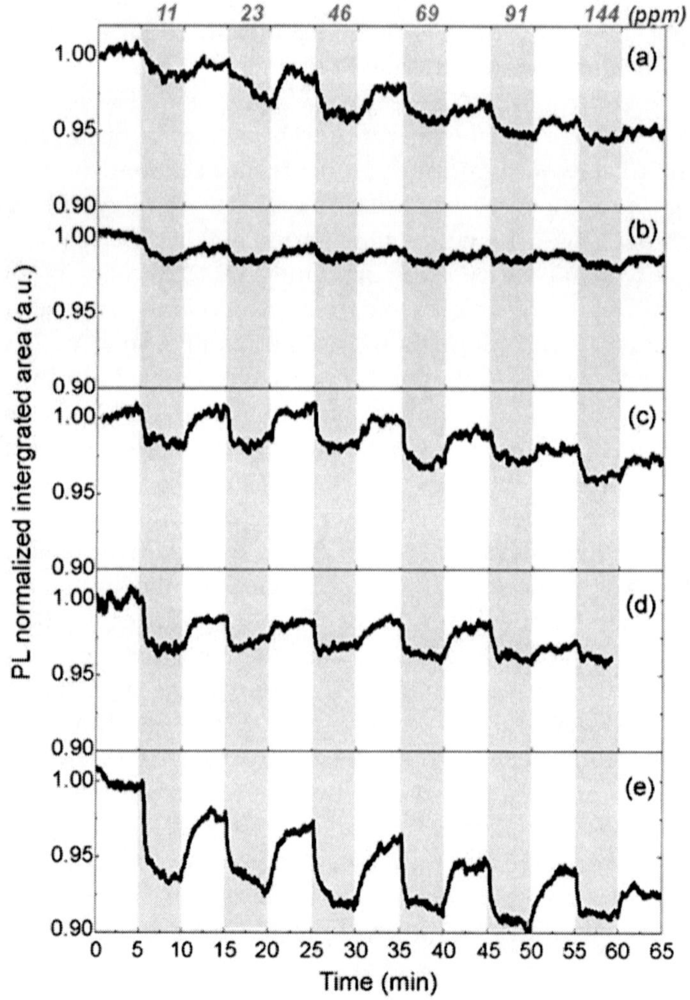

Figure 3.10. Room-temperature NO_2 optical sensing measurements obtained for different ZnO structures exhibiting different morphologies. Larger responses are exhibited by samples with larger specific surface. The PL signal was obtained by integration of the PL spectrum over the band-edge region (380–420 nm). Gas concentrations varied from 11 to 114 ppm (reported on top of the graph), Time intervals of NO_2 exposure are gray-shaded. (Reprinted with permission from Valerini et al 2010. Copyright 2010 Elsevier.)

as electron trappers, such as nitrogen dioxide (NO_2), have a significant effect on its electron-transport properties. Moreover, ZnO is a well-known high-quality optical material, due to its optical transparency in the visible range. Concerning optical sensing by ZnO, applications based on refractive response using ZnO waveguides have been reviewed previously (see Section 3), while optical sensing by PL modification in ZnO seems to have been less investigated.

An interesting study of PL modifications of ZnO nanostructures due to nitrogen dioxide (NO_2) exposure was reported by Valerini and co-workers (2010), who focused on the role of the surface morphology of a set of ZnO samples grown by pulsed laser deposition. These authors demonstrated that NO_2 concentrations down to 3 ppm could be detected at room temperature by monitoring PL quenching in ZnO pencil-like structures having diameters in the range 80–500 nm. The study pointed out the importance of sample morphology on gas-sensing performance; in Figure 3.10, for instance, the quenching of the PL integrated area and the subsequent recovery in air is reported for NO_2 concentration varying in a range from 11 ppm to 114 ppm. Different curves (1–5) correspond to samples having different morphologies. In Figure 3.10, curves 3–5 correspond to samples exhibiting a more structured and porous morphology; in particular, the largest response was obtained from sample 5, which corresponded to morphology with high specific surface. Another remarkable point evidenced were the fast and sharp variations of the PL signal observed when switching between NO_2 and dry-air exposure. This can be seen more clearly in Figure 3.11, where a magnification of curves corresponding to 11-ppm exposure is shown.

Very interestingly, Valerini et al. (2010) found that quenching of the PL integrated area decreased as the gas concentration increased in the range 11–114 ppm for all the samples. Such counterintuitive behavior was interpreted by suggesting that an 11-ppm NO_2 concentration was high enough to almost saturate the available adsorption molecular sites, so further increase of NO_2 concentration did not lead to an increase in response. Additionally, incomplete reversal of the NO_2 adsorption process during sample recovery could lead to the observed decreasing behavior. It is to be noted that such a picture points in the direction of one basic assumption of the Langmuir model, namely, existence of a limit for density of adsorbed molecules, defined by a number of "available" adsorption sites (see the quantity Q_0 defined previously). An experimental run using the best sensor (sample 5) at lower NO_2 concentrations indeed gave a sensitivity of the relative response showing a saturation at higher quencher concentrations, as predicted by the Langmuir formula.

Saturated "Langmuir-like" PL responses have also been observed by Baratto et al. (2009). In this work, films constituted by a disordered net of nanowires having an average length of several micrometers and a width of approximately 300 nm were used as sensing elements. Probably because of the large surface area exhibited by the nanowires (thanks to their high length-to-width ratio) and to the open

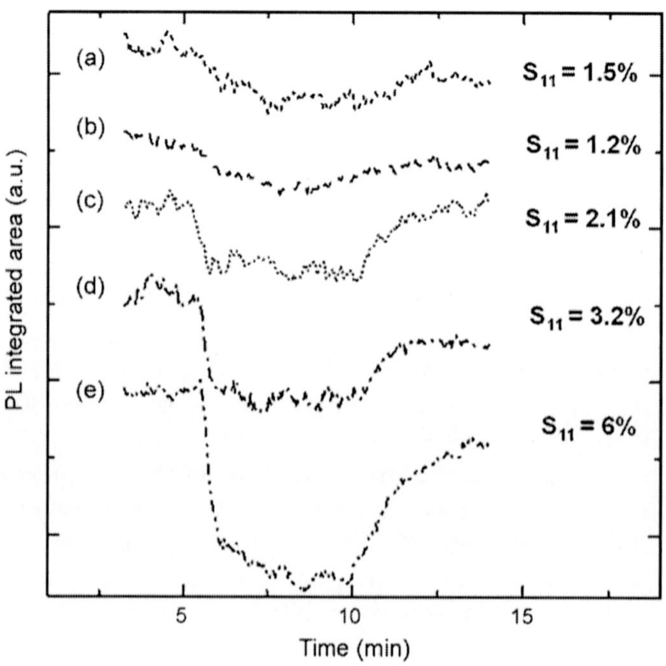

Figure 3.11. Magnification of the dynamic PL response to exposure to 11 ppm of NO_2. The curves are vertically translated for clarity. The relative response (S11) in percent measured for each sample is reported next to the corresponding spectrum. A fast response is observed after exposition to NO_2, with PL intensity reaching a steady-state condition in about 2 min. (Reprinted with permission from Valerini et al. 2010. Copyright 2010 Elsevier.)

network structure featuring large accesses for gas penetration, Baratto and co-workers could detect a response to NO_2 even in the sub-ppm range. Indicating by Φ the (wavelength-integrated) PL spectral area relative to the visible defect-related PL emission, the relative response versus NO_2 concentration (indicated as $[NO_2]$) is shown in Figure 3.12, where the data points were fitted by using the following expression:

$$\frac{\Phi}{\Phi_0} = \left(1 - \frac{\alpha b[NO_2]}{1 + b[NO_2]}\right) \tag{3.30}$$

which can be easily obtained by using Eq. (3.24) (with Φ_0 representing the PL spectral area in the absence of NO_2). The good agreement between Eq. (3.30) and experimental data suggests a PL quenching following the kinetics we indicated as "proportional quenching," i.e., scaling linearly as a function of surface coverage. As discussed previously, in such a case the analysis of time-resolved PL emission is of interest to investigate the possible contribution to quenching from changes in recombination rates ("dynamic") or in the density of emitting centers ("static"). Time-resolved measurements of PL decay of the UV component of ZnO nanorod photoluminescence under exposure to different concentrations of NO_2 showed very slight changes and practically negligible modifications in recombination rates (Lettieri et al. 2012), thus indicating a static-like behavior for the PL quenching. An example is reported in Figure 3.13, where peak-normalized time-resolved PL decay curves are shown for ZnO nanowire samples under exposure to NO_2 flow at different concentrations. The PL signal is defined by integrating the PL spectrum over the wavelength region characterizing the band-edge emission (wavelength range $\Delta\lambda$ = 370–400 nm), similar to the work of Valerini et al. (2010) (the curves obtained for different NO_2 concentrations have been shifted vertically for the sake of clarity). No significant modification of recombination kinetics is noticed, in agreement with previous findings obtained for visible (defect-related) PL emission (Baratto et al. 2009).

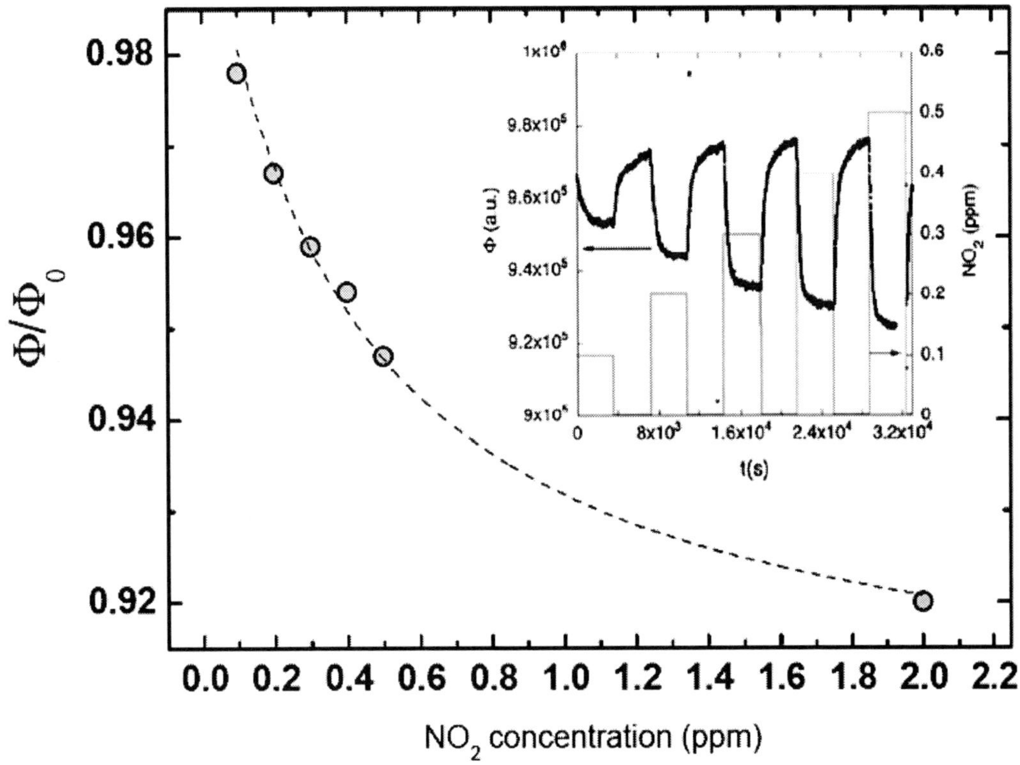

Figure 3.12. Measured ratio Φ/Φ_0 of PL yield versus NO_2 concentration (Φ, spectrally integrated PL intensity in the presence of NO_2; Φ_0, spectrally integrated PL intensity in dry air). Dashed line is best fit using Lagmuir function—see Eq. (3.25). (Reprinted with permission from Baratto et al. 2009. Copyright 2009 Elsevier.)

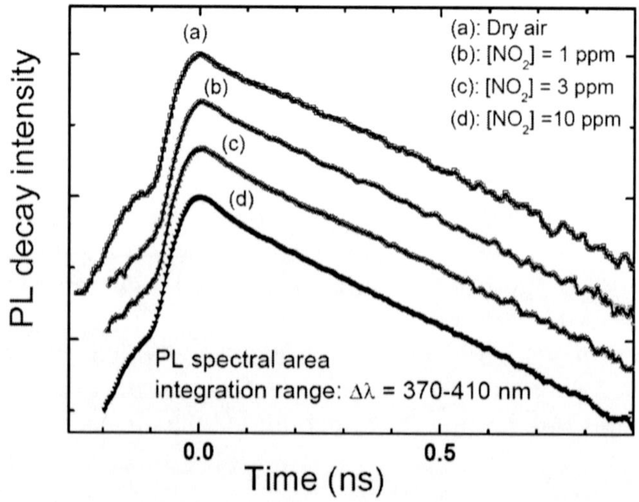

Figure 3.13. Normalized time-resolved PL decay curves for ZnO nanowire samples under exposure to NO_2 flow at different concentrations. The PL signal is obtained as spectral area integrated over the band-edge emission (wavelength range $\Delta\lambda$ = 370–400 nm), similarly to the work of Valerini et al. (2010). The PL curves have been shifted vertically for clarity (Lettieri et al. 2012, unpublished results).

Another very interesting feature worth noting is represented by the response time to NO_2 exposure. The fast response observed by Valerini and co-workers is in close agreement with findings reported by Comini et al. (2007), who compared electrical and PL response to NO_2 and found a typical PL quenching response time at room temperature of the order of 30 s, much faster than the room-temperature electrical response (of the order of 15 min).

The above-described findings and in particular the low sensitivity of recombination dynamics to NO_2 exposure support a picture in which the gas molecules act as static quenchers, partially suppressing the physical mechanisms giving rise to the green luminescence. One possible direct quenching mechanism could lie in chemical "suppression" of radiative states due to interaction with chemisorbed NO_2 molecules on the ZnO surface. It is worth noting that even if such a mechanism has still been not studied in the case of ZnO, evidence supporting it has been reported for the case of SnO_2 surfaces by means of first-principle theoretical calculations of surface band structures; this work will be reviewed in the subsection dedicated to tin oxide.

Another likely origin of "proportional" and static PL quenching may be more "electrostatic" in nature, namely, due to trapping of ZnO electrons from surface defect states to bonds with the adsorbed species. Such a trapping mechanism accounts very well for electrical conductivity quenching under NO_2 exposure. However, in the opinion of the author of this chapter, in the case of optical quenching such a mechanism should be discussed and investigated further. Indeed, the PL quenching mechanism may equally well be interpreted assuming that electrons generated by optical excitation in PL experiments are trapped to NO_2 and become unavailable for radiative recombination with holes. However, it is not completely clear if such an effect should be regarded as static or dynamic: Trapping of photo-generated electrons may be regarded as introduction of an additional nonradiative recombination path and thus introducing some modification of recombination rates, which is indeed not observed in time-resolved PL experiments. Moreover, assuming that electrical and PL quenching result from the same electron-trapping mechanism should lead to similar response times to gas exposure, a fact that is not confirmed by results reported by Comini et al. (2007). A similar argument, based on evidence of different response times for electrical and optical sensors, is also valid for SnO_2 nanostructures, as will be examined later. More investigations are probably needed to clarify the overall picture.

4.3. SILICA NANOSTRUCTURES

Amorphous SiO_2 (or "silica") is an electrical insulator and is therefore not suitable by itself for application in conductometric-based gas sensing. Nevertheless, silica is for various reasons a material of interest for chemical sensing. In fact, micro- or nanostructured silica having very high (or even huge) porosities can be prepared by various routes and loaded by semiconducting or conducting particles to obtain a composite conductometric gas sensor. Another route worth mentioning is the use of porous silica as a coating material on optical fibers for absorbance-based gas sensors. Toward this aim, SiO_2 can be considered an oxide material of interest for gas sensing even if in previous sections we have focused mainly on literature based on transparent semiconducting oxides.

Worth mentioning here is a peculiar application of SiO_2 alone (that is, as an undoped insulator material) for PL-based chemical detection. The application is based on the natural photoluminescence of mesoporous silica, which, due to several kinds of native defects (mainly surface oxygen stoichiometric defects) usually exhibits defect-related luminescence activity in the green-blue region. An interesting and surprising example of "natural" luminescent porous silica is represented by silica shells ("frustules") of diatoms, which are unicellular algae consisting of two valves joined together by a series of silica bands linked along the margins. Frustules are essentially made of silica, and their surfaces are covered by arrays of micropores and mesopores, where the diameter of the pores typically ranges from a few hundred nanometers up to a few micrometers. An example of a porous-rich diatom frustule surface is shown in Figure 3.14, where micropores of about 30 nm diameter distributed on the inner surface of one valve are shown.

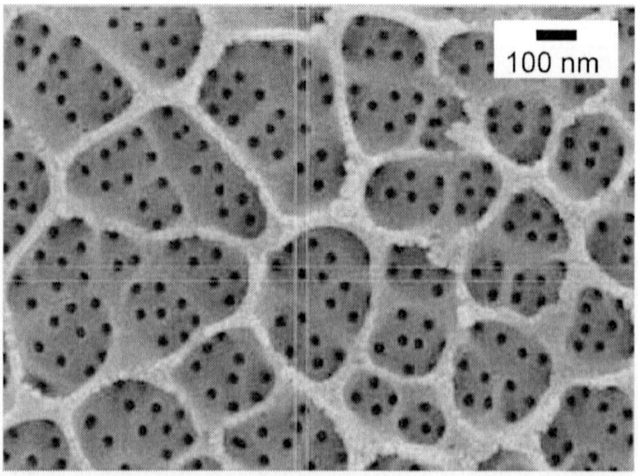

Figure 3.14. Scanning electron microscopy (SEM) viewgraph of a diatom frustule surface. Micropores of about 30 nm diameter are seen. (Reprinted with permission from Lettieri et al 2008b. Copyright 2008 Wiley-VCH.)

Diatom frustules display an efficient visible photoluminescence emission (Butcher et al. 2005), and this fact, together with their highly porous structures, represents the starting point for some PL-based sensing applications.

De Stefano et al. (2005) reported PL modification of various SiO_2 frustules for several gas and organic vapors chemical species. As shown in Figure 3.15, PL quenching was clearly recognized in the case of exposure to NO_2 and acetone, while the opposite effect was obtained after exposing the diatoms to electron-donor substances such as xylene and pyridine. Similar results were reported by Setaro et al. (2007), who discussed PL quenching and enhancement under exposure to NO_2 and CH_4, respectively (see Figure 3.16). These works show that relative responses and gas concentration ranges of sensitivity differ significantly when using different cultures of diatoms with different surface morphologies. Natural availability of a large variety of frustule surface morphologies and porosities, unequaled by availability of artificial mesoporous silica structures, encouraged further investigations on these materials for optical sensing.

These interesting led Lettieri et al. (2008b) to perform a more detailed study of PL quenching mechanisms in silica. In their work, analysis of silica frustules PL modification upon exposure to oxidizing NO_2 was accompanied by infrared spectroscopy and time-resolved PL experiments. Infrared spectroscopy analysis on frustule samples demonstrated that the gas-sensing frustule samples consisted mostly of slightly understoichiometric amorphous silicon dioxide (a-SiO_x), exhibiting –OH surface terminations. Saturated Langmuir-like behavior of PL quenching under NO_2 exposure could be seen, as shown in Figure 3.17, to be like that in

ZnO, even if in the case of SiO_x–NO_2 interactions one cannot invoke mechanisms based on gas-induced trapping of native free electrons.

The picture of static quenching based on suppression of light-emitting centers by NO_2 molecules adsorbing by a Langmuir kinetic mechanism was confirmed by TRPL analysis and by a continuous-wave PL quenching study at variable temperature. In particular, a direct test of Eqs. (3.22) and (3.24) was performed by a Langmuir-based fit of quenching ratios Φ/Φ_0 on the same sample at variable

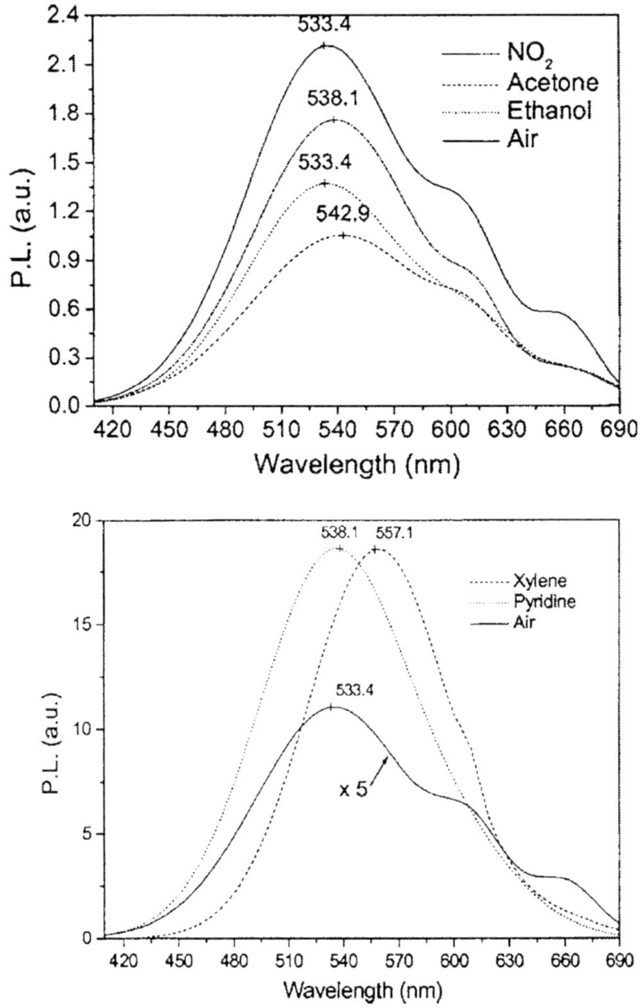

Figure 3.15. Quenching and enhancement of diatoms photoluminescence spectra in the presence of several gaseous substances. The numbers on top of each spectrum indicate the peak position as estimated by numerical fit. (Reprinted with permission from De Stefano et al. 2005. Copyright 2005 American Institute of Physics.)

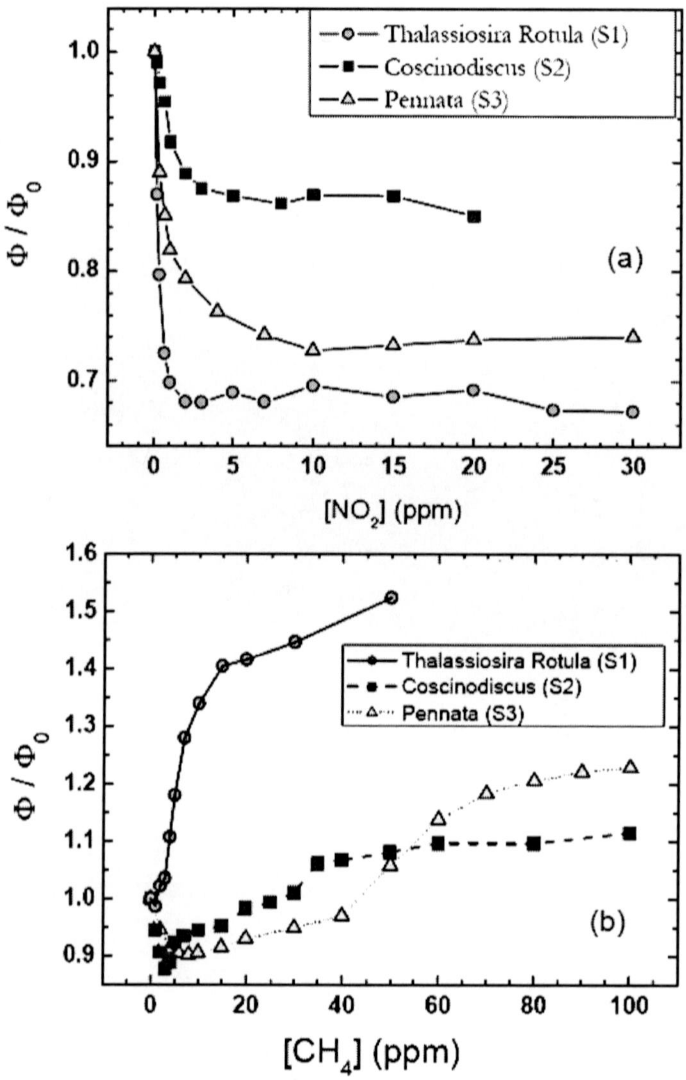

Figure 3.16. Ratio Φ/Φ_0 for spectrally integrated PL emission as a function of gas concentration in test chamber. Upper figure: exposure to NO_2 (PL quenching). Lower figure: exposure to CH_4 (PL enhancing). Gas concentrations are expressed in ppm units. Samples S1 to S3 correspond to silica frustules of different diatom species. (Reprinted with permission from Setaro et al. 2007. Copyright 2007 American Institute of Physics.)

temperature. As predicted by Eq. (3.22), an increase in temperature led to a decrease in the dynamic range of PL quenching ratio, which was indeed observed experimentally, as shown in Figure 3.18. The validity of Eq. (3.22) could thus be probed by plotting the quantity $\ln\left(b\sqrt{k_B T}\right)$ as a function of $1/(k_B T)$ to check

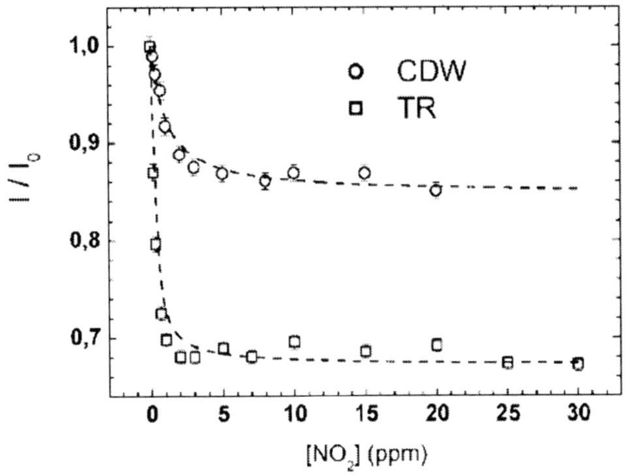

Figure 3.17. Photoluminescence quenching ratio versus NO_2 concentration for frustules of *Thalassiosira rotula* and *Coscinodiscus Wailesii* diatom. The two species are characterized by frustules with different morphologies and specific surfaces, reflecting different responses and saturation toward gas exposures. Dashed lines represent the best-fit curves obtained using the Langmuir model. (Reprinted with permission from Lettieri et al. 2008b. Copyright 2008 Wiley-VCH.)

whether the predicted linear behavior was observed. From the fit, a value of ΔG_{des} = 36 ± 2 kJ mol^{-1} was obtained as free energy of desorption of NO_2 from silica surface. This value is interestingly close to that obtained for NO_2 desorption from

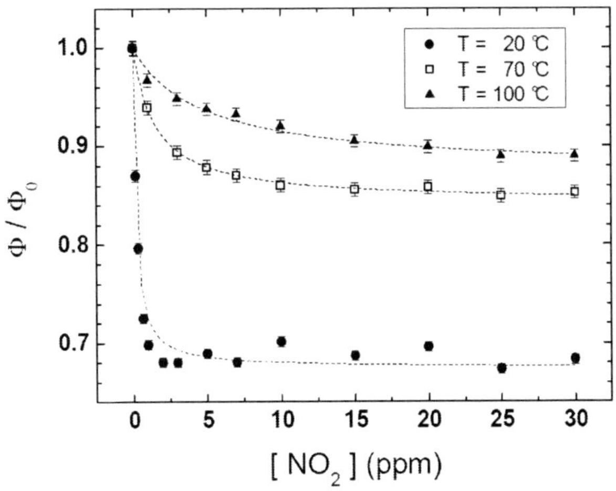

Figure 3.18. Photoluminescence quenching ratio versus NO_2 concentration for *Thalassiosira rotula* diatom frustules obtained at different sample temperatures. Circles, T = 20°C; squares, T = 70°C; triangles, T = 100°C. Dashed lines represent the best-fit curves obtained using the Langmuir model. (Reprinted with permission from Lettieri et al. 2008b. Copyright 2008 Wiley-VCH.)

porous silicon surfaces (23 kJ mol^{-1}) (Geobaldo et al. 2004) and indicative of a surface chemisorption interaction.

Finally, some conclusions about the physical mechanisms involved can be sketched. Concerning the light-emitting centers in silica, low-temperature behavior of PL efficiency (Lettieri et al. 2008b) and previous literature findings (Zhu and Ding 1999; Anedda et al. 2005; De Julián Fernández et al. 2008) suggest that PL emission is related to surface oxygen vacancy states, acting as excited states for radiative transitions. Thus, findings suggest investigations of specific interactions occurring between surface oxygen vacancy states and nitrogen dioxide molecules should be undertaken.

4.4. TIN DIOXIDE NANOSTRUCTURES

Tin dioxide (SnO_2) is a key functional material for many different applications, such as gas sensing, transparent conducting electrodes, catalyst support, dye-based solar cells. SnO_2 is a wide-band-gap semiconductor belonging to the family of *n*-type metal oxides, similar to other metal oxides such as ZnO, In_2O_3, and TiO_2. Because of its large chemoresistive response, SnO_2 is probably the most popular oxide material being used today for preparation of conductometric solid-state gas sensors. Typical applications of SnO_2 (such as those mentioned above) depend strongly on its surface properties, knowledge of which is thus of practical importance. The large interest in the surface physics of SnO_2 (Batzill and Diebold 2005) has been further boosted in the last decade by the improvement in synthesis of quasi-one-dimensional tin dioxide nanostructures. Quasi-1-D SnO_2 nanostructures can be realized by several techniques and demonstrate characteristics such as well-defined single-crystal surfaces, thermal stability, and large surface area (Dai et al. 2002; Liu et al. 2004; Ramgir et al. 2004; Wang et al. 2004). As introduced in Section 2, such characteristics are of extreme interest for realization of stable and sensitive conductometric gas sensors (Comini et al. 2002, 2004; Law et al. 2002).

Due to the importance and widespread use of SnO_2 as a chemical sensing material, literature on the topic of optical sensing by SnO_2 nanostructures is somewhat more extensive than for some other oxides. In particular, empirical and experimental evidence can now be coupled with recently published theoretical calculations, thus allowing interpretation of the gas-induced PL modifications that have been observed in laboratories. Here, after a discussion of the experimental evidence on PL-based sensing in SnO_2 nanostructures (Section 4.4.1), gas-sensing mechanisms and the nature of emitting states in SnO_2 will be discussed (Section 4.4.2), based on both experimental observation and ab-initio calculations. This discussion also emphasizes how, starting from a practical need, such as characterization of optical-based nanostructures, one can also explore basic physics mechanisms, hopefully allowing us to sketch routes to tailor possible final devices.

4.4.1. Gas-Induced PL Quenching Phenomena in SnO_2 Nanowires

Bulk SnO_2 crystals exhibit low photoluminescence efficiency (Jeong et al. 2003), whereas the PL efficiency of SnO_2 nanostructures is usually large. The key factor affecting the light-emission efficiency of SnO_2 nanostructures appears to be their defect composition and therefore their oxygen vacancies density (Chang and Park 2002; Jeong et al. 2003; Calestani et al. 2005; Maestre et al. 2005; Zhou et al. 2006; He and Wu 2006; Luo et al. 2006a, 2006b). PL emissions have been observed in various SnO_2 nanostructures and exhibit some common features regardless of the kind of morphology. For example, PL emission was observed in SnO_2 nanoribbons (Hu et al. 2003), nanotubes (Zhou et al. 2006), and nanobelts and nanowires (Calestani et al. 2005; Cai et al. 2005; He and Wu 2006; Zhou et al. 2006). In all these cases, all the analyzed nanostructures exhibited a broad PL emission peak in the yellow range (1.9–2.1 eV). Considering the wide-band-gap energy of SnO_2 (3.6 eV), the emission band cannot be assigned to interband or to exciton recombination and thus transitions involving defect states within the bandgap have to be invoked.

Literature results indicate a key role for oxygen vacancy (OV) states in these PL emissions. This role can be seen in all cases where oxygen stoichiometry is intentionally modified through annealing treatments (in vacuum and/or under oxygen flow). As a general trend, the PL efficiency is reduced (or eventually almost suppressed) as oxygen vacancies are removed (Jeong et al. 2003; Cai et al. 2005; Luo et al. 2006; Zhou et al. 2006).

OV states may represent either a “direct” or an “indirect” cause of luminescence. For example, OVs lying inside the band gap above the Fermi level may be upper electronic levels decaying radiatively to ground state. In this case, OVs play a “direct” role, meaning that the vacancy states themselves represent the light-emitting centers (LECs) in light-emitting SnO_2 nanostructures. On the other hand, the presence of OV states may lead to light emission “indirectly,” by triggering some specific mechanisms that result in photon emission, such as, for example, the formation of self-trapped excitons in TiO_2 (Tang et al. 1993) or $SrTiO_3$ (De Haart and Blasse 1986). In these latter cases, one can refer to an “indirect” role of the OV states.

Figure 3.19 provides a very simple schematic representation of “indirect” (see Figures 3.19A and 3.19B) versus “direct” (see Figures 3.19C and 3.19D) role of oxygen vacancies (indicated by V_O in the figures) in SnO_2 nanoparticle photoluminescence (see the figure caption for further details). Distinguishing between these two possibilities is important to understanding light emission in

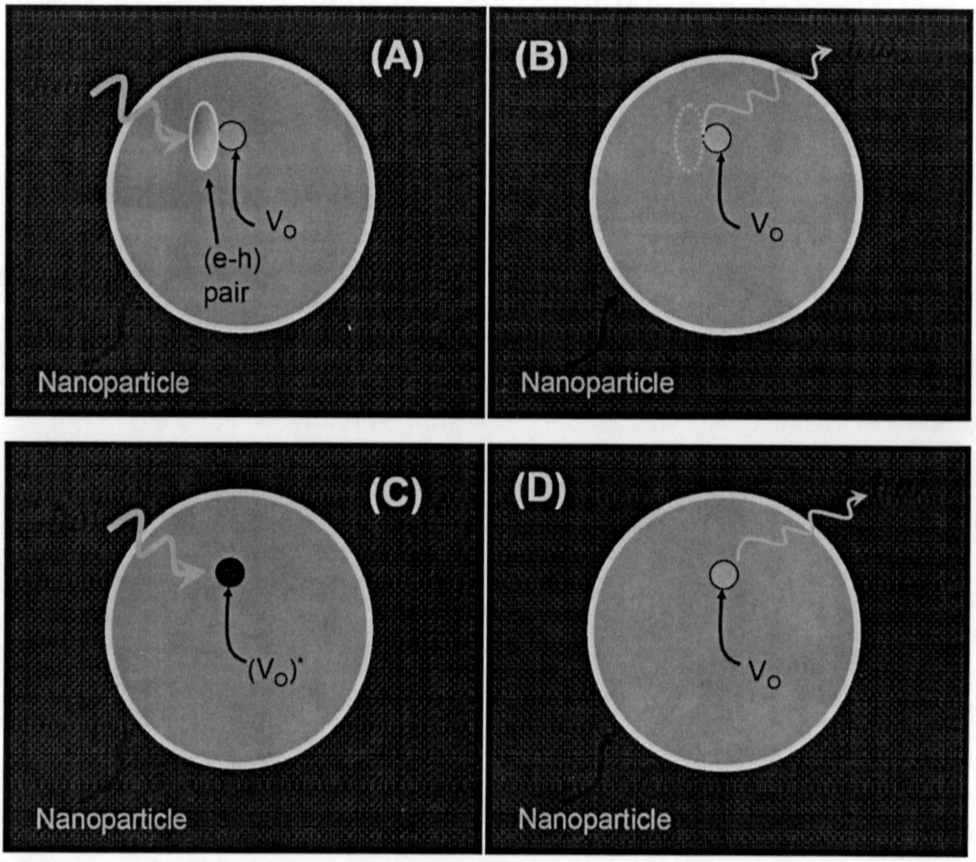

Figure 3.19. Schematic representation of "indirect" (upper panels: A and B) versus "direct" (lower panels: C and D) role of oxygen vacancy (indicated as V_O) in nanoparticle photoluminescence. Upper panels, "indirect role": (A) The presence of V_O locally triggers the formation of radiative states (generally indicated as e–h pair) after photon absorption that (B) decays spontaneously. Here $\hbar\omega$ and $\hbar\omega'$ are the photon energies of absorbed (excited) and emitted photons, respectively. Notice that the vacancy state does not excite optically nor take place actively in the emission process. Lower panels, "direct role": The V_O states are excited by impinging radiation and next recombine radiatively (D). In (C), V_O^* represents an excited state of oxygen vacancy whose radiative recombination to ground state (V_O, panel D) is the process responsible for photoluminescence emission.

SnO_2 nanostructures. Note that in the case of a "direct" role, the OVs are acting as dominant light-emitting centers. Thus, eventual gas-induced "suppression" of OV states would in this case lead to a clear example of static quenching. However, the argument cannot be trivially reversed: That is, observation of a static quenching mechanism does not lead automatically to a statement affirming OV states (or other defects) as LECs in SnO_2. In practice, any eventual observation of static-type quenching should be accompanied to other evidence. In particular, a careful study of chemical interactions between gas molecules and states assumed to represent LECs is helpful to further clarify the process involved. Toward this aim, specific information can be provided by first-principles calculations, such as those reviewed in the next subsection.

As a starting point for the discussion, we report some results obtained by Faglia and co-workers (2005), who studied the adsorption effect of NO_2 on PL response of highly crystalline (tetragonal rutile structure) SnO_2 nanobelts prepared by thermal evaporation. Faglia et al. studied the effect of three different species on visible PL emission of the nanobelts samples, NO_2 (0.3–10 ppm), NH_3 (50 ppm), and CO (10–1000 ppm), in dry and humid synthetic air and under normal ambient pressure condition. Figure 3.20 reports the dynamic response of the peak area toward 1 ppm NO_2 in dry air, relative humidity RH = 70% and RH = 30%, at T = 20°C, while the sample was kept at T = 120°C. One feature that can be highlighted is the fast dynamics of PL response toward NO_2 exposure (about 30 s to reach 90% of steady state), whereas the conductometric response toward the same gas was of the order of several minutes. Another interesting feature was the insensitivity of PL efficiency to humidity modifications, as can also be seen in Figure 3.20. This result is particularly interesting because water vapor is ordinarily an important interfering species in gas sensing, since adsorption of water over metal oxides produces a hydroxylated surface that can act as an electron donor and thereby modify the electrical properties of the oxide (Madou and Morrison 1989). In addition, Faglia and co-workers (2005) also observed that no appreciable effect was produced on PL activity of SnO_2 nanostructures by exposing them to NH_3 and CO (see Figure 3.20B), although these gaseous species usually do affect electrical transport. These results (response time, insensitivity to water, insensitivity to other interfering species) suggest strongly that the physical mechanisms behind the PL response to gas adsorption are not the same as those determining the conductometric response. One may recall that the same conclusion (different optical and electrical mechanisms) could also be also drawn from results reported above for ZnO and SiO_2.

Several studies have paid specific attention to the possible mechanisms of PL quenching in light of the model developed in Section 3. In particular, in a work by Lettieri and co-workers (Lettieri et al. 2008a), both a Langmuir-like "proportional quenching" regime and insensitivity of recombination rates (lifetimes) for NO_2 adsorption on SnO_2 nanostructures were observed. In Figure 3.21 the static PL curves are reported as obtained by excitation of SnO_2 nanobelt samples under flow of NO_2 diluted in nitrogen at various concentrations. It is easily seen that the quenching of PL yield induced by NO_2 exhibits nonlinear behavior as a function of the quencher concentration, as significant variations of the PL efficiency are observed upon exposure to low concentrations of NO_2 (a few ppm) followed by saturation at increasing NO_2 concentrations. Such behavior is typical of the familiar Langmuir-limited "proportional quenching" regime.

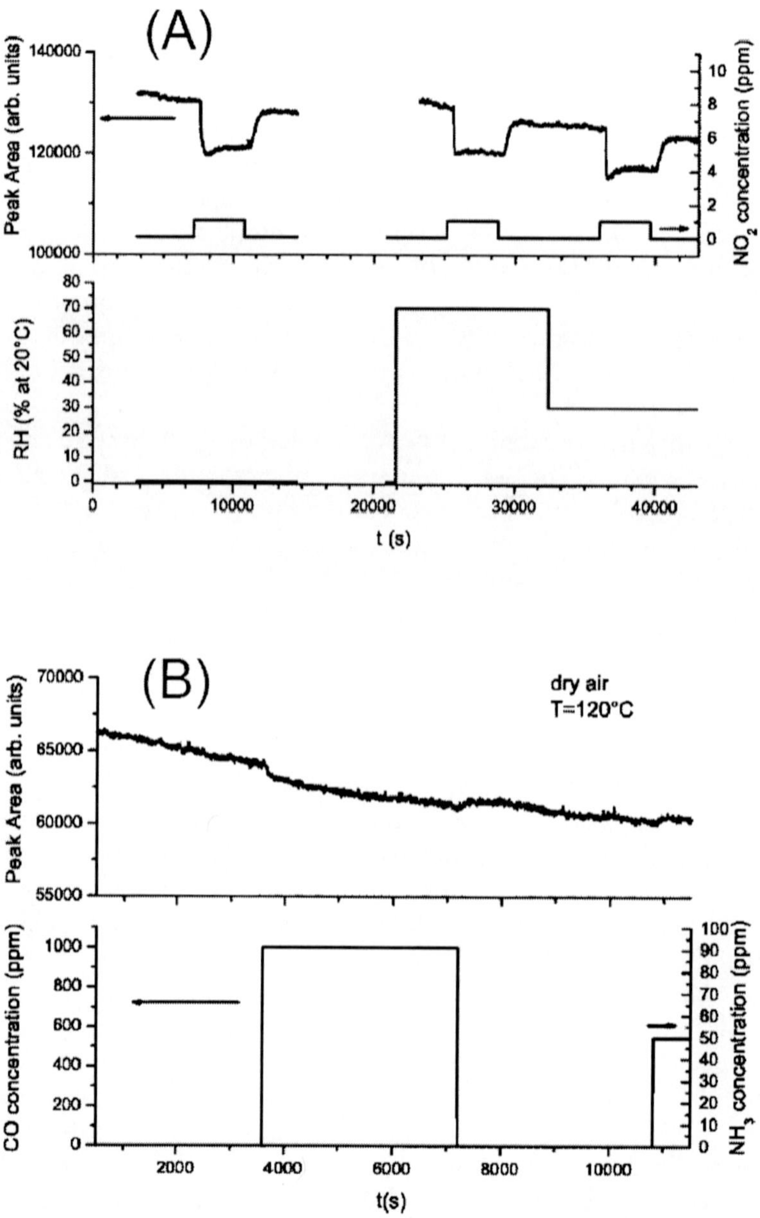

Figure 3.20. (A) Kinetic response of PL peak area intensity of SnO_2 nanowires toward 1 ppm of NO_2 diluted in dry air and for relative humidities of 70% and 30%. (B) Kinetic response of PL peak area under exposure to 1000 ppm of CO and 50 ppm of NH_4. Features to be noted are the fast kinetics and the low sensitivity to humidity changes and to interfering species. (Reprinted with permission from Faglia et al. 2005. Copyright 2005 American Institute of Physics.)

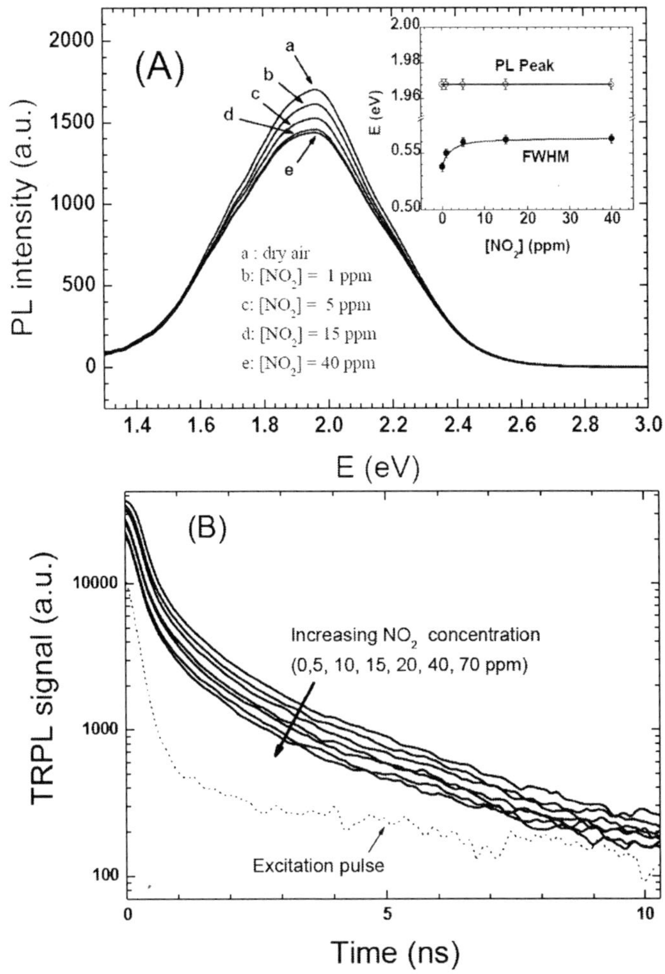

Figure 3.21. (A) CWPL spectra of SnO_2 nanowires obtained at various concentrations of NO_2. Inset shows PL peak energy and full-width at half-maximum of the photoluminescence spectra as a function of NO_2 concentration. (B) Time-resolved PL intensity decay for emission integrated over an interval of 1.6 to 2.4 eV. Inspection of peak-normalized time-dependent curves (not reported here) shows that typical lifetimes of curves were not significantly affected by changes in NO_2 concentration, thus indicating a static-like quenching. Dashed line indicates the temporal profile of the excitation laser pulse. (Reprinted with permission from Lettieri et al. 2008a. Copyright 2008 IOP.)

Time-resolved PL measurements performed by using the third harmonic of a picosecond Nd:YAG laser as excitation source showed PL decay kinetics almost insensitive to NO_2 concentration. Data reported in Figure 3.21 are not peak-normalized to show the intensity quenching effect due to molecular adsorption. However, careful inspection of the time-dependent curves showed that, apart from an amplitude term, the temporal decay dynamics of the different curves are equivalent, i.e., the typical lifetimes of curves were not significantly affected by changes in NO_2 concentration (Lettieri et al. 2008a).

4.4.2. Gas-Sensing Mechanisms and Emitting States in SnO_2

Among the experimental evidence discussed so far, two points seem to be of fundamental relevance: (1) PL sensitivity to NO_2 is ruled by a static quenching mechanism; and (2) "electrically active" states and sensing mechanisms are not equivalent to the "optically active" ones. On this basis, some conclusions can be drawn as follows.

Concerning electron-transport properties, it is commonly accepted that the defect composition is the feature that determines the SnO_2 conductivity. In fact, pure SnO_2 exhibits low formation energies for oxygen vacancies and tin interstitials (Kilic and Zunger 2002). These defects act as shallow donor levels lying below the conduction band (Marley and Dockerty 1965; Ishiguro et al. 1965) and give rise to the high n-type conductivity of nonstoichiometric SnO_2. On the other hand, the correlation between the defect composition and the luminescence properties of SnO_2 has been much less investigated. As discussed previously (Jeong et al. 2003; Luo et al. 2006; Zhou et al. 2006; Prades et al. 2007a), as a general rule, PL emission efficiency increases with increasing number of oxygen vacancy states. PL spectra of samples treated by different annealing processes are shown in Figure 3.22 (Zhou et al. 2006). The intensity of the PL signal coordinates with the number of oxygen defects, as can be seen from the measured stoichiometric ratio [O]/[Sn] (also shown in Figure 3.22). Note that the energy position of the PL spectrum commonly observed in SnO_2 nanostructures (peaking at about 2.0–2.1 eV at room temperature and typically extending over a range of 1.5–2.8 eV; see also Figure 3.21a) is clearly not compatible with radiative transitions involving band-edge states (the energy band gap is about 3.6 eV) or shallow levels. More localized states giving rise to deep energy levels must thus be invoked. We also note that such a hypothesis is also coherent with the observation of different kinetics and mechanisms for PL and conductometric responses to NO_2. In fact, spatially localized states whose energy levels lies close to the Fermi level and well below the conduction band edge are not expected to contribute significantly to electrical transport in a semiconductor, because of their low intrinsic mobility. The typical PL peak at around 2 eV photon energy indicates localized defects as PL-active states, and it is thus reasonable to assume that interaction of these defects with NO_2 molecules may follow kinetics different than the ones leading to modulation of electrical conductivity. Furthermore, such local interactions between the spatially localized defects involved in PL activity and NO_2 molecules may even be quite specific, and thus explain the selectivity of PL response to NO_2 observed by Faglia et al. (2005).

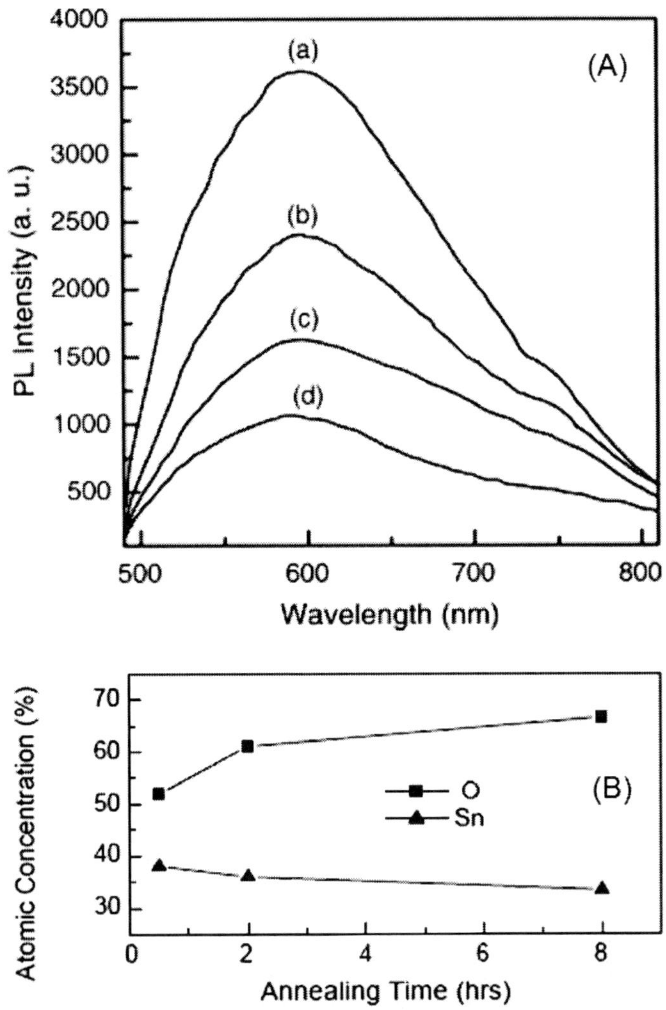

Figure 3.22. (Left) PL spectra of SnO_2 nanowires treated by different annealing processes: (a) as-prepared; (b) 1 h annealing at 800°C; (c) 2 h at 800°C; (d) 4 h at 800°C). The intensity of the PL signal coordinates with the oxygen defects, as can be inferred from measured stoichiometric ratios [O]/[Sn] obtained for different annealing times (right panel). (Reprinted with permission from Zhou et al. 2006. Copyright 2006 Elsevier.)

Taking into account the relation between PL efficiency and number of oxygen defects, similarities observed between PL quenching phenomena in SnO_2 and (formerly discussed) a-SiO_2 (Lettieri et al. 2008) and, in particular, considering the observation of "static" PL quenching features, Lettieri et al. (2008c) suggested an interpretation of SnO_2 PL quenching experiments based on the two following points:

1. Surface OV states may represent the dominant light-emitting centers in SnO_2.
2. The interaction between NO_2 molecules and surface oxygen vacancy states results in suppression of energy levels responsible for PL emission.

Such a hypothesis is immediately compatible with the observation of static quenching dynamics, and also explains the decoupling between PL response and electrical response toward NO_2 adsorption, provided that two assumptions can be demonstrated, namely, that surface OV states represent preferential NO_2 adsorption sites, and that energy levels associated to various surface oxygen vacancies should be compatible with the spectral contribution to PL emission.

Important information about these two issues can be obtained by ab-initio calculations that allow one to extract, for example, total adsorption energies of given molecule on specific defect sites or electron spectrum and level positions of the defects themselves. Ab-initio calculations published in recent years seem to confirm the picture previously discussed based on surface OV states as the direct cause of PL emission and reactivity to NO_2, as will be discussed now.

The first assumption (i.e., that oxygen vacancies represent preferential NO_2 surface adsorption sites) is supported by ab-initio studies performed by Prades and co-workers (2007a, 2007b) of NO_x and SO_2 adsorption onto (110) SnO_2 surfaces. In particular, these authors calculated total adsorption energy (E_{ads}) for NO_x compounds, defined as the energy difference between the state represented by a clean surface and isolated molecule and the energy of the state representing the surface with one adsorbed molecule on it. Thus, a negative E_{ads} means that the adsorption is energetically favorable and may occur spontaneously. The results of the calculations confirmed the role of surface oxygen vacancies as preferential sites for the adsorption of NO_2 molecules. For example, adsorption on oxygen-bridging atoms of a stoichimetric surface (i.e., with no oxygen deficiencies) was found to be largely energetically unfavorable (E_{ads} = +1.51 eV), while adsorption on vacancies of oxygen bridging atoms was found to be very favorable in both 50% reduced and fully reduced SnO_2 surfaces, with adsorption energies of E_{ads} = −2.31 eV and −2.11 eV, respectively (Prades et al. 2007c). In general, for both 50% reduced and fully reduced surfaces, bridging oxygen vacancies were found to be the most relevant adsorption sites for NO_2, followed by in-plane oxygen vacancies.

Interesting information on the question of the energy positions of surface OV levels has also been supplied by ab-initio calculations. Prades et al. (2007c) carried out first-principles calculations based on density functional theory to calculate the energy-level position over the top of the valence band for occupied states associated with both bridging oxygen and in-plane oxygen vacancies. The levels found by these authors, lying at about 1.4 and 0.9 eV over the top of the valence band, allowed interpreting the experimental results of cathodoluminescence spectra as superpositions of different transitions (conduction band to OV states, bulk shallow levels to OV states). However, the approach could not also provide calculation of unoccupied levels associated with OV states. Such calculation was performed by Trani et al. (2008) using the hybrid B3LYP exchange-correlation functional, which is one of the most accurate approximations for the calculation of excited states (Becke 1988). The band structure of a defected SnO_2 (110) surface calculated for both occupied and unoccupied states is reported in Figure 3.23, considering either a bridging oxygen vacancy (left) or an in-plane oxygen vacancy (right). Band diagrams were found to differ substantially with respect to the stoichiometric surface, especially for the intragap states.

The distance of the energies of occupied states from the top of the valence band agree substantially with those reported by Prades et al. (2007c). However, very interesting additional results were obtained by these researchers for unoccupied states. In particular, as a result of the presence of surface OVs, a dispersed band appears below the top of the conduction band, leading to a number of additional possible optical transitions between exited and fundamental surface OV

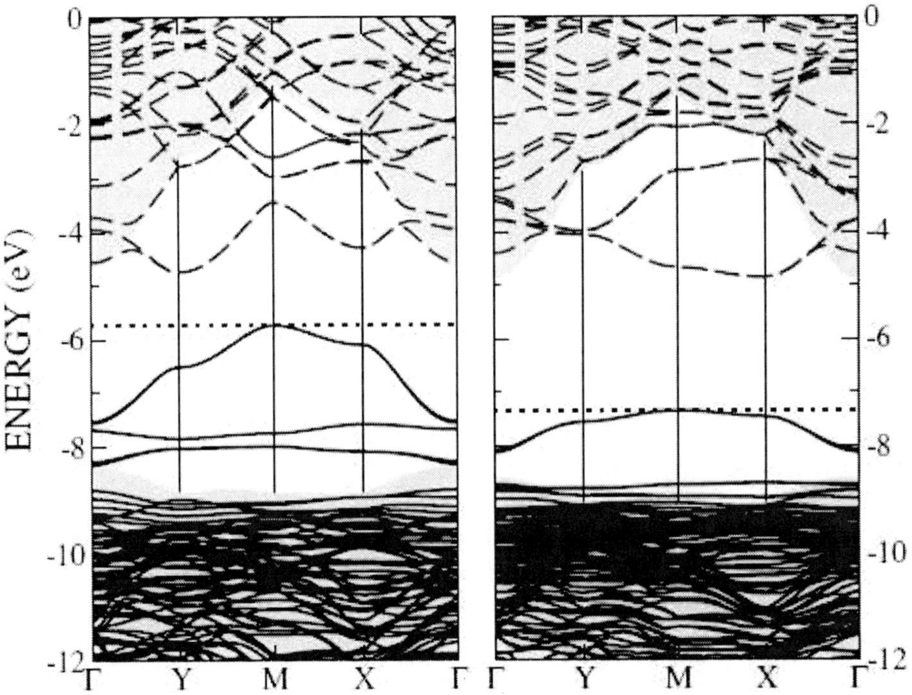

Figure 3.23. Band structure of the SnO_2 (110) surface with an oxygen vacancy at the outermost bridging (left) or in-plane (right) positions. The reference energy is the vacuum level. Occupied and unoccupied bands are reported as solid and dashed lines, respectively. The shaded region corresponds to the projected bulk band structure. (Reprinted with permission from Trani et al. 2008. Copyright 2008 the American Physical Society.)

states ranging between 2.0 to 3.0 eV at points Γ, X, and M for both kinds of surface OV considered.

The works cited so far all focused on the SnO_2 (110) surface, which, however, is not a favorable faceting surface for one-dimensional SnO_2 structures (Batzill and Diebold 2005). The (101) surface, representing one of the most common faceting surfaces of SnO_2 cassiterite (Oviedo and Gillan 2000), is on the contrary the most commonly used lateral facet for SnO_2 nanobelts (Batzill and Diebold 2005) and also one of the favored facets for nanowires prepared by the evaporation–condensation method (Prades et al. 2007a).

An additional study, coupled with experimental optical results and focused on the (101) SnO_2 surface, has been reported by Lettieri et al. (2008c). In this work, calculation of the electron spectrum (band diagram) was performed for stoichiometric and reduced (101) surfaces obtained by removing oxygen atoms at bridging positions (oxygen bridging vacancies) through a hybrid B3LYP functional, including both occupied and unoccupied states. Findings obtained from PL spectra measurements and NO_2-induced PL quenching, performed on single-crystal SnO_2 nanobelt samples (prepared at Sensor Labs, CNR-IDASC, and the University of Brescia, Italy) were also discussed in view of ab-initio calculation results.

Photoluminescence measurements of SnO_2 nanowires obtained at room temperature (in dry air and 40 ppm NO_2) and at low temperature, shown in Figure 3.24, again showed PL quenching, and, thanks to enhanced PL efficiency at low temperature, an additional shoulder in the PL profile was seen at about 2.75 eV (curve C in Figure 3.24A). This spectral feature is more evident in Figure 3.24B, where the PL profile of curve C is shown in logarithmic scale (dashed lines are the results of a two-peak Gaussian best fit).

A calculated band diagram for the SnO_2 (101) surface including bridging oxygen vacancies is reported in Figure 3.25 and, starting from it, valuable information on the photoluminescence activities can be inferred as follows.

When looking at pairs of states whose transition energies are near those observed experimentally, two cases arise. In the first case, the occupied OV electron states above the valence band act as final states for transitions involving excited bulk electrons, in the same manner as proposed by Prades and co-workers (Prades et al. 2007a, 2007c). In particular, transition from bulklike conduction-band states to localized OV states are predicted at the Γ point, which is a valley point for the conduction band and thus represents a region of accumulation for excited electrons. The second case corresponds to transitions involving surface-localized OV states as both initial (excited) state and final state. In this case, the lowest-energy optical transition shows up at the X point, as a direct-gap recombination between electron and hole states well localized around the vacancy site.

According to first-principles energy calculations, both "extended-to-localized" and "localized-to-localized" transitions are thus compatible with experimental findings on SnO_2 nanostructures. It must be noted, however, that since the

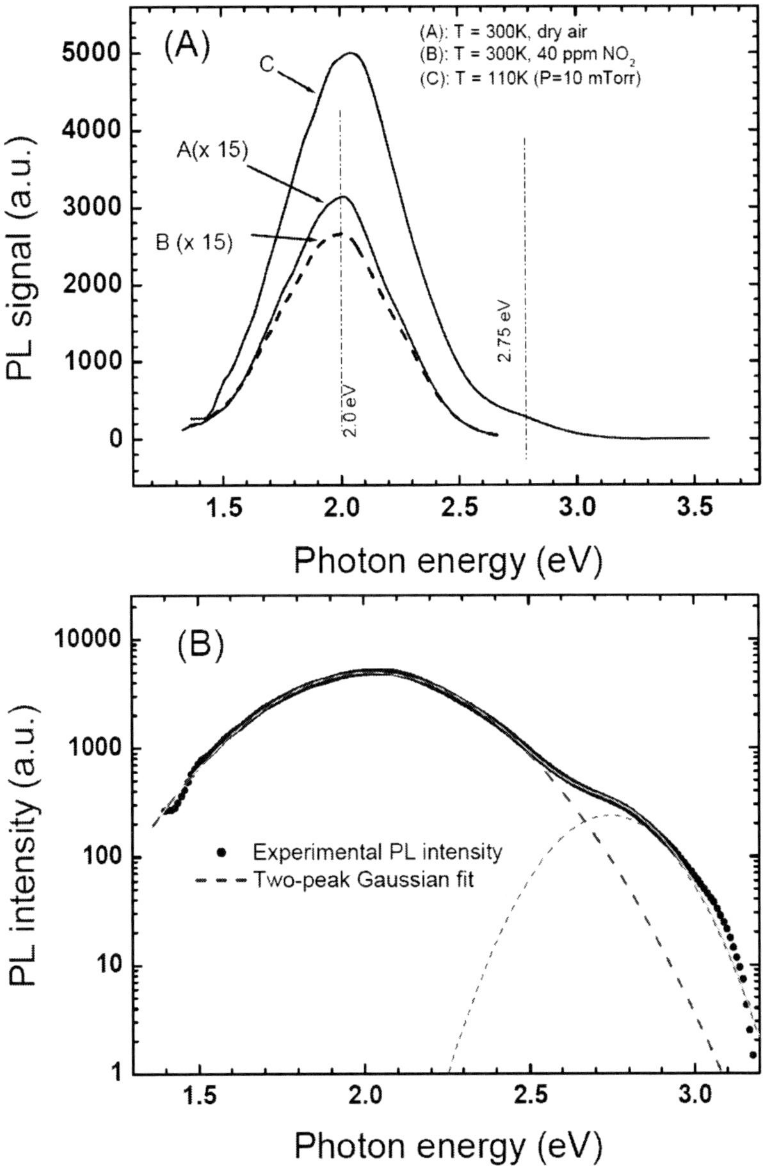

Figure 3.24. (A) PL spectra of SnO_2 nanobelts obtained at temperature T = 300 K in pure dry air atmosphere (curve A), at 40 ppm NO_2 concentration (curve B), and under vacuum conditions (10 mTorr) at temperature 110 K (curve C). The data corresponding to curves A and B have been multiplied for a factor of 15 for better comparison. (B) PL intensity profile of curve C in log scale. The dashed lines represent the Gaussian best fit used to deconvolve the spectrum. (Reprinted with permission from Lettieri et al. 2008c. Copyright 2008 American Institute of Physics.)

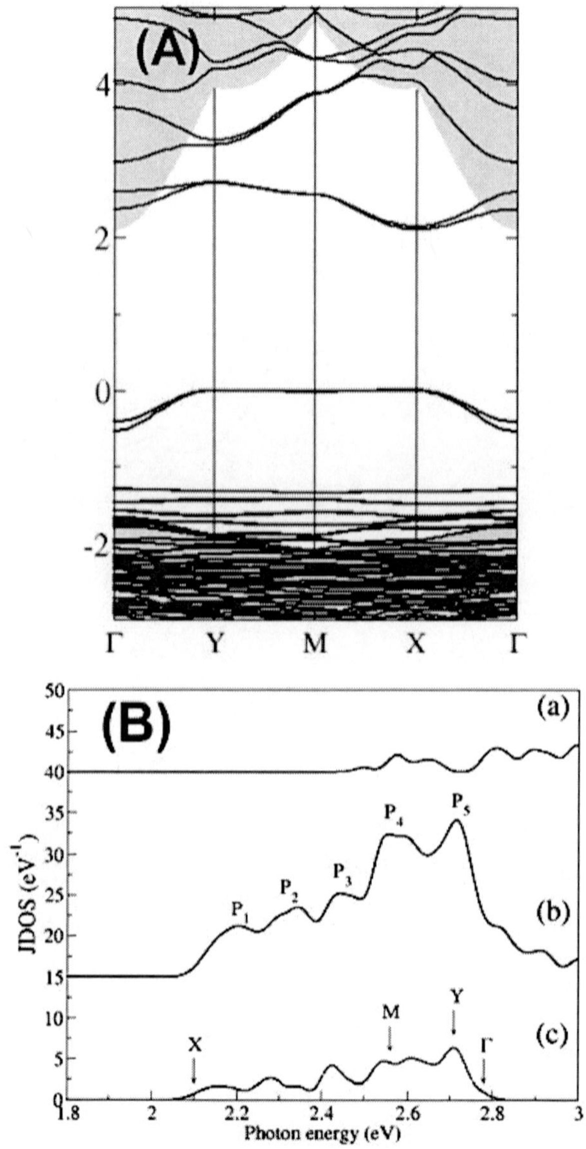

Figure 3.25. (A) Surface bands (solid lines) and projected bulk bands (shaded regions) for the SnO_2 (101) surface. The zero of energy has been taken as the energy of the last occupied state. (B) Computed JDOS for SnO_2 (101) surface. Curve (a), stoichiometric surface. Curve (b), JDOS calculated by summing over all the transitions among occupied to unoccupied states. Curve (c), JDOS calculated for just the first valence to conduction band. The main features in the JDOS have been labeled as P1–P5, while positions of the main Brillouin-zone critical points have been shown explicitly with vertical arrows. (Reprinted with permission from Lettieri et al. 2008c. Copyright 2008 American Institute of Physics.)

oscillator strength of a given transition is proportional to the spatial overlap between the wave functions involved, it is more likely that the second mechanism, which involves functions which are strongly localized near the oxygen vacancy, is the dominant one. Thus, one can propose the reasonable hypothesis that surface OV states (here located at bridging oxygen atom positions) represent *both* excited and fundamental states for optical transitions (recall the [X] and [X*] levels introduced in Section 4)—in other words, suggesting a complete "direct role" for OV states in PL activity. One useful test of this hypothesis is to compute the joint density of states (JDOS) associated to surface bands only. In fact, the spectral shape of the JDOS may give useful indications about the origin of spectral resonances exhibited by PL spectra, because JDOS represents a "spectral measurement" of the number of allowed optical transitions. The result of calculations by Lettieri et al. (2008c) is reported in Figure 3.25. Curve (a) in Figure 3.25B shows the JDOS relative to the transitions between the lowest unoccupied and the highest occupied surface bands for the stoichiometric (101) surface, while curves (b) and (c) show the JDOS plotted for all computed surface vacancy bands and for just the first transition between the top occupied and the lowest unoccupied bands.

As can be seen in Figure 3.25B, curve (a), no surface-related optical transitions are predicted in the range 2.0–2.4 eV in the case of a stoichiometric surface. Also, even if the stoichiometric JDOS is non-null in the range 2.4–2.8 eV, its values are much smaller than those obtained in the same energy range for the case of an oxygen-deficient surface.

Introduction of surface bridging oxygen vacancies leads to the appearance of a number of spectral features corresponding to transitions at the critical points X, M, Y, and Γ. A multipeak Gaussian fit of the JDOS calculated by summing over all surface bands allowed Lettieri et al. (2008c) to discriminate the typical energies of separated spectral features (indicated by P1 to P5), and to obtain energies of 2.2 eV (P1), 2.32 eV (P2), 2.45 eV (P3), 2.56 eV (P4), and 2.77 eV (P5).

It is worth noting that such JDOS resonances are compatible with several cathodoluminescence (CL) or photoluminescence observations reported in different references. For example, peaks at 2.2, 2.37, and 2.75 eV were reported in CL experiments on SnO_2 nanobelts (Prades et al. 2007a). Yuan et al. (2006) reported CL peaks in SnO_2 nanowires at 2.1 and 2.7 eV, and Calestani and co-workers (2005) reported CL peaks at 560 nm (2.2 eV) and a shoulder at 450 nm (2.75 eV). We can notice that the same spectral features (2.1 and 2.75 eV) were also observed by Lettieri and co-workers (2008c) (see also Figure 3.22), while similar peaks (2.64 and 2.21 eV) have been reported in low-temperature PL of SnO_2 nanowires (Luo et al. 2006) and in PL emission of thin SnO_2 films prepared by chemical vapor deposition (2.75 and at 2.3 eV) (Luo et al. 2006a, 2006b). Emission energies of 2.6–2.7 eV have also been reported in the case of SnO_2 nanoparticles (Pal et al. 2008).

In conclusion, the discussed findings reasonably support the interpretation of surface oxygen vacancy states as light-emitting centers. As a consequence, they also strengthen the hypothesis that the observed NO_2-induced static PL quenching acts by means of a suppression of (light-emitting) surface OV states, due to chemical adsorption of NO_2 molecules on oxygen vacancy sites. Such an interpretation appears to be, at least for now, in agreement with most experimental findings. However, further analysis mostly likely to be required to obtain a full understanding of the mechanisms of PL-based NO_2 sensing in SnO_2.

5. CONCLUSIONS

Optical approaches to chemical sensing (or, more shortly, optochemical sensing) exhibit a number of unusual characteristics and potential advantages over traditional sensing based on conductometric devices. Optochemical sensing may involve a broad variety of possible methods, experimental setups, and signal transduction routes, thus intrinsically allowing considerable flexibility in approaches and solutions. Another possible advantage of optochemical approaches is that they can be implemented so as to give multiparametric responses and information. In fact, optical sensors can potentially detect multiple species from differences in the intensity, wavelength, phase, and polarization of the output light signals. Such multiparametric detection features cannot be easily achieved with conductometric sensing systems, which rely on modifications of a single quantity only (electrical resistance). The absence of electrical contacts and enhanced insensitivity to electromagnetic noise are other useful characteristics of optochemical sensing.

In addition to these well-known features of optical sensing, there is another advantage that may be easily overlooked: Investigation of less well explored optical sensing approaches, such as those relying on PL modifications in oxides, can also encourage advancement of basic knowledge in chemophysical processes involved in transduction mechanisms. This particular advantage is exemplified quite effectively by studies on gas-induced PL modifications, such as those that have been discussed in this chapter. Preliminary findings using these approaches have shown interesting characteristics, as in the case of NO_2-induced PL quenching in SnO_2, for which satisfactory or even high sensitivities (Lettieri et al. 2008a) and selectivity toward interfering species (Faglia et al. 2005) have been demonstrated. These results point toward the hypothesis that mechanisms involved in modifications in PL efficiency are different from those giving rise to chemoresistive responses.

From studies performed on various nanomaterials and nanocrystals, such as those reviewed in Section 4, some recurring conclusions seem to emerge, such as the following.

1. A "static" type of quenching mechanism is observed for different oxide nanomaterials (ZnO, SnO_2, SiO_2). Gas–oxide interaction seems to lead to suppression of light-emitting centers ("static" mechanisms) instead of introduction of

competitive nonradiative recombination channels ("dynamic" mechanisms). It has to be shown that such a distinction can be made only by means of time-resolved PL analysis in the presence of gas quencher species; as discussed in Section 3.4, it is unlikely that measurements of PL quenching can by themselves allow distinguishing a specific kind of quenching kinetics.

2. As seen for SnO_2 (Faglia et al. 2005), oxidizing species seem to give much stronger responses than reducing ones. Furthermore, the gas-induced PL response is present in oxides whose PL activity is likely to be ruled by the presence of oxygen defects.
3. Gas-induced PL response is a very active phenomenon in materials with high specific surfaces. Moreover, the kinetics of PL quenching (the variation of PL intensity versus the concentration of quencher gas) seems to indicate a surface-limited phenomenon, well described by the Langmuir adsorption model, where the decrease of PL intensity is linearly proportional to the density of adsorbed molecules at the surface.

These features seem to more or less characterize most of the empirical observations reviewed in this chapter. Thus, a picture that coherently joins those elements can be proposed, assuming that surface OV states may indeed represent the dominant light-emitting centers and that interaction between NO_2 molecules and surface oxygen vacancy states results in suppression of energy levels responsible for PL emission. Such a hypothesis has been proposed explicitly for SnO_2 and tested through ab-initio calculations of surface OV energy positions. The results obtained from first-principles calculations of energy levels associated to surface OVs in SnO_2 indeed seem to support the suggested model, indicating optical transitions whose energies are in agreement with experimental values. Moreover, such an interpretation also indirectly explains the decoupling between PL response and electrical response toward NO_2 adsorption.

At the end of our discussion, we would like to stress that a practical interest lies behind the understanding of gas-induced PL modification phenomena, not limited to performance exhibited by optical (PL-based) devices. Indeed, even if this specific approach proves in the future to be less effective than others, it is still worth exploring for other reasons. One valuable example is the potential selectivity of the approach: Confirmation of the physical picture described above would mean that enhanced selectivity toward oxidizing molecules may in fact be obtained by preparation of oxygen-depleted metal oxide surfaces.

In conclusion, PL-based sensing using metal oxide nanostructures and nanocrystals can be regarded as a relatively unexplored approach evidencing phenomena and physical mechanisms that are not completely familiar in the field of gas sensing. Similar mechanisms have been found to control the PL quenching phenomenon in various nanostructured and porous oxides (such as ZnO, SnO_2, and SiO_2). Experimental analysis and theoretical calculations reported up to now seem to suggest a picture in which oxygen vacancies at surfaces are mostly involved in the PL response, acting as quenchable emitting centers. If confirmed, this picture promises an interesting advance toward realization of selective optical devices, exhibiting enhanced response to oxidizing gas species and reduced response toward interfering reducing species.

Even if further analyses are likely to be required to obtain a full understanding of PL-based gas-sensing phenomena in MOX nanostructures, we think that the evidence so far should encourage studies in this field for both practical and fundamental purposes.

ACKNOWLEDGMENTS

I would like to thank my former and current co-workers for their collaboration and support in studies on nanostructured oxides-based optical sensing, especially A. Setaro, L. Santamaria, D. Pallotti, S. Amoruso, and P. Maddalena. Thanks are also due to colleagues and collaborators who contributed to some of the results reviewed in this contribution: M. C. Carotta, E. Comini, L. De Stefano, G. Faglia, and G. Sberveglieri. A special thank you is due to Prof. P. Maddalena for introducing me to the field of optochemical sensing.

REFERENCES

Ando M., Kobayashi T., and Haruta M. (1997a) Combined effects of small gold particles on the optical gas sensing by transition metal oxide films. *Catal. Today* **36**, 135–141. DOI: 10.1016/S0920-5861(96)00206-4

Ando M., Kobayashi T., Iijima S., and Haruta M. (1997b) Optical recognition of CO and H_2 by use of gas-sensitive Au-Co_3O_4 composite films. *J. Mater. Chem.* **7**, 1779–1783. DOI: 10.1039/a700125h

Ando M., Chabicovsky R., and Haruta M. (2001) Optical hydrogen sensitivity of noble metal-tungsten oxide composite films prepared by sputtering deposition. *Sens. Actuators B* **76**, 13–17. DOI: 10.1016/S0925-4005(01)00561-5

Ando M., Kobayashi T., Iijima S., and Haruta M. (2003) Optical CO sensitivity of Au–CuO composite film by use of the plasmon absorption change. *Sens. Actuators B* **96**, 589–595. DOI: 10.1016/S0925-4005(03)00645-2

Ando M., Steffes H., Chabicovsky R., Haruta M., and Stangl G. (2004) Optical and electrical H_2 and NO_2 sensing properties of Au/$In_xO_yN_z$ films. *IEEE Sensors J.* **4**, 232–236. DOI: 10.1109/JSEN.2004.823656

Anedda A., Carbonaro C.M., Clemente R., Corpino R., and Ricci P.C. (2005) Low temperature investigation of the blue emission in mesoporous silica. *Mater. Sci. Eng. C* **25**, 631–634. DOI: 10.1016/j.msec.2005.07.009

Baratto C., Todros S., Faglia G., Comini E., Sberveglieri G., Lettieri S., Santamaria L., and Maddalena P. (2009) Luminescence response of ZnO nanowires to gas adsorption. *Sens. Actuators B* **140**, 461–466. DOI: 10.1016/j.snb.2009.05.018

Bassani F. and Pastori Parravicini G. (1975) *Electronic States and Optical Transitions in Solids.* Pergamon Press, Oxford, UK.

Batzill M. and Diebold U. (2005) The surface and materials science of tin oxide. *Prog. Surf. Sci.* **79**, 471–554. DOI: 10.1016/j.progsurf.2005.09.002

Becke A.D. (1988) Density-functional exchange-energy approximation with correct asymptotic behaviour. *Phys. Rev. A* **38**, 3098–3100. DOI: 10.1103/PhysRevA.38.3098

Bismuto A., Lettieri S., Maddalena P., Baratto C., Comini E., Faglia G., Sberveglieri G., and Zanotti L. (2006) Room-temperature gas sensing based on visible photoluminescence properties of metal oxide nanobelts. *J. Opt. A: Pure Appl. Opt.* **8**, S585–S588. DOI: 10.1088/1464-4258/8/7/S45

Borner S., Ruter C.E., Voss T., Kip D., and Schade W. (2007) Modelling of ZnO nanorods for evanescent field optical sensors. *Phys. Stat. Sol. (a)* **204**, 3487–3495. DOI: 10.1002/pssa.200723137

Buso D., Busato G., Guglielmi M., Martucci A., Bello V., Mattei G., Mazzoldi P., and Post M.L. (2007) Selective optical detection of H_2 and CO with SiO_2 sol–gel films containing NiO and Au nanoparticles. *Nanotechnology* **18**, 475505. DOI: 10.1088/0957-4484/18/47/475505

Butcher K.S.A., Ferris J.M., Philips M.R., Wintrebertt-Foquet M., Wah J.W.J., Jovanovic N., Vyverman W., and Chepurnov V. (2005) A luminescence study of porous diatoms. *Mater. Sci. Eng. C* **25**, 658–663. DOI: 10.1016/j.msec.2005.06.049

Cai D., Su Y., Chen Y.Q., Jiang J., He Z.Y., and Chen L. (2005) Synthesis and photoluminescence properties of novel SnO_2 asterisk-like nanostructures. *Mater. Lett.* **59**, 1984–1988. DOI: 10.1016/j.matlet.2005.01.045

Calestani D., Zha M., Zappettini A., Lazzarini L., Salviati G., Zanotti L,. and Sberveglieri G. (2005) Structural and optical study of SnO_2 nanobelts and nanowires. *Mater. Sci. Eng. C* **25**, 625–630. DOI: 10.1016/j.msec.2005.07.014

Chaim R., Levin M., Shlayer A., and Esournes C. (2008) Sintering and densification of nanocrystalline ceramic oxide powders: A review. *Adv. Appl. Ceram.* **107**, 159–169. DOI: 10.1179/174367508X297812

Chang S.S. and Park D.K. (2002) Novel Sn powder preparation by spark processing and luminescence properties. *Mater. Sci. Eng. B* **95**, 55–60. DOI: 10.1016/S0921-5107(02)00162-9

Chen Y., Cui X., Zhang K., Pan D., Zhang S., Wang B., and How J.G. (2003) Bulk-quantity synthesis and self-catalytic VLS growth of SnO_2 nanowires by lower-temperature evaporation. *Chem. Phys. Lett.* **369**, 16–20. DOI: 10.1016/S0009-2614(02)01949-8

Chen J., Xu L.N., Li W.Y., and Gou X.L. (2005) Three-dimensional porous copper-tin alloy electrodes for rechargeable lithium batteries. *Adv. Mater.* **17**, 582–586. DOI: 10.1002/adma.200401101

Choudhury B., Shinar R., and Shinar J. (2004) Glucose biosensors based on organic light emitting devices structurally integrated with a luminescent sensing element. *J. Appl. Phys.* **96**, 2949–2954. DOI: 10.1063/1.1778477

Comini E. (2006) Metal oxide nano-crystals for gas sensing. *Anal. Chim. Acta* **568**, 28–40. DOI: 10.1016/j.aca.2005.10.069

Comini E., Faglia G., Sberveglieri G., Pan Z.W., and Wang Z.L. (2002) Stable and highly sensitive gas sensors based on semiconducting oxide nanobelts. *Appl. Phys. Lett.* **81**, 1869–1871. DOI: 10.1063/1.1504867

Comini E., Guidi V., Malagù C., Martinelli G., Pan Z., Sberveglieri G., and Wang Z.L. (2004) Electrical properties of tin dioxide two-dimensional nanostructures. *J. Phys. Chem. B* **108**, 1882–1887. DOI: 10.1021/jp036693y

Comini E., Baratto C., Faglia G., Ferroni M., and Sberveglieri G. (2007) Single crystal ZnO nanowires as optical and conductometric chemical sensor. *J. Phys. D: Appl. Phys.* **40**, 7255–7259. DOI: 10.1088/0022-3727/40/23/S08

Dai Z.R., Gole J.L., Stout J.D., and Wang Z.L. (2002) Tin oxide nanowires, nanoribbons and nanotubes. *J. Phys. Chem. B* **106**, 1274–1279. DOI: 10.1021/jp013214r

Dang H.Y., Wang J., and Fan S.S. (2003) The synthesis of metal oxide nanowires by directly heating metal samples in appropriate oxygen atmospheres. *Nanotechnology* **14**, 738–741. DOI: 10.1088/0957-4484/14/7/308

De Boer J.H. (1953) *The Dynamical Character of Adsorption*. Clarendon Press, Oxford, UK.

De Haart L. and Blasse G. (1986) The observation of exciton emission from rutile single-crystals. *J. Solid State Chem.* **61**, 135–136. DOI: 10.1016/0022-4596(86)90015-0

De Julián Fernández C., Manera M.G., Pellegrini G., Bersani M., Mattei G., Rella R., Vasanelli L., and Mazzoldi P. (2008) Surface plasmon resonance optical gas sensing of nanostructured ZnO films. *Sens. Actuators B* **130**, 531–537. DOI: 10.1016/j.snb.2007.09.065

De Stefano L., Rendina I., De Stefano M., Bismuto A., and Maddalena P. (2005) Marine diatoms as optical chemical sensors. *Appl. Phys. Lett.* **87**, 233902. DOI: 10.1063/1.2140087

Dieguez A., Romano-Rodriguez A., Morante J.R., Nelli P., Sengaletti L., and Sberveglieri G. (1999) Analysis of the thermal oxidation of tin droplets and its implications on gas sensor stability. *J. Electrochem. Soc.* **146**, 3527–3535. DOI: 10.1149/1.1392509

Dikovska A.O., Atanasov P.A., Tonchev S.H., Ferreira J., and Escoubas L. (2007) Periodically structured ZnO thin films for optical gas sensor application. *Sens. Actuators A* **140**, 19–23. DOI: 10.1016/j.sna.2007.05.032

Dimitrov I.G., Dikovska A.O., Atanasov P.A., Stoyanchov T.R., and Vasilev T. (2008) Al-doped ZnO thin films for gas sensors application. *J. Phys.: Conf. Ser.* **113**, 012044. DOI: 10.1088/1742-6596/113/1/012044

Eguchi K. (1992) Optical gas sensors, In: Sberveglieri G. (ed.), *Gas Sensors*. Kluwer, Dordrecht, The Netherlands, 307–328. DOI: 10.1007/978-94-011-2737-0_9

Elmi I., Zampolli S., Cozzani E., Mancarella F., and Cardinali G.C. (2008) Development of ultra-low-power consumption MOX sensors with ppb-level VOC detection capabilities for emerging applications. *Sens. Actuators B* **135**, 342–351. DOI: 10.1016/j.snb.2008.09.002

Faglia G., Baratto C., Sberveglieri G., Zha M., and Zappettini A. (2005) Adsorption effects of NO_2 at ppm level on visible photoluminescence response of SnO_2 nanobelts. *Appl. Phys. Lett.* **86**, 011923. DOI: 10.1063/1.1849832

Fallert J., Hauschild R., Stezl F., Urban A., Wissinger M., Zhou H., Klingshrn C., and Kalt H. (2007) Surface-state related luminescence in ZnO nanocrystals. *J. Appl. Phys.* **101**, 073506. DOI: 10.1063/1.2718290

Geobaldo F., Rivolo P., Borini S., Boarino L., Amato G., Chiesa M., and Garrone E. (2004) Chemisorption of NO_2 at boron sites at the surface of nanostructured mesoporous silicon. *J. Phys. Chem. B* **108**, 18306–18310. DOI: 10.1021/jp046918k

Guo Y., Zhang X.W., and Han G.R. (2006) Investigation of structure and properties of N-doped TiO_2 thin films grown by APCVD. *Mater. Sci. Eng. B* **135**, 83–87. DOI: 10.1016/j.mseb.2006.08.031

Gurlo A., Bârsan N., Ivanovskaya M., Weimar U., and Göpel W. (1998) In_2O_3 and MoO_3–In_2O_3 thin film semiconductor sensors: Interaction with NO_2 and O_3. *Sens. Actuators B* **47**, 92–99. DOI: 10.1016/S0925-4005(98)00033-1

He J.H. and Wu T.H. (2006) Beaklike SnO_2 nanorods with strong photoluminescent and field-emission properties. *Small* **2**, 116–120. DOI: 10.1002/smll.200500210

Horsthuis W.H.G. and Pannekoek R. (1985) ZnO processing for integrated optic sensors. In: *Proceedings of 3rd European Conference on Integrated Optics, Berlin. May 6–8*. Springer Series in Optical Sciences, Vol. **48**. Springer-Verlag, Berlin, 152–157.

Horsthuis W.H.G. (1986) ZnO processing for integrated optic sensors. *Thin Solid Films* **137**, 185–192. DOI: 10.1016/0040-6090(86)90018-0

Hu J., Bando Y., Liu Q., and Golberg D. (2003) Laser-ablation growth and optical properties of wide and long single-crystal SnO2 ribbons. *Adv. Funct. Mater.* **13**, 493–496. DOI: 10.1002/adfm.200304327

Huang M.H., Wu Y.Y., Feick H., Tran N., Weber E., and Yang P.D. (2001) Catalytic growth of zinc oxide nanowires by vapor transport. *Adv. Mater.* **13**, 113–115. DOI: 10.1002/1521-4095(200101)13:2<113::AID-ADMA113>3.0.CO;2-H

Hudson J.B. (1998) *Surface Science*. John Wiley, New York.

Ishiguro K., Sasaki T., Arai T., and Imai I. (1958) Optical and electrical properties of tin oxide films. *J. Phys. Soc. Jpn.* **13**, 296–304. DOI: 10.1143/JPSJ.13.296

Jeong J., Choi S.P., Chang C.I., Shin D.C., Park J.S., Lee B.T., Park Y.J., and Song H.J. (2003) Photoluminescence properties of SnO_2 thin films grown by thermal CVD. *Solid State Commun.* **127**, 595–597. DOI: 10.1016/S0038-1098(03)00614-8

Kan D., Kanda R., Kanemitsu Y., Shimakawa Y., Takano M., Terashima T., and Ishizumi A. (2005) Blue-light emission at room temperature from Ar^+-irradiated $SrTiO_3$. *Nature Mater.* **4**, 816–819. DOI: 10.1038/nmat1498

Kida T., Nishiyama A., Yuasa M., Shimanoe K., and Yamazoe N. (2009) Highly sensitive NO_2 sensors using lamellar-structured WO_3 particles prepared by an acidification method. *Sens. Actuators B* **135**, 568–574. DOI: 10.1016/j.snb.2008.09.056

Kiliç C. and Zunger A. (2002) Origins of coexistence of conductivity and transparency in SnO_2. *Phys. Rev. Lett.* **88**, 095501. DOI: 10.1103/PhysRevLett.88.095501

Kolmakov A., Zhang Y., Cheng G., and Moskovits M. (2003) Detection of CO and O_2 using tin oxide nanowire sensors. *Adv. Mater.* **15**, 997–1000. DOI: 10.1002/adma.200304889

Korotcenkov G. (2007) Metal oxides for solid-state gas sensors: What determines our choice? *Mater. Sci. Eng. B* **139**, 1–23. DOI: 10.1016/j.mseb.2007.01.044

Kundu S. and Biswas P.K. (2005) Synthesis and photoluminescence property of nanostructured sol-gel indium tin oxide film on glass. *Chem. Phys. Lett.* **414**, 107–110. DOI: 10.1016/j.cplett.2005.08.062

Law M., Kind H., Messer B., Kim F., and Yang P. (2002) Photochemical sensing of NO_2 with SnO_2 nanoribbon nanosensors at room temperature. *Angew. Chem. Int. Ed.* **41**, 2405–2408. DOI: 10.1002/1521-3773(20020703)41:13<2405::AID-ANIE2405>3.0.CO;2-3

Lei Y., Zhang L.D., Meng G.W., Li G.H., Zhang X.Y., Liang C.H., Chen W., and Wang S.X. (2001) Preparation and photoluminescence of highly ordered TiO_2 nanowire arrays. *Appl. Phys. Lett.* **78**, 1125. DOI: 10.1063/1.1350959

Lettieri S., Setaro A., Baratto C., Comini E., Faglia G., Sberveglieri G., and Maddalena P. (2008a) On the mechanism of photoluminescence quenching in tin dioxide nanowires by NO_2 adsorption. *New J. Phys.* **10**, 043013. DOI: 10.1088/1367-2630/10/4/043013

Lettieri S., Setaro A., De Stefano L., De Stefano M., and Maddalena P. (2008b) The gas-detection properties of light-emitting diatoms. *Adv. Funct. Mater.* **18**, 1257–1264. DOI: 10.1002/adfm.200701124

Lettieri S., Causà M., Setaro A., Trani F., Barone V., Ninno D., and Maddalena P. (2008c) Direct role of surface oxygen vacancies in visible light emission of tin dioxide nanowires. *J. Chem. Phys.* **129**, 244710. DOI: 10.1063/1.3041775

Lettieri S., Amoruso S., Baratto C., Comini E., Faglia G., Maddalena P., Pallotti D., Santamaria L., and Sberveglieri G. (2012) Unpublished results.

Li C.C, Du Z.F., Li L.M., Yu H.C., Wan Q., and Wang T.H. (2007) Surface-depletion controlled gas sensing of ZnO nanorods grown at room temperature. *Appl. Phys. Lett.* **91**, 032101. DOI: 10.1063/1.2752541

Liao Z.M., Zhang H.Z., Zhou Y.B., Xu J., Zhang J.M., and Yu D.P. (2008) Surface effects on photoluminescence of single ZnO nanowires. *Phys. Lett. A* **372**, 4505–4509. DOI: 10.1016/j.physleta.2008.04.013

Liu P., Lee S., Tracy C.E., Turner J.A., Pitts J.R., and Deb S.K. (2003) Electrochromism of amorphous ruthenium oxide thin films. *Solid State Ionics* **165**, 223–228. DOI: 10.1016/j.ssi.2003.08.044

Liu Y., Dong J., and Liu M. (2004) Well-aligned “nano-box-beams” of SnO_2. *Adv. Mater.* **16**, 353–356. DOI: 10.1002/adma.200306104

Luo S.H., Fan J.Y., Liu W.L., Zhang M., Song Z.T., Liu C.L., Wu X.L., and Chu P.K. (2006a) Synthesis and low-temperature photoluminescence properties of SnO_2 nanowires and nanobelts. *Nanotechnology* **17**, 1695–1699. DOI: 10.1088/0957-4484/17/6/025

Luo S., Chu P.K., Liu W.L., Zhang M., and Lin C.L. (2006b) Origin of low-temperature photoluminescence from SnO_2 nanowires fabricated by thermal evaporation and annealed in different ambients. *Appl. Phys. Lett.* **88**, 183112. DOI: 10.1063/1.2201617

Madou M. and Morrison R. (1989) *Chemical Sensing with Solid State Devices*. Academic Press, San Diego, CA.

Maestre D., Cremades A., and Piqueras J. (2005) Growth and luminescence properties of micro and nanotubes in sintered tin oxide. *J. Appl. Phys.* **97**, 044316. DOI: 10.1063/1.1851602

Marley J.A. and Dockerty R.C. (1965) Electrical properties of stannic oxide single crystals. *Phys. Rev.* **140**, A304–A310. DOI: 10.1103/PhysRev.140.A304

Mazingue T., Escoubas L., Spalluto L., Flory F., Socol G., Ristoscu C., Axente E., Grigorescu S., Mihailescu I.N., and Vainos N.A. (2005) Nanostructured ZnO coatings grown by pulsed laser deposition for optical gas sensing of butane. *J. Appl. Phys.* **98**, 074312. DOI: 10.1063/1.2076442

Mazingue T., Escoubas L., Spalluto L., Flory F., Jacquouton P., Perrone A., Kaminska E., Piotrowska A., Mihailescu I., and Atanasov P. (2006) Optical characterizations of ZnO, SnO_2, and TiO_2 thin films for butane detection. *Appl. Opt.* **45**, 1425–1435. DOI: 10.1364/AO.45.001425

Meskini N., Mattausch H.J., and Hanke W. (1983) Many-body effects in the absorption spectrum of a heteropolar crystal. *Solid State Commun.* **48**, 807–809. DOI: 10.1016/0038-1098(83)91023-2

Modi A., Koratkar N., Lass E., Wei B.Q., and Ajayan P.M. (2003) Miniaturized gas ionization sensors using carbon nanotubes. *Nature* **424**, 171–174. DOI: 10.1038/nature01777

Nguyen P., Hou T., Kong J., Cassell A.M., Quinn R., Li J., Han J., McNeil M., and Meyyappan

M. (2003) Epitaxial directional growth of Indium-doped tin oxide nanowire arrays. *Nanoletters* **3**, 925–928. DOI: 10.1021/nl0342186

Nishizawa K. (1988) Optical chemical sensors. in: Seiyama T. (ed.), *Chemical Sensor Technology*. Kodansha, Tokyo, 237–245.

Okazaki S., Nakagawa H., Asakura S., Tomiuchi Y., Tsuji N., Murayama H., and Washiya M. (2003) Sensing characteristics of an optical fiber sensor for hydrogen leak. *Sens. Actuators B* **93**, 142–145. DOI: 10.1016/S0925-4005(03)00211-9

Oviedo J. and Gillan M.J. (2000) Energetics and structure of stoichiometric SnO_2 surfaces studied by first-principles calculations. *Surf. Sci.* **463**, 93–101. DOI: 10.1016/S0039-6028(00)00612-9

Ozgur U., Alivov Y.I., Liu C., Teke A., Reshchikov M.A., Dogan S., Avrutin V., Cho S.J., and Morkoc H. (2005) A comprehensive review of ZnO materials and devices. *J. Appl. Phys.* **98**, 041301. DOI: 10.1063/1.1992666

Pal U., Perez-Centeno A., and Herrea-Zaldivar M. (2008) Cathodoluminescence defect characterization of hydrothermally grown SnO_2 nanoparticles. *J. Appl. Phys.* **103**, 064301. DOI: 10.1063/1.2885060

Pan Z.W., Dai Z.R., and Wang Z.L. (2001) Nanobelts of semiconducting oxides. *Science* **291**, 1947–1949. DOI: 10.1126/science.1058120

Parr R.G. and Yang W. (1989) *Density-Functional Theory of Atoms and Molecules*. Oxford University Press, New York.

Prades J.D., Arbiol J., Cirera A., Morante J.R., Avella M., Zanotti L., Comini E., Faglia G., and Sberveglieri G. (2007a) Defect study of SnO_2 nanostructures by cathodoluminescence analysis: Application to nanowires. *Sens. Actuators B* **126**, 6–12. DOI: 10.1016/j.snb.2006.10.014

Prades J.D., Cirera A., Morante J.R., Pruneda J.M., and Ordejon P. (2007b) Ab initio study of NO_x compounds adsorption on SnO_2 surface. *Sens. Actuators B* **126**, 62–67. DOI: 10.1016/j.snb.2006.10.040

Prades J.D., Cirera A., and Morante J.R. (2007c) First-principles study of NO_x and SO_2 adsorption onto SnO_2. *J. Electrochem. Soc.* **154**, H675–H680. DOI: 10.1149/1.2742295

Ramgir N.S., Mulla I.S., and Vijayamohanan K.P. (2004) Shape selective synthesis of unusual nanobipyramids, cubes, and nanowires of RuO_2:SnO_2. *J. Phys. Chem. B* **108**, 14815–14819. DOI: 10.1021/jp046859a

Sakai G., Baik N.S., Miura N., and Yamazoe N. (2007) Surface chemistry of nanocrystalline SnO_2: Effect of thermal treatment and additives. *Sens. Actuators B* **126**, 52–55. DOI: 10.1016/j.snb.2006.10.047

Savvate'ev V., Chen-Esterlit Z., Aylott J.W., Choudhury B., Kim C.H., Zou L., Friedl J.H., Shinar R., Shinar J., and Kopelman R. (2002) Integrated organic light-emitting device/fluorescence-based chemical sensors. *Appl. Phys. Lett.* **81**, 4652–4654. DOI: 10.1063/1.1518154

Sberveglieri G., Baratto C., Comini E., Faglia G., Ferroni M., Ponzoni A., and Vomiero A. (2007) Synthesis and characterization of semiconducting nanowires for gas sensing. *Sens. Actuators B* **121**, 208–213. DOI: 10.1016/j.snb.2006.09.049

Seiyama T., Kato A., Fujiishi K., and Nagatani M. (1962) A new detector for gaseous components using semiconductive thin films. *Anal. Chem.* **34**, 1502–1503. DOI: 10.1021/ac60191a001

Sekimoto S., Nakagawa H., Okazaki S., Fukuda K., Asakura S., Shigemori T., and Takahashi

S. (2000) A fiber-optic evanescent-wave hydrogen gas sensor using palladium-supported tungsten oxide. *Sens. Actuators B* **66**, 142–145. DOI: 10.1016/S0925-4005(00)00330-0

Setaro A., Lettieri S., De Stefano L., and Maddalena P. (2007) Highly sensitive opto-chemical gas detection by luminescent marine diatoms. *Appl. Phys. Lett.* **91**, 051921. DOI: 10.1063/1.2768027

Setaro A., Lettieri S., Diamare D., Maddalena P., Malagu C., Carotta M.C., and Martinelli G. (2008) Nanograined anatase titania-based optochemical gas detection. *New J. Phys.* **10**, 053030. DOI: 10.1088/1367-2630/10/5/053030

Shalish I., Temkin H., and Narayanamurti V. (2004) Size-dependent surface luminescence in ZnO nanowires. *Phys. Rev. B* **69**, 245401. DOI: 10.1103/PhysRevB.69.245401

Shi J., Zhu Y., Zhang X., Baeyens W.R.G., and Garcıa-Campana A.M. (2004) Recent developments in nanomaterial optical sensors. *Trends Anal. Chem.* **23**, 351–360. DOI: 10.1016/S0165-9936(04)00519-9

Shukla S.K., Parashar G.K., Mishra A.P., Misra P., Yadav B.C., Shukla R.K., Bali L.M., and Dubey G.C. (2004) Nano-like magnesium oxide films and its significance in optical fiber humidity sensor. *Sens. Actuators B* **98**, 5–11. DOI: 10.1016/j.snb.2003.05.001

Singh S.P., Arya S.K., Pandey P., Malhotra B.D., Saha S., Sreenivas K., and Gupta V. (2007) Cholesterol biosensor based on rf sputtered zinc oxide nanoporous thin film. *Appl. Phys. Lett.* **91**, 063901. DOI: 10.1063/1.2768302

Sysoev V.V., Schneider T., Goschnick J., Kiselev I., Habicht W., Hahn H., Strelcov E., and Kolmakov A. (2009) Percolating SnO_2 nanowire network as a stable gas sensor: Direct comparison of long-term performance versus SnO_2 nanoparticle films. *Sens. Actuators B* **139**, 699–703. DOI: 10.1016/j.snb.2009.03.065

Trani F., Causà M., Ninno D., Cantele G., and Barone V. (2008) Density functional study of oxygen vacancies at the SnO_2 surface and subsurface sites. *Phys. Rev. B* **77**, 245410. DOI: 10.1103/PhysRevB.77.245410

Valerini D., Cretì A., Caricato A.P., Lomascolo M., Rella R., and Martino M. (2010) Optical gas sensing through nanostructured ZnO films with different morphologies. *Sens. Actuators B* **145**, 167–173. DOI: 10.1016/j.snb.2009.11.064

Vayssieres L. (2003) Growth of arrayed nanorods and nanowires of ZnO from aqueous solutions. *Adv. Mater.* **15**, 464–466. DOI: 10.1002/adma.200390108

Wan Q., Li Q.H., Chen Y.J., Wang T.H., He X.L., Li J.P., and Lin C.L. (2004) Fabrication and ethanol sensing characteristics of ZnO nanowire gas sensors. *Appl. Phys. Lett.* **84**, 3654–3656. DOI: 10.1063/1.1738932

Wang C.F., Xie S.Y., Lin S.C., Cheng X., Zhang X.H., Huang R.B., and Zheng L.S. (2004) Glow discharge growth of SnO_2 nano-needles from SnH_4. *Chem. Commun.* **15**, 1766–1767. DOI: 10.1039/b404362f

Wang J.X., Sun X.W., Yang Y., Huang H., Lee Y.C., Tan O.K., and Vayssieres L. (2006) Hydrothermally grown oriented ZnO nanorod arrays for gas sensing applications. *Nanotechnology* **17**, 4995–4998. DOI: 10.1088/0957-4484/17/19/037

Wehrspohn R.B., Zhu M., and Godet C. (1998) Visible photoluminescence and its mechanisms from a-SiO_x : H films with different stoichiometry. *J. Lumin.* **80**, 449–453. DOI: 10.1016/S0022-2313(98)00146-X

Wen J.G., Lao J.Y., Wang D.Z., Kyam T.M., Foo Y.L., and Ren Z.F. (2003) Self-assembly of semiconducting oxide nanowires, nanorods and nanoribbons. *Chem. Phys. Lett.* **372**, 717–722. DOI: 10.1016/S0009-2614(03)00485-8

Wolfbeis O.S. (2002) Fiber-optic chemical sensors and biosensors. *Anal. Chem.* **74**, 2663–2678. DOI: 10.1021/ac020176e

Wolfbeis O.S. (2004) Fiber-optic chemical sensors and biosensors. *Anal. Chem.* **76**, 3269–3284. DOI: 10.1021/ac040049d

Yamazoe N. (1991) New approaches for improving semiconductor gas sensors. *Sens. Actuators B* **5**, 7–19. DOI: 10.1016/0925-4005(91)80213-4

Yu P.Y. and Cardona M. (1995) *Fundamentals of Semiconductors: Physics and Materials Properties.* Springer-Verlag, Berlin. DOI: 10.1007/978-3-642-00710-1

Yu P.Y. and Welber B. (1978) High pressure photoluminescence and resonant Raman study of GaAs. *Solid State Commun.* **25**, 209–211. DOI: 10.1016/0038-1098(78)90214-4

Yuan Z.Y., Colomer J.F., and Su B.L. (2002) Titanium oxide nanoribbons. *Chem Phys. Lett.* **363**, 362–366. DOI: 10.1016/S0009-2614(02)01205-8

Yuan X.L., Lazzarini L., Salviati G., Zha M., and Sekiguchi T. (2006) Cathodoluminescence characterization of SnO_2 nanoribbons grown by vapor transport technique. *Mater. Res. Semicond. Process.* **9**, 331–336. DOI: 10.1016/j.mssp.2006.01.029

Zhou J.X., Zhang M.S., Hong J.M., and Yin Z. (2006) Raman spectroscopic and photoluminescence study of single-crystalline SnO_2 nanowires. *Solid State Commun.* **138**, 242–246. DOI: 10.1016/j.ssc.2006.03.007

Zhu M., Han Y., Wehrspohn R.B., Godet C., and Etemadi R. (1998) The origin of visible photoluminescence from silicon oxide thin films prepared by dual-plasma chemical vapor deposition. *J. Appl. Phys.* **83**, 5386–5393. DOI: 10.1063/1.367407

Zhu Y.C. and Ding C.X. (1999) Investigation on the surface state of TiO_2 ultrafine particles by luminescence. *J. Solid State Chem.* **145**, 711–715. DOI: 10.1006/jssc.1999.8292

CHAPTER 4

Simulation and Modeling of Hydrogen Leak Sensors Based on Optical Fiber Gratings

C. Caucheteur
M. Debliquy
G. Ravet
D. Lahem
P. Megret

1. INTRODUCTION

Gas detection is a key issue today in numerous fields such as environmental, process control, automotive, and chemical industries. Besides classical detection systems that are often cumbersome and expensive, there exists a huge need for miniaturized systems, allowing multiple the measurement points with a single interrogation unit. In that frame, sensors based on optical fibers are particularly interesting, thanks to their ability to provide numerous sensing points.

Among the gases of interest for detection systems, hydrogen has attracted considerable attention because it participates to a wide range of chemical processes and appears during energy production and transport. It is heavily employed in aerospace applications, where hydrogen serves as fuel for rockets. Besides these

DOI: 10.5643/9781606503201/ch4

important fields, a growing market is now beginning to emerge: the use of hydrogen as a fuel source for clean-burning engines and power plants. However, one of the problems in storing, transporting, or handling hydrogen is that it is extremely flammable and very volatile. With its high diffusivity, it can burn and explode in air at a concentration as low as 4%, that is, the lower explosive limit (LEL), with a flame velocity almost 10 times higher than that of natural gas. Therefore, to mitigate the risk linked to its use, versatile and fast hydrogen leak detectors are required in all application fields, especially for low concentrations.

In potentially explosive atmospheres, sensors based on optical fibers offer a set of advantages without equivalent for well-established technologies such as electrochemical sensors, semiconductor sensors, or pellistors (Yu 2002). Indeed, optical fibers are immune to electromagnetic interference and offer good resistance to high temperatures and to chemical corrosion. They are also durable in high-radiation environments. Their small dimensions combined with their light weight and flexibility allow embedding them directly in the structures to monitor without affecting their mechanical resistance. Last but not least, depending on the way the system is set up, optical fibers can yield distributed (the parameter is sensed all along the fiber length) or quasi-distributed (the parameter is sensed at different points cascaded along the fiber length) measurements that can be remotely addressed. It is not intrinsically possible to achieve this with any other sensing method.

Different types of optical fiber gas sensors have been reported in the past and have been adapted to hydrogen sensing. These configurations mainly include evanescent wave sensors (Sekimoto 2000), end-coated optical fiber sensors (Arregui 2003), fiber grating–based sensors (Sutapun 1999), or surface plasmon wave sensors (Bévenot 2002). In all these configurations, the fiber of itself is not sensitive to the gas concentration, so a sensitive layer is required in addition to the optical fiber probe.

- In an evanescent-wave fiber optic sensor, the cladding of the optical fiber is often partially removed and placed in direct contact with a sensitive material. Ideally, this material absorbs the wavelengths carried by the fiber and modifies the absorption as a function of the gas concentration to be detected. In turn, this mechanism drains energy from the fiber core, and the output power at the detector is modified as a function of the light absorption.
- End-coated optical fiber sensors exploit the Fresnel reflection occurring at the interface between the cleaved fiber end and the surrounding medium. Here, a sensitive material deposited on the cleaved fiber end is expected to modify its refractive index as a function of the gas concentration to monitor. As a result, this modifies the reflection conditions at the interface and the reflected power varies with respect to the gas concentration.
- A fiber grating is a permanent and (quasi-)periodic refractive index modulation of the fiber core. It is conventionally obtained through lateral illumination

of the fiber with an interference pattern of ultraviolet light. Depending on the type of grating that is manufactured, the refractive index modulation induces light coupling inside the fiber core or between the core and the cladding. To detect a gas, a grating should be associated to a sensitive layer that induces an environmental stimulus such as a temperature change, a mechanical stress, a pressure change, or a refractive index change.
- Surface plasmon wave sensors exploit a nanometric sheath of noble metal (most often gold or silver) deposited on the cladding surface that, under particular conditions of excitation, leads to the generation of surface plasmon waves. These waves propagate at the interface between the metal and the dielectric (surrounding medium). These waves are strongly sensitive to a change of the surrounding refractive index. Hence, by depositing a sensitive material on the gold sheath (i.e., a material that modifies its refractive index in the presence of the gas to be detected), one can realize an efficient and very sensitive gas sensor.

In all cases, the sensitive layer has to fulfill some practical requirements to obtain an efficient and robust sensor. In particular, it has to yield a fast response (of the order of several seconds or even better) and to be specific to the gas species to be detected, i.e. it should be inert to other gases.

In this chapter, the focus will be made on fiber grating configurations used for hydrogen leak detection. Two configurations will be detailed:

1. Fiber gratings covered with a palladium layer, in which the sensing mechanism is based on the swelling of the palladium coating, resulting in a mechanical stress on the grating. This configuration is used in a nitrogen environment and is mainly dedicated to aerospace applications.
2. Fiber gratings covered with a catalytic sensitive layer made of a ceramic doped with noble metal, resulting in a temperature elevation around the grating. This configuration works in air environments, whatever the relative humidity level and ambient temperature. It is designed for ground applications such as the monitoring of storage places and pipelines.

The remainder of the chapter is organized as follows. Section 2 briefly summarizes the basic concepts of fiber gratings technology. Section 3 focuses on the modeling of hydrogen sensors in a nitrogen environment based on the use of fiber gratings coated with palladium. This model uses a mechanical approach that links the palladium expansion to an axial strain on the grating. Section 4 details the modeling of hydrogen sensors in air environments that use fiber gratings coated with tungsten oxide doped with platinum. Here, an energy-balance approach is used to model the flux exchanged between the optical fiber and the surroundings. Finally, Section 5 draws some conclusions.

2. FUNDAMENTALS OF FIBER GRATINGS

A fiber Bragg grating (FBG) is a periodic and permanent modification of the core refractive index value along the optical fiber axis (Kashyap 1999; Othonos 1999). This modification is usually obtained by transversally exposing the core of a photosensitive optical fiber to an intense interference pattern of ultraviolet light at a wavelength around 240 nm. Indeed, due to the presence of germanium oxide dopants inside the core, an optical fiber is photosensitive (i.e., it benefits from the ability to permanently change its refractive index when exposed to light) in a wavelength band centered around 240 nm. For this reason, continuous-wave frequency-doubled argon-ion lasers emitting at 244 nm or pulsed excimer lasers emitting at 248 nm are most often used to manufacture FBGs. To create an interference pattern, two writing techniques are preferred: the interferometric method (also called the transverse holographic method) and the phase mask technique.

An FBG is defined by several physical parameters, as sketched on Figure 4.1. The grating length L is the optical fiber length along which the refractive index modulation is realized. The periodicity and the amplitude of the refractive index modulation are labeled Λ and δn, respectively. The order of magnitude of these parameters typically varies between 200 and 1000 nm for Λ, from a few millimeters to a few tens of centimeters for L, and from 10^{-5} to 10^{-3} for δn.

Perturbation induces light coupling between two counterpropagating core modes. This coupling occurs for some wavelengths around the Bragg wavelength, defined by

$$\lambda_{\text{Bragg}} = 2n_{\text{eff}}\Lambda \tag{4.1}$$

where n_{eff} is the effective refractive index of the core mode at the Bragg wavelength.

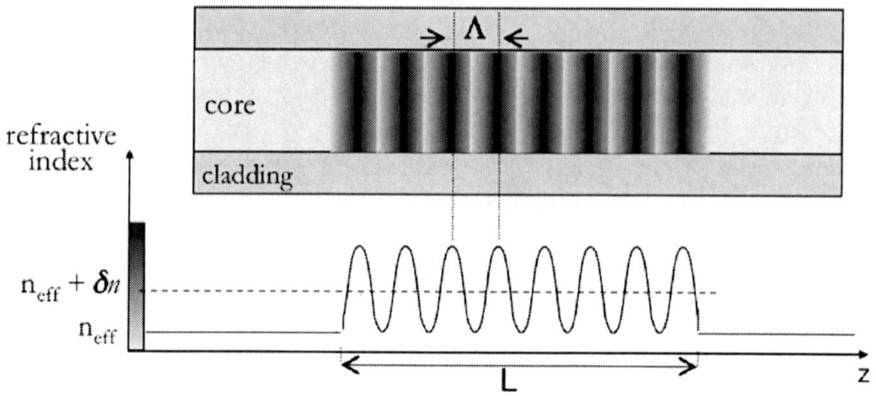

Figure 4.1. Sketch of a fiber Bragg grating.

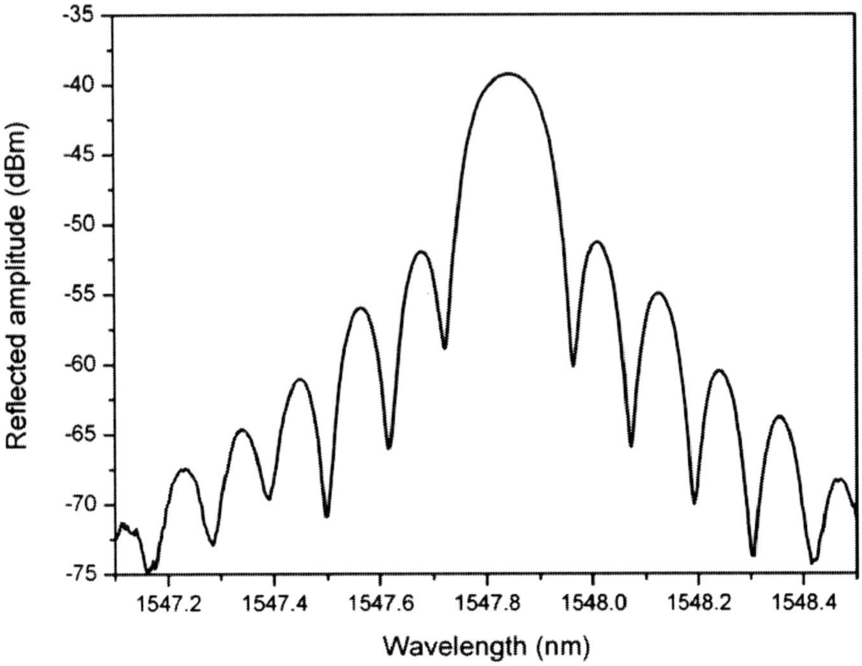

Figure 4.2. Amplitude reflected spectrum of a 1-cm-long FBG.

A uniform FBG acts as a selective mirror in wavelengths around the Bragg wavelength to yield a pass-band reflected amplitude spectrum, as depicted in Figure 4.2 for a 1-cm-long FBG. In practice, at each refractive index discontinuity along the fiber axis, a weak Fresnel reflection is generated. They add in phase at the Bragg wavelength, yielding an important reflection.

In practice, the effective refractive index of the core and the spatial periodicity of the grating are both affected by changes in strain and temperature. In particular, the effective refractive index is modified through the thermo-optic and strain-optic effects, respectively. Hence, from Eq. (4.1), the Bragg wavelength shift $\Delta\lambda_{\text{Bragg}}$ due to strain $\Delta\varepsilon$ and temperature ΔT variations is given by

$$\Delta\lambda_{\text{Bragg}} = 2\left(\Lambda\frac{dn_{\text{eff}}}{dT} + n_{\text{eff}}\frac{d\Lambda}{dT}\right)\Delta T + 2\left(\Lambda\frac{dn_{\text{eff}}}{d\varepsilon} + n_{\text{eff}}\frac{d\Lambda}{d\varepsilon}\right)\Delta\varepsilon \tag{4.2}$$

The first term in Eq. (4.2) represents the effect of temperature on the Bragg wavelength. The shift of the Bragg wavelength due to thermal expansion comes from the modification of the grating spacing and the refractive index. The relative wavelength shift due to a temperature change ΔT can be written as:

$$\frac{\Delta\lambda_{\text{Bragg}}}{\Delta T} = \lambda_{\text{Bragg}}\left(\frac{1}{n_{\text{eff}}}\frac{dn_{\text{eff}}}{dT} + \frac{1}{\Lambda}\frac{d\Lambda}{dT}\right) \tag{4.3}$$

where

$$\frac{1}{n_{\text{eff}}}\frac{dn_{\text{eff}}}{dT}$$

is the thermo-optic coefficient, which is approximately equal to $8.6 \times 10^{-6}\ \text{K}^{-1}$ for a germanium-doped silica core optical fiber and $(1/\Lambda)(d\Lambda/dT)$ is the thermal coefficient of the optical fiber, which is approximately equal to $0.55 \times 10^{-6}\ \text{K}^{-1}$ for silica so that the refractive-index change is by far the dominant effect (Othonos 1999). The order of magnitude of the temperature sensitivity of the Bragg wavelength is 10 pm/°C around 1550 nm.

The second term in Eq. (4.2) represents the effect of longitudinal strain on an optical fiber. It corresponds to a change in the grating periodicity and the strain-optic–induced change in the refractive index. By defining the strain as $\varepsilon = \Delta\Lambda/\Lambda$, the change of the grating periodicity can be expressed by

$$\Lambda_S = \Lambda + \Delta\Lambda = \Lambda(1 + \varepsilon) \tag{4.4}$$

where Λ_S represents the modified grating period after the application of the perturbation.

The elasto-optic effect relates the refractive-index change to the applied strain (Yariv 2003):

$$\Delta n_{ij} = \left(\Delta\frac{1}{n_{\text{eff}}^2}\right)_{ij} = \sum_{i=1}^{6}\sum_{j=1}^{6} p_{ij}\varepsilon_{ij} \tag{4.5}$$

where Δn_{ij} is the change in the electric impermeability tensor, ε_{ij} are the strain components, and p_{ij} are the elements of the elasto-optic tensor. The shape of the elasto-optic tensor, but not the magnitude of the coefficients p_{ij}, can be derived from the symmetry of the material under consideration. In the most general case, there are 36 different coefficients p_{ij}. For the class of isotropic materials to which fused silica glass belongs, this number is reduced to two independent coefficients p_{11} and p_{12} (Yariv 2003). The magnitudes of the individual coefficients p_{ij} are dependent on the material considered. Measurements on single-mode optical fibers at 628 nm yielded values of $p_{11} = 0.113$ and $p_{12} = 0.252$ (Barlow 1983).

Assuming that the grating is strained in the *z* direction only and that the fiber material follows Hooke's law, the Bragg wavelength shift as a result of the applied strain is defined by

$$\Delta\lambda_{\text{Bragg}} = \lambda_{\text{Bragg}}(1 - p_e)\Delta\varepsilon \tag{4.6}$$

$$p_e = \frac{n_{\text{eff}}^2}{2}\left[p_{12} - \upsilon(p_{11} + p_{12})\right] \tag{4.7}$$

where υ is Poisson's ratio.

Substitution of parameters ($p_{11} = 0.113$, $p_{12} = 0.252$, $\upsilon = 0.16$, and $n_{\text{eff}} = 1.482$) in Eqs. (4.6) and (4.7) gives a strain-optic constant $p_e = 0.21$ and an axial strain sensitivity of the Bragg wavelength of 1.2 pm/$\mu\varepsilon$ around 1550 nm.

Figure 4.3 displays the evolution of the Bragg wavelength shift as a function of a temperature change and an axial strain. The evolutions are linear and without hysteresis, which is very important for sensing purposes.

The main advantage of FBG sensors is that the information about the perturbation is wavelength-encoded. This property makes the sensor self-referencing and independent of fluctuating light levels. The system is therefore immune to source power and connector losses that affect many other types of optical fiber sensors. The very low insertion loss and narrow-band wavelength reflection of FBGs offer convenient serial multiplexing along a single-mode optical fiber. There are further advantages of FBGs over conventional electrical strain gauges, such as linearity in response over many orders of magnitude. The small diameter of the optical fiber combined with the small physical length of FBGs make them compatible with applications for which slightly intrusive probes are required, such as in the human body for temperature and/or biochemical monitoring. Furthermore, FBGs can be easily embedded into materials to provide damage detection or internal strain field mapping. FBG sensors are therefore very important components for the development of smart structure technology and for monitoring composite material curing and response. Their association with sensitive materials opens the way to other sensing applications such as gas leak monitoring (e.g., hydrogen, as described in the remaining of the text) or chemical species profiling.

3. HYDROGEN LEAK SENSOR IN NITROGEN ENVIRONMENT USING FBG COVERED BY PALLADIUM

Palladium (Pd) is known for its ability to absorb hydrogen and to expand as hydrogen absorption converts Pd into hydrides PdHx, which has a lower density and a larger specific volume (Lewis 1967). When a Pd coating deposited on

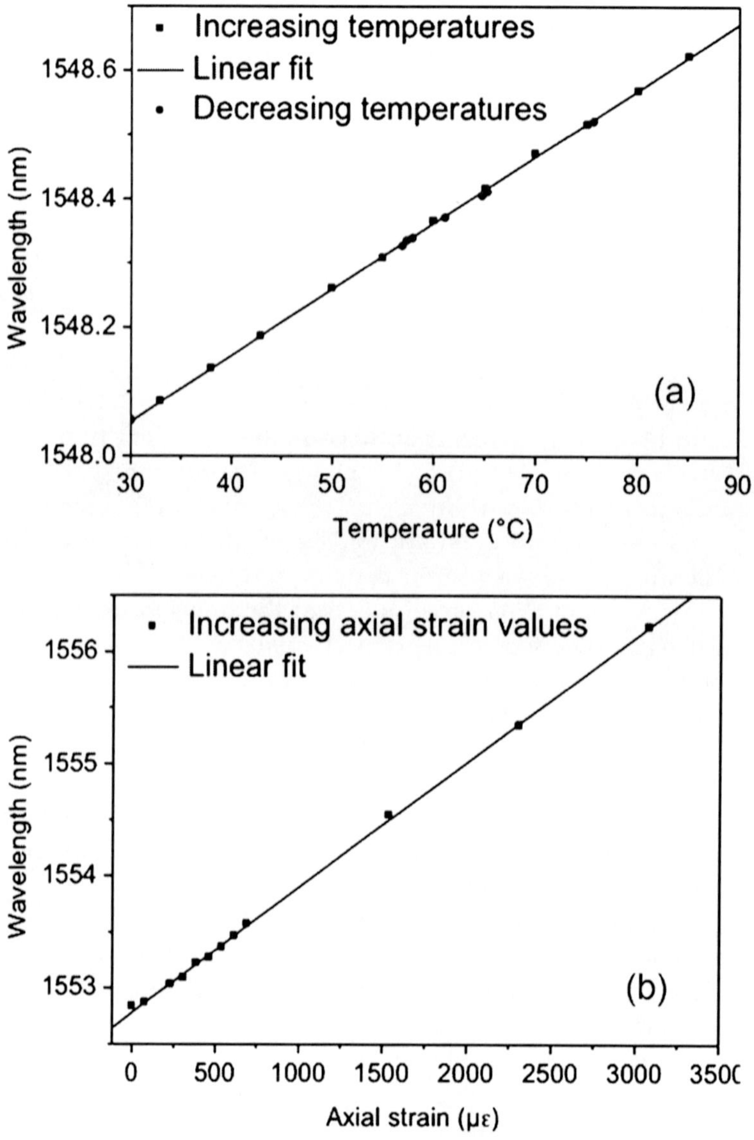

Figure 4.3. Bragg wavelength shift as a function of a temperature change (a) and a mechanical axial strain (b).

a FBG absorbs hydrogen, the mechanical expansion occurring in the Pd layer stretches the optical fiber and causes the grating period to expand and the effective refractive index to change via the elasto-optic effect. As the stress induced by the formation of the hydride phase is directly related to the hydrogen concentration, the shift of the Bragg wavelength can be straightforwardly linked to this concentration. Following this principle, hydrogen sensors in nitrogen environment can be achieved.

3.1. AXIAL STRAIN EFFECT

The behavior in response to the hydrogen concentration can be simulated starting from Eqs. (4.6) and (4.7). The most often encountered quantitative analysis uses the model proposed by Butler (1988) to determine the axial strain in the fiber core. In this model, it is assumed that the relationship between the hydrogen content inside the bulk palladium and the partial pressure of hydrogen outside the bulk follows Sievert's law (Lewis 1967). As a consequence, the change in the dimensions of the Pd coating and the subsequent Bragg wavelength shift can be computed.

The strain in a free-standing bulk Pd is equal to the ratio of the fractional change in the lattice constant of Pd. The relationship between the resulting strain in the Pd coating and its content in hydrogen (C) is given by (Butler 1988; Sutapun 1999)

$$\delta = 0.026C \tag{4.8}$$

C is linked to the partial pressure of hydrogen (p in Torr) according to Sievert's law (Lewis 1967):

$$\sqrt{p} = KC \tag{4.9}$$

where K is Sievert's coefficient and is equal to 350 $\text{Torr}^{1/2}$. Merging Eq. (4.8) and (4.9) yields the following relationship:

$$\delta = 0.026\frac{\sqrt{p}}{K} \tag{4.10}$$

Now that the strain in the Pd layer is known, the strain transmitted to the FBG through the expansion of the Pd coating can be computed. For the sake of simplicity, let us consider only axial strain. If Y_F and Y_{Pd} denote the Young's modulus of the optical fiber and the Pd coating, respectively, and A_F or A_{Pd} their effective areas, the axial strain on the FBG is obtained by balancing the force acting on the FBG due to the Pd expansion [$\varepsilon(Y_F A_F + Y_{Pd} A_{Pd})$] and the force acting on the coating ($\delta Y_{Pd} A_{Pd}$):

$$\varepsilon = \delta\frac{(b^2 - a^2)Y_{Pd}}{a^2 Y_F - (b^2 - a^2)Y_{Pd}} \tag{4.11}$$

where a and b denote the cladding diameter and the total sensor diameter (cladding + Pd coating), respectively.

Substituting Eq. (4.11) into Eq. (4.6) yields the relationship of the Bragg wavelength shift of the Pd-coated FBG as a function of the hydrogen partial pressure:

$$\Delta\lambda_{\text{Bragg}} = 0.026\frac{\sqrt{p}}{K}\left[\frac{(b^2-a^2)Y_{\text{Pd}}}{a^2Y_{\text{F}}-(b^2-a^2)Y_{\text{Pd}}}\right]0.78\lambda_{\text{Bragg}} \tag{4.12}$$

Using this equation, one can compute the Bragg wavelength shift as a function of the hydrogen concentration. Figure 4.4 compares the sensor response to the hydrogen concentration and its sensitivity for different geometries (different cladding and coating thicknesses). The reduction of the cladding diameter can be obtained experimentally by etching the silica fiber with hydrofluoric acid. A cladding reduction weakens the mechanical resistance of the fiber, so we have limited our simulations to half of the fiber thickness. The following values were chosen for the Young's modulus: $Y_F = 69$ GPa and $Y_{Pd} = 121$ GPa. The Bragg wavelength was chosen to be 1550 nm. It can be seen that the sensitivity increases when the fiber diameter decreases or when the coating thickness increases. However, as the latter also increases the response time, due to hydrogen diffusion inside the coating,

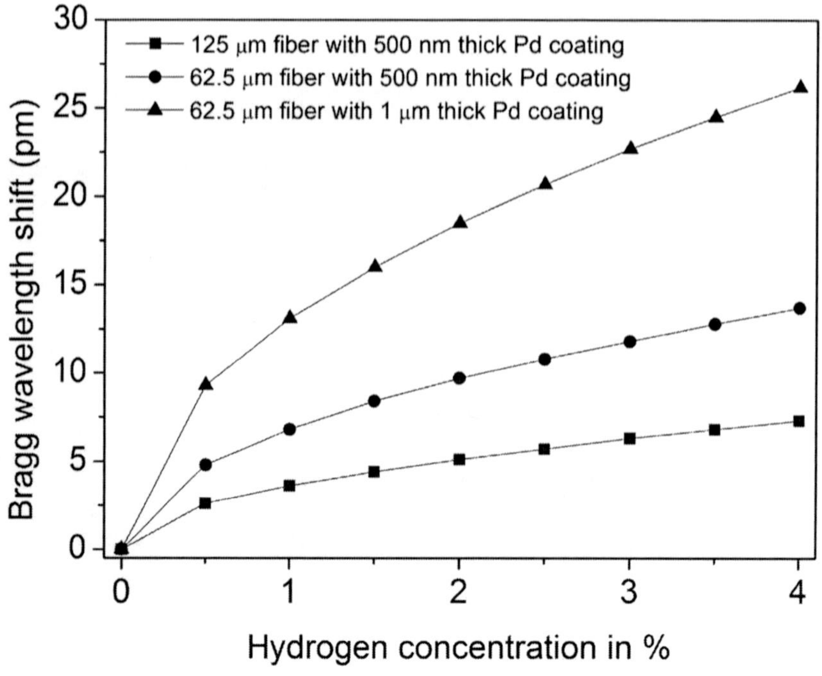

Figure 4.4. Influence of the sensor geometry on its sensitivity to hydrogen.

it cannot be too important. A few hundreds of nanometers remain a good compromise, leading to a sensitivity of the Bragg wavelength equal to several picometers per percent hydrogen concentration below the explosion threshold. These numerical figures are in good agreement with results reported experimentally (Sutapun 1999; Villatoro 2005; Buric 2007).

3.2. TEMPERATURE EFFECT

The sensor response is also affected by surrounding temperature changes leading to two contributions: the thermal effect on the optical fiber itself [cf. Eq. (4.3)] and the stress due to the thermal expansion of the Pd coating. Because it interferes with the response of the sensor to the hydrogen concentration, the thermal expansion of the Pd coating has to be modeled as well. In practice, the thermal expansion coefficient of Pd is equal to $11 \times 10^{-6}/°C$, about 20 times higher than that of silica, which is equal to $0.55 \times 10^{-6}/°C$. The thermal expansion of the Pd coating impacts the Bragg wavelength in a similar way as its expansion due to the hydrogen absorption. Therefore, from Eq. (4.12), one can derive the following relationship that links the Bragg wavelength shift as a function of the thermal stress:

$$\Delta\lambda_{\text{Bragg}} = \left[\frac{(b^2 - a^2)Y_{\text{Pd}}}{a^2 Y_{\text{F}} - (b^2 - a^2)Y_{\text{Pd}}}\right]\alpha_{\text{Pd}}\, \Delta T \lambda_{\text{Bragg}} \tag{4.13}$$

where α_{Pd} is the thermal expansion coefficient of Pd.

Using a 125-μm optical fiber and a 500-nm-thick Pd coating, Eq. (4.13) yields a Bragg wavelength shift equal to 0.467 pm/°C, which can be neglected in comparison to the intrinsic temperature sensitivity of the Bragg wavelength, which is close to 10 pm/°C.

4. HYDROGEN LEAK SENSOR IN AIR ENVIRONMENT USING FBG COVERED BY TUNGSTEN OXIDE DOPED WITH PLATINUM

When using platinum-doped tungsten trioxide nanolamellae as sensitive layer deposited on FBGs, the following chemical reactions occur with hydrogen in the presence of oxygen (in air):

$$WO_3 + H_2 \leftrightarrow WO_2 + H_2O \tag{4.14}$$

$$WO_2 + \frac{1}{2}O_2 \leftrightarrow WO_3 \tag{4.15}$$

The net reaction corresponds to the combustion of hydrogen on the surface of the sensitive coating that acts as a catalyst for this exothermic reaction. The reaction heat of hydrogen combustion in normal conditions is 57.8 kcal/mol. So, when exposed to hydrogen in air environment, the temperature around the FBG increases, which results in a shift of its Bragg wavelength.

As demonstrated by Debliquy (2009), different environmental parameters such as ambient temperature and relative humidity level influence the reaction. Two other parameters are also of great influence: the efficiency of the catalyst on the one hand and light coupling from the fiber core to the sensitive layer on the other hand. Such light coupling can be obtained through the use of a radiative fiber grating that couples the core mode into cladding modes. It provides an activation energy that can stimulate the reaction. The latter requires activation energy equal to 0.15 eV. These influences can be simulated as explained below.

The temperature reached by the FBG (and therefore the Bragg wavelength shift) results from the balance between the thermal losses and the energy released by the reaction. Figure 4.5 summarizes the different losses and energy fluxes for a simple fiber without any protection. The positive contribution to the balance (Φ_{reac}) comes from the heat released at the surface of the fiber due to the reaction. The losses can be divided into three components: (1) Φ_{conv} results from the exchanges with ambient air due to convection; (2) Φ_{rad} comes from heat exchanges due to radiation between the fiber and the ambient; and (3) Φ_{cond} is due to thermal conduction along the axis of the fiber. As explained below, this term is negligible compared to the first two.

In practice, the Biot number is defined by

$$\mathrm{Bi} = \frac{kD}{\lambda_{glass}} \tag{4.16}$$

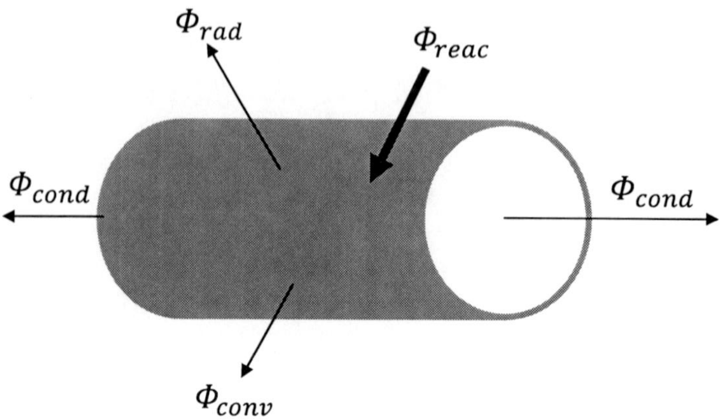

Figure 4.5. Thermal exchanges existing between the optical fiber and the surroundings.

where k is the exchange coefficient resulting from convection and radiation (typically 50 W/m^2 K), D is the fiber diameter, and λ_{glass} is the thermal conductivity of the glass (≈ 1 W/m K).

For a solid in contact with a fluid and exchanging heat with it, the Biot number represents the ratio of the heat transfer resistance through the solid and the heat transfer resistance through the surrounding fluid. When the Biot number is small (<0.1), the main resistance results from the exchange with the fluid, so the temperature gradient in the solid can be neglected. In the hydrogen sensor configuration, the fiber diameter is 125 μm and Bi is ~6×10^{-3}. Therefore, the temperature can be considered uniform in the exposed zone.

At thermal equilibrium, in the presence of hydrogen reacting on its surface, the fiber will stabilize at a higher temperature than the ambient temperature according to the thermal balance:

$$\Phi_{\text{reac}} = \Phi_{\text{conv}} + \Phi_{\text{rad}} + \Phi_{\text{cond}} \tag{4.17}$$

The total energy released on a section of length L of the optical fiber is equal to its losses due to convection, radiation with the ambient, and conductivity along the fiber. The different terms of Eq. (4.17) are detailed hereafter.

4.1. REACTION ON THE FIBER

The power density (W/m^2) released by the exothermic reaction is given by

$$\Phi_{\text{reac}} = -\upsilon\,\Delta H_r \tag{4.18}$$

where υ (in mol/s m^2) is the reaction rate on the catalyst surface and $-\Delta H_r$ is the reaction heat released by the combustion on the catalyst. In air, $-\Delta H_r = 57.8$ kcal/mol or 2.41×10^5 J/mol. The reaction rate depends on the concentration of the reactants (hydrogen and oxygen) at the surface of the catalyst. The concentration at the surface is imposed by the mass transfer mechanism. If diffusion is not limiting, the reaction rate is given directly by

$$\upsilon = k_r \prod_i C_i^{n_i} \tag{4.19}$$

where k_r is the reaction constant, which depends on temperature; C_i is the concentration of reactant i in mol/m^3, and n_i is the partial order of reactant i. For the reaction of combustion of hydrogen in air, oxygen is in large excess and its concentration is assumed to remain at its usual concentration in air. Because the partial order of hydrogen is 1, Eq. (4.19) degenerates to

$$v = k_r C_{\mathrm{H}} \tag{4.20}$$

k_r depends on the temperature according to the Arrhenius relationship,

$$k_r = k_0 e^{-E_a/RT} \tag{4.21}$$

where k_0 is the reaction coefficient for infinite temperature (m/s), E_a is the activation energy (J/mol), R is the gas constant (J/mol K), and T is the absolute temperature where the reaction takes place (here, the surface of the catalyst). For hydrogen combustion on platinum/ceramic catalyst, E_a is 1738 J/mol (Appel 2004).

This term shows that when the reaction takes place, the temperature increase due to the heat release by the reaction itself accelerates the reaction. This avalanche effect is limited by diffusion of the reactant, since hydrogen has to diffuse to the surface before reacting. A concentration gradient appears. The diffusion rate of hydrogen is proportional to the gradient according to Fick's law:

$$\overrightarrow{\Phi_i} = -D_i \vec{\nabla} C_i \tag{4.22}$$

where $\overrightarrow{\Phi_i}$ is the flux of the diffusing species i (mol/s m^2) and D_i is the diffusion coefficient of species i (m^2/s).

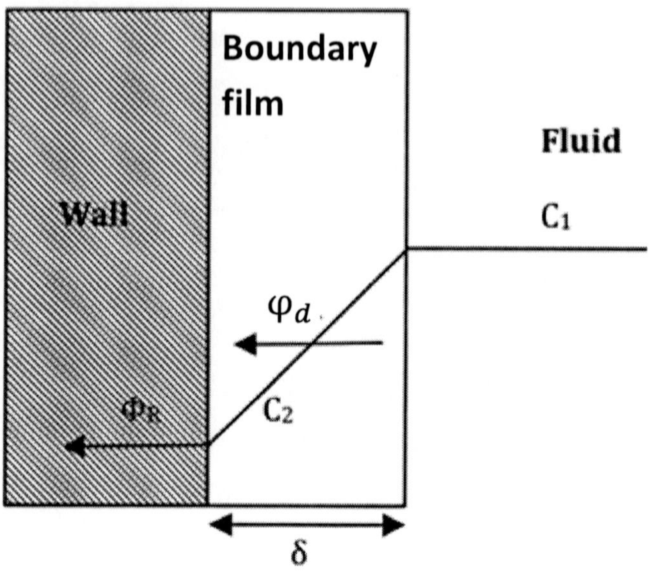

Figure 4.6. Mass transfer according to the film theory.

When the reaction is rapid, the concentration at the surface is close to 0 and hydrogen diffusion to the surface reaches its maximal value. The reaction rate is then imposed by the diffusion rate, which does not depend on the efficiency of the catalyst and which is proportional to the hydrogen concentration.

Using the film theory (Perry 1984), the hydrogen concentration and its diffusion rate to the surface can be calculated. As for thermal exchanges, it is supposed that the exchange between the fluid and the surface happens in a limited region, i.e., a film near the surface, as depicted in Figure 4.6. The diffusion rate is given by the resistance to mass transfer of this film. The matter flux (in mol/s m^2) due to diffusion is then given by

$$\Phi_d = k_d(C_1 - C_2) \tag{4.23}$$

k_d depends on the geometry of the system and is given by the following relationship, which uses the nondimensional Schmidt number Sc:

$$\mathrm{Sc} = \frac{L_{\mathrm{ref}} k_d}{d} \tag{4.24}$$

where L_{ref} is a reference length and d is the diffusion coefficient. For the cylindrical geometry of the fiber, L_{ref} corresponds to its diameter. Sc depends on the geometry as well as on the type of flow (laminar or turbulent flow). For the system considered here, Sc is 9.0.

When the system is stationary, the flux of hydrogen that reacts and disappears (υ) is equal to that coming by diffusion (Φ_d), so that the following relationship can be written:

$$k_d(C_{\mathrm{H}}^F - C_{\mathrm{H}}^S) = k_r C_{\mathrm{H}}^S \tag{4.25}$$

with C_{H}^F the nominal hydrogen concentration in air and C_{H}^S the concentration at the surface.

Combining Eqs. (4.18), (4.20), and (4.25) yields the following relationship for the heat released by the unit surface:

$$\Phi_{\mathrm{reac}} = \frac{k_r k_d}{k_r + k_d} C_{\mathrm{H}}^F \tag{4.26}$$

So, the total heat flux for a length L of the optical fiber (diameter D) is

$$\Phi_{\mathrm{reac}} = \pi DL \frac{k_r k_d}{k_r + k_d} C_{\mathrm{H}}^F = \pi DLK_{\mathrm{reac}} C_{\mathrm{H}}^F \tag{4.27}$$

In practice, the reaction coefficient k_r depends on the catalyst through the specific surface and the density of active sites, which are linked to the synthesis of the catalytic layer. Also, the presence of other species in the gas mixture can play a significant role as they compete with the reactants. Hence, if the surface is already occupied by other species such as adsorbed water molecules, this coefficient decreases. The surface occupied by water molecules depends on the relative humidity and on the temperature according to an adsorption isotherm.

When combustion of hydrogen takes place on the optical fiber, the temperature increases very rapidly, leading to the desorption of the water molecules and the release of the active sites. As already explained, the reaction rate becomes rapidly limited by diffusion. This implies that the sensitivity is not affected by humidity but that the necessary hydrogen concentration to ignite the reaction will be higher for a higher relative humidity level. This effect can be simulated by choosing an adequate adsorption isotherm for water vapor. In our model, a simple Langmuir isotherm is used. In practice, the reaction coefficient k_0 is reduced by the proportion of active surface occupied by water molecules. This, in turn, implies that k_0 becomes

$$k_{0,\mathrm{eff}} = k_0\left(1 - \frac{p_{\mathrm{H_2O}}}{p_{\mathrm{H_2O}} + K_{HT}}\right) \tag{4.28}$$

where $p_{\mathrm{H_2O}}$ is the partial pressure of water and K_{HT} is the adsorption coefficient of water molecules:

$$K_{HT} = k_H e^{-E_{\mathrm{ads}}/RT} \tag{4.29}$$

E_{ads} is the adsorption heat of the water molecules. In this case, no particular interaction is expected between the catalyst and water and therefore E_{ads} is close to the condensation heat and approximated to this value. Equation (4.28) shows that when the temperature increases, K_{HT} increases, so that the active sites are progressively released.

$p_{\mathrm{H_2O}}$ is linked to the relative humidity by the relationship

$$p_{\mathrm{H_2O}} = \frac{\mathrm{RH}\cdot p_{\mathrm{sat}}(T)}{100} \tag{4.30}$$

where RH is the relative humidity (%) and $p_{\mathrm{sat}}(T)$ is the vapor pressure of water at temperature T.

Usually, the chemical reactions are activated by thermal agitation so that k_r follows an Arrhenius law. However, when the reaction can be initiated by another form of energy, such as electromagnetic radiation, another term needs to be included in the coefficient k_r. This happens, for instance, when some light intensity is coupled into the sensitive layer. In these conditions, Eq. (4.21) becomes

$$k_r = k_0 e^{-E_a/RT} + k_i I \tag{4.31}$$

for which k_i depends on the wavelength of the radiation and I is the light intensity coupled into the sensitive layer.

Hence, coupling sufficiently energetic radiation into the sensitive layer allows us to increase the reaction rate and, as a consequence, to decrease the detection threshold of the sensor.

4.2. CONVECTION LOSSES

The flux due to convection losses can be expressed by

$$\Phi_{\text{conv}} = \pi D L K_c (T - T_a) \tag{4.32}$$

where K_c is the exchange coefficient with the fluid around the coated optical fiber (air in this case), T is the temperature reached by the optical fiber, and T_a is the ambient temperature.

K_c is calculated using the dimensional Nüsselt number defined by

$$\text{Nu} = \frac{K_c D}{\kappa} \tag{4.33}$$

where κ is the thermal conductivity of the fluid (air in this case). Note that, because the concentration of hydrogen is generally low for the application, its effect on the thermal conductivity or other physical properties is neglected.

The Nüsselt number depends on the fluid properties (viscosity, thermal conductivity, density), the geometry, and the type of convection mechanism (natural or forced). In this case, natural convection in air is considered and the fiber can be modeled as a horizontal tube of diameter D at uniform temperature in the zone considered. Under these conditions, the Nüsselt number is given by (Churchill 1983)

$$\sqrt{\text{Nu}} = \sqrt{\text{Nu}_0} + \left\{ \frac{\text{Gr}\,\text{Pr}/300}{\left[1 + (0.5/\text{Pr})^{9/16}\right]^{16/9}} \right\}^{1/6} \tag{4.34}$$

This relationship is valid for $10^{-4} \le \text{Gr Pr} \le 4 \times 10^{14}$, $0.022 \le \text{Pr} \le 7640$, and for a constant wall temperature. The Grashof and Prandl numbers are given by

$$\text{Gr} = \frac{g\beta D^3 \Delta T}{\mu^2} \tag{4.35}$$

$$\text{Pr} = \frac{\mu}{e} \tag{4.36}$$

respectively, where g is the acceleration due to gravity, β is the buoyancy coefficient (for perfect gases, $\beta = 1/T$), ΔT is the temperature difference between the surface and the fluid, μ is the kinematic viscosity, and ε is the effusivity of air. For gases, Pr is 0.7.

In the literature, the Rayleigh number Ra, which is the product of Gr and Pr, is often used. Due to the fact that Ra is involved in a relationship with a small exponent [see Eq. (4.34)], the coefficient will not vary a lot in the temperature range considered. It turns out that Nu ≈ 0.6, so the order of magnitude for K_c is 30 W/m^2 °C.

4.3. RADIATION LOSSES

The flux lost by radiation can be expressed by

$$\Phi_{\text{rad}} = \pi D L \varepsilon_s \sigma (T^4 - T_a^4) \tag{4.37}$$

where ε_s is the emissivity of the surface of the material (here WO_3/Pt and $\varepsilon_s = 0.2$) and $\sigma = 5.6710^{-8}$ W/m^2·K^4 is Stefan's constant (according to Stefan's law of emission of hot bodies). T and T_a are the absolute temperatures of the fiber and the ambient, respectively. To obtain a relationship of the same order as Eq. (4.32), Eq. (4.37) can be rewritten as

$$\Phi_{\text{rad}} = \pi D L K_r (T - T_a) \tag{4.38}$$

with

$$K_r = \frac{\varepsilon_s \sigma (T^4 - T_a^4)}{T - T_a} \tag{4.39}$$

The order of magnitude of K_r is 10 W/m^2 °C.

4.4. CONDUCTION LOSSES ALONG THE AXIS OF THE FIBER

This flux is defined by

$$\Phi_{\text{cond}} = \pi \frac{D^2}{2} K_l (T - T_a) \tag{4.40}$$

where K_l is a coefficient related to the conduction through the fiber along its axis. In practice, the conduction through the thin layer of absorbing material is neglected. K_l can be calculated by assuming that the rest of the fiber behaves as a winglet. In doing so, the order of magnitude for K_l is 65 W/m^2 °C.

4.5. SUM OF THE VARIOUS CONTRIBUTIONS

Now that every term of Eq. (4.17) has been defined, the balance between the different fluxes can be made to determine the temperature reached by the optical fiber. Combining Eqs. (4.27), (4.32), (4.38), and (4.40) yields the following relationship:

$$\pi DLK_{\text{reac}} C_{\text{H}}^F = \pi DLK_c (T - T_a) + \pi DLK_r (T - T_a) + \pi \frac{D^2}{2} K_l (T - T_a) \tag{4.41}$$

or, dividing by πDL,

$$K_{\text{reac}} C_{\text{H}}^F = (K_c + K_r + K_i')(T - T_a) \tag{4.42}$$

with

$$K_i' = \frac{D}{2L} K_l \tag{4.43}$$

The temperature variation is then given by

$$T - T_a = \frac{K_{\text{reac}} C_{\text{H}}^F}{K_c + K_r + K_i'} \tag{4.44}$$

Because L is generally less than a few centimeters and D is 125 μm, K_i' is small in comparison to K_c and K_r and can therefore be neglected.

According to Eq. (4.44), if K_{reac} were constant, the response of the sensor would be perfectly linear. Unfortunately, as already explained, K_{reac} may depend strongly on the hydrogen concentration and the presence of interfering gases such as water vapor, which affects the linearity of the sensor response.

It must be noticed that all material physical properties (viscosity, density, conductivity, etc.) used in the different relationships above are taken at the film temperature and correspond to an arithmetical mean between the surface temperature and the fluid temperature.

4.6. SIMULATION RESULTS

Using the model described above, one can predict the behavior of the sensor under various environmental conditions. The parameters that affect the reaction can be classified into extrinsic and intrinsic ones. Intrinsic parameters are the physical properties of the materials that cannot be changed (conductivity, diffusion coefficient, etc.), and k_0, k_i, k_H, which are unknown and depend on the properties of the catalyst. For our model, k_0 and k_H were adjusted by fitting the model to the experimental results reported by Debliquy (2009). The following values were therefore chosen: k_0 = 5000 m/s and k_H = 150,000 Pa^{-1}. k_0 was modified to include the influence of the efficiency of the catalyst on the reaction.

As already mentioned, the main extrinsic parameters are the ambient temperature (T_a), the relative humidity level (RH), and the intensity of the light coupled out of the fiber (I). In the following, each contribution (k_0, T_a, RH, or I) is simulated independently of the others.

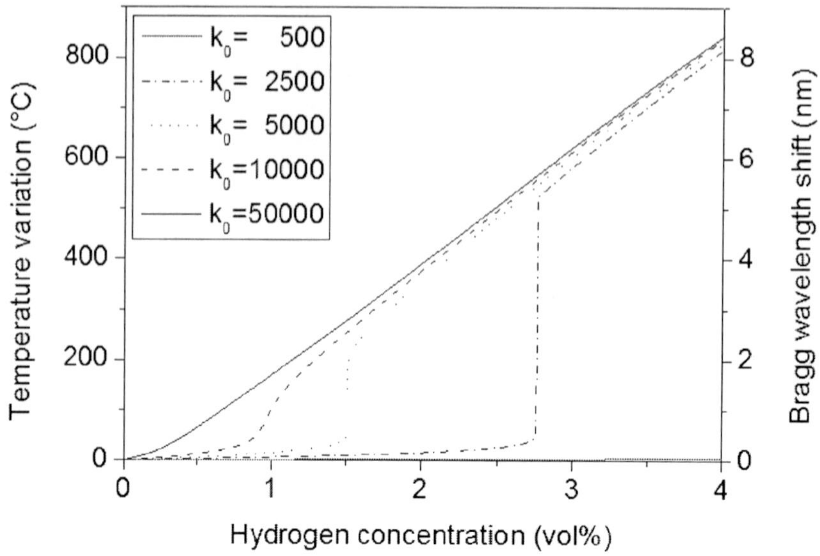

Figure 4.7. Simulated response of the hydrogen sensor for different efficiencies of the catalyst.

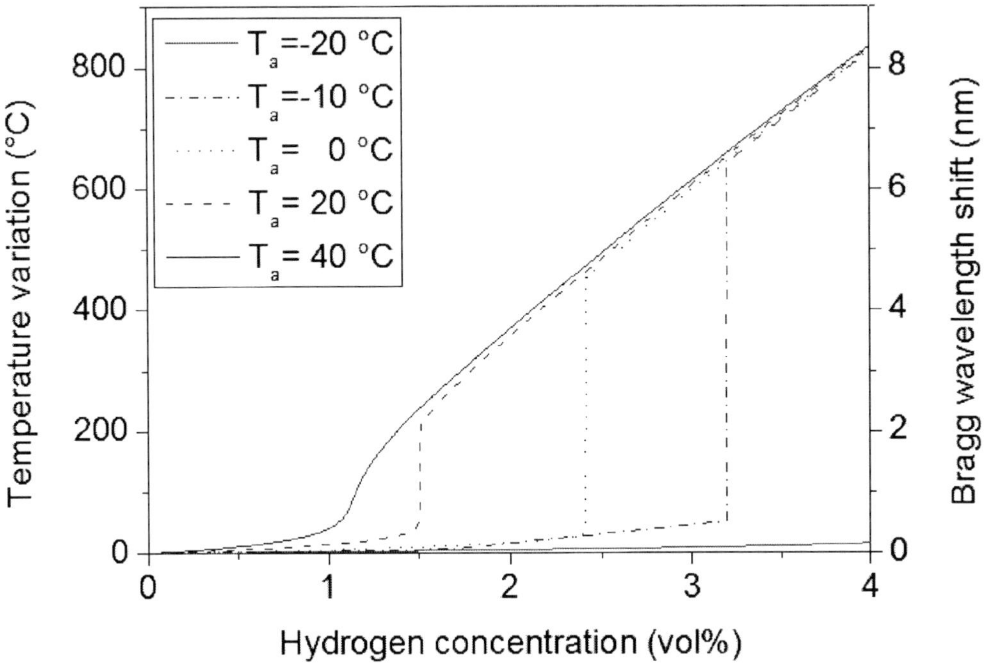

Figure 4.8. Simulated response of the hydrogen sensor for different ambient temperatures.

For the simulations, the concentration of hydrogen was chosen in the range 0–4%, which corresponds to the normal detection range because 4% is the lower explosive limit.

The first set of simulations takes into account the influence of the efficiency of the catalyst. For these simulations, T_a = 20°C, RH = 0%, and I = 0, while k_0 is varied from 500 to 50,000 m/s. Figure 4.7 depicts the temperature variation of the optical fiber as a function of the hydrogen concentration in air. It shows the presence of a threshold in terms of hydrogen concentration whose value decreases as k_0 increases. It also shows that the sensitivity of the sensor does not depend on the efficiency of the catalyst, provided that the concentration has reached the threshold value.

Figure 4.8 depicts the influence of the ambient temperature on the sensor response. Here, RH = 0%, I = 0, k_0 = 5000 m/s, and T_a varies between −20°C and +40°C. Once again, as found experimentally, the ambient temperature variations do not play a major role in the sensitivity but do affect the threshold value. At low temperatures, the sensor may be completely insensitive in the normal concentration range (0–4%). This is again explained by the fact that the reaction has to be ignited by a temperature increase sufficient to cause an avalanche effect.

Keeping all other parameters constant (I = 0, T_a = 20°C, and k_0 = 5000 m), the influence of the relative humidity has also been investigated numerically.

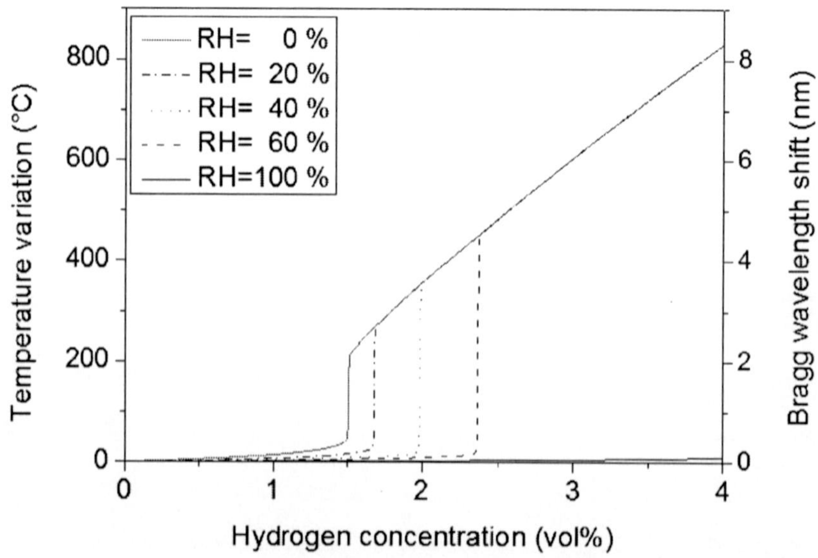

Figure 4.9. Simulated response of the hydrogen sensor for different relative humidity levels.

Figure 4.9 presents the results obtained for RH varying between 0% and 100%. A threshold increase appears as RH increases. Depending on other parameters, for humidity values reaching 100%, the response can be even completely inhibited below the LEL. This is again explained by the fact that the reaction has to be initiated by a sufficient temperature increase. It has to be mentioned that, in practi-

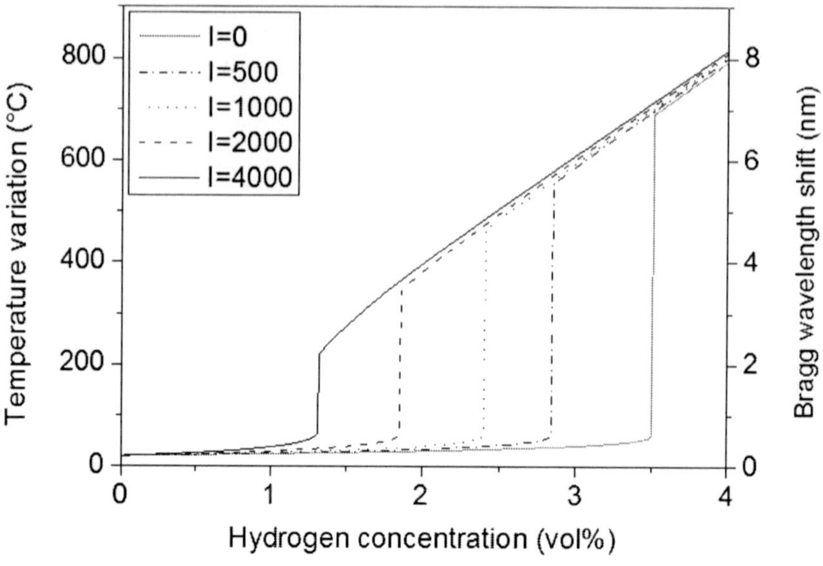

Figure 4.10. Simulated response of the hydrogen sensor for different light intensities coupled to the sensitive layer.

cal cases, when the reaction takes place, the temperature increases very rapidly, leading to desorption of the water molecules and release of the active sites, and thus to a situation similar to the case of RH = 0.

The last simulation results display the influence of the light coupling. Here, T_a = 20°C, RH K_c 0%, k_0 K_c 5000 m, while I (in arbitrary units) is varied from 0 to 4000. Figure 4.10 shows that the threshold decreases drastically from ~3.5% hydrogen concentration down to ~1.3% when the light intensity coupled to the sensitive later increases. This result confirms the interest in using a radiative grating that, in addition to the probe grating, couples light energy from the fiber core to the fiber cladding, as in the hybrid configuration reported by Caucheteur (2008).

All these simulation results are consistent with experimental results reported by Caucheteur (2008) for uniform fiber Bragg gratings covered by dip coating with a sensitive layer made of WO_3 doped with Pt. The numerically obtained sensitivity (~200°C per percent hydrogen below the LEL) is also in good agreement with that reported experimentally by Caucheteur (2008).

5. CONCLUSIONS

This chapter has summarized the modeling concepts for two hydrogen-sensing configurations that use fiber Bragg gratings. The first configuration exploits the elasto-optic effect and is based on the use of a palladium coating on a FBG. It swells in the presence of hydrogen, which strains the grating and induces a Bragg wavelength shift. A simple model based on the strain transfer from the palladium coating to the fiber core can be used to predict the behavior of such a sensor. Working in a nitrogen environment, it has a sensitivity of the order of a few picometers per percent hydrogen below the LEL.

The second configuration makes use of a ceramic doped with a noble metal as sensitive layer. Here, an exothermic reaction occurs, which elevates the temperature around the FBG and shifts its Bragg wavelength. The model used for simulation purposes is based on flux exchanges between the optical fiber and the surrounding medium. The sensor behavior can then be predicted as a function of some external parameters such as the ambient temperature or the relative humidity level of the air. This model is consistent with results reported experimentally and confirms that, once a threshold is reached, the sensitivity of the sensors ranges around 200°C (or 20 nm in terms of Bragg wavelength shift) per percent hydrogen concentration.

REFERENCES

Appel C., Mantzaras J., Schaeren R., Bombach R., and Inaen A. (2004) Catalytic combustion of hydrogen-air mixtures over platinum: Validation of hetero/homogeneous chemical reaction schemes. *Clean Air* **5**, 21–44. DOI: 10.1615/InterJEnerCleanEnv.v5.i1.20

Arregui F.J., Matias I.R., and Claus R.O. (2003) Optical fiber gas sensors based on hydrophobic alumina thin films formed by the electrostatic self-assembly monolayer process. *IEEE Sensors J.* **3**, 56–61. DOI: 10.1109/JSEN.2003.810109

Barlow A. and Payne D. (1983) The stress-optic effect in optical fibers. *J. Quant. Electron.* **19**, 834–839. DOI: 10.1109/JQE.1983.1071934

Bévenot X., Trouillet A., Veillas C., Gagnaire H., and Clément M. (2002) Surface plasmon resonance hydrogen sensor using an optical fibre. *Meas. Sci. Technol.* **13**, 118–124. DOI: 10.1088/0957-0233/13/1/315

Buric M., Chen K.P., Bhattarai M., Swinehart P.R., andMaklad M. (2007) Active fiber Bragg grating hydrogen sensors for all-temperature operation. *IEEE Photonic Technol. Lett.* **19**, 255–257. DOI: 10.1109/LPT.2006.888973

Butler M.A. and Ginley D.S. (1988) Hydrogen sensing with palladium-coated optical fibers. *J. Appl. Phys.* **64**, 3706–3712. DOI: 10.1063/1.341414

Caucheteur C., Debliquy M., Lahem D., and Mégret P. (2008a) Catalytic fiber Bragg grating sensor for hydrogen leak detection in air. *IEEE Photonic Technol. Lett.* **20**, 96–98. DOI: 10.1109/LPT.2007.912557

Caucheteur C., Debliquy M., Lahem D., and Mégret P. (2008b) Hybrid fiber gratings coated with a catalytic sensitive layer for hydrogen sensing in air. *Opt. Express* **16**, 16854–16859. DOI: 10.1364/OE.16.016854

Churchill S.W. (1983) Free convection around immersed bodies, In: Fritz W. (ed.), *Heat Exchanger Design Handbook*. Hemisphere, London, UK.

Debliquy M., Lahem D., Caucheteur C., and Mégret P. (2009) Finding hydrogen leaks by means of the fiber Bragg gratings technology, In: Lethien C. (ed.), *Optical Fiber New Developments*. Intech, e-book.

Kashyap R. (1999) *Fiber Bragg Gratings*. Academic Press, San Diego, CA.

Lewis F.A. (1967) *The Palladium Hydrogen System*. Academic Press, London, UK.

Othonos A. and Kalli K. (1999) *Fiber Bragg Gratings: Fundamentals and Applications in Telecommunications and Sensing*. Artech House, Norwood, MA.

Perry R.H. and Green D. (1984) *Perry's Chemical Engineers' Handbook*. McGraw-Hill, New-York.

Sekimoto S., Nakagawa H., Okazaki S., Fukuda K., Asakura S., Shigemori T., and Takahashi S.A (2000) Fiber-optic evanescent-wave hydrogen gas sensor using palladium-supported tungsten oxide. *Sens. Actuators B* **66**, 142–145. DOI: 10.1016/S0925-4005(00)00330-0

Sutapun B., Tabib-Azar M., and Kazemi A. (1999) Pd-coated elastooptic fiber optic Bragg grating sensors for multiplexed hydrogen sensing. *Sens. Actuators B* **60**, 27–34. DOI: 10.1016/S0925-4005(99)00240-3

Villatoro J., Luna-Moreno D., and Monzon-Hernandez D. (2005) Optical fiber hydrogen sensor for concentrations below the lower explosive limit. *Sens. Actuators B* **110**, 23–27. DOI: 10.1016/j.snb.2005.01.009

Yariv A. and Yeh P. (2003) *Optical Waves in Crystals: Propagation and Control of Laser Radiation*. John Wiley, New York.

Yu F.T.S., Yin S., and Yu Y.T.S. (2002) *Fiber Optic Sensors*. Marcel Dekker, New York. DOI: 10.1201/9780824744571

CHAPTER 5

Simulation and Modeling of Surface Plasmon Resonance–Based Fiber Optical Sensors

Banshi D. Gupta
Rajan Jha

1. INTRODUCTION AND HISTORICAL BACKGROUND OF SURFACE PLASMONS

The phenomenon of surface plasmon resonance (SPR) originated about a century ago, when Zenneck formulated a surface wave solution to Maxwell's equations and demonstrated the presence of a surface electromagnetic wave at the interface of two media, one a "lossy" dielectric and the other a "loss-free" one (Zenneck 1907). According to Zenneck, it was the imaginary part of the dielectric function of the lossy medium that was responsible for the binding of the electromagnetic wave to the interface. Later, it was reported that the field amplitudes of the surface waves varied inversely as the square root of the horizontal distance from the source dipole (Sommerfeld 1909). Furthermore, the field decayed exponentially with distance perpendicular to the interface. However, the real progress on SPR phenomena was made when the existence of surface plasmon waves at a metal–dielectric interface was demonstrated (Ritchie 1957). The excitation of surface plasmons at the metal–dielectric interface was also observed using electrons

DOI: 10.5643/9781606503201/ch5

(Powell and Swan 1960). At nearly the same time, the coupling of electromagnetic radiation to surface plasmons was also shown (Stern and Ferrell 1960). Further, the dispersion relations for electromagnetic surface waves at metal surfaces were also derived. Later, to enable the coupling of electromagnetic light waves with surface electromagnetic waves, the attenuated total reflection (ATR) prism coupling method was devised (Otto 1968), as shown in Figure 5.1a. The coupling was carried out using evanescent waves. Because of the finite gap between the prism base and the metal layer, the device was more suited to delicate surfaces, which one would not like to be damaged or touched by the prism. It was therefore more useful for the study of single crystal surfaces. This configuration was later modified for the excitation of surface plasmons and is now known as the Kretschmann configuration, as shown in Figure 5.1b (Kretschmann and Reather 1968).

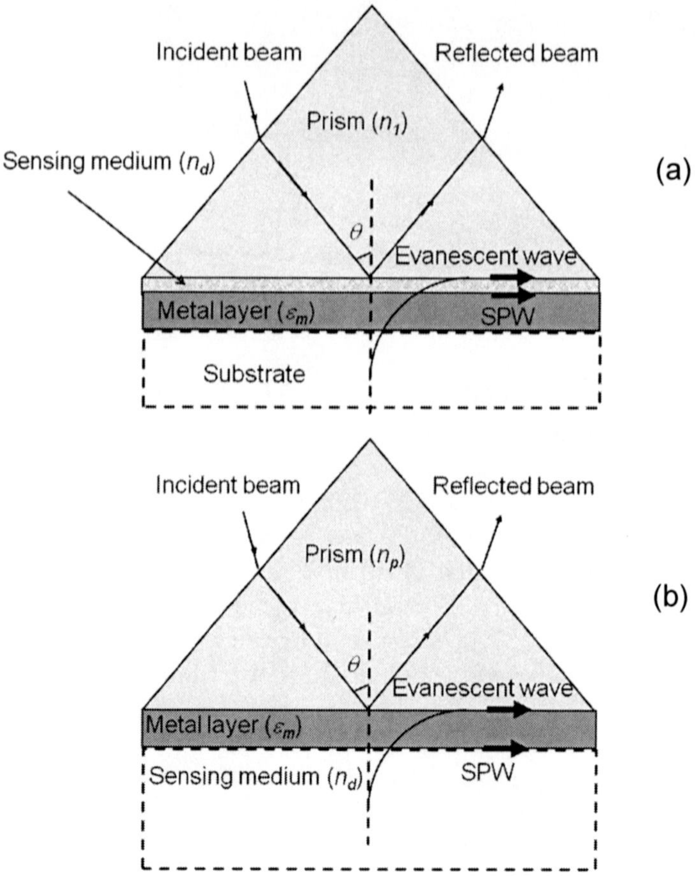

Figure 5.1. (a) Otto and (b) Kretschmann configurations for excitation of surface plasmons at metal–dielectric interface.

In order to understand surface plasmons, we need to understand some of the fundamental equations that govern the propagation of surface plasmons. In a homogeneous and isotropic medium with charge density $\rho = 0$ and current density $\vec{J} = 0$, the wave equations for electric and magnetic fields are given by the following Maxwell equations:

$$\nabla^2 \vec{E} - \frac{\mu\varepsilon}{c^2}\frac{\partial^2 \vec{E}}{\partial t^2} = 0 \tag{5.1}$$

$$\nabla^2 \vec{H} - \frac{\mu\varepsilon}{c^2}\frac{\partial^2 \vec{H}}{\partial t^2} = 0 \tag{5.2}$$

where c is the speed of light and μ and ε are permeability and permittivity, respectively. The solutions to these equations are plane waves, and for the electric field the solution is given by

$$\vec{E}(\vec{r},t) = \vec{E_0} e^{i(\vec{k}.\vec{r} - \omega t)} \tag{5.3}$$

Equation (5.3) describes an electric field that propagates in the $\hat{k}$ direction with angular frequency ω. The magnitudes of $\hat{k}$ and ω are related by the following dispersion relation:

$$k^2 = \frac{\omega^2}{c^2}\mu\varepsilon \tag{5.4}$$

For light in free space, the wave vector (k_0) is given by $k_0 = 2\pi/\lambda_0$; λ_0 is the wavelength in free space. In the case of a conductor, there is a dense assembly of negatively charged free electrons and also an equally charged positive ion lattice. Since positive ions have very large mass compared to electrons, the positive ions can be replaced by a positive constant background. If an external field is applied to the conductor, movement of electrons takes place and the density of the free electrons is locally reduced. The free electrons are no longer screened by the positive background. Instead, they are attracted by the positive ion background, resulting in their movement to the positive region. At the same time, the Coulomb repulsion among the moving free electrons produces motion in the opposite direction. The resultant of these two motions in opposite directions sets up the longitudinal oscillations known as "plasma oscillations" among the free electrons. A quantum of the plasma oscillation is called as plasmon. This type of coherent oscillation is supported at the metal–dielectric interface, and these charge density oscillations along the metal–dielectric interface are known as surface plasmon oscillations. The quantum of these oscillations is referred to as a surface plasmon wave (SPW) or a surface plasmon mode. A surface plasmon wave is transverse-magnetic (TM) polarized, with an electric field decaying exponentially in the metal as well as in the dielectric medium. As a result, the field has its maximum at the metal–dielectric interface. Using Maxwell's equation, it can be shown that the SPW propagation constant (k_{sp}) is continuous through the metal–dielectric interface and is given by (Ritchie et al. 1968; Reather 1998)

$$k_{sp} = \frac{\omega}{c}\sqrt{\frac{\varepsilon_m \varepsilon_d}{\varepsilon_m + \varepsilon_d}} \tag{5.5}$$

where ε_m and ε_d represent the dielectric constants of the metal and dielectric medium, respectively.

Since the dielectric constant of the dielectric medium is weakly dispersive compared to that of the metal, the metal's dispersive nature plays an important role in determining the characteristics of the surface plasmon. At the interface, a trapped surface mode tied to the oscillating surface charge density propagates. For such charges to be sustained, the electric field normal to the interface must change sign across the interface. This means that for a guided mode, the real parts of the dielectric constants of the two media must be of opposite sign. The dielectric constant of the dielectric medium is usually positive, which implies that the real part of the dielectric constant of metal must be negative, and this is indeed the case. Moreover, from the free electron theory in metal, there is a limiting frequency known as the plasma frequency (ω_p), above which the metal is no longer metallic. The following equation gives the variation of dielectric constant with the wavelength of light in a metal as given by the Drude model:

$$\varepsilon_m(\lambda) = \varepsilon_m' + i\varepsilon_m'' = 1 - \frac{\lambda^2 \lambda_c}{\lambda_p^2(\lambda_c + i\lambda)} \tag{5.6}$$

where ε_m' and ε_m'' are the real and imaginary parts of the dielectric constant of the metal, respectively, and λ_p and λ_c denote the plasma wavelength and collision wavelength, respectively. The existence of surface plasmons at a metal–dielectric interface is confined to the region longer than a certain critical wavelength, which depends on the plasma frequency and is specific to certain metal. For surface plasmon–active metals, the critical wavelength lies in the ultraviolet or visible region.

For a given frequency, the wave vector of a surface plasmon is always greater than that of a photon in the medium bounding metal; this means that one cannot use incident plane waves to couple to surface plasmon modes. To couple, the wave vector of the excitation light should be increased to the level that it becomes equal to the wave vector of the surface plasmon. When these two wave vectors become equal, the excitation light transfers its energy to the surface plasmons.

To increase the wave vector of excitation radiation, one uses a prism, grating, or waveguide to couple the light to plasmons formed at the metal–dielectric interface. Usually, the metallic layer used is either silver or gold along with copper and aluminum. Gold and silver have been mainly used in sensing utilizing surface plasmon phenomena. Gold is chemically more stable than silver, which is prone to oxidation. In the past, aluminum was used as a plasmon-active metal for sensing studies (Krane and Reather 1976). It has been shown that aluminum may provide a sharper surface plasmon resonance curve than Ag, Au, and Cu (Sharma and Gupta 2007). Practically speaking, Al also has poor chemical stability, which makes it difficult to realize a reliable sensor for practical applications. However, a new structure of bimetallic layers (with a few nanometers of Au as the outer layer) may not only protect the thicker Ag layer from oxidation but also may enhance the overall performance compared to a single layer of Au (Sharma and Gupta 2007).

2. EXCITATION OF SURFACE PLASMONS AND COUPLING TECHNIQUES

The maximum propagation constant of a light wave at frequency ω propagating through air is given by

$$k = \frac{\omega}{c} \tag{5.7}$$

Since $\varepsilon_m < 0$ and $\varepsilon_d > 0$, for a given frequency and the same polarization state, the propagation constant of a surface plasmon in Eq. (5.5) is greater than that of the light wave in air (see Figure 5.2). Hence, direct light cannot excite surface plasmons at a metal–dielectric interface, which are referred to as nonradiative surface plasmons. As mentioned above, to excite surface plasmons, the momentum and hence the wave vector of the excitation light in the dielectric medium should be increased. The extra momentum can be imparted if the coupling of light wave is carried out using a prism, waveguide, or grating (Homola et al. 1999). We discuss these below.

2.1. PRISM COUPLING

In prism coupling, a thin film of metal is deposited on the base of a high-refractive-index prism and the dielectric is kept on the other side of the metal film. The p-polarized light is incident on the prism–metal interface through one of the faces of the prism. The reflected light through the other face of the prism is recorded using a detection system. When the angle of incidence of the light is greater than the critical angle, an evanescent wave is produced at the interface, which propagates along the interface. The evanescent wave excites a TM-polarized charge density wave, a

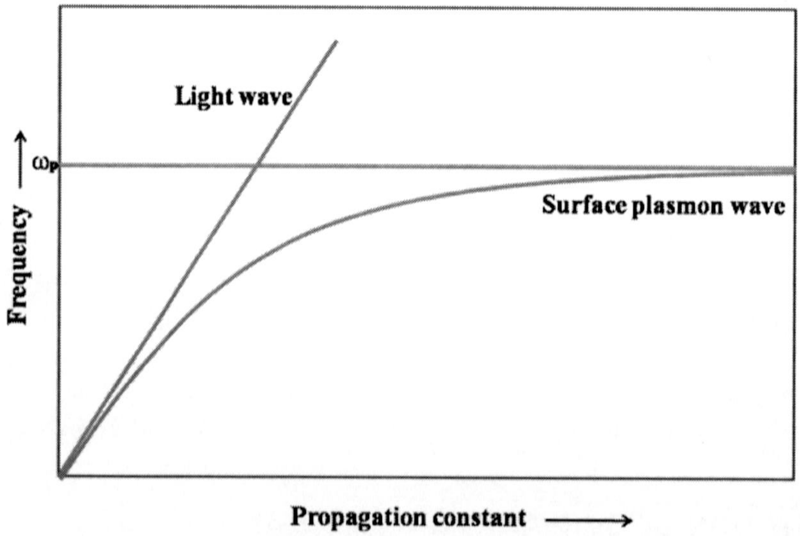

Figure 5.2. Dispersion curves for direct light wave and surface plasmon wave.

surface plasmon wave, along the metal–dielectric interface. Although these waves travel parallel to the surface, they also extend evanescently into the dielectric, and as soon as the propagation constants of an SPW and an evanescent wave match, a sharp dip in reflected light intensity is observed. This matching of propagation constants is known as the resonance condition and is given as

$$k_0 n_1 \sin\theta_{\mathrm{SPR}} = k_0 \left(\frac{\varepsilon_m' n_d^2}{\varepsilon_m' + n_d^2} \right)^{1/2} \tag{5.8}$$

The term on the left-hand side of Eq. (5.8) is the propagation constant of the evanescent wave, which is generated under attenuated total reflectance (ATR) of the light incident at a resonance angle θ_{SPR} through a light-coupling prism of refractive index n_1. The right-hand term is the real part of the SPW propagation constant (k_{sp}), with ε_m' as the real part of the metal dielectric constant (ε_m) and n_d as the refractive index of the sensing (dielectric) medium. The excitation of surface plasmons occurs when the wave vector of the propagation constant of the evanescent wave exactly matches that of the surface plasmons of similar frequency and state of polarization.

In prism-based SPR sensing, the angular interrogation is performed by keeping the incident-light wavelength constant and varying the angle of incidence of the light beam. According to Eq. (5.8), when there is a change in refractive index of the sensing or the dielectric medium, the value of θ_{SPR} has to change in order to satisfy the resonance condition, because all other parameters are constant at a fixed wavelength. This shift in resonance angle at a given wavelength can be conveniently monitored under angular interrogation with standard prisms such as silica, BK7, SF6, etc. BIAcore and other manufacturers have used prism coupling for design of sensors based on surface plasmon resonance with high resolution (Karlsson and Stahleberg 1995). Generally, SPR measurements with conventional prisms do not allow detection in the infrared wavelength region, which requires attention due to its many environmental, medical, and security applications. Moreover, the SPR-based structures in IR have substantially different parameters of excitation and support certain advantages for sensing applications in terms of high probe depth (Nelson et al. 1999) and more accurate determination of SPR dip (Johansen et al. 2000). SPR sensing in the near-IR with a silicon prism has showed interesting results for further studies in the IR regime (Jha and Sharma 2009a, 2010). In this view, chalcogenide glass based on the Group VI chalcogen elements (sulfur, selenium, and tellurium) is the potential candidate for designing SPR sensors to operate in IR region, owing to their transparent behavior in the near- to mid-IR region (Bureau et al. 2004). Chalcogenide glasses are formed by adding materials such as arsenic, gallium, germanium, or antimony to the above-mentioned chalcogenide elements and have been widely used for various applications (Paivasaari et al. 2007), including sensing (Jha and Sharma 2009b, 2011). An Al-based chalcogenide glass sensor has been found to have enhanced performance (Jha and Sharma 2009c).

2.2. WAVEGUIDE COUPLING

The process of exciting SPR using a waveguide is similar to that in the Kretschmann configuration. A light wave is guided by either a single or a multilayer (slab or channel) waveguide having a metal overlayer. After entering, it evanescently penetrates through the metal layer. Provided that the surface plasmon and the guided mode can be phase-matched, the incident light excites surface plasmons at the outer interface of the metal layer. The use of optical waveguides in exciting SPR sensors has some attractive features, including simple control of the optical path in the sensor system, and advantages in terms of ruggedness and compactness due to the availability of the latest microfabrication techniques. One of the first reports on integrated optical waveguide SPR sensors was that of Lambeck and co-workers (Lambeck 1992). Since then, various groups have contributed to the development of SPR sensors utilizing waveguide coupling, owing to the obvious advantages as mentioned above. SPR-based integrated waveguides using slabs (Homola et al. 1997) and channels (Mouvet et al. 1997) have been reported. Various possibilities for tuning the operating range of the sensor have been explored, such as a high-refractive-index overlayer (Suzukia et al. 2005) and complex multilayer structures (Weiss et al. 1996). However, all the approaches that introduced additional layers were found to yield less sensitive SPR sensing devices because of a relatively lower concentration of electromagnetic field in the analyte. Recently, a detailed design principle and propagation characteristics of channel Bragg-plasmon-coupled-waveguides (Srivastava et al. 2010), and highly accurate and sensitive SPR sensors based on channel photonic crystal waveguides, have been reported (Srivastava et al. 2011).

2.3. GRATING COUPLING

Excitation of surface plasmons can also be achieved using the diffraction of light by an optical grating. Consider a case in which the metal–dielectric interface is periodically modified, and if the light wave is incident from a dielectric medium on a metallic grating, then the incident optical wave is diffracted and forms a series of beams (sometimes called orders of the grating) at different angles. These diffracted waves couple with the surface plasmon wave if the propagation constant of the diffracted wave matches the propagation constant of the surface plasmon wave. The theoretical modeling of grating-based SPR sensing structures is more complex than that for planar prism-based systems (Moharam and Gaylord 1986). Further, one of the major drawbacks of grating-based SPR systems is that the light beam is incident through the sample solution and therefore the analyte and flow cell need to be optically transparent, which sometimes is difficult to handle and operate in practice. However, in contrast to prism-based coupling techniques, accurate control of the thickness of the plasmon-active metal layer is not required in grating-based SPR sensors. The chief advantage of grating-coupled SPR sensors is that they can be produced by mass replication technologies such as injection molding and hot embossing. These technologies have opened the door for producing low-cost, high-throughput SPR platforms for label-free monitoring of biomolecular interactions (Dostálek et al. 2005). Gold-based SPR grating sensors have been reported for the monitoring of biomolecular interactions in aqueous environments (Cullen and Lowe 1990).

3. *N*-LAYER MODEL FOR DIFFERENT CONFIGURATIONS

3.1. PRISM-BASED ANGULAR INTERROGATION

To calculate an expression for the amplitude reflection coefficient from a multilayer structure consisting of a prism, metal, and a sensing or dielectric layer for a p-polarized incident beam, a matrix method for the N-layer model is used (Gupta and Sharma 2005). The layers are assumed to be stacked along the vertical direction (Figure 5.3).

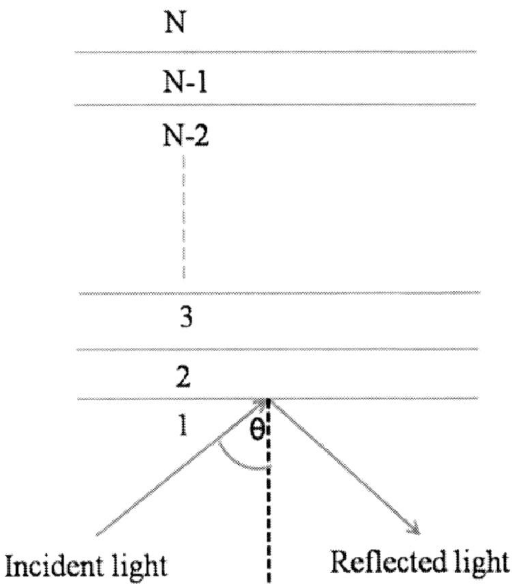

Figure 5.3. *N*-layer model.

The arbitrary medium layer is defined by thickness d_k, dielectric constant ε_k, permeability μ_k, and refractive index n_k. The tangential fields at the first boundary, $Z = Z_1 = 0$, are related to those at the final boundary, $Z = Z_{N-1}$ by

$$\begin{bmatrix} U_1 \\ V_1 \end{bmatrix} = M \begin{bmatrix} U_{N-1} \\ V_{N-1} \end{bmatrix} \tag{5.9}$$

where U_1 and V_1 are the tangential components of the electric and magnetic fields, respectively, at the boundary of the first layer. U_{N-1} and V_{N-1} are the corresponding fields at the boundary of the *N*th layer. Here, *M* is known as the characteristic matrix of the combined structure and is given by

$$M = \prod_{k=2}^{N-1} M_k = \begin{bmatrix} M_{11} & M_{12} \\ M_{21} & M_{22} \end{bmatrix} \tag{5.10}$$

with

$$M_k = \begin{bmatrix} \cos\beta_k & (-i\sin\beta_k)/q_k \\ -iq_k\sin\beta_k & \cos\beta_k \end{bmatrix} \tag{5.11}$$

where

$$q_k = \left(\frac{\mu_k}{\varepsilon_k}\right)^{1/2} \cos\theta_k = \frac{\left(\varepsilon_k - n_1^2 \sin^2\theta_1\right)^{1/2}}{\varepsilon_k} \tag{5.12}$$

and

$$\beta_k = \frac{2\pi}{\lambda} n_k \cos\theta_k \left(z_k - z_{k-1}\right) = \frac{2\pi d_k}{\lambda}\left(\varepsilon_k - n_1^2 \sin^2\theta_1\right)^{1/2} \tag{5.13}$$

If there are three layers, then $N = 3$. Therefore, accordingly, n_1 is the refractive index of the prism material, n_2 is the refractive index of the metal layer, and n_3 is the refractive index of the sensing layer. Thus, transfer matrices M_k are calculated as per the above equations for different layers. The different matrices are multiplied to obtain the final transfer matrix M of the multilayer setup to calculate the amplitude reflection coefficient (r_p) for a p-polarized incident wave as follows:

$$r_p = \frac{\left(M_{11} + M_{12}q_4\right)q_1 - \left(M_{21} + M_{22}q_3\right)}{\left(M_{11} + M_{12}q_3\right)q_1 + \left(M_{21} + M_{22}q_3\right)} \tag{5.14}$$

In this equation, q_1 is the corresponding term for the prism and q_3 is the corresponding term for the sensing layer. Both these terms are calculated from Eq. (5.12) by inputting their corresponding refractive index values (i.e., n_1 and n_3), dielectric constants (i.e., ε_1 and ε_3) and incident-angle values. Finally, reflectance (R_p) for p-polarized light as a function of angle of incidence is obtained using the following equation:

$$R_p = \left|r_p\right|^2 \tag{5.15}$$

3.2. FIBER-BASED WAVELENGTH INTERROGATION

In the case of fiber optic SPR technique, the coupling prism used in the basic SPR theory is replaced by the core of a multimode optical fiber. A spectral interrogation method is generally used in fiber optic SPR sensors because all the guided modes are launched in the fiber. In this section we will discuss the N-layer model under the following different launching conditions:

1. Simple geometry: on-axis excitation
2. Meridional and skew rays: off-axis excitation
3. Diffuse source as a light source

3.2.1. Simple Geometry: On-Axis Excitation

In all the cases we shall discuss, highly multimoded plastic-clad silica (PCS) fibers are used. For calculating the normalized transmitted power using the N-layer model, it is assumed that the cladding from a small portion of the fiber is removed and the unclad core is coated with a SPR-active metal layer, which is further surrounded by a dielectric sensing layer. The light from a polychromatic source (in the case of spectral interrogation) is launched into one of the ends of the optical fiber, and the spectrum of the light transmitted after passing through the SPR sensing region is detected at the other end.

The amplitude reflection coefficient for a p-polarized incident beam is calculated as explained in the previous section for the N-layer model. The N-layer model is very useful because it can be applied to a system containing any number of layers. Also, the calculations are very easy, accurate, and precise, due to the absence of approximations. The reflectance (R_p) for p-polarized light is given by Eq. (5.15). However, to calculate the power transmitted through the optical fiber, one has to consider all guided rays launching for simulations. In this kind of launching, the light from a collimated source is focused onto the endface of the fiber at the axial point with the help of a microscope objective (lens). The angular power distribution of all-guided rays launching depends on the angle (θ) of the ray with the normal to the core–cladding interface and is given by (Snyder and Love 1991; Gupta et al. 1993)

$$dP \propto \frac{n_1^2 \sin\theta \cos\theta}{\left(1 - n_1^2 \cos^2\theta\right)^2} d\theta \tag{5.16}$$

where n_1 is the refractive index of the fiber core. In order to determine the effective transmitted power, the reflectance (R_p) for a single reflection is raised to the power of the number of reflections the specific propagating angle undergoes with the core–cladding interface. Therefore, for p-polarized light, the generalized expression for the normalized transmitted power, P_{trans}, in a fiber optic SPR sensor is

$$P_{\text{trans}} = \frac{\displaystyle\int_{\theta_{cr}}^{\pi/2} R_p^{N_{\text{ref}}(\theta)} \frac{n_1^2 \sin\theta \cos\theta}{\left(1 - n_1^2 \cos^2\theta\right)^2} d\theta}{\displaystyle\int_{\theta_{cr}}^{\pi/2} \frac{n_1^2 \sin\theta \cos\theta}{\left(1 - n_1^2 \cos^2\theta\right)^2} d\theta} \tag{5.17}$$

where

$$N_{ref}(\theta) = \frac{L}{2\rho \tan\theta} \tag{5.18}$$

and

$$\theta_{cr} = \sin^{-1}\left(\frac{n_{cl}}{n_1}\right) \tag{5.19}$$

In Eq. (5.18), N_{ref} represents the total number of reflections performed by a ray making angle θ with the normal to the core–metal layer interface in the sensing region, while L and ρ represent the length of the exposed sensing region and the fiber core radius, respectively. In Eq. (5.19), θ_{cr} is the critical angle of the given fiber and n_{cl} is the refractive index of the fiber cladding. To find the resonance wavelength, a curve between the transmitted output power and the wavelength is plotted.

3.2.2. Meridional and Skew Rays: Off-Axis Excitation

If the collimated beam is focused at some point other than the axial point on the input face of the fiber, as shown in Figure 5.4, then both meridional and skew rays will be present in the fiber. This has been modeled and simulated (Dwivedi et al. 2007). The transmitted power in the presence of skew rays is given by the following relation:

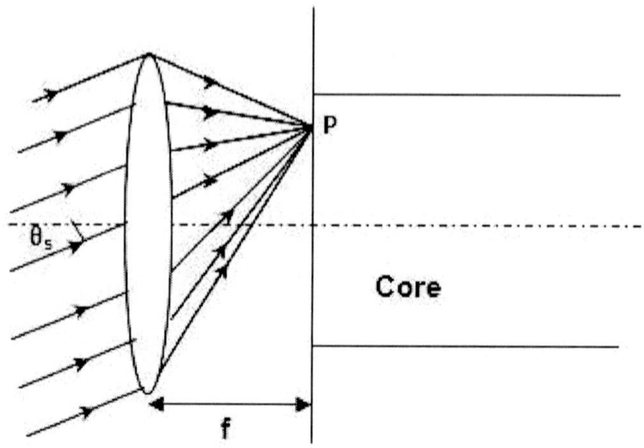

Figure 5.4. Off-axis excitation.

$$P_{\text{trans}} = \frac{\displaystyle\int_0^{\alpha_{\max}} \int_{\theta_{cr}}^{\pi/2} R_p^{N_{\text{ref}}(\theta,\alpha)} \frac{n_1^2 \sin\theta\cos\theta}{\left(1-n_1^2\cos^2\theta\right)^2} \cos\theta_s \, d\theta \, d\alpha}{\displaystyle\int_0^{\alpha_{\max}} \int_{\theta_{cr}}^{\pi/2} \frac{n_1^2 \sin\theta\cos\theta}{\left(1-n_1^2\cos^2\theta\right)^2} \cos\theta_s \, d\theta \, d\alpha} \tag{5.20}$$

where

$$N_{\text{ref}}(\theta,\alpha) = \frac{L}{2\rho\cos\alpha\tan\theta} \tag{5.21}$$

Here, α is the skewness angle in the sensing region and θ_s is the angle the collimated beam makes with the axis of the lens. For meridional rays, α is zero.

3.2.3. *Diffuse Source as a Light Source*

In this case we discuss the launching of light in the fiber using a diffuse source such as a light-emitting diode (LED). The power distribution inside the fiber for the diffuse source is given by (Snyder and Love 1991; Gupta and Singh 1994a)

$$dP \propto n_1^2 \sin\theta\cos\theta \, d\theta \tag{5.22}$$

For SPR-based fiber optic sensors utilizing a diffuse source, there will be both meridional and skew rays. The power transmitted through the fiber is given by (Singh et al. 2010)

$$P_{\text{trans}} = \frac{\displaystyle\int_0^{\pi/2} d\alpha \int_{\theta_{cr}}^{\pi/2} R_p^{N_{\text{ref}}(\theta,\alpha)} n_1^2 \sin\theta\cos\theta \, d\theta}{\displaystyle\int_0^{\pi/2} d\alpha \int_{\theta_{cr}}^{\pi/2} n_1^2 \sin\theta\cos\theta \, d\theta} \tag{5.23}$$

where α is the skewness angle of the ray, R_p is the reflection coefficient as discussed in the previous section, and N_{ref} is the number of reflections a ray undergoes with the core–cladding interface in the sensing region. Equation (5.23) includes the transmission of both skew and meridional rays. For propagation of meridional rays only inside the fiber, α is equal to zero and the power distribution inside the fiber is given by Eq. (5.17). The number of reflections for the guided ray with angles (θ, α) in length L of the sensing region is given by Eq. (5.21).

4. SENSING PRINCIPLE OF SPR: PERFORMANCE PARAMETERS

The sensing principle of SPR sensors is based on Eq. (5.8). For a given frequency of the light source and the dielectric constant of the metal film, one can determine the dielectric constant (ε_d) of the sensing layer adjacent to the metal layer by knowing the value of either the resonance angle or the resonance wavelength in the cases of angular interrogation and spectral interrogation, respectively.

The performance of a SPR sensor is analyzed with the help of two parameters: sensitivity and detection accuracy or signal-to-noise ratio (SNR). For the best performance of the sensor, both these parameters should be as high as possible. The sensitivity of a SPR sensor depends on how much the resonance angle or resonance wavelength shifts with a change in refractive index of the sensing layer. If the shift is large, the sensitivity is large. Figure 5.5 shows a plot of reflectance as a function of angle of the incident light beam for sensing layers with refractive indices n_d and $n_d + \delta n_d$. An increase in refractive index by δn_d shifts the resonance curve by $\delta\theta_{\text{res}}$ angle (Figure 5.5a) or $\delta\lambda_{\text{res}}$ (Figure 5.5b). The sensitivity of a SPR sensor with angular interrogation is defined as

$$S_n = \frac{\delta\theta_{\text{res}}}{\delta n_d} \qquad (5.24)$$

The sensitivity of SPR-based sensors has been an important issue since the beginning of research in this area (Yeatman 1996; Homola 1997). Enhancement of the sensitivity of SPR-based sensors has been achieved by changing the shape of the probe (Verma and Gupta 2008; Verma et al. 2007, 2008) or by adding a high-refractive-index layer over the metal layer (Lahav et al. 2008; Bhatia and Gupta 2011; Verma et al. 2011). The detection accuracy or the SNR of a SPR sensor depends on how accurately and precisely the sensor can detect the resonance angle and, hence, the refractive index of the sensing layer. This accuracy of detection of the resonance angle depends further on the width of the SPR curve. The narrower the SPR curve, the higher is the detection accuracy. Therefore, if $\delta\theta_{0.5}$ is the angular width of the SPR response curve, corresponding to 50% reflectance, the detection accuracy of the sensor can be assumed to be inversely proportional to $\delta\theta_{0.5}$. The SNR of a SPR sensor with angular interrogation is thus defined as

$$\text{SNR} = \frac{\delta\theta_{\text{res}}}{\delta\theta_{0.5}} \qquad (5.25)$$

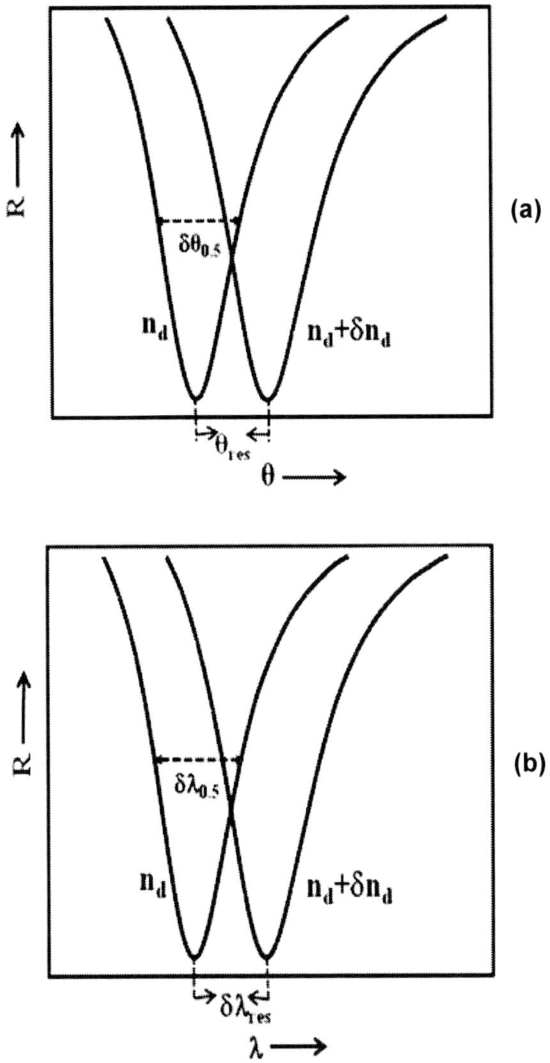

Figure 5.5. (a) Angular interrogation and (b) spectral interrogation for SPR system.

Another important parameter of an SPR sensor is its operating range, or the values of the parameters that can be measured by the sensor. While the operating range of intensity measurement–based SPR sensors is naturally limited due to the limited width of the SPR dip, the operating range of angular and wavelength interrogation–based SPR sensors may be made much wider. In principle, the operating range of these sensors is determined by the detection system, more specifically, by the angular or spectral range covered by the optical system–angular position detector array or spectrum analyzer, respectively. In other words, it is the range of the sensing-layer refractive index (n_d) that a sensor can detect for a given angular or wavelength range.

5. FIBER OPTIC SPR SENSORS

Prism-based SPR sensing devices have a number of drawbacks. Among these are their bulky size and the presence of various (moving) optical and mechanical parts, which makes them difficult to optimize and commercialize on a large scale. Further, they cannot be used for remote sensing. SPR probes, however, can be miniaturized by using optical fibers rather than prisms. One of the important advantages of using optical fiber is the small diameter of its core, which allows it to be used in very small areas. Other advantages of fiber optic SPR sensing configurations are simplified optical design and the capability for remote sensing. In fiber optic SPR sensors, the coupling prism used in the basic SPR concept is replaced by the core of a multimode optical fiber. Currently, optical fiber SPR probes present the highest level of miniaturization of SPR devices, allowing for chemical and biological sensing in inaccessible locations where the mechanical flexibility and the ability to transmit optical signals over a long distance make the use of optical fiber very attractive.

The fiber optic SPR sensor configuration comprises a high-refractive-index fiber core, a metal layer, and a low-refractive-index dielectric (sensing medium) as shown in Figure 5.6. Below we discuss the various constituents of the sensor.

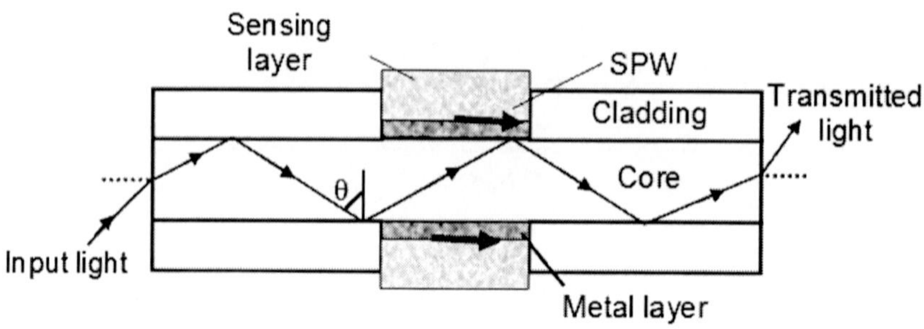

Figure 5.6. Illustration of a typical fiber optic SPR sensor. (Reprinted with permission from Sharma et al. 2007a. Copyright 2007 IEEE.)

5.1. FIBER CORE

For theoretical modeling of a SPR sensor utilizing wavelength interrogation, the refractive indices of the optical fiber, metal layer, and other layers, if present, are required. We generally consider a step-index multimode silica core fiber. The wavelength dependence of the refractive index (n_1) of silica is given by the following Sellmeier dispersion relation:

$$n_1(\lambda) = \sqrt{1 + \frac{a_1\lambda^2}{\lambda^2 - b_1^2} + \frac{a_2\lambda^2}{\lambda^2 - b_2^2} + \frac{a_3\lambda^2}{\lambda^2 - b_3^2}} \tag{5.26}$$

where λ represents the wavelength in micrometers and a_1, a_2, a_3, b_1, b_2, and b_3 are Sellmeier coefficients. The Sellmeier relation is valid for the 0.21- to 2.2-μm wavelength range. The values of these coefficients can be obtained from the literature (Sharma and Gupta 2006a). If a dopant is added in the core of the fiber, then the values of the Sellmeier coefficients change depending on the nature of the doping materials.

5.2. METAL LAYER

For the dispersion relation in the metal, we use the Drude model given by Eq. (5.6). Values of the plasma and collision wavelengths for different SPR-active metals can be obtained from the literature (Reather 1998; Gupta and Sharma 2005; Sharma and Gupta 2007).

5.3. SENSING MEDIUM

The dielectric medium in contact with the metal layer on the other side of the fiber core acts as the sensing medium in a fiber optic SPR sensor. The resonance angle or resonance wavelength is very sensitive to variation in the refractive index (or dielectric constant) of the sensing medium. An increase in the refractive index of the sensing medium increases the resonance angle or the wavelength. To know the refractive index of the sensing medium very accurately, the resonance parameter should be determined with utmost care. The refractive index of the sensing medium can also be changed indirectly, for example, by some chemical reaction due to an analyte. Therefore, SPR has been widely used for analyzing various chemical reactions or parameters such as bimolecular interactions, interactions between low-molecular-weight molecules, conformational changes of proteins, and sensing of pesticides and other biological entities. In all these cases a change in refractive index occurs, and hence these changes can be detected using SPR.

6. EVOLUTION OF FIBER OPTIC SPR SENSORS

The first use of fiber optic SPR sensors was proposed for the measurement of sucrose concentration in aqueous solution (Villuendas and Pelayo 1990). The sensitivity and dynamic range of the sensor were also reported. Later, Jorgenson and Yee (1993) reported theoretical as well as experimental work on a fiber optic chemical sensor based on SPR by fabricating the probe on a highly multimode optical fiber and using a white light source instead of a monochromatic one. They used the wavelength interrogation technique and formed a SPR-sensing probe by using a conventional polymer-clad silica fiber. The cladding was partly removed and a SPR-active metal layer was deposited symmetrically around the exposed unclad section of the fiber. This approach allowed construction of a miniaturized optical fiber SPR probe with limited interaction area, with a length of about 10 mm. The performance of the sensor was studied as a function of refractive index of the sensing medium, metal film thickness, and length of the sensing probe.

Fiber optic SPR sensors were also explored to monitor the deposition of a multi-layered cadmium arachidate Langmuir-Blodgett film (Mar et al. 1993). Shifts in resonance wavelength were observed as the number of monolayers was increased. Thus the film thickness could be determined from knowledge of the change in resonance wavelength. In other research, a theoretical approach was applied to provide a better understanding of the surface plasmon excitation in fiber optic sensors (Esteban et al. 2002). The asymmetric metal coating gives multiple resonance dips in the spectrum due to varying film thickness (Monzón-Hernández et al. 2004). The multiple dips increase the dynamic range of the sensor. In some cases, multilayers have been used as the SPR probe (Gupta and Sharma 2005). Recently, different theoretical schemes such as doping of the fiber core with different dopants (Sharma et al. 2007, Jha and Badenes 2009), a tapered fiber optic probe (Verma et al. 2007, 2008), and a U-shaped fiber optic probe (Verma and Gupta 2008) have been proposed and simulated to enhance the performance of the fiber optic SPR sensor.

7. OTHER PROBES: SENSITIVITY ENHANCEMENT

Depending on the application, various types of SPR-based fiber optic probes have been proposed. These probes have been designed by judiciously altering and optimizing the doping material, numerical aperture, core diameter, shape, size, etc., of the optical fiber to enhance the overall sensor performance. Here, we shall discuss doped optical fiber, tapered optical fiber, U-shaped optical fiber, and long-range optical fiber SPR probes.

7.1. DOPED OPTICAL FIBER PROBE

One of the parameters on which the SPR condition depends is the refractive index of the material of the fiber core. Therefore, if an optical fiber fabricated using

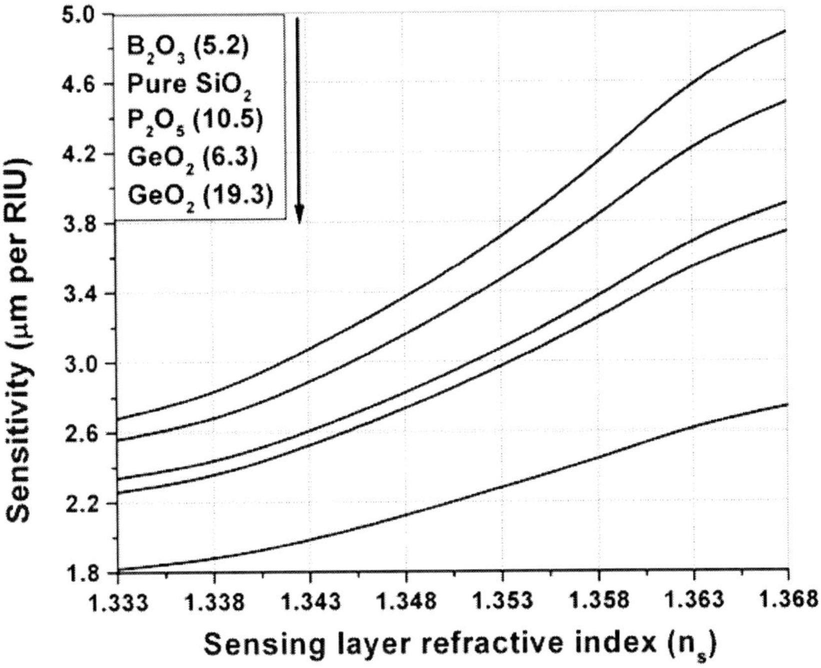

Figure 5.7. Variation of sensitivity with refractive index of the sensing medium. The core is of silica, while the dopants are GeO_2 (19.3), GeO_2 (6.3), P_2O_5 (10.5), and B_2O_3 (5.2). Numbers in brackets are molar concentrations of dopant in mol%. (Reprinted with permission from Sharma et al. 2007b. Copyright 2007 Elsevier.)

different dopants in the core is used, then it will either enhance or tune the performance of the sensor. In general, a pure silica fiber core is used for carrying out any analysis or experiment on SPR-based sensors. Different types of dopants, such as germanium oxide (GeO_2), boron oxide (B_2O_3), and phosphorus pentoxide (P_2O_5), are generally used for doping the fiber core. The dopant affects the sensitivity of the sensor while the detection accuracy remains independent of the doping material.

Figure 5.7 shows the variation of sensitivity with the refractive index of the sensing medium for various dopants and their concentrations (Sharma et al. 2007b). The simulations were carried out using the simple geometry described in Section 3.2.1. The sensitivity increases by almost 50% when the core is doped with B_2O_3 (5.2) as compared to GeO_2 (19.3). Therefore, fiber core doping with B_2O_3 is more useful than doping with GeO_2 or P_2O_5. Moreover, sensitivity of the SPR-based fiber optic sensor at a particular molar concentration (13.5) of B_2O_3 and GeO_2 is shown in Figure 5.8 (Jha and Badenes 2009). The sensitivity of B_2O_3-doped fiber shows better sensitivity for all values of sensing medium refractive index because, for a given refractive index of the sensing medium, the resonance

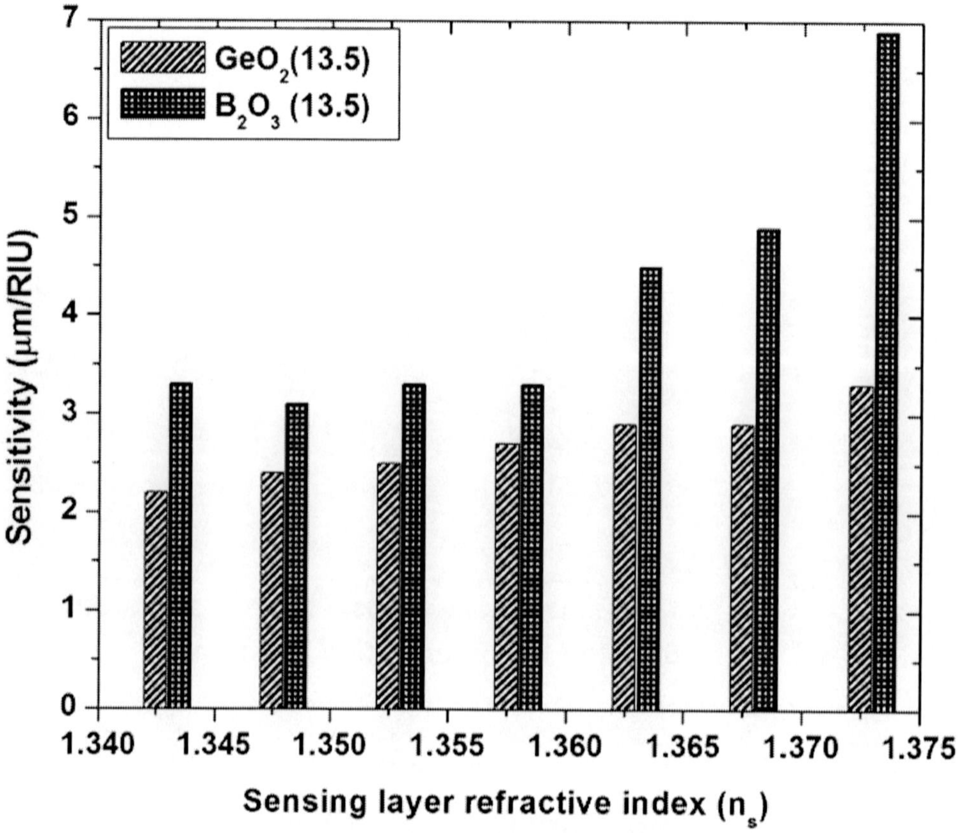

Figure 5.8. Variation of sensitivity with different values of sensing-layer refractive index. (Reprinted with permission from Jha and Badenes 2009. Copyright 2009 Elsevier.)

angle shifts more in the case of B_2O_3-doped fiber. However, doping the fiber core with a mixture of GeO_2 and B_2O_3 is also a very effective way to tune sensor performance. Figure 5.9 shows the results for a mixed case in which both the sensitivity and SNR are maximum (Jha and Badenes 2009). Thus SPR-based fiber optic sensors utilizing a mixture of dopants (GeO_2 and B_2O_3) may be advantageous in tuning the performance of the sensor to meet the requirements of the application.

7.2. TAPERED OPTICAL FIBER PROBE

Figure 5.10 shows a tapered fiber optic probe in which the diameter of the core first decreases linearly up to some length and then increases to its initial value along the axis of the fiber (Verma et al. 2008). Different taper profiles can be

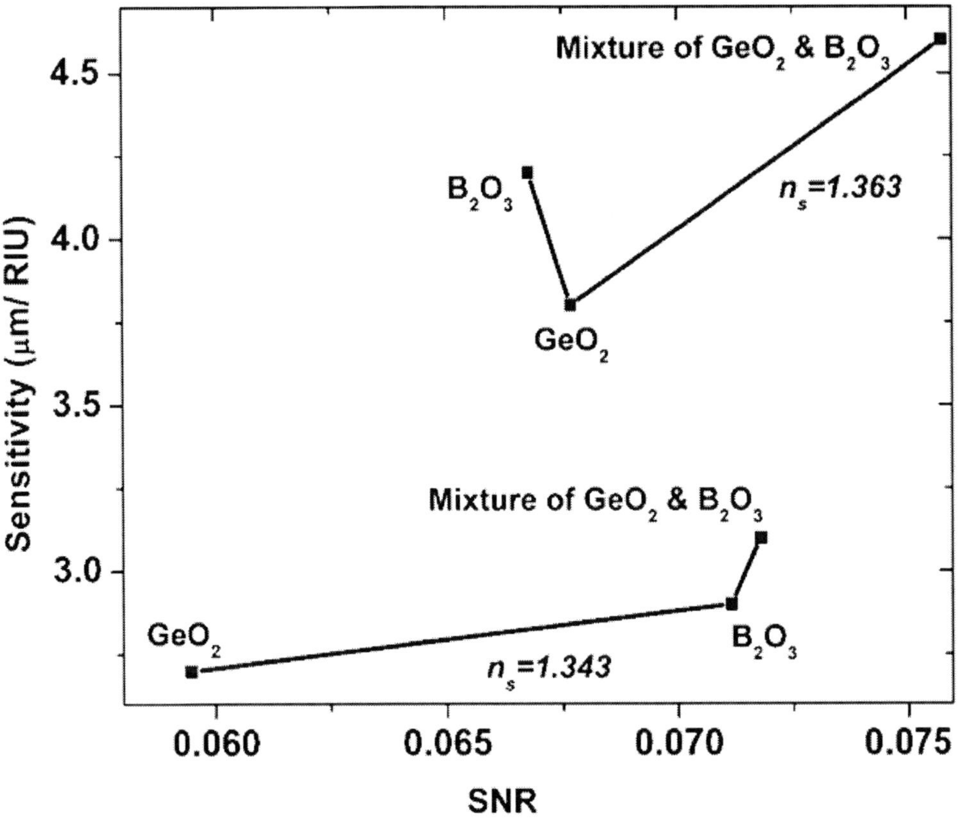

Figure 5.9. Variation of sensitivity and SNR for GeO_2, B_2O_3, and a mixture of GeO_2 and B_2O_3. (Reprinted with permission from Jha and Badenes 2009. Copyright 2009 Elsevier.

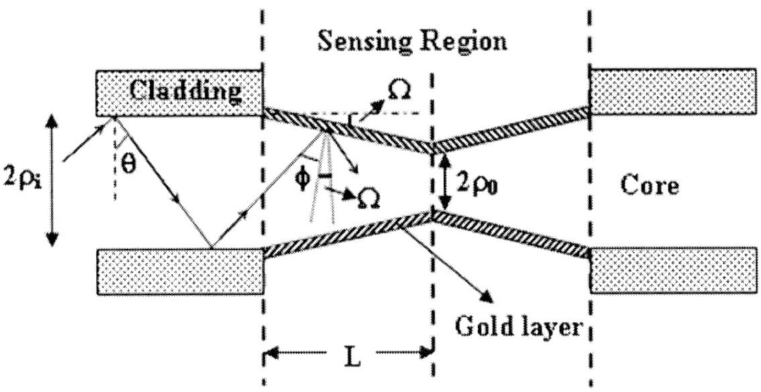

Figure 5.10. Tapered optical fiber. (Reprinted with permission from Verma et al. 2008. Copyright 2008 Elsevier.)

generated during fabrication. The method uses a gas burner and a pulling machine. The following taper profiles, $\rho(z)$, have been considered in the literature (Gupta and Singh 1994b; Verma et al. 2008).

Linear:
$$\rho(z) = \rho_i - \frac{z}{L}(\rho_i - \rho_0) \tag{5.27}$$

Exponential linear:
$$\rho_e(z) = (\rho_i - \rho_0)\left[e^{(-z/L)} - \frac{z}{L}(e^{-1})\right] + \rho_0 \tag{5.28}$$

Parabolic:
$$\rho_P(z) = \left[\rho_i^2 - \frac{z}{L}(\rho_i^2 - \rho_0^2)\right]^{1/2} \tag{5.29}$$

where ρ_i and ρ_0 are the radii of the taper at the input and output ends, respectively, $\rho(z)$ is the taper radius at a distance z from the input end of the taper, and L is the length of the taper. The guided rays in the uniform core fiber enter the sensing tapered region as shown in Figure 5.10. For adiabatic (slow) tapering, the angle θ of the ray transforms into angle $\phi(z)$ in the tapered region according to the following relation:

$$\phi(z) = \cos^{-1}\left[\frac{\rho_i \cos\theta}{\rho(z)}\right] - \tan^{-1}\left(\frac{\rho_i - \rho_0}{L}\right) \tag{5.30}$$

The second term on right-hand side is the taper angle (Ω). For the range of incident angles (θ_1, θ_2), the transmitted power at the output end of the fiber is given by

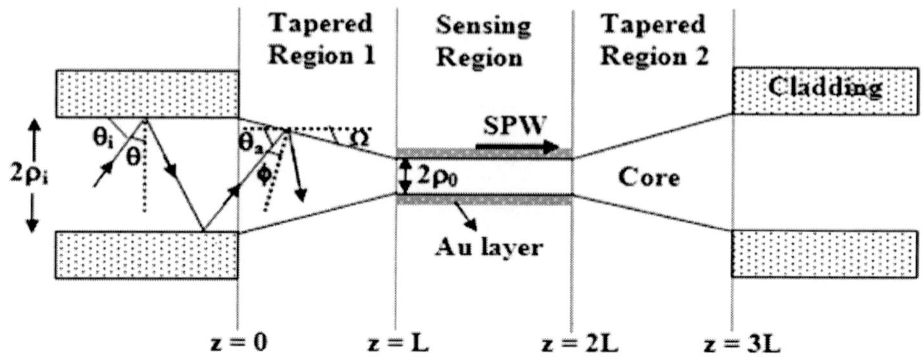

Figure 5.11. Optical fiber sandwiched between two tapered ends. (Reprinted with permission from Verma et al. 2007. Copyright 2007 IEEE.)

$$P_{trans} = \frac{\int_0^L dz \int_{\phi_1(z)}^{\phi_2(z)} R_P^{N_{ref}(\theta,z)} \frac{n_1^2 \sin\theta\cos\theta}{(1-n_1^2\cos^2\theta)^2} d\theta}{\int_0^L dz \int_{\phi_1(z)}^{\phi_2(z)} \frac{n_1^2 \sin\theta\cos\theta}{(1-n_1^2\cos^2\theta)^2} d\theta} \tag{5.31}$$

The expressions for $\phi_1(z)$ and $\phi_2(z)$ can be found by substituting the value of θ in Eq. (5.30). For all the guided rays in the fiber, $\theta_1 = \sin^{-1}(n_{cl}/n_1)$ and $\theta_2 = \pi/2$. Further, N_{ref} for the ray of angle θ and core radius $\rho(z)$ is given as

$$N_{ref}(\theta, z) = \frac{L}{2\rho(z)\tan(\theta+\Omega)} \tag{5.32}$$

For this type of sensor, sensitivity increases quite significantly with taper ratio. The use of an exponential-linear taper profile instead of a linear or parabolic taper profile further improves the sensitivity. Therefore, high taper ratio with an exponential-taper profile is recommended to design a highly sensitive fiber optic SPR probe (Verma et al. 2008). Moreover, a SPR probe of uniform core (with metallic coating) sandwiched between two unclad tapered fiber regions, as shown in Figure 5.11, has also been analyzed (Verma et al. 2007). Figure 5.12 shows

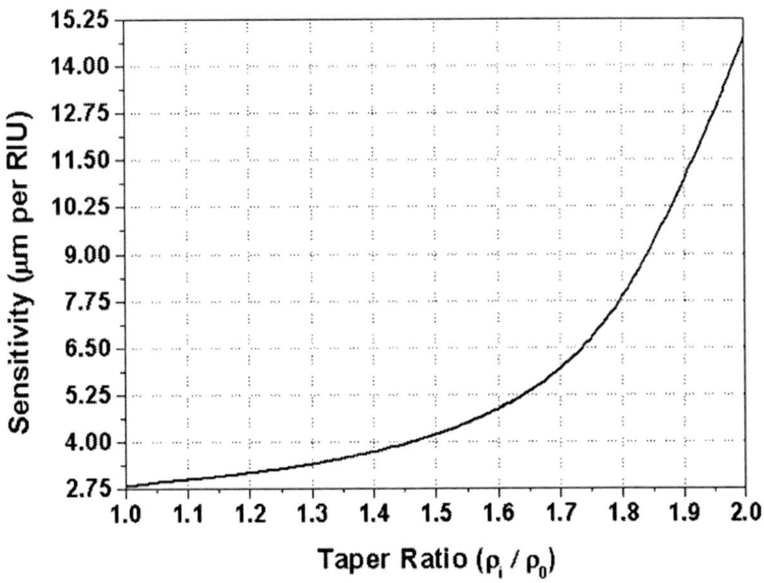

Figure 5.12. Variation of sensitivity (in μm per RIU) of SPR-based fiber optic sensor with taper ratio. (Reprinted with permission from Verma et al. 2007. Copyright 2007 IEEE.)

the variation of sensitivity of this kind of sensor with taper ratio. A sensitivity enhancement of more than 5 times in comparison to conventional fiber optic SPR sensors is observed for taper ratio equal to 2.0.

7.3. U-SHAPED OPTICAL FIBER PROBE

The sensitivity of a SPR-based optical fiber probe can also be increased if the probe is bent and given a U-shape (Verma and Gupta 2008). Its analysis can be made simpler if a two-dimensional approach is considered. For sensing, the outer surface of the bent region is considered to be coated with metal. Further, all the guided rays and their electric vectors (p-polarized light) are assumed to be confined in the plane of bending. For such a probe, angle θ in the straight region transforms to angles $\phi(h)$ and $\delta(h)$ at the outer and the inner surfaces of the U-region and are given by

$$\phi(h) = \sin^{-1}\left[\frac{(R+h)}{(R+2\rho)}\sin\theta\right] \tag{5.33}$$

and

$$\delta(h) = \sin^{-1}\left[\frac{(R+h)}{R}\sin\theta\right] \tag{5.34}$$

respectively. Here h is the distance from the core cladding interface (of the inner surface of the U-region) to the point at which the ray strikes the entrance face of the bending region, and R is the bending radius of the inner circular region as shown in Figure 5.13.

For the range of ray angles θ_1–θ_2, the corresponding range of angles for the outer surface of the U-region alters to $\phi_1(h)$–$\phi_2(h)$ inside the U-region. These angles are obtained from Eq. (5.33) after substituting $\theta_1 = \sin^{-1}(n_2/n_1)$ and $\theta_2 = \pi/2$ (the range of angles in the straight region of the U-shaped probe). The inner surface of the U-region is used to guide light by means of total internal reflection. Thus for all the guided rays, the power transmitted at the output end of the fiber is given by

$$P_{\text{trans}} = \frac{\int_0^{2\rho} dh \int_{\phi_1(h)}^{\phi_2(h)} R_p^{N_{\text{ref}}} \frac{n_1^2 \sin\theta\cos\theta}{\left(1-n_1^2\cos^2\theta\right)^2} d\theta}{\int_0^{2\rho} dh \int_{\phi_1(h)}^{\phi_2(h)} \frac{n_1^2 \sin\theta\cos\theta}{\left(1-n_1^2\cos^2\theta\right)^2} d\theta} \tag{5.35}$$

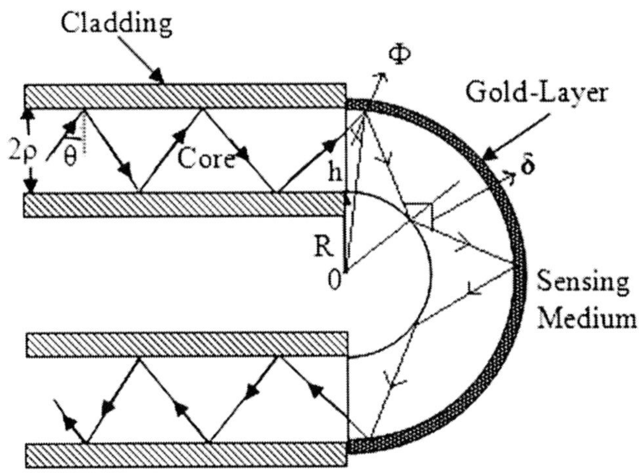

Figure 5.13. U-shaped probe. (Reprinted with permission from Verma and Gupta 2008. Copyright 2008 IOP.)

where

$$N_{\text{ref}} = \frac{L}{8\rho}\left\{\cot\theta + \cot\left[\left(\frac{R+2\rho}{R}\right)\theta\right]\right\} \tag{5.36}$$

L is the length of the sensing region (metal-coated) of the probe. The simulation predicted an increase in the sensitivity of the sensor with a decrease in the bending radius. The maximum sensitivity of the SPR sensor with a U-shaped probe is reported to be about 25 times that of a conventional straight fiber optic SPR sensor (Verma and Gupta 2008).

7.4. LONG-RANGE SURFACE PLASMON RESONANCE

The sensitivity of the SPR probe can be increased by increasing the number of layers in the sensor structure. The use of more layers enhances the number of degrees of freedom for designing a sensor with high sensitivity. For example, introducing a dielectric layer between the prism base or the fiber and the thin metal layer increases the sensitivity of the sensor (Nenninger et al. 2001). The refractive index of the sensing medium on the other side of the metal layer is kept close to the refractive index of the introduced dielectric layer (Slavik and Homola 2006). In this kind of structure, two modes that are bound to both metal layer surfaces are formed. These modes have fields that decay exponentially outside the structure.

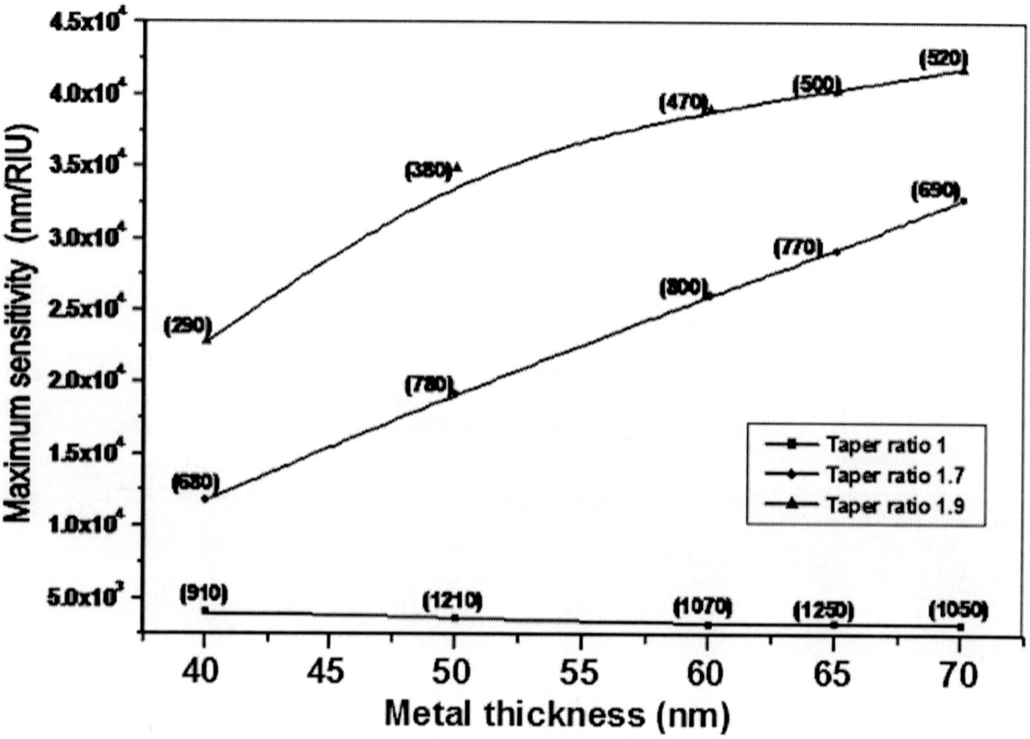

Figure 5.14. Variation of maximum sensitivity of the sensor with metal-layer thickness for different Teflon thicknesses and different taper ratios. Numbers in brackets indicate the Teflon thickness (in nm) that gives maximum sensitivity. (Reprinted with permission from Jha et al. 2008. Copyright 2008 Springer-Verlag.)

The mode having low loss is termed the long-range surface plasmon (LRSP), and the other, having high loss and decaying rapidly, is called the short-range surface plasmon (SRSP). Generally, the sensing probe consists of a symmetrical structure of a thin metal (gold) layer with Teflon AF-1600 on one side and a sensing medium of nearly the same refractive index on the other side. The other side of the Teflon layer is in contact with the fiber core. The SPR spectrum of such a probe has two dips having different sensitivities (Rajan et al. 2007). A probe similar to that in Figure 5.11 has also been analyzed by introducing a Teflon AF-1600 layer between the metal layer and the fiber core (Jha et al. 2008). All possible combinations of metal and Teflon thicknesses and taper ratio have been considered for maximizing the sensitivity of the sensor. The sensitivity of the tapered LRSP probe with taper ratio 1.9 is reported to be about 4 times that without a Teflon layer, while it is more than 15 times that of a uniform-core (without tapering) SPR probe as shown in Figure 5.14 (Jha et al. 2008).

8. SUMMARY

In this chapter we have presented modeling and simulations of various designs of fiber optic SPR sensors. In addition, we have also considered different kinds of light launching in the fiber. The performance of these sensors may be affected if the sensors are exposed to high-temperature environments. Variation in temperature is known to affect the dielectric constants of the fiber, metal, and any other layer in the fiber optic SPR probe. The temperature variation may affect the sensitivity, signal-to-noise ratio, and operating range of the SPR sensor. Therefore, the effect of temperature on the performance of SPR sensors has been analyzed theoretically by taking into account the temperature dependencies of the dielectric constants in the model (Sharma and Gupta 2006a). Apart from studying the effect of ambient temperature on the performance of the SPR sensor, an SPR-based fiber optic temperature sensor has also been designed and simulated using a film of large thermo-optic coefficient over the metal film (Sharma and Gupta 2006b). The modeling of SPR sensors has also been reported for telecommunication wavelengths using a Bragg structure (Das and Jha 2009). SPR sensors utilizing metal nanoparticles (also called localized surface plasmon resonance, or LSPR) are also attracting attention these days. The LSPR technique has shown a lot of promise. The modeling and simulation of LSPR-based fiber optic sensors for selecting an appropriate light source has recently been reported (Srivastava et al. 2009). In addition, a temperature sensor utilizing metal nanoparticles also shows a lot of promise (Srivastava and Gupta 2010). In all, the modeling and simulations help a lot in optimization of the crucial factors and parameters of the probe to achieve best performance of the sensor after its fabrication.

ACKNOWLEDGMENT

The authors would like to acknowledge the contributions of research scholars, project staff, and students whose research work has helped us in writing this chapter. We would also like to convey our sincere thanks to various publishing houses for allowing us to reprint the figures.

REFERENCES

Bhatia P. and Gupta B.D. (2011) Surface-plasmon-resonance-based fiber-optic refractive index sensor: Sensitivity enhancement. *Appl. Opt.* **50**, 2032–2036. DOI: 10.1364/AO.50.002032

Bureau B., Zhang X.H., Smektala F., Adam J., Troles J., Ma H., Boussard-Pledel C.,

Lucas J., Lucas P., Le Coq D., Riley M.R., and Simmons J.H. (2004) Recent advances in chalcogenide glasses. *J. Non-Crystalline Solids* **345–346**, 276–283. DOI: 10.1016/j.jnoncrysol.2004.08.096

Cullen D.C. and Lowe C.R. (1990) A direct surface plasmon-polariton immunosensor: Preliminary investigation of the non-specific adsorption of serum components to the sensor interface. *Sens. Actuators B* **1**, 576–579. DOI: 10.1016/0925-4005(90)80276-6

Das R. and Jha R. (2009) On the modal characteristic of surface plasmon polaritons at a metal Bragg interface at optical frequencies. *Appl. Opt.* **48**, 4904–4908. DOI: 10.1364/AO.48.004904

Dostálek J., Homola J., and Miler M. (2005) Rich information format surface plasmon resonance biosensor based on array of diffraction gratings. *Sens. Actuators* B **107**, 154–161. DOI: 10.1016/j.snb.2004.08.033

Dwivedi Y.S., Sharma A.K., and Gupta B.D. (2007) Influence of skew rays on the sensitivity and signal-to-noise ratio of a fiber-optic surface-plasmon-resonance sensor: A theoretical study. *Appl. Opt.* **46**, 4563–4569. DOI: 10.1364/AO.46.004563

Esteban O., Alonso R., Navarrete C., and Gonzalez-Cano C.A. (2002) Surface plasmon excitation in fiber-optics sensors: A novel theoretical approach. *J. Lightwave Technol.* **20**, 448–453. DOI: 10.1109/50.988993

Gupta B.D. and Sharma A.K. (2005) Sensitivity evaluation of a multi-layered surface plasmon resonance-based fiber optic sensor: A theoretical study. *Sens. Actuators B* **107**, 40–46. DOI: 10.1016/j.snb.2004.08.030

Gupta B.D., Sharma A., and Singh C.D. (1993) Evanescent wave absorption sensors based on uniform and tapered fibers: A comparative study of their sensitivities. *Int. J. Optoelectron.* **8**, 409–418.

Gupta B.D. and Singh C.D. (1994a) Evanescent-absorption coefficient for diffuse source illumination: Uniform and tapered-fiber sensors. *Appl. Opt.* **33**, 2737–2742. DOI: 10.1364/AO.33.002737

Gupta B.D. and Singh C.D. (1994b) Fiber-optic evanescent field absorption sensor: A theoretical evaluation. *Fiber Integrat. Opt.* **13**, 433–443. DOI: 10.1080/01468039408202251

Homola J. (1997) On the sensitivity of surface plasmon resonance sensors with spectral interrogation. *Sens. Actuators B* **41**, 207–211. DOI: 10.1016/S0925-4005(97)80297-3

Homola J., Ctyroky J., Skalsky´ M., Hradilova´ J., and Kolarova (1997) A surface plasmon resonance based integrated optical sensor. *Sens. Actuators B* **38–39**, 286–290. DOI: 10.1016/S0925-4005(97)80220-1

Homola J., Yee S.S., and Gauglitz G. (1999) Surface plasmon resonance sensor: Review. *Sens. Actuators* B **54**, 3–15. DOI: 10.1016/S0925-4005(98)00321-9

Jha R. and Badenes G. (2009) Effect of fiber core dopant concentration on the performance of surface plasmon resonance based fiber optic sensor. *Sens. Actuators A* **150**, 212–217. DOI: 10.1016/j.sna.2009.01.004

Jha R. and Sharma A.K. (2009a) SPR based infrared detection of aqueous and gaseous media with silicon substrate. *Europhys. Lett.* **87**, 10007. DOI: 10.1209/0295-5075/87/10007

Jha R. and Sharma A.K. (2009b) Chalcogenide glass prism based SPR sensor with Ag-Au bimetallic nanoparticle alloy in infrared wavelength region. *J. Opt. A: Pure Appl. Opt.* **11**, 045502. DOI: 10.1088/1464-4258/11/4/045502

Jha R. and Sharma A.K. (2009c) High performance sensor based on surface plasmon resonance with chalcogenide prism and aluminum for detection in infrared. *Opt. Lett.* **34**, 749–751. DOI: 10.1364/OL.34.000749

Jha R. and Sharma A.K. (2010) Design of silicon based plasmonic biosensor chip for human blood group detection. *Sens. Actuators B* **145**, 200–204. DOI: 10.1016/j.snb.2009.11.060

Jha R. and Sharma A.K. (2011) Design considerations for plasmonic-excitation based optical detection of liquid and gas media in infrared. *Sens. Actuators A* **165**, 271–275. DOI: 10.1016/j.sna.2010.10.021

Jha R., Verma R.K., and Gupta B.D. (2008) Surface plasmon resonance-based tapered fiber optic sensor: Sensitivity enhancement by introducing a Teflon layer between core and metal layer. *Plasmonics* **3**, 151–156. DOI: 10.1007/s11468-008-9068-9

Johansen K., Arwin H., Lundstrom I., and Liedberg B. (2000) Imaging surface plasmon resonance sensor based on multiple wavelengths: Sensitivity considerations. *Rev. Sci. Instrum.* **71**, 3530–3538. DOI: 10.1063/1.1287631

Jorgenson R.C. and Yee S.S. (1993) A fiber-optic chemical sensor based on surface plasmon resonance. *Sens. Actuators B* **12**, 213–220. DOI: 10.1016/0925-4005(93)80021-3

Karlsson R. and Stahleberg R. (1995) Surface plasmon resonance detection and multispot sensing for direct monitoring of interactions involving low-molecular-weight analytes and for determination of low affinities. *Anal. Biochem.* **228**, 274–280. DOI: 10.1006/abio.1995.1350

Krane K.J. and Reather H. (1976) Measurements of surface plasmon dispersion in aluminum and indium. *Phys. Rev. Lett.* **37**, 1355–1357. DOI: 10.1103/PhysRevLett.37.1355

Kretschmann E. and Reather H. (1968) Radiative decay of non-radiative surface plasmons excited by light. *Z. Naturforsch.* **23**, 2135–2136.

Lahav A., Auslender M., and Abdulhalim I. (2008) Sensitivity enhancement of guided-wave surface-plasmon resonance sensors. *Opt. Lett.* **33**, 2539–2541. DOI: 10.1364/OL.33.002539

Lambeck P.V. (1992) Integrated opto-chemical sensors. *Sens. Actuators B* **8**, 103–116. DOI: 10.1016/0925-4005(92)85016-P

Mar M., Jorgenson R.C, Letellier S., and Yee S.S. (1993) In-situ characterization of multilayered Langmuir-Blodgett films using a surface plasmon resonance fiber optic sensor. In: Proceedings of 15th Annual International Conference IEEE, 1551–1552. DOI: 10.1109/IEMBS.1993.979277

Moharam M.G. and Gaylord T.K. (1986) Rigorous coupled-wave analysis of metallic surface-relief gratings. *J. Opt. Soc. Am. A* **3**, 1780–1787. DOI: 10.1364/JOSAA.3.001780

Monzón-Hernández D., Villatoro J., Talavera D., and Luna-Moreno D. (2004) Optical-fiber surface-plasmon resonance sensor with multiple resonance peaks. *Appl. Opt.* **43**, 1216–1220.

Mouvet C., Harris R.D., and Maciag C. (1997) Determination of simazine in water samples by waveguide surface plasmon resonance. *Anal. Chim. Acta* **338**, 109–117. DOI: 10.1016/S0003-2670(96)00443-6

Nelson B.P., Frutos A.G., Brockman J.M., and Corn R.M. (1999) Near-infrared surface plasmon resonance measurements of ultrathin films. *Anal. Chem.* **71**, 3928–3934. DOI: 10.1021/ac990517x

Nenninger G.G., Tobiska P., Homola J., and Yee S.S. (2001) Long range surface plasmons for high resolution surface plasmon resonance sensors. *Sens. Actuators B* **74**, 145–151. DOI: 10.1016/S0925-4005(00)00724-3

Otto A. (1968) Excitation of nonradiative surface plasma waves in silver by the method of frustrated total reflection. *Z. Phys.* **216**, 398–410. DOI: 10.1007/BF01391532

Paivasaari K., Tikhomirov V.K., and Turunen J. (2007) High refractive index chalcogenide glasses for photonic crystal application. *Opt. Express* **15**, 2336–2340; DOI: 10.1364/OE.15.002336

Powell C.J. and Swan J.B. (1960) Effect of oxidation on the characteristic loss spectra of aluminum and magnesium. *Phys. Rev.* **118**, 640–643. DOI: 10.1103/PhysRev.118.640

Rajan, Sharma A.K., and Gupta B.D. (2007) Fiber optic sensor based on long-range surface plasmon resonance: A theoretical analysis. *J. Opt.: Pure Appl. Opt.* **9**, 682–687. DOI: 10.1088/1464-4258/9/7/020

Reather H. (1998) *Surface Plasmons on Smooth and Rough Surfaces and on Gratings.* Springer-Verlag, Berlin.

Ritchie R.H. (1957) Plasma losses by fast electrons in thin films. *Phys. Rev.* **106**, 874–881. DOI: 10.1103/PhysRev.106.874

Ritchie R.H., Arakawa E.T., Cowan J.J., and Hamm R.N. (1968) *Phys. Rev. Lett.* **21**, 1530–1533. DOI: 10.1103/PhysRevLett.21.1530

Sharma A.K. and Gupta B.D. (2005) On the sensitivity and signal-to-noise ratio of a step-index fiber optic surface plasmon resonance sensor with bimetallic layers. *Opt. Commun.* **45**, 159–169; DOI: 10.1016/j.optcom.2004.10.013

Sharma A.K and Gupta B.D. (2006a) Theoretical model of a fiber optic remote sensor based on surface plasmon resonance for temperature detection. *Opt. Fiber Technol.* **12**, 87–100. DOI: 10.1016/j.yofte.2005.07.001

Sharma A.K. and Gupta B.D. (2006b) Influence of temperature on the sensitivity and signal-to-noise ratio of a fiber-optic surface plasmon resonance sensor. *Appl. Opt.* **45**, 151–161. DOI: 10.1364/AO.45.000151

Sharma A.K. and Gupta B.D. (2007) On the performance of different bimetallic combinations in surface plasmon resonance based fiber optic sensors. *J. Appl. Phys.* **101**, 093111. DOI: 10.1063/1.2721779

Sharma A.K., Jha R., and Gupta B.D. (2007a) Fiber optic sensors based on surface plasmon resonance: A comprehensive review. *IEEE Sens. J.* **7**, 1118–1129. DOI: 10.1109/JSEN.2007.897946

Sharma A.K., Jha R., and Gupta B.D. (2007b) Influence of different dopants on the performance of a fiber optic SPR sensor. *Opt. Commun.* **274**, 320–326. DOI: 10.1016/j.optcom.2007.02.030

Singh S., Verma R.K., and Gupta B.D. (2010) LED based fiber optic surface plasmon resonance sensor. *Opt. Quant. Electron.* **42**, 15–28. DOI: 10.1007/s11082-010-9418-7

Slavik R. and Homola J. (2006) Simultaneous excitation of long and short range surface plasmons in an asymmetric structure. *Opt. Commun.* **259**, 507–512. DOI: 10.1016/j.optcom.2005.09.011

Snyder A.W. and Love J.D. (1991). *Optical Waveguide Theory.* Chapman & Hall, London.

Sommerfeld A. (1909) Propagation of waves in wireless telegraphy. *Ann. Physik* **28**, 665–736. DOI: 10.1002/andp.19093330402

Srivastava S.K. and Gupta B.D. (2010) Simulation of a localized surface plasmon resonance based fiber optic temperature sensor. *J. Opt. Soc. Am. A* **27**, 1743–1749. DOI: 10.1364/JOSAA.27.001743

Srivastava S.K., Verma R.K., and Gupta B.D. (2009) Theoretical modeling of a localized surface plasmon resonance based intensity modulated fiber optic refractive index sensor. *Appl. Opt.* **48**, 3796–3802. DOI: 10.1364/AO.48.003796

Srivastava T., Das R., and Jha R. (2010) Design considerations and propagation characteristics of channel Bragg-plasmon-coupled-waveguides. *Appl. Phy. Lett.* **97**, 213104. DOI: 10.1063/1.3518922

Srivastava T., Das R., and Jha R. (2011) Highly accurate and sensitive surface plasmon resonance sensor based on channel photonic crystal waveguides. *Sens. Actuators B* **157**, 246–252. DOI: 10.1016/j.snb.2011.03.057

Stern E.A. and Ferrell R.A. (1960) Surface plasma oscillations of a degenerate electron gas. *Phys. Rev.* **120**, 130–136. DOI: 10.1103/PhysRev.120.130

Suzuki A., Kondoh J., Matsui Y., Shiokawa S., and Suzuki K. (2005) Development of novel optical waveguide surface plasmon resonance (SPR) sensor with dual light emitting diodes. *Sens. Actuators B* **106**, 383–387. DOI: 10.1016/j.snb.2004.08.021

Verma R., Gupta B.D., and Jha R. (2011) Sensitivity enhancement of a surface plasmon resonance based biomolecules sensor using graphene and silicon layers. *Sens. Actuators B* **160**, 623–631. DOI: 10.1016/j.snb.2011.08.039

Verma R.K. and Gupta B.D. (2008) Theoretical modeling of a bi-dimensional U-shaped surface plasmon resonance based fiber optic sensor for sensitivity enhancement. *J. Phys. D* **41**, 095106. DOI: 10.1088/0022-3727/41/9/095106

Verma R.K., Sharma A.K., and Gupta B.D. (2007) Modelling of tapered fiber-optic surface plasmon resonance sensor with enhanced sensitivity. *Phot. Tech. Lett.* **19**, 1786–1788. DOI: 10.1109/LPT.2007.906840

Verma R.K, Sharma A.K., and Gupta B.D. (2008) Surface plasmon resonance based tapered fiber optic sensor with different taper profiles. *Opt. Commun.* **281**, 1486–1491. DOI: 10.1016/j.optcom.2007.11.007

Villuendas F. and Pelayo J. (1990) Optical fibre device for chemical sensing based on surface plasmon excitation. *Sens. Actuators A* **23**, 1142–1145. DOI: 10.1016/0924-4247(90)87104-Q

Weiss M.N., Srivastava R., Groger H., Lo P., and Luo S.F. (1996) A theoretical investigation of environmental monitoring using surface plasmon resonance waveguide sensors. *Sens. Actuators A* **51**, 211–217.

Yeatman E.M. (1996) Resolution and sensitivity in surface plasmon microscopy and sensing. *Biosens. Bioelectron.* **11**, 635–649. DOI: 10.1016/0956-5663(96)83298-2

Zenneck J. (1907) Uber die Fortpflanztmg ebener elektro-magnetischer Wellen langs einer ebenen Leiterflache und ihre Beziehung zur drahtlosen Telegraphie. *Ann. Physik* **23**, 846–866.

CHAPTER 6

Fiber Optic Sensor Operating in a Microfluidic Device

A Finite-Element Analysis

G. Louarn
M. Kanso
T. Makiabadi

1. INTRODUCTION

Today, many microbiological analysis laboratories use microdevices that integrate one or several functions; these devices are also called "lab-on-a-chip." The success of these systems is due mainly to their automation and miniaturization, which makes possible measurements with a single device, while analyzing and handling samples very quickly. In biotechnology, these samples often consist of living organisms. To preserve the studied targets, it is important to be able to manage them in a small volume of fluid flowing through channels with micrometer or nanometer dimensions. This manipulation of fluids is referred to as microfluidics.

Driven mainly by the recent development of biotechnology, microfluidics has become an important sector in fluid mechanics, and research on "lab on a chip" and sensitive substrates has been progressing rapidly. This evolution is due mainly to strong demand for handling samples in small quantities, and then to

DOI: 10.5643/9781606503201/ch6

increase the surface/volume ratio during surface chemical reactions. In this way, biosensors coupled with microfluidic devices make it possible to automate screening and then increase the capability to perform complex analyses on cellular or molecular entities (Wang et al. 2003; Minc and Viovy 2004; Berthier and Silberzan 2006; Huang et al. 2006).

In the same time, surface plasmon resonance (SPR) has shown great potential for studies of biomolecular interactions, allowing real-time analysis without the use of labeled molecules. SPR is now one of the most widely used technologies for observations of biomolecular interaction; it is widely used in many sensors that operate in varying areas such as gas detection or bio- and chemical sensing (Homola et al. 2002; Homola 2003). During the last 20 years, several companies have commercialized SPR sensor equipment. Generally, these devices are constructed of an optical system made of a transducer that analyzes the (bio)chemical activity and transfers this information to the optical setup; and an electronic system, which allows for data processing. As illustrated in Figure 6.1, SPR biological targets are generally transported through a microfluidic system by a buffer fluid or a carried fluid. Then, when the transducing media (ligands) react with the target molecules present in the analyte, the refractive index of the analyte changes and this change is determined by optical interrogation.

To perform this interrogation, usually the optical system is detected using an optical structure with a thin metallic layer coated on a prism coupler (Kretschmann configuration; Kretschmann and Raether 1968). The aim of this coupler is to excite and interrogate the surface plasma waves (SPWs) generated on the metal–analyte interface. The excitation of SPWs by the incident optical wave results in the resonant transfer of energy into the SPW, which can be evidenced by the resonant absorption of incident light. Because this absorption is resonant, the intensity of the reflected light dips to a sharp minimum. When the refractive index of the analyte

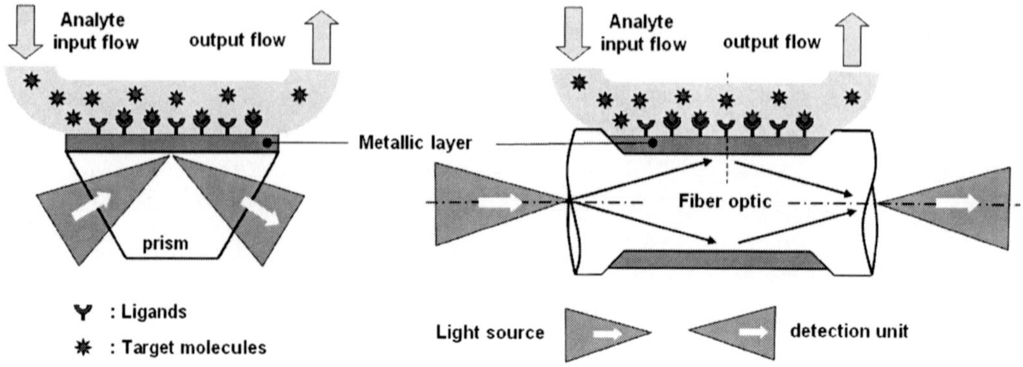

Figure 6.1. Schematic illustration of conventional SPR: Kretschmann configuration (left) and fiber optic configuration (right).

changes, different conditions will be required to satisfy the coupling condition, and the position of the minimum will change. This shift in the resonance location can be monitored by angle, wavelength (angular or wavelength interrogation), or reflection intensity changes. It is possible to track this shift with a great degree of accuracy, thus offering a very high limit of detection.

SPR is now an established technique (Phillips and Cheng 2007), but in order to satisfy the need for miniaturization and integration, recent developments in fiber optics and waveguide-based SPR devices are still under development. The fiber optic SPR (FO-SPR) device is not very different from a standard prism SPR system. The realization of FO-SPR sensors is based on a similar concept, specifically, that a chemical ligand is fixed to the metallic fiber surface. Then, the analyte flows in the microfluidic channel and the target molecules bind to the ligand, creating a thicker layer that can be detected by a tiny change in refractive index. The most important differences concern the axisymmetric geometry and the fact that reflections occur frequently on the fiber optic surface (Figure 6.1b). This complicates the theoretical treatment of the system.

The performance of sensors is generally discussed in terms of limit of detection, dynamic range, detector sensitivity, and sensitivity of the analysis for the species to be detected. The *limit of detection* describes the minimum concentration of analyte that can be reliably detected. The *detector sensitivity,* characterizing device performance, is the change in signal per unit value of the sensor response, whereas the *sensitivity* is the performance of the sensor in response to changing concentrations of analyte. This latter parameter is determined from the slope of the linear region of the response curve. Finally, the *dynamic range* of the sensor for the specific analyte under study and a measure of the chemical reversibility of analyte binding to the coating are important to complete the figures of merit of the sensor under investigation.

The knowledge of these figures of merit is of benefit to those trying to compare SPR designs, materials, and detection techniques. For instance, choices of couplers (prism or fiber optic) and interrogation modes are really important because they substantially affect these four parameters of performance. In this way, such SPR sensors can be optimized through the determination of ideal geometric parameters and knowledge of the best working conditions. Experimental parametric studies or numerical simulations of SPR response curves seem to be two particularly interesting approaches.

The purpose of this chapter is to present how, from a numerical approach, it is possible to optimize the performance of a fiber optic sensor based on surface plasmon resonance in actual operating conditions. In order to evaluate appropriate values of parameters, an experimental setup was developed, and from these experimental data, numerical simulations were carried out. This led us to study and to model (1) the flow and the transport of mass by diffusion and advection of target molecules inside a microfluidic cell; (2) the reactions of hybridization of target molecules with ligands grafted on the sensitive surface; and (3) the sensitivity and the detection limit of our system.

2. THEORY—GOVERNING EQUATIONS

2.1. GENERAL CONSIDERATIONS

From the SPR effect, the hybridization of target molecules on the surface induces a variation of the optical transmission in the optical fiber (Homola 2003; Gupta and Sharma 2007; Balaa et al. 2007; Kanso et al. 2008). As a result of this hybridization, a gradient of concentration of the target molecules is created in an area near the surface of the fiber, known as the concentration boundary layer (see Section 2.5). Then, diffusion phenomena emerge, which are modulated by the circulation of analyte. Quickly, equilibrium of concentrations near the surface is established, which is visualized by the curves called sensorgrams (see Section 2.3).

Practically, the biological targets are diluted in a liquid (carrier or buffer solution) to be guided toward the reactive surface, where they are immobilized during the time of measurement. Then, various forces operate on these targets. On a microscopic scale, we have to take into account the diffusion effect, which describe the natural, rapid, and random movement of the particles (Brownian motion), the advection effect (forced convection) of the particles in the moving fluid, and the mechanisms of hybridization (i.e., adsorption and desorption) of the particles on the surface of the sensor.

Chemical surface reactions (hybridization, for instance), like the phenomena of adsorption and desorption, are often accounted for by using the Langmuir model. All these phenomena have to be implemented in the numerical model to simulate the light transmission through the optical fiber by the SPR effect (Myszka and Norton 1998; Karlsson 2006; Gervais and Tensen 2006).

When a fluid flows in a pipe, it encounters a viscous resistance which creates a pressure drop. Depending on the importance of this pressure loss, the flow system can be either laminar or turbulent. For constant boundary conditions imposed on the flow, Reynolds showed in 1883 that it is possible to anticipate the type of flow (laminar or turbulent), depending on the value of a dimensionless number now called the Reynolds number, usually denoted by Re:

$$\mathrm{Re} = \frac{\rho \times u \times L}{\eta} \tag{6.1}$$

where u is the velocity of the fluid (m/s), ρ is the mass density (kg/m^3), L is the characteristic size of the channel (m), and η is the viscosity of the fluid (Pa s).

When Re is low, the flow lines are stationary and the flow is called laminar. On the other hand, when Re is high, the stream becomes unsteady and the flow is called turbulent. The transition from a laminar to a turbulent regime usually occurs when Re is about 1000–2000.

The main difference between the macroscopic and microscopic systems is that in most cases, the macroscopic flow is turbulent while the microscopic flow is laminar. In biotechnology and microchemistry, the character of the laminar flow is often very high. The flow rate is very low, so it is generally stable. This stability is related to the value of the Reynolds number, which decreases as the system size decreases.

With our experimental cell, the channels have a characteristic dimension of about 300 μm, and we worked with fluid velocities in the range of 1–50 mm/s. The calculation gives, for a water viscosity of 10^{-3} Pa and a density of 1000 kg/m^3, a Reynolds number less than 10.

This Reynolds number indicates that the turbulence is absent in our microfluidic system. This assumption greatly simplifies the modeling, because the flow can then be modeled by the well-known Navier-Stokes equations.

A second important consideration is to rule on the viscous behavior of the analyte. In a homogeneous fluid in which each element of the fluid glides parallel to the other without mixing, the shear flow in the fluid is given by Newton's law:

$$\vec{\tau} = -\eta \cdot \vec{\nabla}\mathbf{u} \tag{6.2}$$

Usually, $\vec{\tau}$ is the shearing stress tensor (N/m^2) and $\mathbf{u}$ is the velocity vector. For a unidirectional flow with a velocity profile $u(y)$ in the direction normal to y (Hagen Poiseuille type), Newton's equation reduces to

$$\tau = -\eta \times \frac{\partial u}{\partial y} \tag{6.3}$$

where η is the dynamic viscosity of the analyte (Pa s).

If this viscosity is independent of the velocity gradient, the fluid is called Newtonian. Most biological fluids, such as blood or concentrated fluids of macromolecules or polymers, exhibit non-Newtonian behavior. This is especially true in the transient state. In the present case, however, SPR-based biosensors generally have to work in aqueous medium and with low concentrations of analyte, so we will limit our analysis to incompressible fluids, homogeneous and Newtonian.

2.2. NAVIER-STOKES EQUATIONS

Under laminar flow condition, the Navier-Stokes equations describe how the velocity, pressure, temperature, and density of a moving fluid are related. The equations include the effects of viscosity on the flow. The equations are a set of coupled differential equations which consist of a time-dependent continuity equation for conservation of mass, conservation of momentum, and time-dependent conservation of energy. For a Newtonian fluid with gravity as the only force acting on the volume field, the equations governing the flow can be expressed as follows:

$$\frac{\partial \rho}{\partial t} + \rho \nabla \cdot \mathbf{u} = 0 \tag{6.4}$$

$$\rho \frac{D\mathbf{u}}{Dt} = \rho\left(\mathbf{u} \cdot \nabla \cdot \mathbf{u} + \frac{\partial u}{\partial t}\right) = -\nabla p + \rho \mathbf{g} + \eta \Delta \mathbf{u} \tag{6.5}$$

where $\mathbf{u}$, ρ, t, p, and $\mathbf{g}$, are the velocity vector, the density, the time, the pressure, and the gravitational acceleration, respectively. The differential symbol D/Dt is different than the usual $\partial/\partial t$ and is used to indicate the following sum of partial derivatives:

$$\frac{D}{Dt} = \frac{\partial}{\partial t} + u\frac{\partial}{\partial x} + v\frac{\partial}{\partial y} + w\frac{\partial}{\partial z}$$

The three components of the velocity are the u component in the x direction, the v component in the y direction, and the w component in the z direction. The term on the left side of Eq. (6.5) represents the total, convective, and time-dependent acceleration. The right side of the equation is a summation of body forces such as gravity and divergence of stress (pressure and viscous effects). For laminar and incompressible flow, the density and the viscosity are assumed to be constant in the equations. These equations represent a complete set of equations for four parameters: the pressure and the three velocity components. For nonviscous flow, the last term of Eq. (6.5) is null. Usually, the gravitational term in Eq. (6.5) is neglected. Thus, if we consider an incompressible and Newtonian fluid, neglecting the gravity term, the equations can be simplified. The continuity equation is given by

$$\nabla \cdot \mathbf{u} = 0 \tag{6.6}$$

The momentum equation becomes

$$\rho \frac{\partial \mathbf{u}}{\partial t} = -\vec{\nabla} p + \eta \Delta \mathrm{u} \tag{6.7}$$

Note that if changes in temperature occur (which will not be discussed in this chapter), the equation of conservation of energy is added to complete the system of equations above.

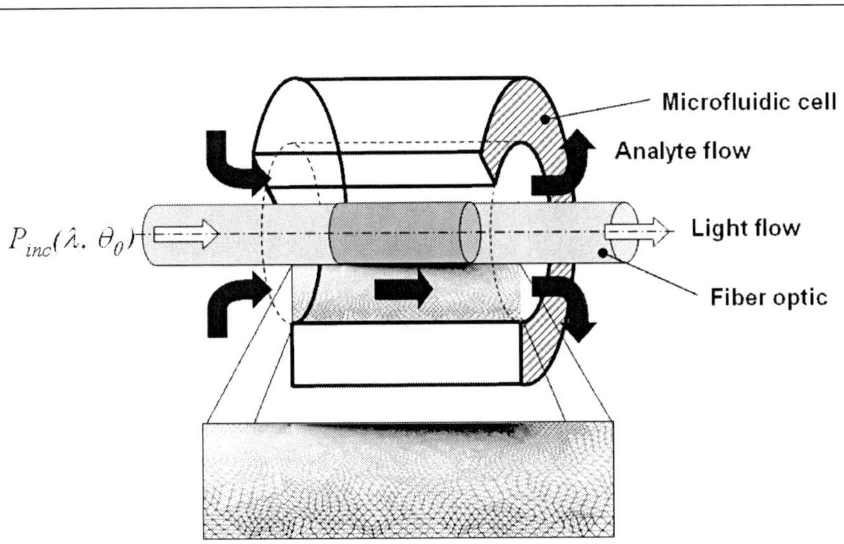

Figure 6.2. Sketch of the microfluidic cell. By using axial symmetry, the 2-D model can be solved instead of a full 3-D model, saving considerable memory and computation time and thus making the modeling an efficient tool for solving this problem.

2.3. LAMINAR FLOW BETWEEN FIXED PARALLEL PLATES

Three types of microchannels are usually used in microfluidic studies: circular or rectangular pipes and flow between parallel plates. Hagen-Poiseuille defined the velocity profile for these three forms. In our case, we consider the flow of the analyte in the vicinity of the sensitive area as a parallel flow between two parallel plates (see Figure 6.2), the wall of the microfluidic channel and the fiber optic, separated by a distance $2d$ (Figure 6.3). The analyte is still regarded as a homogeneous and incompressible Newtonian fluid, in laminar flow along the x axis.

To the extent that the two surfaces are large compared to the distance between the two plates, and considering the symmetry around the axis, an axisymmetric approach can be developed, to reduce the study of the flow in a plane (Figure 6.2).

The continuity equation [$\nabla \cdot \mathbf{u} = 0$ with $\mathbf{u}(u,v,w) = \mathbf{u}(u,0,0)$] imposes $\partial u/\partial x = 0$. In addition, the projections of the momentum equation (6.7) give

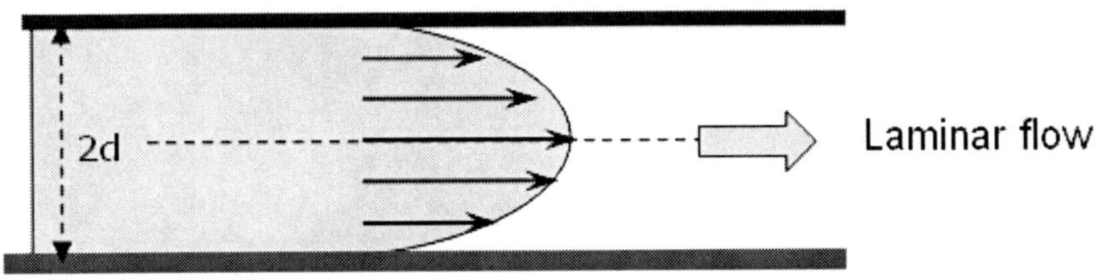

Figure 6.3. Velocity profile in a cross section of a microfluidic cell.

$$\left.\begin{aligned} 0 &= -\frac{\partial p}{\partial x} + \eta\,\Delta u \\ 0 &= \frac{\partial p}{\partial y} \\ 0 &= \frac{\partial p}{\partial z} \end{aligned}\right\} \qquad (6.8)$$

The velocity profile in the channel is given by solving the following differential equation:

$$\eta \frac{d^2u}{dy^2} = \frac{\partial p}{\partial x} \qquad (6.9)$$

with η and $\partial p/\partial x$ as constants and $u(d) = u(-d) = 0$ as boundary limits (zero speed at the walls). Finally, for laminar flow between two parallel plates, we find the velocity profile (Figure 6.3) given by

$$u(y) = u_{\max}\left[1 - \left(\frac{y}{d}\right)^2\right] \qquad (6.10)$$

with $u_{\max}$ the maximum velocity:

$$u_{\max} = -\frac{1}{2\mu}\left(\frac{\partial P}{\partial x}\right) = \frac{3}{2}\overline{U} \qquad (6.11)$$

where $\overline{U}$ is the mean velocity of the flow, defined by

$$\overline{U} = \frac{1}{S}\iint \vec{V}\, d\vec{S} \qquad (6.12)$$

with S the section of the channel. These equations show that the velocity profile is a quadratic function which depends only on y.

2.4. DIFFUSION-ADVECTION EQUATIONS IN A MICROFLUIDIC CHANNEL

In a biological analyte, target molecules are diluted. Following their injection into the microfluidic cell, they can move in the direction of the surface of interaction, and then associate/disassociate themselves with a ligand, or leave the fluidic cell. As mentioned, these molecules can be considered as nanoparticles or microparticles in a carrier fluid in motion. The Navier-Stokes theory is then incomplete and requires the inclusion of diffusion (Brownian motion) and forced convection. For this, it becomes important to present these transport mechanisms.

Two mechanisms of transport targets coexist: (1) forced convection or advection which involves moving a set of molecules, (2) diffusion corresponding to a succession of elementary displacements in the system. The diffusion flux is proportional to the concentration gradient (∇c). We can deduce the mass flux by diffusion $\boldsymbol{J}$ [Eq. (6.13)] using Fick's law:

$$\boldsymbol{J} = -D\nabla c \tag{6.13}$$

with D the diffusion coefficient (m^2/s). The flow balance equation then leads to Fick's second law:

$$\frac{\partial c}{\partial t} = \nabla \cdot (D\nabla c) + S \tag{6.14}$$

with S a source term or a loss of concentration. For instance, in the case we will discuss in the rest of this chapter, if a hybridization or dissociation reaction occurs somewhere in the microfluidic channel, the concentration of target molecules can locally increase or decrease. We return to this term afterward.

The balance of the particle flux in the cell depends on the convection and diffusion flux (at this step, the hybridization reaction is ignored). The equation that governs the motion of the liquid is then given by the advection-diffusion equation as follows:

$$\frac{\partial c}{\partial t} + \mathbf{u} \cdot \nabla c = \nabla \cdot (D\nabla c) \tag{6.15}$$

where c is the analyte concentration, D is the diffusion coefficient given by Fick's equation, and $\mathbf{u}$ is the velocity of the analyte as given by Eq. (6.7).

Analytically, if we substitute the velocity term from Eq. (6.10), where in a microfluidic system the diffusion term along x is negligible compared to the convection, the equation of advection-diffusion becomes

$$\frac{\partial c}{\partial t} = D\left(\frac{\partial^2 c}{\partial y^2}\right) - u_{\max}\left[1 - \left(\frac{y}{d}\right)^2\right]\frac{\partial c}{\partial x} \tag{6.16}$$

Note that we have not solved the model by an analytical method but by the finite-element method. This analytical development only allows here a good understanding of physical mechanisms which must be considered to resolve the problem.

2.5. BIOCHEMICAL REACTION AND LANGMUIR ADSORPTION MODEL

In a microfluidic device, a molecular reaction can be achieved in two different ways: One is called homogeneous, when the ligands and analytes are mixed in the volume of the cell; the second is known as heterogeneous, when the ligands are anchored on the surface of interaction and the analytes are in the solution.

Heterogeneous reactions are widely used in microsystems (biochips, biosensors, etc.). They are often more favorable than the homogeneous reactions for two main reasons: (1) because the functionalized surface can be reused after "washing" of the reaction cell, and (2) because heterogeneous reactions allow generally simple detection of hybridization (ligand–analyte) on the surface of the sensitive area. Thus it is necessary to go through two steps: immobilization of the ligand on the sensitive surface and then injection of the analyte. In this chapter, we limit the discussion to heterogeneous reactions.

Basically, the objective of a surface plasmon resonance sensor, and more generally of a biochip, is the analysis and recognition of target macromolecules (DNA, proteins, etc.). The biorecognition can be based on a mechanism called "key-and-lock." Biomolecular reaction can be described as the association of two molecules to form a "complex," such as two complementary single-stranded DNA molecules (Glasser 1993; Christensen 1997; Karlsson and Falt 1997; Tsoi and Yang 2002). Among the various models of heterogeneous molecular reactions, we have chosen to implement the Langmuir model. This model uses the concepts of association and/or dissociation of a complex on the surface, and determines the affinity of a ligand (L) for an analyte (A).

The Langmuir model assumes that the reactions involve a ligand L and an analyte (or target) A which react solely with each other to form an organic layer LA. This layer is grafted onto the surface of the metal layer by noncovalent bonds (hydrogen bonds, van der Waals bonds, or Coulomb interactions). This model of adsorption is widely used to describe adsorption of particles on a surface when the number of adsorption sites on the surface is fixed, meaning that each site can adsorb only one particle and the interactions between particles are neglected.

$$\mathrm{L} + \mathrm{A} \underset{k_{\mathrm{diss}}}{\overset{k_{\mathrm{ass}}}{\rightleftarrows}} \mathrm{LA} \tag{6.17}$$

If the interaction is not limited by mass transport, then the complex formation rate follows the following law:

$$\frac{d[\mathrm{LA}]}{dt} = k_{\mathrm{ass}}.[\mathrm{A}].[\mathrm{L}] - k_{\mathrm{diss}}[\mathrm{LA}] \tag{6.18}$$

where [A] is the ligand concentration, [L] is the analyte concentration, and [LA] is the ligand–analyte complex concentration. k_{ass} and k_{diss} are the association and dissociation constants of the hybridization reaction, respectively.

Equation (6.18) can be simplified by assuming that the analyte concentration remains constant (due to continuous injection) and the effective concentration of the free ligands [L] is the difference between the initial concentration Γ_0 and the concentration of the complex at the wall, Γ. It can be written as follows:

$$\frac{d\Gamma}{dt} = k_{ass}C_0(\Gamma_0 - \Gamma) - k_{diss}\Gamma \tag{6.19}$$

To simplify the equation, we note that Γ = [LA], C_0 = [A], and Γ_0 = [L].

Solving the Eq. (6.19) leads to

$$\frac{\Gamma}{\Gamma_0} = \frac{k_{ass}C_0}{k_{ass}C_0 + k_{diss}}[1 - e^{-(k_{ass}C_0 + k_{diss})t}] \tag{6.20}$$

From this relation, the kinetics of surface concentration can be sketched as shown in Figure 6.4.

If a buffer solution is injected in place of the analyte, only the dissociation mechanism of the complex formed on the surface occurs (Figure 6.4), and the Langmuir equation is reduced to the simpler form

$$\frac{d\Gamma}{dt} = -k_{diss}\Gamma \tag{6.21}$$

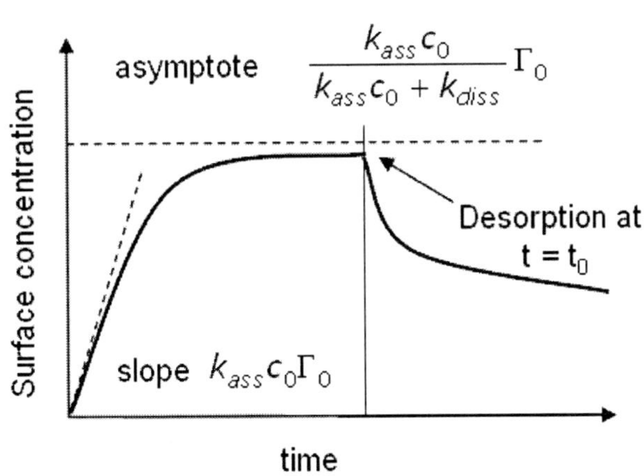

Figure 6.4. Surface concentration of the ligand–analyte complexes as a function of the time (kinetic curves) obtained from Eqs. (6.20) and (6.22).

From this latter equation, the dissociation kinetics are then given by

$$\frac{\Gamma}{\Gamma_a} = e^{-k_{\text{diss}}(t-t_a)} \tag{6.22}$$

with Γ_a the surface concentration when the buffer is injected:

$$\Gamma_a = \frac{k_{\text{ass}}C_0}{k_{\text{ass}}c_0 + k_{\text{diss}}}\Gamma_0 \tag{6.23}$$

More generally, in a microfluidic cell, the analyte is transported toward the surface by convection and diffusion, as it is injected in continuous flow. The ligand–analyte reaction, especially its kinetics, depends not only on the reaction rates on the surface but also on the processes of mass transport near the surface (Figure 6.5).

During the transport process, if the diffusion is very slow, the concentration of the analyte near the surface will decrease during the reaction. As already mentioned, a boundary layer of concentration will form. Then, the overall kinetics of the reaction are controlled by the diffusion flux, and the rates of adsorption and desorption are not predominant (Marie et al. 2002). The concentration C_0 of the Langmuir model is then replaced by a variable concentration C_{surf} [Eq. (6.24)]. The hybridization rate Γ is then calculated by solving the equation

$$\frac{d\Gamma}{dt} = k_{\text{ass}}C_{\text{surf}} \times (\Gamma_0 - \Gamma) - k_{\text{diss}} \times \Gamma \tag{6.24}$$

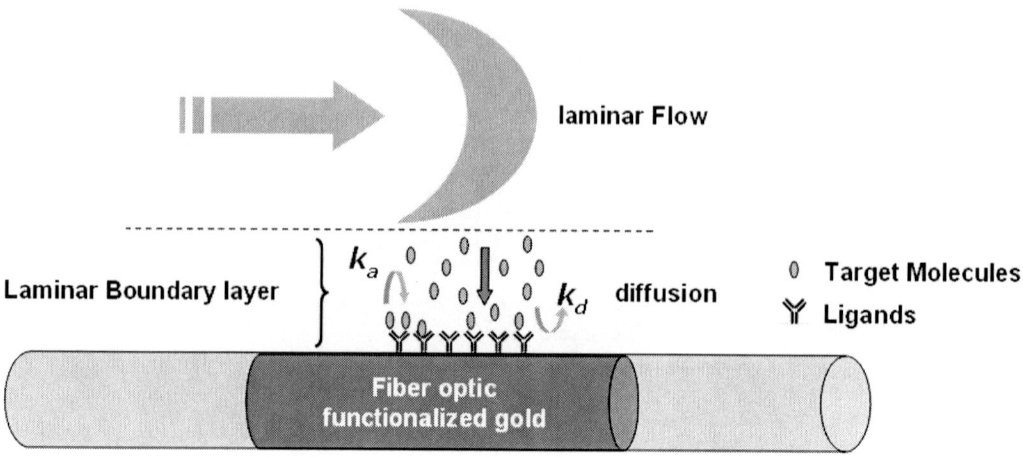

Figure 6.5. Mass transfer coefficients and phenomenon of concentration boundary layer in the vicinity of the sensitive surface.

where C_{surf} is the concentration of the analyte at the surface. Now, the coupling between the surface reaction and the transport mechanisms appears via C_{surf}. To determine the concentration C_{surf}, the balance of the masses on the surface should be introduced, taking into account the depletion of concentration of target molecules near the surface. In this way, the mechanism of advection-diffusion is used (buffer fluid flowing through the microchamber). Concentration and the gradient of concentration in the fluid is obtained by solving the usual advection-diffusion equation:

$$\frac{\partial c}{\partial t} + \mathbf{u} \cdot \nabla c = \nabla \cdot (D \nabla c) \tag{6.25}$$

Equations (6.24) and (6.25) are not independent. The coupling between the two equations is realized by Fick's law:

$$\frac{d\Gamma}{dt} = -D \nabla C\big|_{\text{surf}} \tag{6.26}$$

Then, the surface concentration of hybridized targets can be computed as a function of the analyte concentration near the surface, using Eq. (6.27):

$$\Gamma = \frac{D \nabla C\big|_{\text{surf}} + k_{\text{ass}} C_{\text{surf}} \Gamma_0}{k_{\text{ass}} C_{\text{surf}} + k_{\text{diss}}} \tag{6.27}$$

Concerning the boundary condition at the active surface, the rate of the reaction at the surface is balanced with the flow of the analyte and the concentration of the adsorbed species as described by the following equation:

$$\mathbf{n} \cdot (-D \nabla c + c \times v) = -k_{\text{ass}} C_{\text{surf}} \times (\Gamma_0 - \Gamma) + k_{\text{diss}} \times \Gamma \tag{6.28}$$

Here, the right side of the equation is the source term S (gain or loss of concentration), already introduced in Eq. (6.14), and $\mathbf{n}$ is the normal vector to the wall. Solving the problem is then complicated because we have to solve simultaneously equations describing the strong coupling between the hybridization reactions and the mass transport effects (convection-diffusion).

2.6. SURFACE PLASMON RESONANCE ABSORPTION

When the analyte reacts with the ligands attached to the surface, the refractive index at the sensor surface increases. Then, any modification of the concentration C_{surf} on the surface will produce a change in the condition of surface plasmon resonance, which will be manifested by a change in the SPR signal. By this way, we are able to follow the concentration variations and the hybridization kinetics in the vicinity of the surface (Davis and Wilson 2000; Homola 2003).

In order to simulate the SPR curves, a numerical model established from Fresnel's equations, where the system analyte/gold/fiber optic can be modeled by a multilayer approach. Surface plasmon resonance is a charge-density oscillation, which can exist at the interface between metallic and dielectric media. This electronic oscillation is associated with a surface electromagnetic wave, a so-called surface plasma wave (SPW), of which the electric field is maximum at the interface and decreases evanescently into both media. Under certain specific conditions, the energy carried by photons arriving at the metal/dielectric interface can be transferred to the surface plasmons. If this transfer is achieved, part of the light is absorbed and the phenomenon is called surface plasmon resonance.

More precisely, considering a system of three superposed layers: the first (layer 1) corresponds to a dielectric medium (fiber optic) of refractive index n_0, the second (layer 2) is a metallic layer characterized by a dielectric function ε_m, and the third (layer 3) represents a dielectric medium (analyte or the hybridized layer) with a dielectric constant ε_s (Figure 6.6).

The classical SPR coupling condition is obtained by equaling the tangential component of the wave vector k_x of the incident light and is propagated in layer 1 with the propagation constant β_{sp} of the surface plasma waves, present at the layer 2/layer 3 interface. If the incident beam is propagated in layer 1 with an incident angle θ with respect to the normal of the interface, and λ is the vacuum wavelength ($\lambda = 2\pi c/\omega$), then the coupling condition can be written as

$$\beta_{sp} = k_x \Rightarrow \left(\frac{\omega}{c}\right)\sqrt{\frac{\varepsilon_m \varepsilon_s}{\varepsilon_m + \varepsilon_s}} = \sqrt{\varepsilon_0}\,\frac{\omega}{c}\sin\theta \tag{6.29}$$

The values of the dielectric functions, ε_0 and ε_m defined above, were determined previously. For ε_0, the analytic formula of the fused silica dielectric constant at room temperature should be given by the Schott glass formula (power-series approximation to the Sellmeier), which is valid for the wavelength range of interest and lies between 500 and 1200 nm (Wolfe 1978; Tropf et al. 1995).

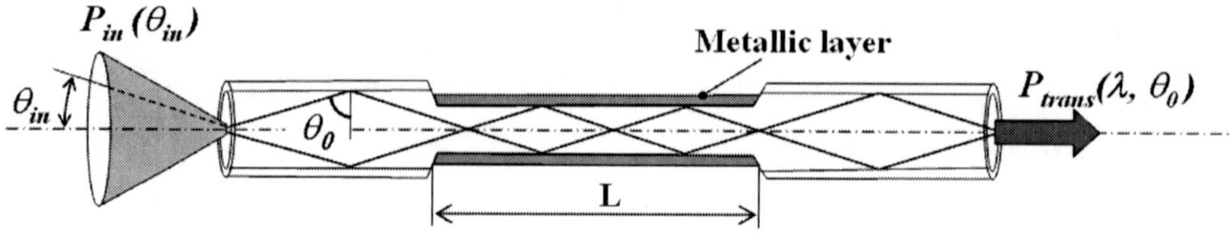

Figure 6.6. Schematic illustration of the light path in the fiber optic SPR sensor.

The dielectric function, ε_m, corresponding to a gold film, was obtained following the approach developed by Etchegoin et al. (2006). The frequency-dependent optical properties of gold in the visible/near-UV range can be represented by an analytical formula (6.30) with three main components: the contribution of a Drude model and two contributions related to the interband transitions $G_1(\omega)$ and $G_2(\omega)$:

$$\varepsilon_{\mathrm{Au}}(\omega) = \varepsilon_\infty - \frac{\omega_p^2}{\omega(\omega + i \cdot \omega_\tau)} + G_1(\omega) + G_2(\omega) \tag{6.30}$$

where ω_p is the plasma frequency, ω_τ is the damping frequency, and ε_∞ is the high-frequency dielectric constant. The critical-point transitions, $G_1(\omega)$ and $G_2(\omega)$, were used as described by Leng et al. (1998).

The response curves of the fiber optic–based sensor are computed, and then the location of the lowest transmitted power and the width of the dip are extracted. To achieve this goal, we developed modeling for the reflectance R of the light in the optical fiber, based on the Fresnel transfer matrix formalism. We calculate the transmission of the sensing fiber by computing the multiple reflections of the incident ray in the fiber. Figure 6.6 is a representation of a light ray propagating through the fiber.

A matrix method developed for analysis of thin-film optical filters has been used to compute the reflection coefficients. This part is based on the calculation of the light propagation through a multilayer medium (Yeh 1988).

The resulting transmitted power P_{trans} through the complete fiber is then calculated by integrating the product of the light reflectance R^N by the angular power distribution $\boldsymbol{P}(\theta_0)$. The whole light reflectance is obtained by adding 50% of light reflectance for a p-polarization, R_p, with 50% of light reflectance for s-polarization, R_s:

$$P_{\mathrm{trans}}(\lambda, \varepsilon_N) \propto \left[\int_{\theta_{\mathrm{crit}}}^{\pi/2} \left[R_p(\lambda, \theta_0, \varepsilon_N) \right]^{N_{\mathrm{ref}}(\ell,\phi,\theta)} P(\theta_0)\, d\theta_0 + \ldots \right.$$
$$\left. \ldots + \int_{\theta_{\mathrm{crit}}}^{\pi/2} \left[R_s(\lambda, \theta_0, \varepsilon_N) \right]^{N_{\mathrm{ref}}(\ell,\phi,\theta)} P(\theta_0)\, d\theta_0 \right] \tag{6.31}$$

where $\theta_{\mathrm{crit}} = \sin^{-1}(n_{\mathrm{cl}}/n_0)$ is the critical fiber angle, n_{cl} is the refractive index of the cladding, and $N_{\mathrm{ref}} = L/(D \cdot \tan\theta_0)$ is the number of reflections within the active area of the sensor, with D and L representing the core diameter and the length of the active area, respectively (Figure 6.6).

The response curves represent the transmittance $T(\lambda)$ through the fiber as a function of the wavelength. The transmittance is defined as the ratio between the transmitted power P_{sens} through the fiber, which is immersed into the analyte, and the transmitted power P_{ref} obtained for a blank medium:

$$T(\lambda)=\frac{P_{\text{sens}}(\lambda)}{P_{\text{ref}}(\lambda)}=\left\{\frac{\int_{\theta_{\text{crit}}}^{\pi/2}\left[\left(R_p^{\text{sens}}\right)^N+\left(R_s^{\text{sens}}\right)^N\right]P(\theta)\,d\theta}{\int_{\theta_{\text{crit}}}^{\pi/2}\left[\left(R_p^{\text{ref}}\right)^N+\left(R_s^{\text{ref}}\right)^N\right]P(\theta)\,d\theta}\right\} \tag{6.32}$$

Numerical calculations for theoretical SPR curves were performed using Matlab software. For these calculations, the dielectric constant of fused silica was determined analytically using experimental data which were obtained from a handbook of optics (Wolfe 1978; Tropf et al. 1995). Finally, for the numerical calculations, the length of the sensing area, L, the gold thickness, d_m, and the core diameter of the fiber optic, D, were 20 mm, 50 nm, and 300 μm, respectively.

3. FINITE-ELEMENT MODELING

In order to model the experiments, calculation of the optical transmission had to be coupled to the mass transport. The equations used to solve the liquid circulation in a microfluidic system, generally derived from the Navier-Stokes equations, are associated with simplifying assumptions, particularly about the geometry of the channels. To resolve the partial derivative equations (PDEs) and for the

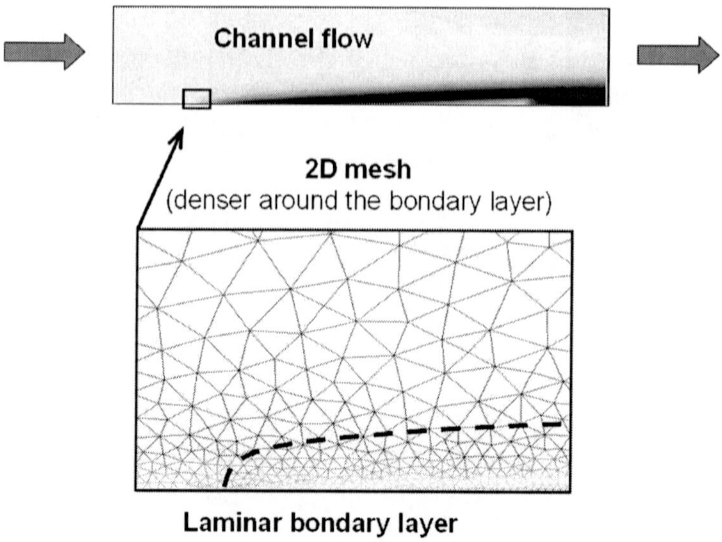

Figure 6.7. Gradient of the analyte concentration near the active surface, also called the concentration boundary layer (modeling obtained from finite elements analysis), and detail of the mesh around the starting point of the sensitive surface.

discretization of space, the Comsol Multiphysics software (version 3.5a) was used. The coupling with the optical measurements was implemented under Matlab 7. Figure 6.4 presents the experimental and simulation curves of the light transmission at 600 nm versus time (injection of a water/glycerol solution).

To model our problem, we proceeded as follows: (1) determination of the physical model, the geometry of the problem, and the boundary conditions; (2) creation of the mesh, discretization of the area according to simple geometric elements, and checking the stability of results, particularly in the concentration boundary layer; and (3) computation of the required variables, the concentration field and the surface concentration, and the hybridization rate that characterize the functionalized surface.

The modeling of a 2-D geometry was chosen (Figure 6.6). This choice was motivated by several reasons, one being the symmetry of the axisymmetric cell that allows this approximation. The second is related to the computing time. In addition, to reflect the Peclet criterion, the mesh size should not exceed a few micrometers in the boundary layer. This requires using very fine meshes; more than 90,000 meshes (200,000 degrees of freedom) was necessary for the 2-D geometry (see Figure 6.7).

In these simulations, a laminar flow was considered. A Hagen-Poiseuille type of profile was implemented [Eq. (6.10)], with the mean velocity of the flow ranging from 1 to 50 mm/s), and a channel width $2d$ = 300 μm. This velocity was estimated by dividing the rate given by the pump through the channel section in the cell. We considered water as a carrier fluid, and simulations were performed with density ρ = 1000 kg/m^3 and viscosity η = 10^3 Pa s.

The equation of diffusion/advection was then introduced, reflecting the mass transport of microparticles in the carrier fluid [Eq. (6.15)]. The diffusion coefficient is generally between 10^{-11} and 10^{-9} m^2/s for organic molecules in water. We chose an arbitrary value of 10^{-9} m^2/s, close to the conventionally accepted coefficient for proteins.

The hybridization rate on the active surface (or concentration) was then extracted versus time from these simulations, and the results were used directly to calculate the curves of transmitted power by the fiber optic with the model presented above. Indeed, from simulations, changes in concentration at the sensitive surface can be extracted (advection/diffusion model). The average concentration was then obtained by integrating the concentration over the entire length of the sensitive area, and a mean refractive index of the surface was estimated from $\bar{n}_s = r \times n_s$ with

$$r = \frac{\bar{c}}{c_0} = \frac{1}{c_0}\int_0^L c \cdot d\ell \qquad (6.33)$$

4. QUANTIFICATION OF THE REACTION BY SPR

The optical effects induced by surface plasmon resonance and the partial derivative equations representing mass transport of analytes were calculated simultaneously, taking into account the geometry of the cell. In this configuration, the channels had a dimension of approximately 320 μm, and the speed of circulation U of analytes was about 1–50 mm/s (William and Zimmerman 2005).

To monitor real-time changes of the surface, the SPR signal versus time is drawn on a graph known as a sensorgram (Figure 6.8). A typical sensorgram includes three phases: (1) During injection of the analyte, the signal increases as the analyte interacts with the immobilized ligand, and as the molecular association events are predominant on the dissociation events. At steady state (events of association and dissociation equivalent), the signal is then stable. (2) During the postinjection phase, the flow of analyte is replaced by the buffer. As the dissociation occurs, the signal decreases. (3) The regeneration phase (injection of a buffer solution) is used to separate all the analyte. Posttreatments of sensorgrams allow quantifying the reaction and determining the molecular constants of association, k_{ass}, and dissociation, k_{diss}. The ratio k_{diss}/k_{ass} is the affinity constant of the interaction, k_{aff} (Lahmani et al. 2007).

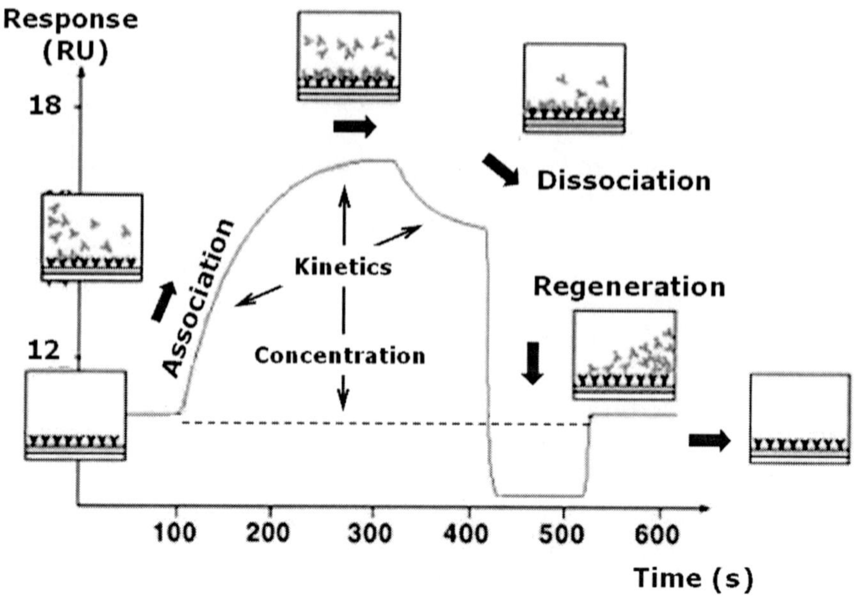

Figure 6.8. Sensorgram showing the evolution of the SPR signal over time. Changes in the signal, expressed in resonance units (RU), allow performing kinetic analysis, to calculate the association (k_{ass}) and dissociation (k_{diss}) constants of the interaction, and the concentration of the analyte.

With SPR signal R and for the Langmuir model, the equation of the kinetic reaction can be expressed as

$$\frac{dR}{dt} = k_{ass}c_{surf}(R_{max} - R) - k_{diss}R \tag{6.34}$$

where R is the response of the biosensor (RU), which can be converted to a material quantity per unit of area. During the dissociation step, we will obtain

$$\frac{dR}{dt} = -k_{diss}R \tag{6.35}$$

To determine the constants k_{ass} and k_{diss}, two methods can be used, either at steady state or by a kinetic analysis.

- **Analysis at steady state (dR/dt = 0)**

At steady state, Eq. (6.34) becomes

$$k_{diss}R_{eq} = k_{ass}c_{surf}\left(R_{max} - R_{eq}\right) \tag{6.36}$$

With $k_{aff} = k_{ass}/k_{diss}$, the affinity constant of the interaction ligand/analyte, we can write

$$\frac{R_{eq}}{c_{surf}} = k_{aff}(R_{max} - R_{eq}) \tag{6.37}$$

If the process is not limited by diffusion, then $c_{surf} \approx C_0$. It will be possible to determine C_0 if k_{aff} is known, or we will able to calculate k_{aff} if C_0 is known.

- **Kinetic analysis**

The kinetic analysis is a real-time analysis, away from the steady state. In this case, Eq. (6.34) becomes

$$\frac{dR}{dt} = k_{ass}C_0R_{max} - (k_{ass}C_0 + k_{diss})R \tag{6.38}$$

This equation describes the variation of the sensor response over time. When the experimental conditions are satisfied (flow), the values of k_{ass} and k_{diss} can be determined by fitting.

5. EXPERIMENTAL PROCEDURE

The microfluidic cell, designed for this study, is shown in Figure 6.9. It measures the optical responses of the sensor according to the flow and the concentration of the analyte versus time. Multimode fibers were used. First, the optical cladding was removed from 2 cm of the central part of the fiber. A thin gold layer was deposited on the bare zone by thermal evaporation. To obtain a homogeneous film, we used a mechanical system coupled to a stepper motor that allows rotation of the fiber (Balaa et al. 2007; Makiabadi et al. 2011).

The experimental setup consists of a source of white light connected to the input of the optical fiber. This fiber is placed in a glass capillary tube where the analyte flows. A miniature optical fiber spectrometer is connected at the output of the fiber in order to measure the light transmission through the fiber (Kanso et al. 2007). In this way, the response of the sensor versus the wavelength of the input light (550–1100 nm) can be recorded as a function of the hybridization of the analyte on the surface.

The microfluidic system consists of a roller pump, channels with an internal diameter of 1 mm, and a cell crossed by the optical fiber sensor where the expected hybridization occurs on the gold surface during the flow of the analyte.

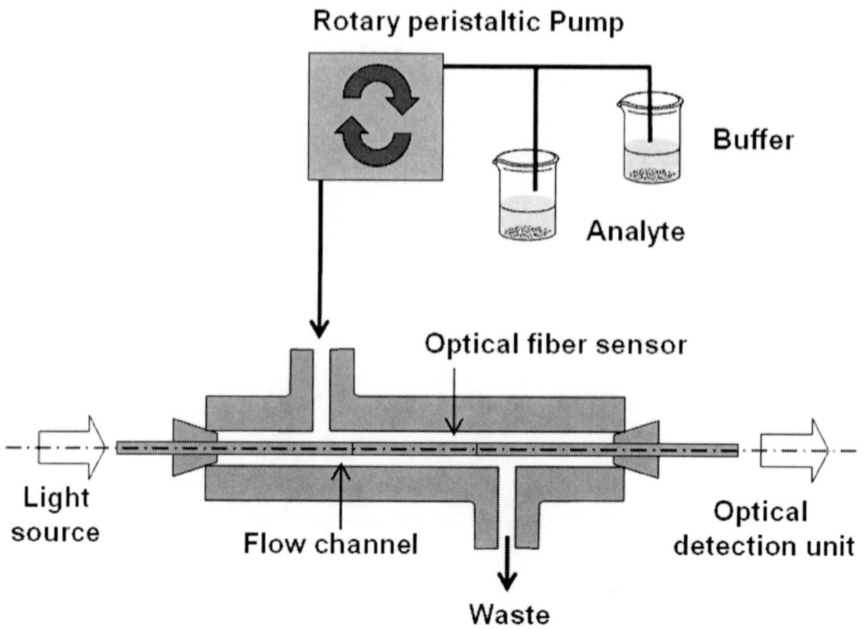

Figure 6.9. Schematic illustration of the fiber optic SPR setup used in this study.

6. NUMERICAL RESULTS

In order to study the response of our sensor according to the biological layer formed by the hybridization of the analyte on the target ligands, this response as a function of the dielectric constant and thickness of the organic was modeled. In particular, we modeled a multilayer system composed of the core of the fiber, the thin metal film, the biological layer, and the carrier fluid solvent.

Figure 6.10 shows the evolution of the SPR response of the fiber optic sensor according to the thickness of the biological layer deposited on the sensitive area. These simulations were obtained with a gold layer of 50 nm. We set the index n_s = 1.3335 to characterize the carrier liquid of target molecules, because most biological studies take place in aqueous media. A linear shift of the surface plasmon resonance is observed with increasing thickness of the organic layer. We note that the shifts in resonance wavelength are significant for a small variation in the thickness of the organic layer, indicating good sensitivity of the sensor relative to surface reactions. This sensitivity will be even better for molecules with higher molecular weight.

On the other hand, Figure 6.11 shows, for a wavelength set to 616 nm, changes in light transmission through the fiber as a function of the thickness of

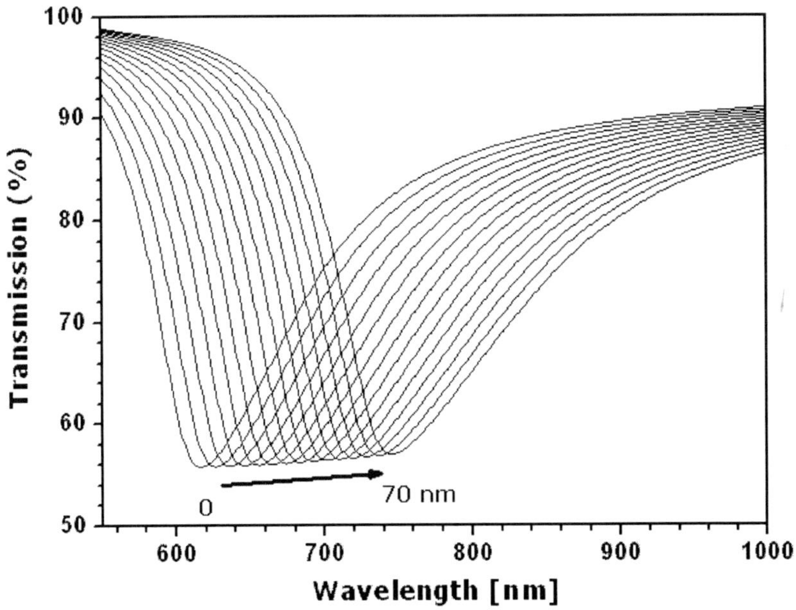

Figure 6.10. Simulated SPR response curves as a function of the thickness of the biologic layer attached on the gold surface (refractive index of the buffer n_s = 1.3335, thickness of the gold thin film = 50 nm, refractive index of the biologic layers n_{bio} = 1.45).

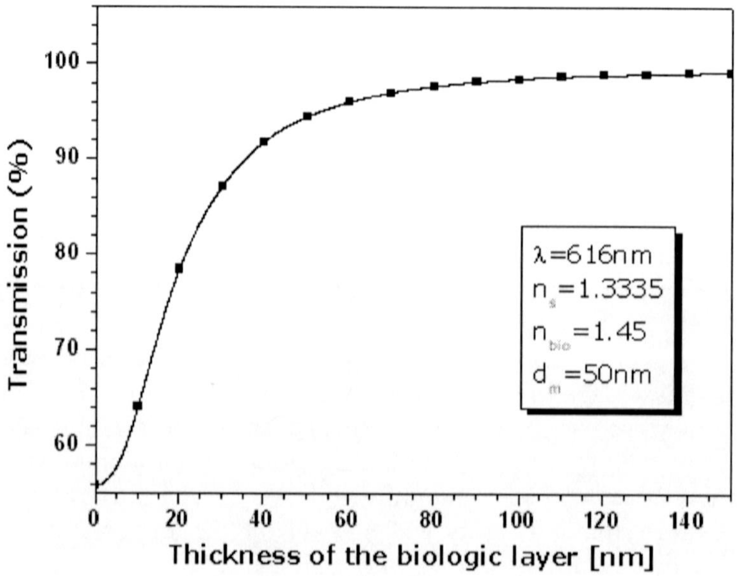

Figure 6.11. Transmitted power (λ = 616 nm) in the fiber optic according to the thickness of the organic layer on the surface (n_s = 1.335, n_{bio} = 1.45).

the biological layer. A saturation of SPR signal when the thickness of the deposited layer is greater than 60 nm is observed. Thus, in this study we will stay in the linear range of the curve.

To take into account the hybridization rate Γ during the formation of the organic layer, the total thickness of the organic layer was set to d_T, and the refractive index was set to ε_m. An intermediate index of the biological layer (n_m), from the carrier medium (solvent) and the partially hybridized layer, can be expressed by the following equation:

$$n_m(t) = \Gamma(t) \times \sqrt{\varepsilon_{\text{bio}}} + \left[1 - \Gamma(t)\right] \times \sqrt{\varepsilon_s} \tag{6.39}$$

where ε_{bio} and ε_s are the dielectric constants of the organic layer and the carrier fluid, respectively. Figure 6.12 depicts the variation of resonance wavelength as a function of Γ.

In this section, the study evaluates the effect of the organic layer thickness on the fiber optic SPR sensor responses. In the next section, the kinetics between biologic reactions and the refractive index changes of the organic layer will be introduced.

Taking into account the ligand–analyte interaction in our numerical model requires the introduction of the values of kinetic parameters (k_{ass} and k_{diss}). These parameters contribute to describe the boundary condition and the molecular response to the sensing area. For this, we used the experimental results obtained by Zimmermann et al. (1999) on the hybridization of protein kinase inhibitor (PKI) protein grafted on the gold surface, and the human X chromosome protein kinase (PrKX) protein as analyte. From this study, values needed for our numerical model could be extracted, and numerical investigations of the SPR responses of our fiber optic SPR sensor were performed. In particular, values of k_{ass} and k_{diss} were extracted from these curves and modeling was performed with 3.8×10^5 $M^{-1}s^{-1}$ and 5.8×10^{-3} s^{-1}, respectively.

Figure 6.13 presents the kinetic response of the refractive index (n_m) versus time for different values of analyte concentration. We note that the changes in the refractive index of the biological layer are proportional to the concentration of the analyte in the injected solution. It varies as a function of the hybridization rate on the surface. However, after a sufficiently long time, balance between the two processes of association and dissociation occurs, and the steady state is observed (see Figure 6.14). In this case, the refractive index of the biological layer has a constant value which corresponds to the maximum of the hybridization rate of the surface (under these conditions). This maximum depends on the value of the concentration of the analyte injected.

In the simulation, the thickness of the biological layer was set to $d_T = 20$ nm. At the initial time $t = 0$, the analyte concentration was zero. After an injection time of 280 s, another solution (buffer) was injected to wash the sensitive area. In this case, the reaction kinetics depends only on k_{diss} and on the value of the concentration in the previous steady state. The equation of the coupling takes the form

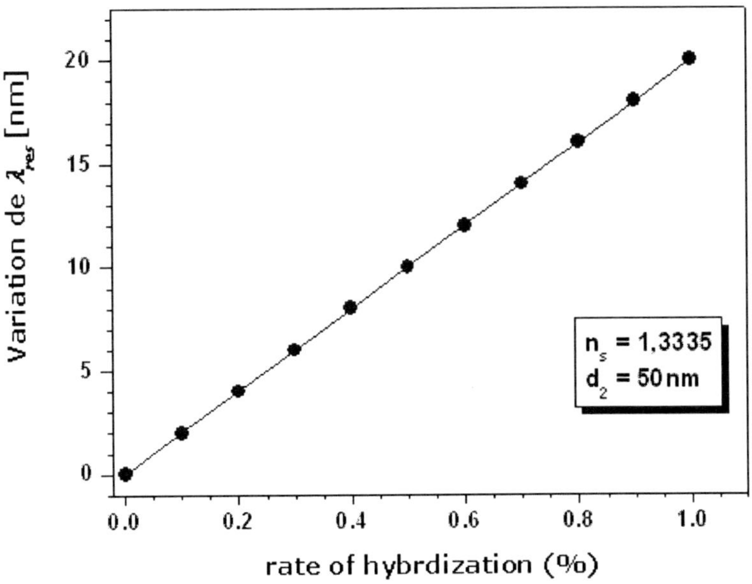

Figure 6.12. Variation of the plasmon resonance wavelength as a function of the hybridization rate on the surface. Organic layer thickness fixed at 15 nm and a refractive index ranging from 1.335 to 1.41.

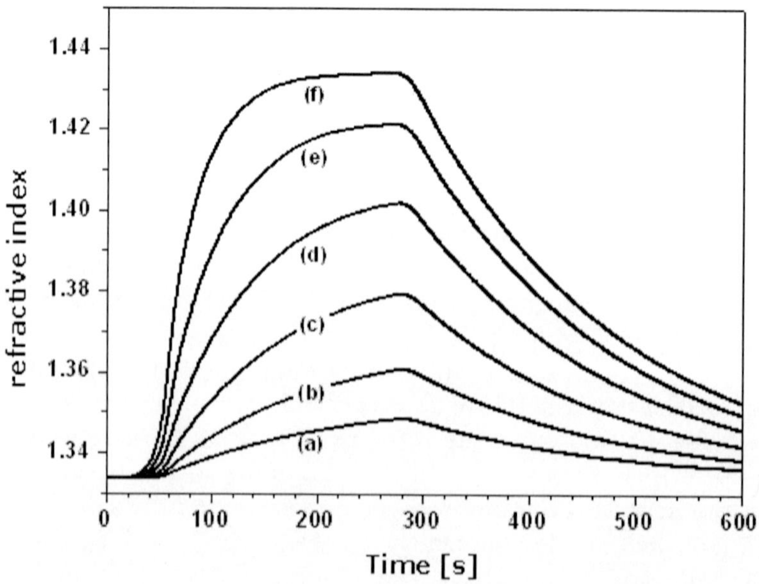

Figure 6.13. Refractive index variation of the hybridized layer of protein PKI (ligand) with protein kinase X (analyte) versus time as a function of analyte concentration: (a) 3 nM; (b) 6 nM; (c) 12 nM; (d) 24 nM; (e) 48 nM; (f) 96 nM. (U = 6 mm/s; $D = 10^{-10}$ m^2/s; $k_{ass} = 3.8 \times 10^5$ M^{-1} s^{-1} and $k_{diss} = 5.8 \times 10^{-3}$ s^{-1}).

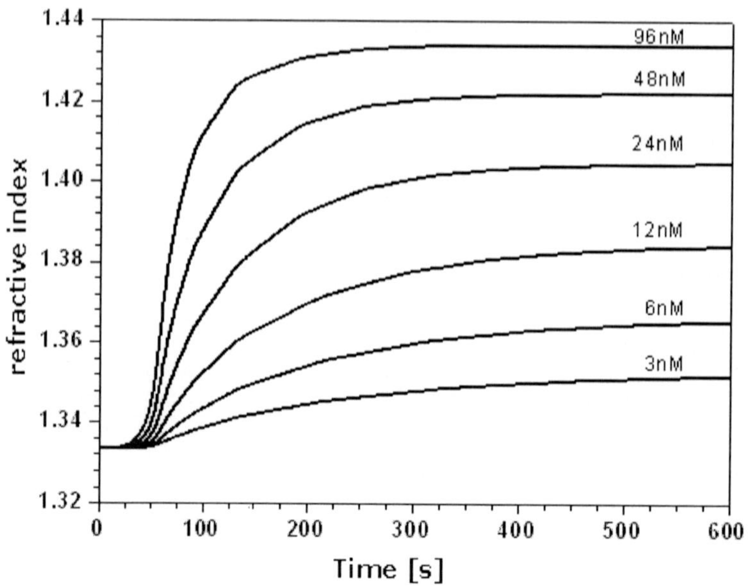

Figure 6.14. Refractive index variation of the organic layer on the gold surface as a function of analyte concentration. (ε_{bio} = 1.45; ε_s = 1.3335; U = 6 mm/s; $D \approx 10^{-10}$ m^2/s, $k_{ass} = 3.8 \times 10^5$ M^{-1} s^{-1} and $k_{diss} = 5.8 \times 10^{-5}$ s^{-1}.)

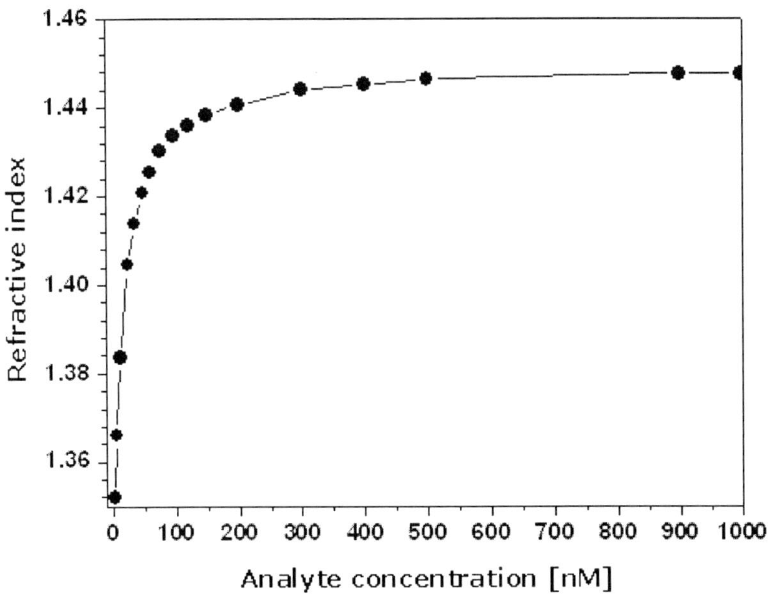

Figure 6.15. Refractive index of the organic layer at steady state as a function of analyte concentration.

$$-D\nabla c + c \cdot \mathbf{u} = -k_{\text{diss}}\Gamma \tag{6.40}$$

To consider this step in the numerical simulations, we performed simulations for a period of 320 s, after the final steady state of the first injection. A decrease of the refractive index from the stationary state is shown in Figure 6.13. This decrease reflects the process of dissociation of the hybridized complex.

For instance, Figure 6.15 shows the evolution of the refractive index of the biological layer as a function of the concentration of the analyte (PrKX), between 0 and 1000 nM. For low concentrations, the refractive index increases linearly versus the concentration, while for high concentrations, greater than 200 nM, the index has a constant value. In other words, beyond 200 nM, the small change in the index shows the saturation of active sites (ligands). This defines the operating range for each experiment, in order to avoid saturation of the signal of the sensor.

In Figure 6.16, the position of the plasmon resonance peak versus the concentration of the analyte is reported. We note the nonlinear behavior of the response above 20–30 nM. From this curve we can still estimate a theoretical sensitivity of our sensor as about 0.6 nM/nm.

$$S_{\text{capt}} = \frac{\Delta c}{\Delta \lambda_{\text{res}}} \qquad (\text{nM/nm}) \tag{6.41}$$

In the same way, it is possible to evaluate the sensitivity of the sensor in terms of mass. The equivalent mass to a layer thickness of 20 nm can be estimated as

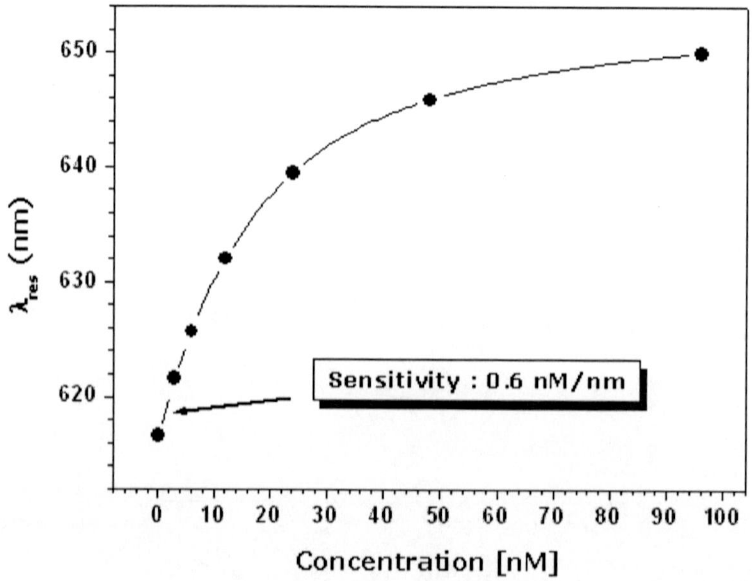

Figure 6.16. Plasmon resonance wavelength as a function of analyte concentration (values obtained after 250 s).

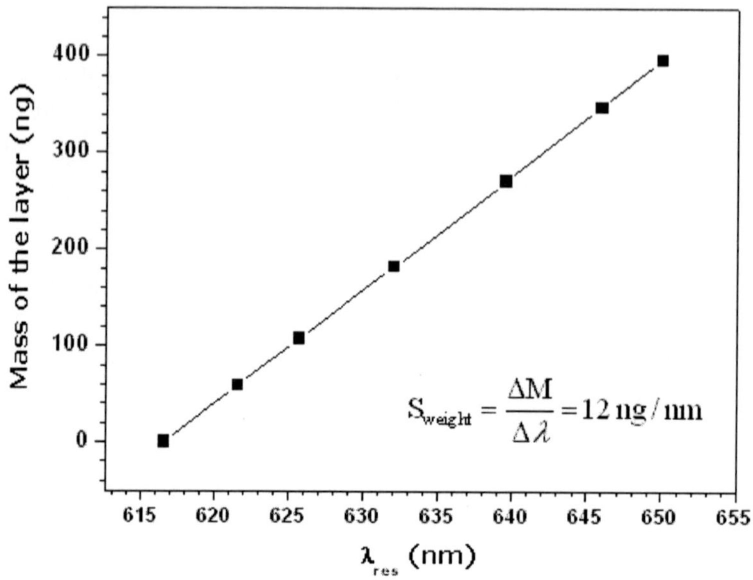

Figure 6.17. Mass changes as a function of plasmon resonance wavelength.

follows. If a fiber with a core of 300 μm and a sensitive area of 20 mm length is considered, taking into account a density of 1.2 g/cm^3, we obtain a mass for a complete layer of 450 ng. Figure 6.17 presents the position of the plasmon resonance as a function of the deposited mass. A sensitivity of this mass sensor of about 12 ng/nm can then be evaluated.

$$S_{\text{mass}} = \frac{\Delta M}{\Delta\lambda_{\text{res}}} (\text{ng/nm}) = 12 \text{ ng/nm} \qquad (6.42)$$

7. CONCLUSION

Based on this modeling, we believe that this numerical model can be used to interpret the experimental results obtained with a fiber optic sensor in a microfluidic environnement. This approach shows the importance of taking into account the convection, diffusion, and reaction kinetics of hybridization. For example, Figure 6.18 shows the modeled kinetic response of the sensor for four different molecular interactions: (a) GST-PKI/PKA-C, (b) IgG/protein A, (c) GST-PKI/PrKX, and (d) GST-PrKX/RI, where the parameters k_{ass} and k_{diss} are, respectively, (1.5 ×

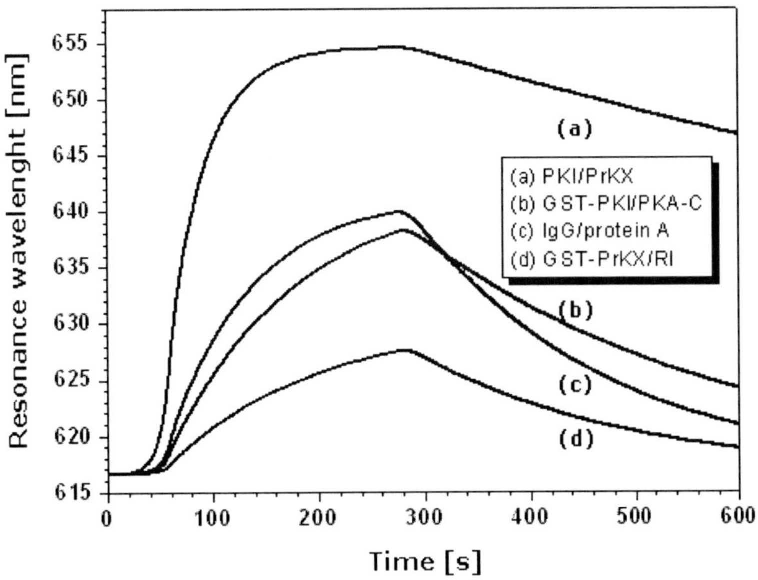

Figure 6.18. Kinetic responses of a fiber optic SPR sensor for four different ligand–analyte combinations (k_{ass} and k_{diss}): (a) $k_{ass} = 1.5 \times 10^6$ M^{-1} s^{-1}; $k_{diss} = 7.6 \times 10^{-4}$ s^{-1}; (b) $k_{ass} = 2.42 \times 10^5$ M^{-1} s^{-1}; $k_{diss} = 3.52 \times 10^{-3}$ s^{-1}; (c) $k_{ass} = 3.8 \times 10^5$ M^{-1} s^{-1}; $k_{diss} = 5.8 \times 10^{-3}$ s^{-1}; (d) $k_{ass} = 1.1 \times 10^5$ M^{-1} s^{-1}; $k_{diss} = 5.4 \times 10^{-4}$ s^{-1}. (C_0 = 25 nM; U = 6 mm/s; $D = 10^{-10}$ m^2/s; d_{bio} = 20 nm; ε_{bio} = 1.45; ε_s = 1.3335.)

10^6 M^{-1} s^{-1}, 7.6 × 10^{-4} s^{-1}), (2.42 × 10^5 M^{-1} s^{-1}, 3.52 × 10^{-4} s^{-1}), and (3.8 × 10^5 M^{-1} s^{-1}, 5.8 × 10^{-4} s^{-1}) (Zimmermann et al. 1999; Catimel et al. 1997). Note that we set the concentration of analyte at 25 nM, the flow velocity to 6 mm/s, and the diffusion constant of proteins to 10^{-10} m^2/s. These examples show the capability of the simulation of a SPR fiber optic sensor to measure different molecular interactions with different kinetics and different affinity constants.

Finally, in this work, responses of an optical fiber SPR sensor were studied with the presence of an analyte subjected to a laminar flow. This flow was carried out in a microfluidic cell, where the fiber sensor was placed. This study allowed us to evaluate the effectiveness and the sensitivity of this sensor in an actual operational configuration. The molecular interaction of PKI protein with PrKX protein was used as a model. A sensor sensitivity of 1.7 nm/nM was determined, with a mass detection limit of about 12 ng.

In conclusion, this work revealed that an optical fiber SPR sensor coupled to a microfluidic cell is able to perform biological analysis, whether quantitative or qualitative.

REFERENCES

Balaa K., Kanso M., Cuenot S., Minea T., and Louarn G. (2007) Experimental realization and numerical simulation of wavelength-modulated fiber optic sensor based on surface plasmon resonance—SPR. *Sens. Actuators B* **126**, 198–203. DOI: 10.1016/j.snb.2006.11.026

Berthier J. and Silberzan P. (2006) *Microfluidics for Biotechnology.* Artech House, London, UK.

Catimel B., Nerrie M., Lee F.T., Scott A.M., Ritter G., Welt S., Old L.J., Burgess A.W., and Nice E.C. (1997) Kinetic analysis of the interaction between the monoclonal antibody A33 and its colonic epithelial antigen by the use of an optical biosensor: A comparison of immobilisation strategies. *J. Chromatogr. A* **776**, 15–30. DOI: 10.1016/S0021-9673(97)00087-3

Christensen L.L. (1997) Theoretical analysis of protein concentration determination using biosensor technology under conditions of partial mass transport limitation. *Anal. Biochem.* **249**, 153–164. DOI: 10.1006/abio.1997.2182

Davis T.M. and Wilson W.D. (2000) Determination of the refractive index increments of small molecules for correction of surface plasmon resonance data. *Anal. Biochem.* **284**, 348–353. DOI: 10.1006/abio.2000.4726

Etchegoin P.G., Ru E.C.L., and Meyer M. (2006) An analytic model for the optical properties of gold, *J. Chem. Phys.* **125**, 164705. DOI: 10.1063/1.2360270

Gervais T. and Tensen K.F. (2006) Mass transport and surface reactions in microfluidic systems. *Chem. Eng. Sci.* **61**, 1102–1121. DOI: 10.1016/j.ces.2005.06.024

Glasser R.W. (1993) Antigen-antibody and mass transport by convection and diffusion to a surface, a two-dimensional computer model of binding and dissociation kinetics. *Anal. Biochem.* **213**, 152–161. DOI: 10.1006/abio.1993.1399

Gupta B.D. and Sharma A.K. (2007) Sensitivity evaluation of a multi-layered surface Plasmon resonance-based fiber optic sensor: A theoretical study. *Sens. Actuators B* **107**, 40–46. DOI: 10.1016/j.snb.2004.08.030

Homola J., Yee S.S., and Myszka D. (2002) Surface plasmon resonance biosensors. In: Ligler F. and Taitt C. (eds.), *Optical Biosensors: Present and Future.* Elsevier, Amsterdam, 207–251.

Homola J. (2003) Present and future of surface plasmon resonance biosensors. *Anal. Bioanal. Chem.* **377**, 528–539. DOI: 10.1007/s00216-003-2101-0

Huang S.C., Lee G.B., Chien F.C., Chen S.J., Chen W.J., and Yang M.C. (2006) A microfluidic system with integrated molecular imprinting polymer films for surface plasmon resonance detection. *J. Micromech. Microeng.* **16**, 1251–1257. DOI: 10.1088/0960-1317/16/7/019

Kanso M., Cuenot S., and Louarn G. (2007) Roughness effect on the SPR measurements for an optical fibre configuration: Experimental and numerical approaches. *J. Opt. A: Pure Appl. Opt.* **9**, 586–592. DOI: 10.1088/1464-4258/9/7/008

Kanso M., Cuenot S., and G. Louarn (2008) Sensitivity of optical fiber sensor based on surface plasmon resonance: Modeling and experiments. *Plasmonics* **3**, 49–57. DOI: 10.1007/s11468-008-9055-1

Karlsson R. and Falt A. (1997) Experimental design for kinetic analysis of protein-protein interactions with surface plasmon resonance biosensors. *J. Immunol. Methods* **200**, 121–133. DOI: 10.1016/S0022-1759(96)00195-0

Karlsson R., Katsamba P.S., Nordin H., Pol E., and Myszka D.G. (2006) Analyzing a kinetic titration series using affinity biosensors. *Anal. Biochem.* **349**, 136–147. DOI: 10.1016/j.ab.2005.09.034

Kretschmann E. and Raether H. (1968) Radiative decay of nonradiative surface plasmons excited by light. *Z. Naturforsch. A* **23**, 2135–2136.

Lahmani M., Boisseau P., and Houdy P (2007) *Les Nanosciences, Tome 3: Nanobiotechnologies et Nanobiologie.* Belin, France, chap. 9.

Leng J., Opsal J., Chu H., Senko M., and Aspnes D.E. (1998) Analytic representations of the dielectric functions of materials for device and structural modelling. *Thin Solid Film* **313**, 132–136. DOI: 10.1016/S0040-6090(97)00799-2

Makiabadi T., Le Nader V., Kanso M., and Louarn G. (2011) Comprehensive study of an optical fiber plasmonic microsensor in a microfluidic device. *Eur. Phys. J. Appl. Phys.* **56**, 13704. DOI: 10.1051/epjap/2011100223

Marie R., Jensenius H., Thaysen J., Christensen C.B., and Boisen A. (2002) Adsoption kinetics and mechanical properties of thiol-modified DNA-oligos on gold investigated by microcantilever sensors. *Ultramicroscopy* **91**, 29–36. DOI: 10.1016/S0304-3991(02)00079-7

Minc N. and Viovy J.L. (2004) Microfluidique et applications biologiques: Enjeux et tendances. *C. R. Phys.* **5**, 565–575. DOI: 10.1016/j.crhy.2004.04.003

Myszka D.G. and Norton T.A. (1998) CLAMP: A biosensor kinetic data analysis program. *Trends Biochem. Sci.* **23**, 149–150. DOI: 10.1016/S0968-0004(98)01183-9

Phillips K.S. and Cheng Q. (2007) Recent advances in surface plasmon resonance based techniques for bioanalysis. *Anal. Bioanal. Chem.* **387**, 1831–1840. DOI: 10.1007/s00216-006-1052-7

Tropf W.J., Thomas M.E., and Haris T.J. (1995) Properties of crystals and glasses. In M. Bass (ed.), *Handbook of Optics Vol. II,* 2nd ed. McGraw-Hill, New York, chap. 33.

Tsoi P.Y. and Yang M. (2002) Kinetic study of various binding modes between human DNA polymerase β and different DNA substrates by surface-plasmon-resonance biosensor. *Biochem. J.* **361**, 317–325. DOI: 10.1042/0264-6021:3610317

Wang Y., Vaidaya B., Farquar H.D., Stryjewski W., Hammer R.P., McCarley R.L., and Soper S.A. (2003) Microarrays assembled in microfluidic chips fabricated from poly(methyl methacrylate) for the detection of low-abundant DNA mutations. *Anal. Chem.* **73**, 1130–1140. DOI: 10.1021/ac020683w

Wolfe W.L. (1978) Properties of optical materials. In: Driscoll W.G. (ed.), *Handbook of Optics. Optical Society of America.* McGraw-Hill, New York, sec. 7.

Yeh P. (1988) *Optical Waves in Layered Media.* John Wiley, New York, 102–117.

Zimmermann B., Chiorinis J.A., Ma Y., and Kotin R.M. (1999) PrKX is a novel catalytic subunit of the CAMP-dependent protein kinase regulated by the regulatory subunit type I. *J. Biol. Chem.* **274**, 5370–5378. DOI: 10.1074/jbc.274.9.5370

Zimmerman W.B.J. (2005) *Process Modelling and Simulation with Finite Element Methods.* Series on Stability, vol. 15. World Scientific, Singapore.

CHAPTER 7

Novel Long-Period Fiber Grating Sensor Based on Dual-Peak Resonance and SPR

Zhengtian Gu

1. INTRODUCTION

A long-period fiber grating (LPFG) is a kind of fiber device with photo-induced periodic modulation of the refractive index in the core, which was first reported in 1995 (Vengsarkar et al. 1996a). The periodicity of an LPFG is typically in the range of 100–1000 μm. An LPFG can couple the fundamental core mode to the co-propagating cladding modes, producing a series of discrete attenuation peaks in the transmission spectrum. By virtue of advantages such as ease of fabrication, low insertion loss, low-level back reflection, and compactness, LPFGs have been used increasingly over the last few years in communications (Vengsarkar et al. 1996a, 1996b) and sensing applications (Bhatia and Vengsarkar 1996; Patrick et al. 1998; James and Tatam 2003).

Because the field distributions and effective refractive index (ERI) of the cladding modes are easily affected by the surrounding medium, LPFGs are very suitable for refractive index sensing (Lee et al. 1997; Tong et al. 2002; Chong et al. 2004). However, this kind of LPFG sensor is sensitive only to a surrounding medium whose refractive index is less than that of the silica cladding, and in addition, it is not species-specific, which limits the extent of its applications. Recently,

DOI: 10.5643/9781606503201/ch7

an LPFG chemical sensor whose cladding is coated with a sensitive thin film has been attracting much interest and attention. N. D. Rees coated an organic-material thin film on the cladding of an LPFG using Langmuir-Blodgett technique, and studied the influence of the thickness of the overlay material on the LPFG response (Nicholas et al. 2002). Ignacio et al. (2005, 2006), Wang et al. (2005), and Cusano et al. (2006) studied the resonance shift and the mode transition in high-refractive-index-coated long-period gratings. The theoretical analysis as well as experimental data suggest that the sensitivity of LPFG-based chemical/biosensors can be improved by appropriate choice of the cladding mode order, the film thickness, and the refractive index, and there is an optimum overlay thickness for each combination of ambient and overlay refractive indices. Based on this coated LPFG structure, Pilla et al. (2005), Gu et al. (2006), and Jesus et al. (2007) designed optical chemo-, gas-, and pH sensors. The coated LPFG sensors overcome the limitation of noncoated LPFGs that the refractive index of the external medium must be less than or equal to that of the cladding, and has the advantages of specification of chemical analyte and broad response range.

In 1999, Shu et al. (1999) found that when the grating period of an LPFG was short (~100 μm), dual resonance peaks of the higher cladding mode appeared during the writing process of the LPFG. They also found that the dual resonance wavelengths shifted in opposite directions with variation of the environmental refractive index. Based on this dual peak effect, Shu and Huang (1999) fabricated a kind of high-sensitivity liquid concentration sensor, for which the sensitivity of the LP_{015} cladding mode is 20 times higher than that of the scheme with conventional grating period (400 μm). If the dual-peak resonance effect is utilized in an LPFG coated with a sensitive thin film, the LPFG not only has high sensitivity to refractive index, due to dual peak resonance, but also enlarges the sensing application range which can be applied in monitoring the solution concentration, gas concentration, and so on. When this kind of LPFG is brought into contact with the external surroundings, the change of refractive index of the thin film leads to variation of the interval between the dual resonance peaks. This coated LPFG based on dual peak resonance is first discussed in this chapter.

Surface plasmon resonance (SPR) is one of the promising optical techniques with potential applications in various fields (Flavio et al. 2005; Chen et al. 2005; Wang and Knoll 2006; Shin et al. 2007). SPR sensors have advantages such as flexibility, low cost, and small size (Lotierzo et al. 2004; Gupta and Sharma 2005). The principle of SPR sensors is that the surface plasmon wave (SPW), which can be excited at the interface between a metal film and an absorbing medium, is extremely sensitive to tiny changes in the refractive index (RI) of the absorbing medium. SPR sensors have been developed into three types: prism-coupled, integrated optical waveguide–coupled, and optical fiber–coupled sensors (Dostalek et al. 2001; Ho et al. 2001; Cao et al. 2006). In recent years, considerable attention has focused on SPR sensors based on metal-coated LPFGs, which are extremely sensitive to small changes in refractive index (Iga et al. 2005; Tang et al. 2006), thus allowing them to be used as ambient RI sensors. Integrated LPFGs with nanolayers of noble metals such as gold and silver have been receiving attention owing to their potential applications for chemical and biological sensing (Xu et al. 2005; Iadicicco et al. 2010). Some biosensor schemes have been realized based on SPR phenomena (Tang et al. 2006). Iadicicco et al. (2010) reported on the sensing performance of an arc-induced LPFG coated with a nanoscale gold coating. This type of sensor makes the tailoring of sensing performance easier and provides a valid platform for chemical sensing applications.

Though metal-coated LPFGs based on SPR have been applied in RI sensors, little theoretical analysis has been done on this kind of LPFG. In their early days, many scientists adjusted temperature and stress by heating the metal film to shift the resonance wavelength without theoretical explanation (Duhem et al. 1999; Costantini et al. 1999). Zhang (2004) and Wei (2003) reported a solution to the complex characteristic equation of an LPFG coated with a thick metal film, and analyzed the mechanism of the shift in resonance wavelength. Further, He (2006) studied the coupling and transmitted characteristics of an LPFG coated with a thin metal film, which provided a theoretical foundation for application. A SPR-based LPFG sensor coated with two kinds of film--a metal film and a sensing film—is presented and studied theoretically in this chapter. This kind of sensor not only possesses sensitive characteristics toward the surroundings of the sensing film, it also makes use of the high sensitivity of SPR in the interface between the metal film and the sensing film.

A novel LPFG sensor based on dual peak resonance (DPR) and surface plasmon resonance (SPR) is presented for high-sensitivity online monitoring of the environment. For a gas sensor, the sensitive film coated on the cladding is exposed to the surrounding gas and interacts with it, which induces a tiny variation of the film refractive index and thus results in a shift of the attenuation peak in the transmission spectrum. Corres et al. (2008) pointed out that the maximum resonance wavelength shift was guaranteed by selecting optimal sensitive film thickness. A high-sensitivity optical film LPFG sensor should respond to tiny variations of film optical parameters when sensitive films are exposed to the ambient (bio)chemical material. In order to bring the variation of the wavelength separation of the dual resonance peaks into play to the utmost, the film optical parameters and the grating structural parameters must be optimized to improve the sensor sensitivity.

Corres et al.'s (2008) experimental result was in accordance with the theoretical simulation; however, there existed some deviation in the resonance wavelength. Corres et al. thought that the deviation might result from a number of theory approximations and the neglect of the depth of the attenuation bands. In our opinion, the influence of material dispersion is an important factor in this deviation of the resonance wavelength. Iadicicco et al. (2010) pointed out that the LPFG sensor exhibited a shielding effect for analyte RI ranging from 1.32 to 1.40, which indicated that this sensor was insensitive to the specified analyte RI. In fact, the sensitivity of the coated LPFG sensor is closely related to the film optical parameters and grating parameters. Therefore, these parameters must be optimized to handle the shielding problem. Moreover, material dispersion can exert influence on sensitivity of the metal-coated LPFG sensor. In optical fiber communication waveband, the metal material dispersion is particularly obvious, and it cannot be ignored. Thus the influence of material dispersion cannot be neglected while optimizating the design of a metal-coated LPFG sensor.

In this chapter, the operation principle, sensing characteristics, and design optimization for an LPFG sensor based on DPR and SPR are introduced. First, the dual peak resonance of the coated LPFG is studied. The method of determining the dual-peak resonance wavelengths is described, and the relations between the resonance wavelength and the grating period for lower- and higher-order cladding modes are discussed. Then the chapter compares the transmission characteristics of coated and noncoated LPFGs, and points out the advantages of the coated LPFG for a refractive index sensor. In the coated LPFG, the distance between the two resonance wavelengths changes apparently with variation of the refractive index of the surrounding medium. Further, the influence of the film optical parameters and the grating structural parameters on the transmission characteristics is studied. These parameters may exert different effects on the dual peak separation, which should be considered in optimizing the design of a coated LPFG sensor.

In addition, the complex characteristic equation of the metal-coated LPFG based on SPR is set up. In solving the complex characteristic equation, the numerical overflow is resolved through approximate deformation and scaling implementation in the Bessel function. The simulation for the field distribution indicates that the proportion of the even mode's intensity in the core is much larger than that of the odd mode's intensity. In addition, there is a jumping phenomenon in certain regions of the film parameters which must be discarded in actual applications of this LPFG sensor.

Based on the above analysis, the LPFG sensor based on DPR and SPR is set up. By analyzing the relationship between the sensitivity and the film optical parameters and the fiber grating parameters, the optimal design parameters of the LPFG sensor can be obtained. Data simulation shows that the resolution of the refractive index of this LPFG sensor based on DPR and SPR is predicted to be 10^{-8}, which means it could be used as a refractive index sensor with high sensitivity.

Finally, the influence of the material dispersion of the core, cladding, metal film, and sensing films on the resonance characteristics of presented LPFG sensor is analyzed. The results show that the core and cladding material dispersion must be considered simultaneously in the theoretical model. Further, the sensitivity of the presented LPFG sensor as a function of the regional distribution of the structural parameters is discussed with consideration of material dispersion or not, and the contour line of the sensor sensitivity provides accurate and valid means for the optimal design of sensor sensitivity.

2. DUAL PEAK RESONANCE IN COATED LPFG

2.1. COUPLED THEORETICAL ANALYSIS OF COATED LPFG

The structure of a coated LPFG sensor may be regarded as a triple-clad LPFG model. The thin film and the surroundings are the second and the third claddings, respectively. Figure 7.1a shows a structural diagram of this kind of coated LPFG sensor. Figure 7.1b shows the refractive index profile of the triple-clad LPFG sensor. The refractive indices of the core, cladding, sensing film, and surroundings (gas) are n_1, n_2, n_3, and n_4, respectively; a_1, a_2, and a_3 are the radii of the core, cladding, and sensing film, respectively. The value of $(a_3 - a_2)$ equals the film thickness h_3. In the study, n_1, n_2, a_1, and a_2 of the fiber parameters are 1.4681, 1.4628, 4.15, and 62.4 μm, respectively.

The coupled mode equations that describe co-propagating interactions in an LPFG can be simplified to (Erdogan 1997)

$$\frac{dA^{co}}{dz} = i\kappa_{11-11}^{co-co}A^{co} + \frac{i}{2}\sum_{\nu}\kappa_{1\nu-11}^{cl-co}A_{\nu}^{cl} \times \sum_{m=1}^{+\infty}\left[A_m \exp\left(-i2\delta_{1\nu-11}^{cl-co}z\right)\right] \tag{7.1}$$

$$\sum_{\nu}\left[\frac{dA_{\nu}^{cl}}{dz} = +i\kappa_{1\nu-11}^{cl-co}A^{co} \times \sum_{m=1}^{+\infty}\left[A_m \exp\left(-i2\delta_{1\nu-11}^{cl-co}z\right)\right]\right] \tag{7.2}$$

where A^{co} is the amplitude of the core mode, A_{ν}^{cl} is the amplitude of the νth cladding mode, and k_{11-11}^{co-co} and $k_{1\nu-11}^{cl-co}$ are coupling constants for the core-mode–core-mode coupling and core-mode–cladding-mode coupling, respectively. They are

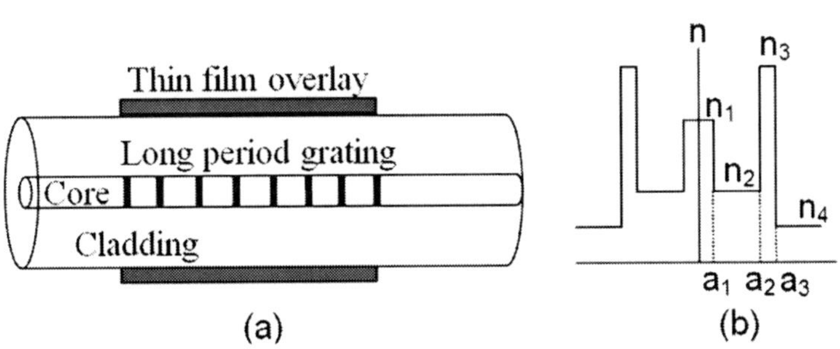

Figure 7.1. Triple-cladding model of coated long-period fiber grating sensor: (a) structural diagram; (b) refractive index profile.

related to the radial and azimuthal electromagnetic field components of the core mode and the cladding modes. $\delta_{1\nu-11}^{cl-co}$ is a small-detuning parameter between the core mode and the νth cladding mode:

$$\delta_{1\nu-11}^{cl-co} = \frac{1}{2}\left(\beta_{11}^{co} - \beta_{1\nu}^{cl} - \frac{2\pi}{\Lambda}\right) \tag{7.3}$$

$\delta_{1\nu-11}^{cl-co} = 0$ is the phase-matching condition of LPFG coupling between the fundamental core mode propagating in the fiber and the co-propagating cladding modes.

$$\beta_{11}^{co} - \beta_{1\nu}^{cl} - 2\pi/\Lambda = 0 \tag{7.4}$$

According to the expression $\beta = k_0 \cdot n_{eff}$, the phase-matching condition can also be expressed as

$$n_{eff,\,co}(\lambda) - n_{eff,\,cl}^{\nu}(\lambda) = \frac{\lambda}{\Lambda} \qquad \nu = 1, 2, 3, \ldots \tag{7.5}$$

where $n_{eff,\,co}(\lambda)$ is the effective refractive index for the fundamental core mode, which can be obtained by solving the core-mode eigenvalue equation (Gloge 1971); $n_{eff,\,cl}^{\nu}(\lambda)$ is the effective index for the νth cladding mode, which can be obtained by solving the cladding-mode eigenvalue equation (Tsao 1992). The effective refractive index of the fundamental core mode and the cladding mode is a function of the wavelength. The wavelength λ which satisfies Eq. (7.5) is called the resonance wavelength. When the grating period is very short, two dual-resonance wavelengths may satisfy Eq. (7.5). There are different dual-resonance wavelengths for different cladding modes.

Assuming that the center of the grating is the origin of the z axis, the boundary conditions for a long-period grating L in length are $A^{co}(z = -L/2) = 1$ and $A_{\nu}^{cl}(z = -L/2) = 0$. The transmissivity through the grating is defined as

$$T = \frac{A^{co}(L/2)}{A^{co}(-L/2)} \tag{7.6}$$

2.2. DETERMINATION OF DUAL RESONANCE WAVELENGTHS

In order to determine the dual peak resonance wavelengths and the corresponding mode order for a given grating period, a function $\Delta n_{eff}(\lambda)$ based on the phase-matching condition [Eq. (7.5)] is defined as follows:

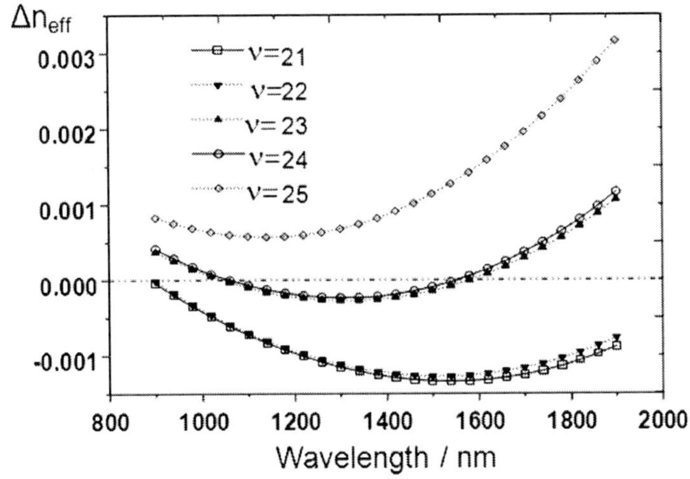

Figure 7.2. Plot of Δn_{eff} as a function of light wavelength. (Reprinted with permission from Gu et al. 2011. Copyright 2011 IOP.)

$$\Delta n_{\text{eff}}(\lambda) = n_{\text{eff, co}}(\lambda) - n^{\nu}_{\text{eff, cl}}(\lambda) - \lambda/\Lambda \tag{7.7}$$

By establishing the relations between Δn_{eff} and wavelength for different cladding modes, the resonance wavelengths can be determined by the roots of the equation $\Delta n_{\text{eff}}(\lambda) = 0$; that is, the solutions to Eq. (7.7) are dual resonance wavelengths. Figure 7.2 shows the curves of Δn_{eff} versus λ for higher cladding modes with a grating period of 142 μm. From Figure 7.2, dual resonance peaks appear for $\nu = 23, 24$, and the corresponding resonance wavelengths can also be obtained. However, there is no dual peak resonance effect for the other higher cladding mode in the given wavelength range.

For lower cladding modes, the relations between resonance wavelength and grating period for ν = 1–4 cladding modes are shown in Figure 7.3a. Considering the common observed wavelength range of 900–1700 nm, the fiber and grating parameters are selected to control the resonance wavelength in the range mentioned above. The parameters used in Figure 7.3 are a_1 = 2.625 μm, a_2 = 15 μm, h_3 = 100 nm, n_3 = 1.47, $\sigma = 2 \times 10^{-4}$ and $n_1 - n_2 = 0.0082$. From Figure 7.3, it can be deduced that the dual-peak resonance phenomenon may appear in the lower-order modes under certain conditions. However, the dual resonance wavelengths are beyond the range of 900–1700 nm for commonly used NIR fiber optic spectrometers. This is why we cannot observe the dual resonance phenomenon of the lower-order modes with usual experimental apparatus.

Figure 7.3b shows the relations between the resonance wavelength and the grating period for ν = 15–25 cladding modes. It is clear that there is dual peak resonance in the wavelength at 900–1900 nm. The smaller the grating period,

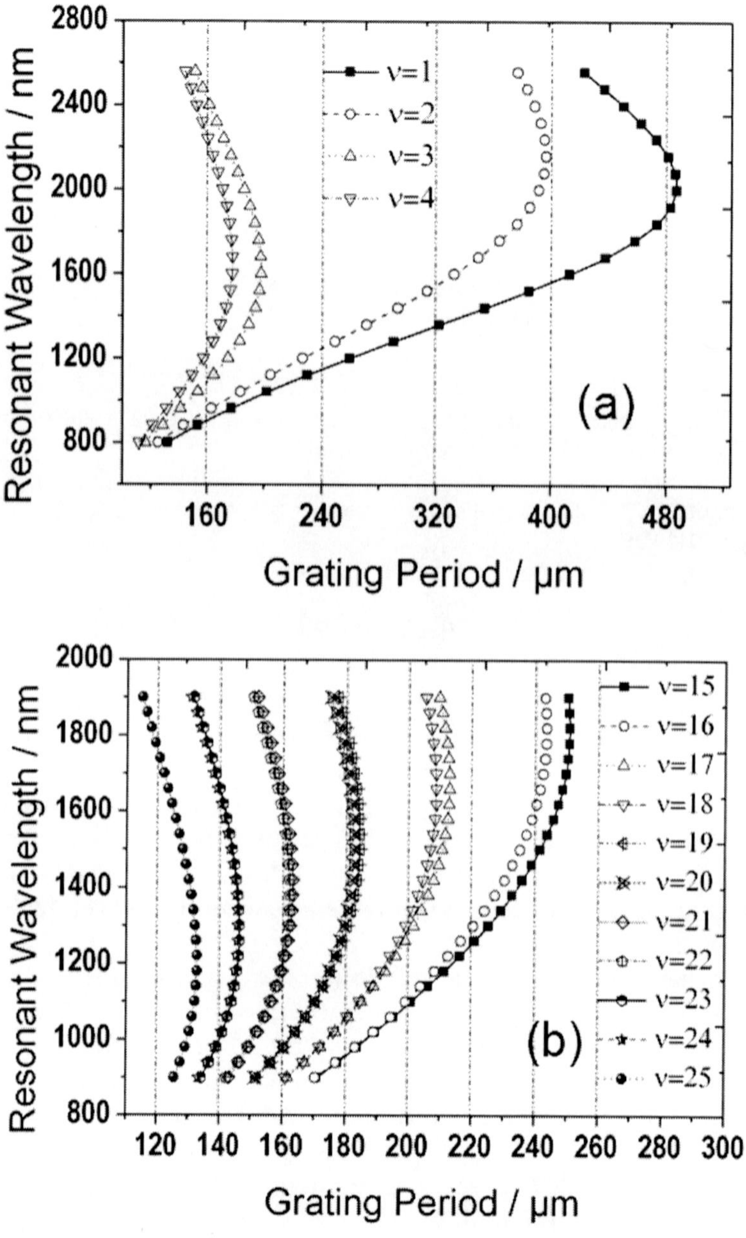

Figure 7.3. Dependence of resonance wavelength on grating period for cladding modes: (a) ν = 1–7; (b) ν = 15–25. (Reprinted with permission from Gu et al. 2011. Copyright 2011 IOP.)

the higher the cladding mode order corresponding to the dual peak resonance is, and accordingly, the smaller the dual resonance wavelengths are. In addition, the dual resonance peak wavelengths of the ν = 22, 23, 24, 25 cladding mode are in the available wavelength range of LPFGs, and their dual peak separations are larger. These cladding modes are very suitable as observed objects in a dual peak resonance sensor. According to the characteristics of the cladding mode at different root order numbers of the eigenvalue equation of the LPFG, the odd order mode represents the HE mode, and the even order mode represents the EH mode. Considering that the coupling strength of the HE mode is much stronger than that of the EH mode (Erdogan 1997), the transmission of dual resonance peaks for the $\nu = 23$ mode ($HE_{1,12}$) is discussed below.

2.3. TRANSMISSION CHARACTERISTICS OF DUAL RESONANCE LPFG

2.3.1. Transmission Property Comparison between Coated LPFG and Noncoated LPFG

In contrast to the noncoated LPFG, the coated LPFG is overlaid by a sensitive thin film, which is very sensitive to variations of the surroundings. For convenience of comparison, these two kinds of LPFGs are used as solution sensors, and the small variation of the film refractive index of the coated LPFG is assumed to be negligible in the surroundings. Figures 7.4a and 7.4b show the transmission spectra of the $HE_{1,12}$ mode of the coated and noncoated LPFGs under different refractive indices of the surroundings. The calculation parameters are listed in Table 7.1. It can be seen that the dual resonance wavelengths shift in opposite directions with variation of the environmental refractive index, and the amplitude of the resonance peak also changes with the environmental refractive index. In particular, the change of wavelength intervals between the dual resonance peaks in the coated LPFG is much larger than that in the noncoated LPFG. Thus, the sensitivity of the coated LPFG sensor based on dual peak resonance is much greater than that of the noncoated LPFG sensor.

In the following, we discuss the influence of the film parameters and the grating parameters on the dual resonance peak characteristics, which relates directly to the sensitivity of the coated LPFG sensor.

2.3.2 Transmission of Dual Resonance Peaks Under Different Film Parameters

First, the transmission characteristics of the dual resonance peaks of the coated LPFG with different film parameters are analyzed. Figure 7.5 shows the transmission spectra of the $HE_{1,12}$ mode under different film refractive indices. It can be seen that the dual resonance peaks shift in the opposite direction with variation

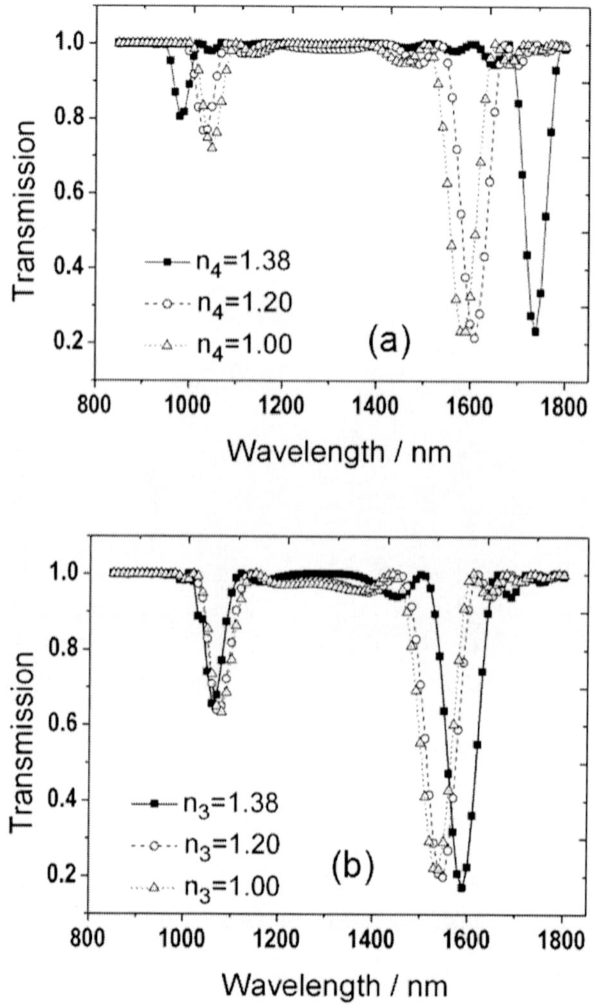

Figure 7.4. Transmission spectra of $HE_{1,12}$ of LPFG for different environment refractive indices: (a) coated LPFG; (b) noncoated LPFG. (Reprinted with permission from Gu et al. 2011. Copyright 2011 IOP.)

Table 7.1. Parameters of fiber core and claddings for calculation

	Triple-clad structure	Double-clad structure
Core	n_1 = 1.4681; r_1 = 4.15 nm	n_1 = 1.4681; r_1 = 4.15 nm
First cladding	n_2 = 1.4628; r_2 = 62.4 nm	n_2 = 1.4628; r_2 = 62.4 nm
Second cladding	n_3 = 1.57; r_3 = 62.58 nm	
Surroundings	n_4 = 1.0	n_3 = 1.0

Source: Reprinted with permission from Gu et al. 2011. Copyright 2011 IOP.

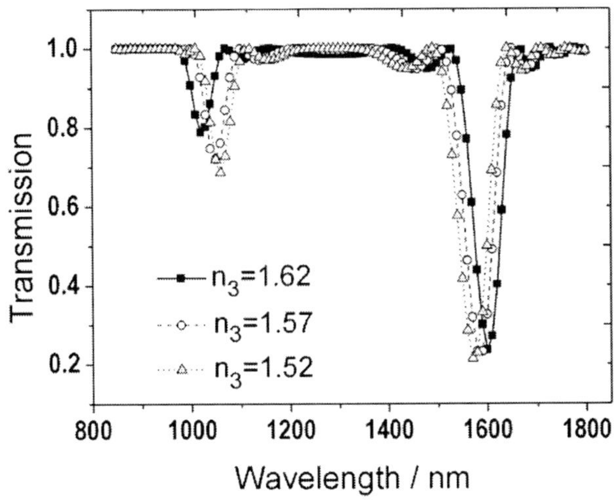

Figure 7.5. Transmission spectra of $HE_{1,12}$ for different film refractive indices: h_3 = 180 nm, σ = 4 $\times 10^{-}$, L = 1 cm, Λ = 142 μm. (Reprinted with permission from Gu et al. 2011. Copyright 2011 IOP.)

of the film refractive index. The peak separation increases with the film refractive index. The influence of the refractive index on the peak separation is similar to that in the noncoated LPFG liquid sensor (Shu and Huang 1999).

Figure 7.6 shows the transmission spectra of the $HE_{1,12}$ mode under different film thicknesses. It is clear that the dual resonance peaks shift in the opposite

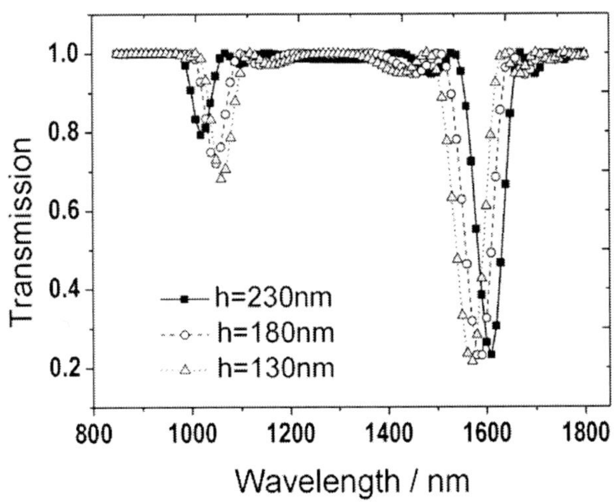

Figure 7.6. Transmission spectra of $HE_{1,12}$ for different film thicknesses: n_3 = 1.57, $\sigma = 4 \times 10^{-4}$, L = 1 cm, Λ = 142 μm. (Reprinted with permission from Gu et al. 2011. Copyright 2011 IOP.)

direction with variation of the film thickness. The peak separation increases with film thickness. From the similarity of Figures 7.5 and 7.6, we can also conclude that the film refractive index and thickness influence the wavelength shift of the dual resonance peaks, and both of them can enhance the sensitivity, which can be significantly improved by an optimum combination of film refractive index and film thickness for a specific LPFG. The optimization design is discussed in detail in Section 4.

2.3.3. Transmission of Dual Resonance Peaks Under Different Grating Parameters

Further, the transmission characteristics of the dual resonance peaks of the coated LPFG with different grating structural parameters are analyzed. Figure 7.7 shows the transmission spectra of the $HE_{1,12}$ mode under different index average modulations. It can be seen that the dual resonance peaks approach each other with an increase of average modulation. This means that when the grating is being written in a fiber, the refractive index of the core increases and the dual resonance peaks get closer together with an increase of UV exposure. This result is in good agreement with the experimental results reported by Shu et al. (1999). In addition, the average modulation also has a large influence on the amplitude and the width of the dual resonance peaks.

Figure 7.8 shows the transmission spectra of the $HE_{1,12}$ mode under different grating lengths. It is found that the grating length has no effect on the dual

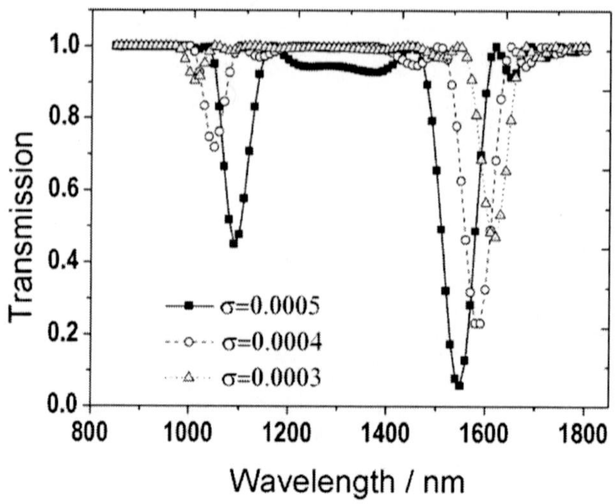

Figure 7.7. Transmission spectra of $HE_{1,12}$ for different index average modulations: n_3 = 1.57, h = 180 nm, L = 1 cm, Λ = 142 μm. (Reprinted with permission from Gu et al. 2011. Copyright 2011 IOP.)

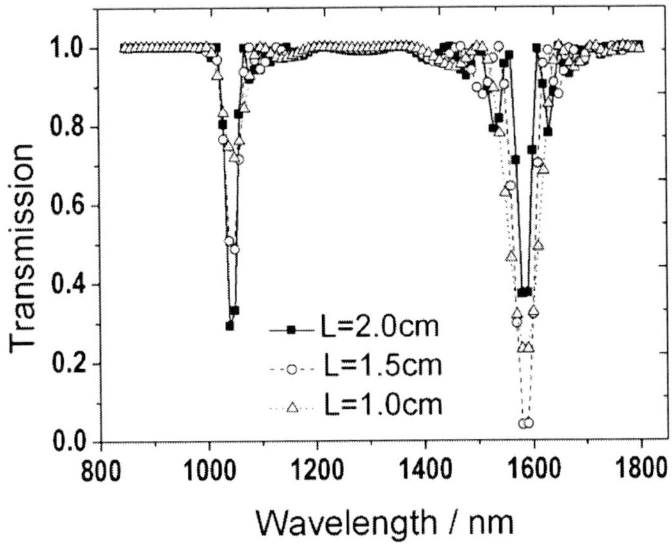

Figure 7.8. Transmission spectra of $HE_{1,12}$ for different grating lengths: $n_3 = 1.57$, $h = 180$ nm, $\sigma = 4 \times 10^{-4}$, $\Lambda = 142$ μm. (Reprinted with permission from Gu et al. 2011. Copyright 2011 IOP.)

resonance peak position, but it has an obvious effect on the bandwidth and the amplitude of the dual resonance peaks. So, the sensitivity of the sensor based on dual peak resonance is not affected by the grating length. However, for a sensor based on variation of transmissivity, it is necessary to choose the proper grating length to make the bandwidth of the resonance peak as narrow as possible,.

The transmission spectra of the $HE_{1,12}$ mode under different grating periods are shown in Figure 7.9. This figure indicates that the grating period exerts a significant influence on the dual peak positions. A tiny variation of grating period will produce a large change in the peak separation. This characteristic is very suitable for high-sensitivity stress sensors and temperature sensors.

From the above results, it can be seen that the film optical parameters and the grating structural parameters influence the position, the amplitude, and the width of the dual resonance peaks. Further, these parameters may exert different effects on the dual peak separation. For instance, the variation of dual peak separation when the film refractive index is in the range 1.57–1.62 is greater than in the range 1.52–1.57 (see Figure 7.5). This is exactly the problem one must be concerned about in optimizing the design of a coated LPFG sensor.

3. MODEL ANALYSIS OF SPR-BASED LPFG WITH METAL COATING

Figure 7.10 shows a structural diagram of the five-layer single-mode LPFG sensor. The refractive indices of the core, cladding, metal film, sensing film, and

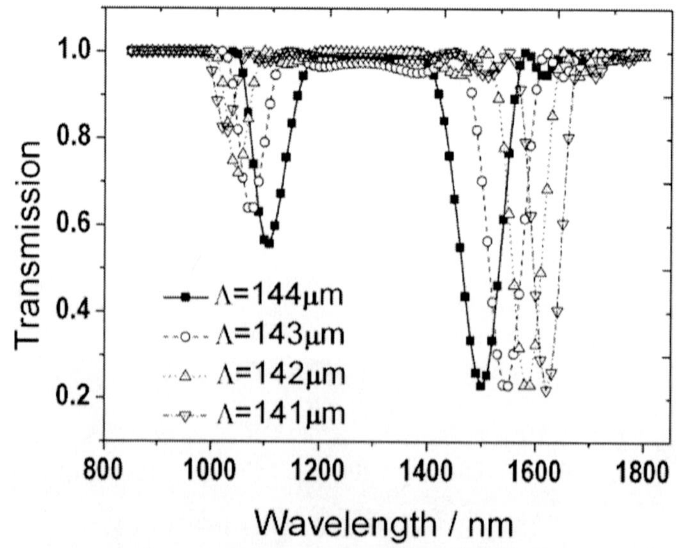

Figure 7.9. Transmission spectra of $HE_{1,12}$ for different grating periods: n_3 = 1.57, h = 180 nm, L = 1 cm, σ = 4 × 10^{-4}. (Reprinted with permission from Gu et al. 2011. Copyright 2011 IOP.)

surroundings are n_1, n_2, n_3, n_4, and n_5, respectively; a_1, a_2, a_3, and a_4 are the radii of the core, cladding, metal film, and sensing film, respectively. The value of $(a_3 - a_2)$ equals the metal film thickness h_3, and the value of $(a_4 - a_3)$ equals the sensing film thickness h_4. The refractive index modulation σ is of order of magnitude 10^{-4}.

The chemical sensing dielectric film is considered under a weak absorption model. The imaginary part of the refractive index is very small and is neglected in the following discussion. The metal film is considered under a strong absorption model, due to its larger imaginary part. Also, the metal film must be very thin because the penetration depth of the electromagnetic wave in the metal

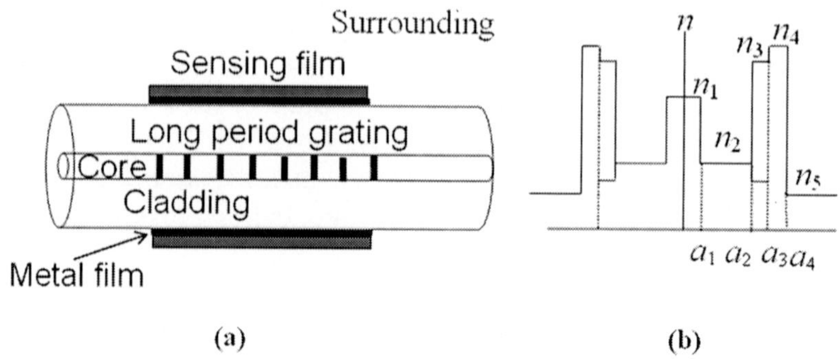

Figure 7.10. Five-layer long-period fiber grating thin film sensor model: (a) structural diagram; (b) refractive index profile.

film is of the order of the wavelength magnitude, and the sensing film should respond to the external surroundings.

3.1. ESTABLISHING THE COMPLEX CHARACTERISTIC EQUATION

Based on the theory of a planar waveguide coated with a metal film, the characteristic equation is the same as that of a common planar waveguide. The difference between them is the feature of their refractive indices. The refractive index of the metal film is a complex number. So, Wei et al. (2003), Zhang et al. (2004), and Xu et al. (2005) set up the complex characteristic equation based on the theory mentioned above. Their works provide the theoretical foundation for establishing of the complex characteristic equation foundation of the five-layer LPFG.

The vector expression for the electromagnetic field is derived from the relation between the electromagnetic vectors $\{\vec{e},\vec{h}\}$ and the Debye potentials $\{\Psi_a,\Phi_a\}$ (Tsao 1992). According to the continuity of the tangential components e_ϕ, e_z, h_ϕ, and h_z across the boundaries, a linear homogeneous system of equations with 16 variables can be obtained:

$$M_{16\times16}\left[C_1,A_1,C_2,D_2,A_2,B_2,C_3,D_3,A_3,B_3,C_4,D_4,A_4,B_4,D_5,B_5\right]^T=0 \qquad (7.8)$$

where $M_{16\times16}$ is the coefficient matrix. In order to ensure a nontrivial solution for Eq. (7.8), the determinant of the matrix $M_{16\times16}$ has to be evaluated as zero, which is just the eigenvalue equation $\det\{M_{16\times16}\}=0$.

Based on the operation rule of the block matrix, the complex characteristic equation may be expressed as follows, which is convenient for programming computation.

$$\begin{aligned} f(n_{\text{eff}}) = \det[&M(u_4a_4)\cdot M(u_4a_3)^{-1}\cdot M(u_3a_3)\cdot M(u_3a_2)^{-1}\cdot M(u_2a_2)\cdot M(u_2a_1)^{-1}\cdot M(u_1a_1)\\ &-M(w_5a_4)]=0 \end{aligned} \qquad (7.9)$$

where

$$M(u_ia_i)=\begin{bmatrix} -\dfrac{u_i^2}{n_i^2}\cdot J_l(u_ia_i) & -\dfrac{u_i^2}{n_i^2}\cdot Y_l(u_ia_i) & 0 & 0\\ 0 & 0 & u_i^2\cdot J_l(u_ia_i) & u_i^2\cdot Y_l(u_ia_i)\\ -\dfrac{\sigma_2}{n_i^2\cdot a_i}\cdot J_l(u_ia_i) & -\dfrac{\sigma_2}{n_i^2\cdot a_i}\cdot Y_l(u_ia_i) & -u_i\cdot J_l'(u_ia_i) & -u_i\cdot Y_l'(u_ia_i)\\ -u_i\cdot J_l'(u_ia_i) & -u_i\cdot Y_l'(u_ia_i) & \dfrac{\sigma_1}{a_i}\cdot J_l(u_ia_i) & \dfrac{\sigma_1}{a_i}\cdot Y_l(u_ia_i) \end{bmatrix}$$

$$M(u_i a_{i-1})^{-1}$$

$$= \frac{n_i^2}{u_i^6} \cdot \frac{1}{JJYY} \begin{bmatrix} u_i^4 \cdot Y_l'(u_i a_{i-1}) & \frac{u_i^3 \cdot \sigma_1 \cdot Y_l(u_i a_{i-1})}{n_i^2 \cdot a_{i-1} \cdot \mu} & 0 & -\frac{u_i^5 \cdot Y_l(u_i a_{i-1})}{n_i^2} \\ -u_i^4 \cdot J_l'(u_i a_{i-1}) & -\frac{u_i^3 \cdot \sigma_1 \cdot J_l(u_i a_{i-1})}{n_i^2 \cdot a_{i-1} \cdot \mu} & 0 & \frac{u_i^5 \cdot J_l(u_i a_{i-1})}{n_i^2} \\ \frac{u_i^3 \cdot \sigma_2 \cdot Y_l(u_i a_{i-1})}{n_i^2 \cdot a_{i-1}} & -\frac{u_i^4 \cdot Y_l'(u_i a_{i-1})}{n_i^2} & -\frac{u_i^5 \cdot Y_l(u_i a_{i-1})}{n_i^2} & 0 \\ -\frac{u_i^3 \cdot \sigma_2 \cdot J_l(u_i a_{i-1})}{n_i^2 \cdot a_{i-1}} & \frac{u_i^4 \cdot J_l'(u_i a_{i-1})}{n_i^2} & \frac{u_i^5 \cdot J_l(u_i a_{i-1})}{n_i^2} & 0 \end{bmatrix}$$

$$M(u_1 a_1) = \begin{bmatrix} -\frac{u_1^2}{n_1^2} \cdot J_l(u_1 a_1) & 0 & 0 & 0 \\ 0 & u_1^2 \cdot J_l(u_1 a_1) & 0 & 0 \\ -\frac{\sigma_2}{n_1^2 \cdot a_1} \cdot J_l(u_1 a_1) & -u_1 \cdot J_l'(u_1 a_1) & 0 & 0 \\ -u_1 \cdot J_l'(u_1 a_1) & \frac{\sigma_1}{a_1} \cdot J_l(u_1 a_1) & 0 & 0 \end{bmatrix}$$

$$M(w_5 a_4) = \begin{bmatrix} 0 & 0 & -\frac{w_5^2}{n_5^2} \cdot K_l(w_5 a_4) & 0 \\ 0 & 0 & 0 & -w_5^2 \cdot K_l(w_5 a_4) \\ 0 & 0 & -\frac{\sigma_2}{n_5^2 \cdot a_4} \cdot K_l(w_5 a_4) & -w_5 \cdot K_l'(w_5 a_4) \\ 0 & 0 & -w_5 \cdot K_l'(w_5 a_4) & \frac{\sigma_1}{a_4} \cdot K_l(w_5 a_4) \end{bmatrix}$$

where $u_1 = k\sqrt{n_1^2 - n_{\text{eff}}^2}$, $u_2 = k\sqrt{n_2^2 - n_{\text{eff}}^2}$, $u_3 = k\sqrt{n_3^2 - n_{\text{eff}}^2}$, $u_4 = k\sqrt{n_4^2 - n_{\text{eff}}^2}$, $w_5 = k\sqrt{n_{\text{eff}}^2 - n_5^2}$. If $\text{real}(n_3) < n_{\text{eff}}^2$, $u_3 = k\sqrt{n_{\text{eff}}^2 - n_3^2}$; $JJYY = J_l(u_i a_{i-1}) \cdot Y_l'(u_i a_{i-1}) - Y_l(u_i a_{i-1}) \cdot J_l'(u_i a_{i-1})$; $k = 2\pi/\lambda$, $\sigma_1 = \beta jl/\omega\mu$, $\sigma_2 = \beta jl/\omega\varepsilon_0$; where J_l, Y_l, K_l denote the usual Bessel function, Neumann function, and Hankel function, respectively; where J_l', Y_l', K_l' are the derivatives of usual Bessel function, Neumann function, and Hankel function, respectively; and $l = 1$ (Erdogan 1997).

Based on the theory of a planer waveguide coated with metal film, the root is a better approximation when the imaginary part of ε_3 is not considered. Thus, the approximate root is taken as a zero-order approximation and the imaginary part of ε_3 is taken as a perturbation to revise the above results. The best approximate root is obtained by an iterative method.

The complex root is expressed as $n_{\text{eff}} = n_{\text{eff},\,r} + i n_{\text{eff},\,i}$. For convenience in resolving the root easily, let the expression $y(n_{\text{eff},\,r}, n_{\text{eff},\,i}) = 0$, that is,

$$y(n_{\text{eff},\,r}, n_{\text{eff},\,i}) = f(n_{\text{eff},\,r}, n_{\text{eff},\,i}) f(n_{\text{eff},\,r}, n_{\text{eff},\,i})^* = 0 \tag{7.10}$$

where $f(n_{\text{eff},\,r}, n_{\text{eff},\,i})^*$ denotes the conjugation of $f(n_{\text{eff},\,r}, n_{\text{eff},\,i})$. So, $y(n_{\text{eff},\,r}, n_{\text{eff},\,i})$ is real and $y(n_{\text{eff},\,r}, n_{\text{eff},\,i}) \geq 0$. The complex root can be obtained by the Nelder-Mead method (Mathews and Fink 2008). The two variables $n_{\text{eff},\,r0}$ and $n_{\text{eff},\,i0}$ are the complex roots corresponding to the expression $y(n_{\text{eff},\,r}, n_{\text{eff},\,i}) = 0$.

3.2. EQUATION SOLUTION METHOD

The fiber is described by the following parameters: a_1 = 4.15 μm, a_2 = 62.4 μm, $a_3 = a_2 + h_3$, $a_4 = a_3$ + 200 nm; n_1 = 1.4681, n_2 = 1.4628, n_3, n_4 = 1.57, n_5 = 1.0. The refractive index modulation σ is 4×10^{-4}, and the grating period Λ is 450 μm.

3.2.1. Processing Method for Infinity Problem

When the complex argument of the Bessel function is large, the Bessel function can be replaced by approximate formulas. The approximate formulas are given as follows (Liang 2004):

$$Y_l(x) \approx \sqrt{\frac{2}{\pi \cdot x}} \cos\left(x - \frac{\pi}{2} \cdot l - \frac{\pi}{4}\right) \qquad J_l(x) \approx \sqrt{\frac{2}{\pi \cdot x}} \sin\left(x - \frac{\pi}{2} \cdot l - \frac{\pi}{4}\right) \tag{7.11}$$

So the expression $J(u_3 a_2) \cdot Y'(u_3 a_2) - Y(u_3 a_2) \cdot J'(u_3 a_2)$ is deduced as follows:

$$\left.\begin{array}{l} J = J_1(x), J' = J_0(x) - \dfrac{J_1(x)}{x} \\ Y = Y_1(x), Y' = Y_0(x) - \dfrac{Y_1(x)}{x} \end{array}\right\} \Rightarrow \left\{\begin{array}{l} J \cdot Y' = Y_0(x) \cdot J_1(x) - \dfrac{Y_1(x) \cdot J_1(x)}{x} \\ Y \cdot J' = Y_1(x) \cdot J_0(x) - \dfrac{Y_1(x) \cdot J_1(x)}{x} \end{array}\right\} \Rightarrow J \cdot Y' - Y \cdot J'$$

$$= \frac{2}{\pi \cdot x} \tag{7.12}$$

Hence,

$$J_l(u_3 a_2) \cdot Y_l'(u_3 a_2) - Y_l(u_3 a_2) \cdot J_l'(u_3 a_2) = \frac{2}{\pi \cdot u_3 \cdot a_2} \tag{7.13}$$

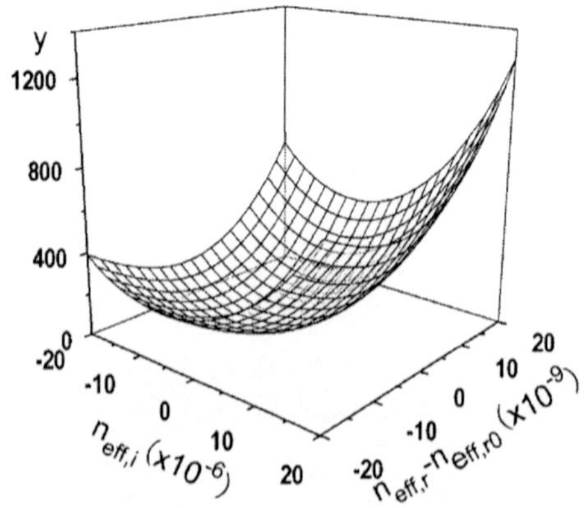

Figure 7.11. Value of the characteristic equation $y(n_{eff,r}, n_{eff,i})$ around the complex root of HE_{12}.

Figure 7.11 shows the value of the function $y(n_{eff,r}, n_{eff,i})$ in the vicinity of the complex root of the HE_{12} mode of the LPFG coated with a gold film in the complex plane. The calculating is described by the following parameters: $a_3 = a_2 + 100$ nm, $n_3 = 0.559 + i * 9.81$. The result shows that the complex root of the HE_{12} mode is $1.462565855223 + i *4.5048347 \times 10^{-6}$. The zero-order root is 1.462567423373 without considering the imaginary part of n_3. The difference between the zero-order root and the real part of the complex root is small. The complex root is surely around the zero root, which proves the rationality and feasibility of this method.

Figures 7.12a and 7.12b show the real parts and imaginary parts of 1–5 cladding modes' complex roots changing with wavelength, respectively. From Figure 7.12, the real parts decrease with the wavelength, and the imaginary parts increase with the wavelength corresponding to the same cladding mode. When the wavelength is given, the real parts increase with the mode ordinal number. In addition, the absolute values of the imaginary parts change from a magnitude of 10^{-6} to 10^{-5}. The change law is consistent with reported results (Zhang and Shi 2004). The metal film thickness is only 20 nm in the LPFG considered in this chapter, so the electromagnetic field in the metal is oscillatory and strong. In the work of Zhang and Shi (2004), the metal film is so thick that it is regarded as the surrounding layer, in which the field power falls sharply because the depth of penetration of the electromagnetic field in the metal film is within a distance of one wavelength. Therefore, these results are rationally explained, which proves that the processing method and results are correct and rational.

3.2.2. Processing Method for Overflow Problem

A numerical overflow problem may arise in the Bessel function of an LPFG coated with aluminum or nickel film, because both the real and imaginary parts of the metal refractive index are so large that the imaginary variation of the Bessel

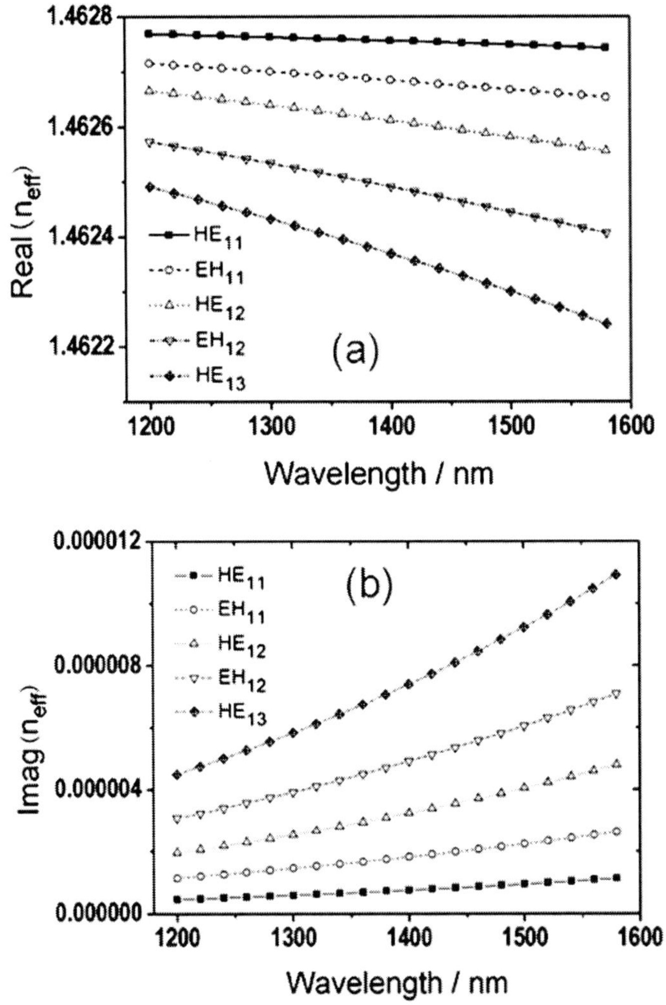

Figure 7.12. Relationship between the real and imaginary parts of the effective index of 1–5 cladding modes and the wavelengths: (a) real part; (b) imaginary part.

function is tremendous. When the program runs, some data are infinite and cannot be recognized by the computer. However, the influence of the variation of the real part on the Bessel function value is small. So, the Bessel function cannot be applied directly by MATLAB until the Bessel function is scaled.

If the Bessel function is divided by a factor A, the complex characteristic equation can be rewritten as

$$f(n_{\text{eff}}) = \det[M(u_4a_4)\cdot M(u_4a_3)^{-1}\cdot\frac{M(u_3a_3)}{A}\cdot[A\cdot M(u_3a_2)^{-1}]\cdot M(u_2a_2)\cdot M(u_2a_1)^{-1}$$
$$\cdot M(u_1a_1) - M(w_5a_4)] = 0 \tag{7.14}$$

where A is a common factor which is determined by Bessel function values. $M(u_3a_3)/A$ means that each matrix element is divided by the same factor, and $\left[A \cdot M(u_3a_2)^{-1}\right]$ means that each matrix element is multiplied by the same factor.

The trigonometric function expansion equations for the Bessel function are

$$\sin z = \frac{1}{2*i}(e^{iz} - e^{-iz}) = \frac{1}{2*i}\left[e^{-y}(\cos x + i\cdot\sin x) - e^{y}(\cos x - i\cdot\sin x)\right]$$
$$\cos z = \frac{1}{2}(e^{iz} + e^{-iz}) = \frac{1}{2}\left[e^{-y}(\cos x + i\cdot\sin x) + e^{y}(\cos x - i\cdot\sin x)\right] \qquad (7.15)$$

where $z = x + i*y$. Equation (7.15) is composed of a sine, a cosine function, and an exponential function. The numerical overflow problem resulting from the real and imaginary variation of the Bessel function cannot occur in the sine and cosine functions, but may occur in the exponential function. Thus, the exponential function should be transformed by dividing it by a common factor A. We have shown that $A = e^{300}$ is suitable for solving characteristic roots in LPFGs coated with aluminum film.

Figure 7.13 shows the value of the function $y(n_{\mathrm{eff},r}, n_{\mathrm{eff},i})$ in the vicinity of the complex root of the HE_{13} mode of an LPFG coated with aluminum film in the complex plane. The calculation is described by the following parameters: $a_3 = a_2 + 20$ nm, $n_3 = 1.38 + i * 16.4$. The result shows that the complex root of the HE_{13} mode is $1.462278831016 + i * 1.0993666 \times 10^{-6}$. The zero-order root is 1.462746896683 without considering the imaginary part of n_3. The difference

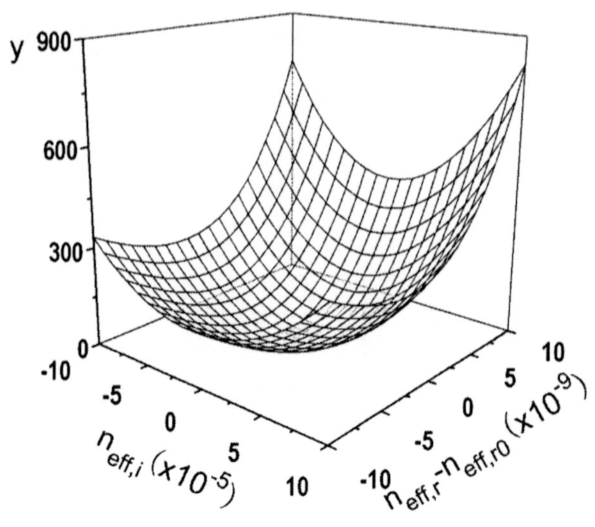

Figure 7.13. Value of the characteristic equation $y(n_{\mathrm{eff},r}, n_{\mathrm{eff},i})$ around the complex root of HE_{13}.

between the zero-order root and the real part of the complex root is small, so this processing method can resolve the numerical overflow problem effectively.

3.2.3. Jump Phenomenon of Effective Index

In the resolution of the complex characteristic equation using the above method, an influence of sensitive films on the complex effective index is observed. Figures 7.14a and 7.14b show the change of the real and imaginary parts of the effective

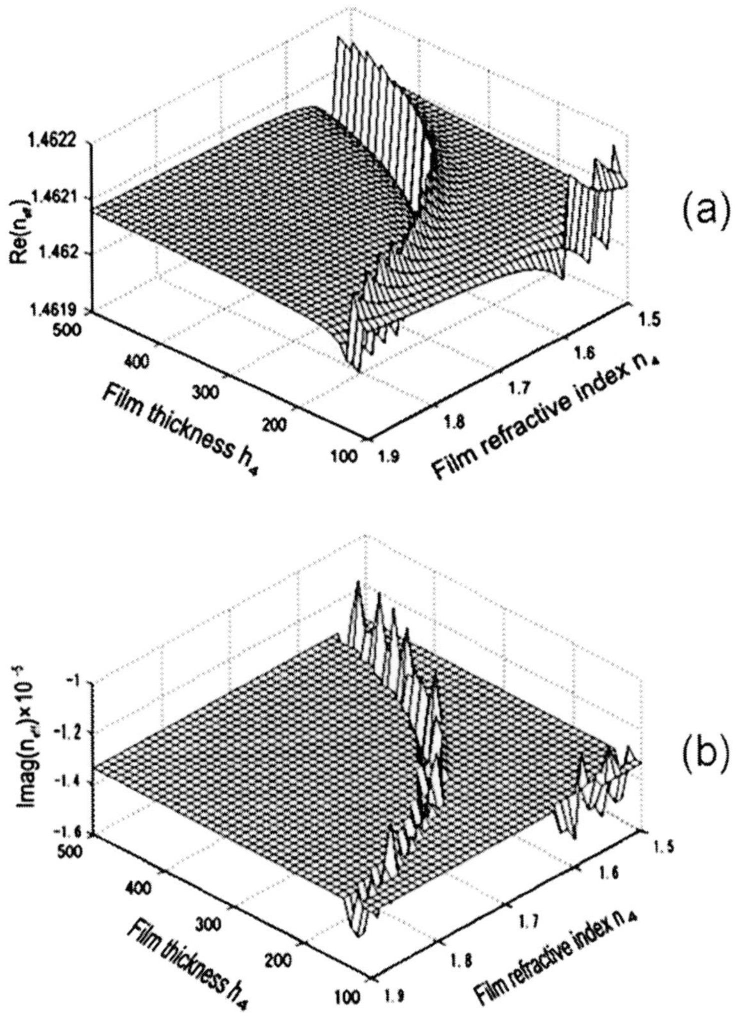

Figure 7.14. Real and imaginary parts of HE_{13} cladding mode effective index change with refractive index and thickness of sensitive film: (a) real part; (b) imaginary part.

index for the cladding mode HE_{13} with the sensitive film's refractive index and film thickness when the metal film parameters are fixed. It can be seen that the real and imaginary parts of the effective index change monotonically with the refractive index and film thickness of sensitive films in most cases. However, there are special regions of combinations of the sensitive film refractive index and film thickness in which the real and imaginary parts of the effective index present irregular jump changes. This may result in the disappearance of the resonance wavelength and its corresponding attenuation peak from the transmission spectrum (Nicholas 2002). Therefore these combinations of parameters of sensitive films must be avoided in designing actual sensors.

3.3. INTENSITY PROFILE OF CLADDING MODE

Figure 7.15 shows the local intensity profiles of the cladding modes (ν = 1–6) in a Ag-coated LPFG. It is evident that the low-order even modes (EH modes, ν = 2, 4, 6) have a larger proportion of intensity localized in the core than the low-order odd modes (HE modes, ν = 1, 3, 5). It can be concluded that the coupling between the core mode and the EH modes may be stronger than that between the core mode and the HE modes, which is contrary to the situation of air cladding.

The intensity is damped to zero rapidly in the Ag film in Figure 7.15. It is seen that the electromagnetic field exists to a very small depth of the Ag film. This observation supports the view that the Ag film should be thin. Otherwise, there will be no electromagnetic field in the sensing film. Figures 7.16a and 7.16b show the three-dimensional distribution of the intensity of the HE_{13} and EH_{13} modes along the radial section. This is in accord with the situation in Figure 7.15.

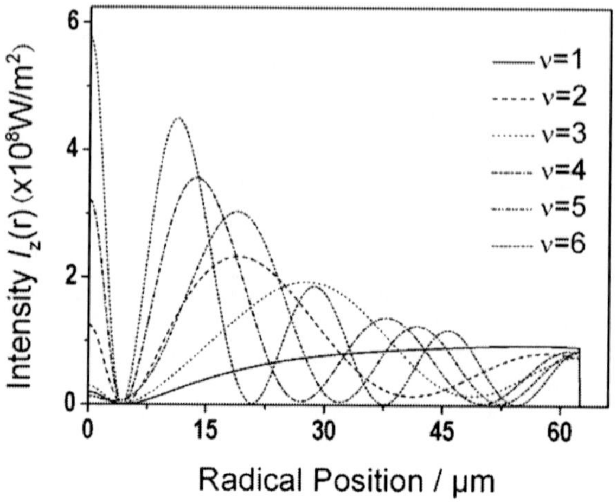

Figure 7.15. Plots of local intensity I_z as a function of radius for six lowest-order $l = 1$ cladding modes.

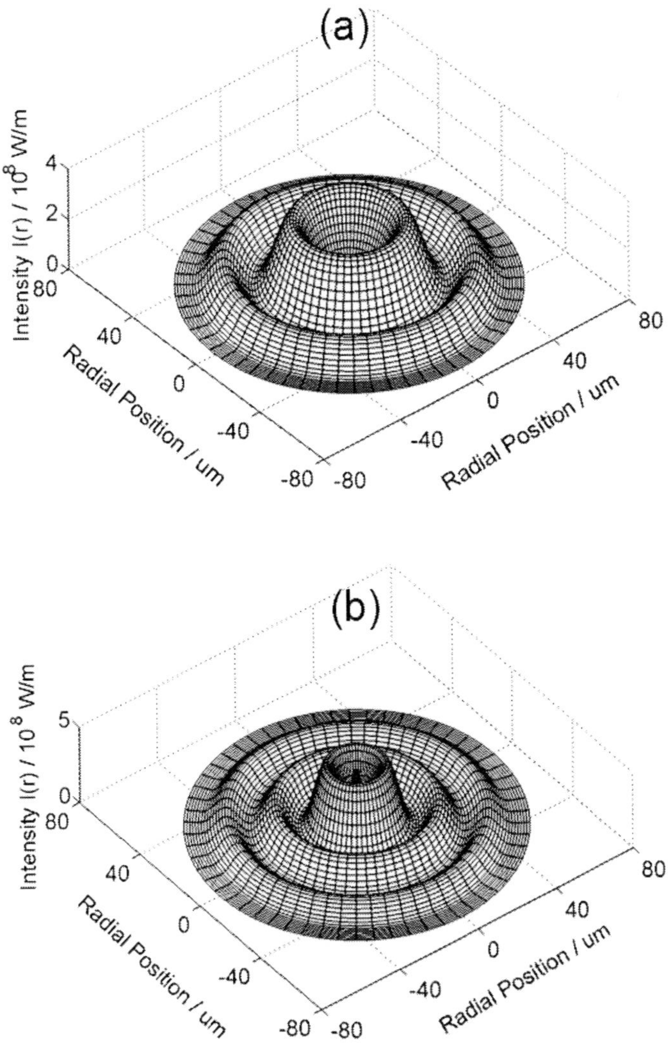

Figure 7.16. The power distribution of the two lowest-order $l = 1$ cladding modes: (a) HE_{13} mode; (b) EH_{13} mode.

4. OPTIMIZATION OF COATED LPFG SENSOR BASED ON DPR AND SPR

4.1. DEFINITION OF SENSOR SENSITIVITY

When a LPFG sensor based on dual peak resonance and surface plasmon resonance is put into the surrounding gas, the sensitive film is exposed to the surrounding gas and interacts with each other, which induces a tiny variation of the film refractive index and thus results in a shift of the attenuation peaks in the transmission spectrum. The sensitivity of an LPFG which indicates the shifting degree of the interval between the dual resonance wavelengths changing with the refractive index of thin film is expressed as

$$S_n = \frac{(\lambda_2' - \lambda_1') - (\lambda_2 - \lambda_1)}{\lambda_2 - \lambda_1} \Big/ \frac{\Delta n_4}{n_4} \tag{7.16}$$

where $\lambda_2 - \lambda_1$ and $\lambda_2' - \lambda_1'$ are the intervals between the dual resonance wavelengths before and after interaction between the sensor and the analyte. S_n is the ratio of the rate of change of the interval between dual resonance wavelengths to the change rate of the refractive index of the thin film. S_n reflects the detective sensitivity because the rate of change of the refractive index of the thin film is related directly to the analyte concentration. Obviously, S_n is a function of the refractive index and the thickness of the thin film, grating parameters, and the incident wavelength. In order to bring the variation of the wavelength separation of the dual resonance peaks into play to the utmost, the film optical parameters and the grating structural parameters must be optimized to improve the sensor sensitivity. By analyzing the response of the sensitivity to these parameters, the optimum film parameters and grating parameters can be obtained.

4.2. OPTIMIZATION OF DESIGN OF LPFG SENSORS

4.2.1. Optical Parameters of the Sensitive Film

Figure 7.17a shows that the sensitivity S_n of the coated LPFG changes with the refractive index n_4 (1.5–1.8) and the thickness h_4 (100–400 nm) of the sensing film (called the 2nd film in Figure 7.17) when the grating parameters ($\sigma = 4 \times 10^{-4}$, $L = 1$ cm, $\Lambda = 143$ μm) are given. The contour of $S_n = 40$ under these conditions is shown in Figure 7.17b. From Figure 7.17a, there is a great difference in S_n for various combinations of n_4 and h_4, and the value of S_n is of order of magnitude 10^2 for some combinations, from which we can select suitable materials according to the optimum n_4 and coat the film with optimum thickness h_4. S_n in the blank area of Figure 7.17a is 0, which shows that the dual resonance wavelengths cannot appear for these combinations of n_4 and h_4 as grating parameters are given. So the parameters in the blank area should be avoided in designing this kind of sensor. The data simulations indicate that the sensitivity S_n of the sensor is 234.73, while the sensing film refractive index n_4 is 1.535 and the film thickness h_4 is120 nm.

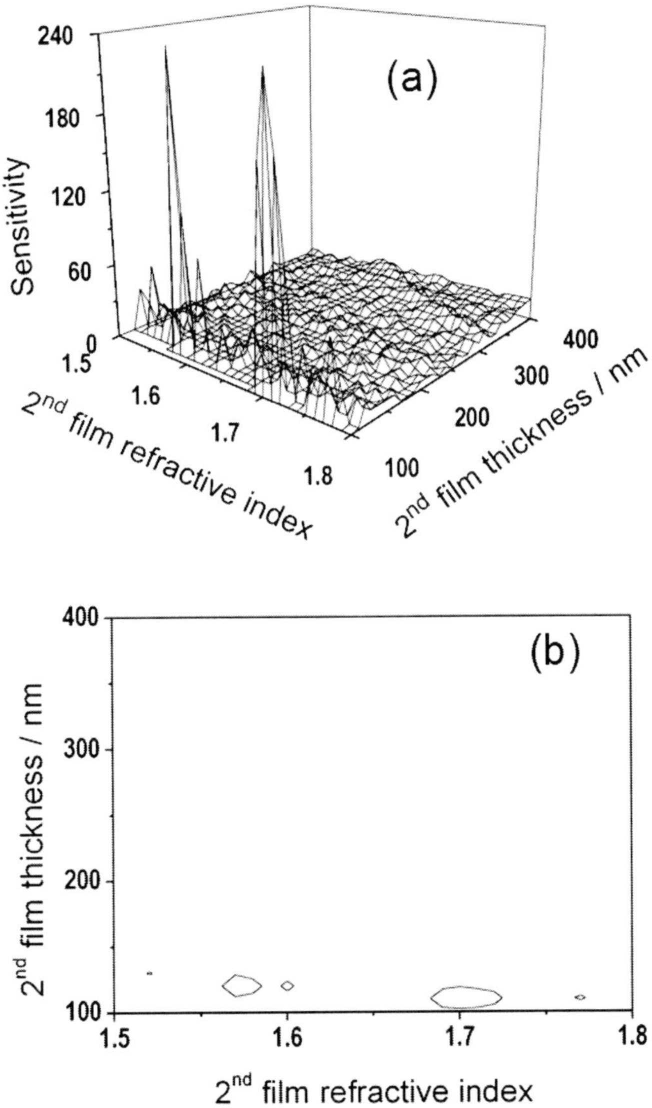

Figure 7.17. Dependence of S_n on h_4 for n_4 ranging from 1.5 to 1.8 μm (h_4 = 100–400 nm) and relevant contour line of S_n = 20: (a) sensitivity relation; (b) contour line.

4.2.2. Optical Parameters of the Metal Film

Figure 7.18a shows that the sensitivity S_n of the coated LPFG changes with the average index modulation σ (3×10^{-4} to 5×10^{-4}) and the metal film thickness h_3 (20–60 nm) of the metal film (called the 1st film in Figure 7.18) when the sensing

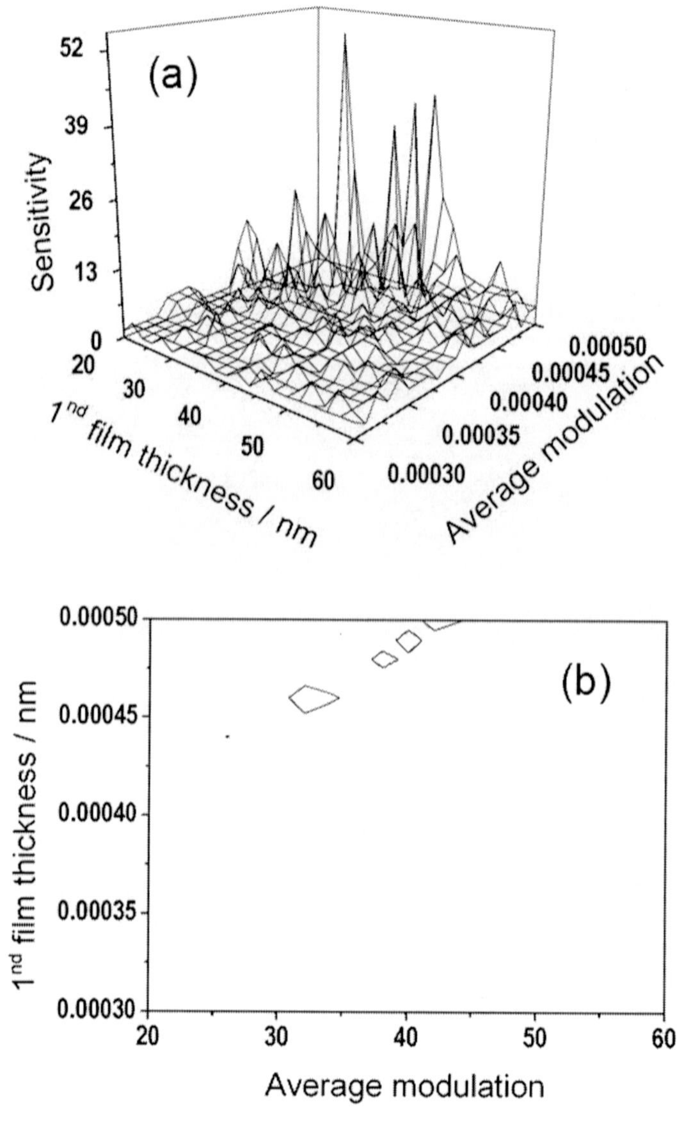

Figure 7.18. Dependence of S_n on h_3 for σ ranging from 0.0003 to 0.0005 (h_3 = 20–60 nm) and relevant contour line of S_n = 20: (a) sensitivity relation; (b) contour line.

film and the other grating parameters (h_4 = 200 nm, L = 1 cm, Λ = 145 μm) are given. The contour of S_n = 20 under these conditions is shown in Figure 7.18b. From Figure 7.18a, there is a great difference in S_n for various combinations of h_3 and σ, and the value of S_n is of order of magnitude 10^1 for some combinations, from which we can coat the film with optimum thickness h_3 and select a suitable exposure time for grating writing according to the optimum σ.

4.2.3. Structure Parameters of the Fiber Grating

The dual resonance wavelengths are also influenced by the grating parameters of the LPFG, which directly influences the sensitivity S_n. Figure 7.19a shows the relations between the sensitivity S_n and the average index modulation σ (2×10^{-4} to 6×10^{-4}) and the sensing film thickness h_4 (200–400 nm) when the other grating

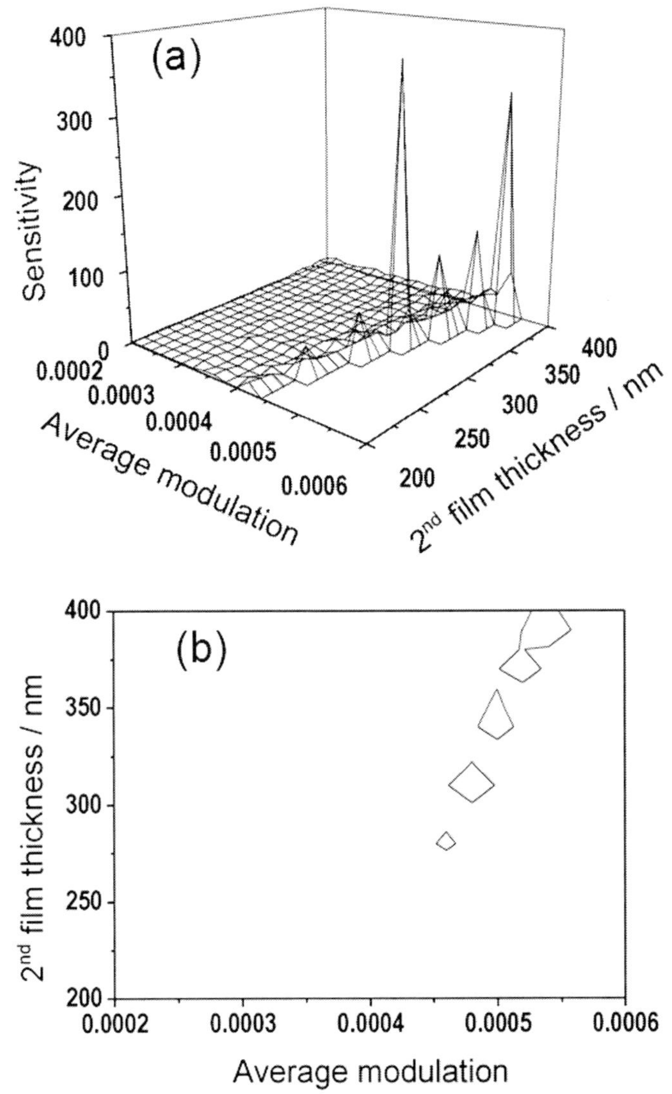

Figure 7.19. Dependence of S_n on h_4 for σ ranging from 0.0002 to 0.0006 (h_4 = 100–300 nm) and relevant contour line of S_n = 40: (a) sensitivity relation; (b) contour line.

parameters are given (n_4 = 1.54, L = 1 cm, Λ = 145 μm). The contour of S_n = 40 corresponding to Figure 7.19a is shown in Figure 7.19b. The result indicates that high sensitivity can be obtained by controlling the average index modulation and overlaying a proper sensing film thickness.

Figure 7.20a shows the sensitivity for various combinations of grating period Λ and sensing film thickness when the other grating parameters are given (n_4 = 1.54, L = 1 cm, σ = 0.0004). The contour of S_n = 40 corresponding to Figure 7.20a is shown in Figure 7.20b. It can be seen that high sensitivity can be obtained by adjusting the grating period controlled by the step length of the moving platform and controlling the sensing film thickness by sol-gel technology.

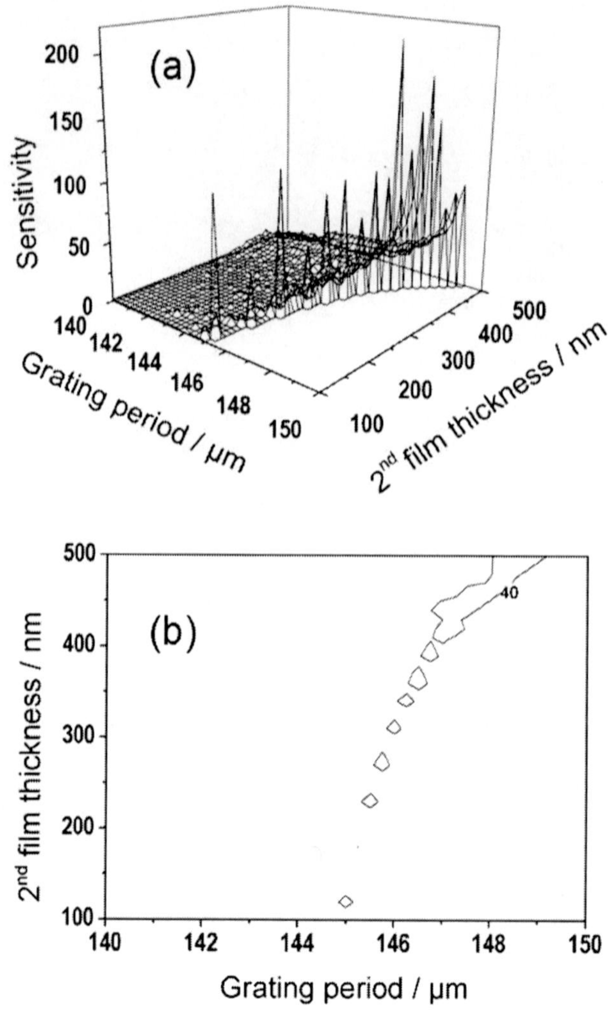

Figure 7.20. Dependence of S_n on h_4 for Λ ranging from 141 to 151 μm (h_4 = 100–500 nm) and relevant contour line of S_n = 40: (a) sensitivity relation; (b) contour line.

Table 7.2. Typical film optical parameters, grating parameters, and relevant sensitivities and resolutions

h_3 (nm)	n_4	h_4 (nm)	σ (× 10^{-4}), L (cm), Λ (μm)	λ (nm)	S_n	σ_n
20	1.535	120	4.0, 1.0, 143.0	1131.7/1499.9	2.34×10^2	5.80×10^{-8}
20	1.605	110	4.0, 1.0, 143.0	1131.5/1500.2	1.72×10^2	8.25×10^{-8}
32	1.540	200	4.6, 1.0, 143.0	1286.1/1331.2	5.22×10^1	2.29×10^{-7}
42	1.540	200	5.0, 1.0, 143.0	1277.9/1335.8	4.06×10^1	2.97×10^{-7}
20	1.540	310	4.8, 1.0, 145.0	1296.9/1322.1	3.73×10^2	3.18×10^{-8}
20	1.540	390	5.4, 1.0, 145.0	1280.5/1328.7	3.17×10^2	3.79×10^{-8}
20	1.540	230	4.0, 1.0, 145.5	1288.6/1318.3	1.18×10^2	1.01×10^{-7}
20	1.540	420	4.0, 1.0, 147.3	1235.4/1314.1	2.09×10^2	5.96×10^{-8}

Table 7.2 lists several different kinds of film optical parameters, grating parameters, and relevant sensitivities and resolutions. The resolution of refractive index of the sensing thin film is expressed as $\sigma_n = S_n^{-1} n_4 (\delta\lambda / \Delta\lambda)$, where $\Delta\lambda$ represents the peak separation and $\delta\lambda$ denotes the wavelength resolution, which is 0.01 nm for commonly used spectrometers. The results from Table 7.2 indicate that the minimum resolution of the refractive index of this LPFG sensor is predicted to be 10^{-8} when the metal and sensing film parameters and the grating parameters are determined appropriately. In addition, the sensitivity to the refractive index of the surroundings of a three-layer structural LPFG sensor is 4.8×10^{-5} (Shu and Huang 1999), so the sensitivity of the five-layer structural LPFG sensor presented here is much higher than that of the three-layer structural LPFG sensor.

Compared with the three-layer or four-layer structural LPFG sensor, the five-layer structural LPFG sensor based on DPR and SPR is more flexible. By selecting suitable sensitive films and adjusting the film optical parameters and grating parameters according to the nature of the analyte, this sensor can provide high sensitivity. In practical application, of course, the values of the designed grating and film parameters must be feasible. For example, the grating length must not be too long in consideration of fiber toughness, and the refractive indices of films should work well with the gas-sensing materials. In addition, the operating wavelength should be around the resonance wavelength to observe the maximum response.

5. DISPERSION IN METAL-COATED LPFG SENSORS

5.1. DISPERSION EXPRESSION

There are two kinds of dispersion mechanisms in a fiber (Li and Cui 2002): wavelength dispersion and modal dispersion. Modal dispersion exists mainly in multimode fibers. Wavelength dispersion in a single-mode optical fiber consists of material dispersion and waveguide dispersion. Theoretical research (Li et al. 2000) showed that the waveguide dispersion is very small and can be neglected when the fiber is far from cutoff mode. Usually, the low-order cladding modes can be regarded as being far from cutoff state for given optical parameters (Deng et al. 2009). Therefore, only material dispersion is considered here.

Given that a single-mode fiber uses a conventional GeO_2-doped SiO_2 core and pure SiO_2 cladding, the refractive index of the core is 0.0050–0.0055 higher than that of the cladding (He 2005). When the wavelength changes between 0.2 and 4.0 μm, the approximate formula for the refractive index of GeO_2-doped fiber cladding is expressed as (Li and Cui 2002)

$$n_2^2 - 1 = \frac{0.6961663\lambda^2}{\lambda^2 - (0.068403)^2} + \frac{0.4079426\lambda^2}{\lambda^2 - (0.1162414)^2} + \frac{0.8974794\lambda^2}{\lambda^2 - (9.89616)^2} \tag{6.17}$$

The refractive index of the fiber core increases by σ after UV irradiation, and the approximate formula of the refractive index of the core is expressed as $n_1 = n_2 + 0.0052 + \sigma$.

In the following section, an LPFG coated with a nano-scaled gold film is studied. According to the discrete data (Palik 1985), the fitting curve is obtained by using the least-squares method, and approximate formulas for the gold refractive index n_3 and the extinction coefficient k_3 are derived respectively:

$$n_3(\lambda) = 1.2101\times10^{11}\cdot\lambda^2 + 2.3051\times10^5\cdot\lambda - 0.0927 \tag{7.18}$$

$$k_3(\lambda) = \begin{cases} -7.3724\times10^{11}\cdot\lambda^2 + 9.7521\times10^6\cdot\lambda - 2.1726 & 826.5\ \text{nm} < \lambda < 1239.8\ \text{nm} \\ 0\cdot\lambda^2 - 2.6875\times10^7\cdot\lambda - 42.0901 & 1239.8\ \text{nm} \le \lambda < 1265.1\ \text{nm} \\ 5.9296\times10^{10}\cdot\lambda^2 + 5.9581\times10^6\cdot\lambda - 0.4503 & 1265.1\ \text{nm} \le \lambda < 2066.3\ \text{nm} \end{cases} \tag{7.19}$$

Sensitive film materials include mainly metallic oxides and organic compounds. TiO_2 is a typical metallic oxide with high gas-sensing ability, so it is selected as a sensing film in the following discussions. The approximate dispersion equation for a TiO_2 thin film from the fitting of discrete data (Palik 1985) is derived as

$$n_4^2 = \frac{2.441\times10^7}{\lambda^2\times10^{20} - 0.803\times10^7} + 5.913 \tag{7.20}$$

5.2. MATERIAL DISPERSION INFLUENCE ON RESONANCE CHARACTERISTICS

5.2.1. Core and Cladding Dispersion

To study the influence of material dispersion on coated LPFGs, the dispersion effect of the fiber core and cladding must first be considered. In this chapter, the radii a_1 and a_2 of the core and the cladding of the fiber are 4.15 and 62.4 μm, respectively. The refractive indices n_1, n_2, and n_3 of the core, the cladding, and the metal film are 1.4521, 1.4469, and 0.559 + i * 9.81, respectively, without considering the dispersion. The average index modulation σ and the fiber grating length L are 4×10^{-4} and 1 cm, respectively. The sensitive film refractive index n_4 and thickness h_4 are 1.57 and 200 nm, respectively, and the gold film thickness h_3 is 50 nm.

The resonance wavelength shift caused by the dispersion of the fiber core and cladding is simulated. In order to determine the resonance wavelengths corresponding to the cladding mode orders, a detuning function Δn based on the phase-matching condition is defined as

$$\Delta n = n_{\text{eff, co}}(\lambda) - n^{\text{v}}_{\text{eff, cl}}(\lambda) - \frac{\lambda}{\Lambda} \qquad (7.21)$$

The resonance wavelengths can be determined by the roots of the equation $\Delta n = 0$. Figure 7.21a shows the curves of Δn versus λ of the tenth cladding mode (EH_{15}) when the grating period is 330 μm. The lines b, c, d, and e in Figure 7.21a present the curves of Δn versus λ with regard to no dispersion, core dispersion, cladding dispersion, and the joint action of core and cladding dispersion, respectively. It can be seen that the resonance wavelengths are located at points C and D when the core and cladding dispersion are taken into consideration separately. Compared with the resonance wavelength without regard to material dispersion, they both shift toward shorter wavelength, and the shifts are very large. Moreover, it is evident that the lines b and e overlap, which indicates that there is little influence on the EH_{15} cladding mode when considering the core dispersion together with the cladding dispersion. This result can be explained as follows: The detunings corresponding to point O are located at point A and point B when considering core and cladding dispersion separately. The values of detuning corresponding to point A and point B are 0.001222 and −0.001165, respectively. The detunings Δn_A and Δn_B are counterbalanced when considering fiber core dispersion together with cladding dispersion, and the sum of Δn_A and Δn_B is close to zero. Thus the resonance wavelength returns to its original position, which corresponds to the case without considering dispersion.

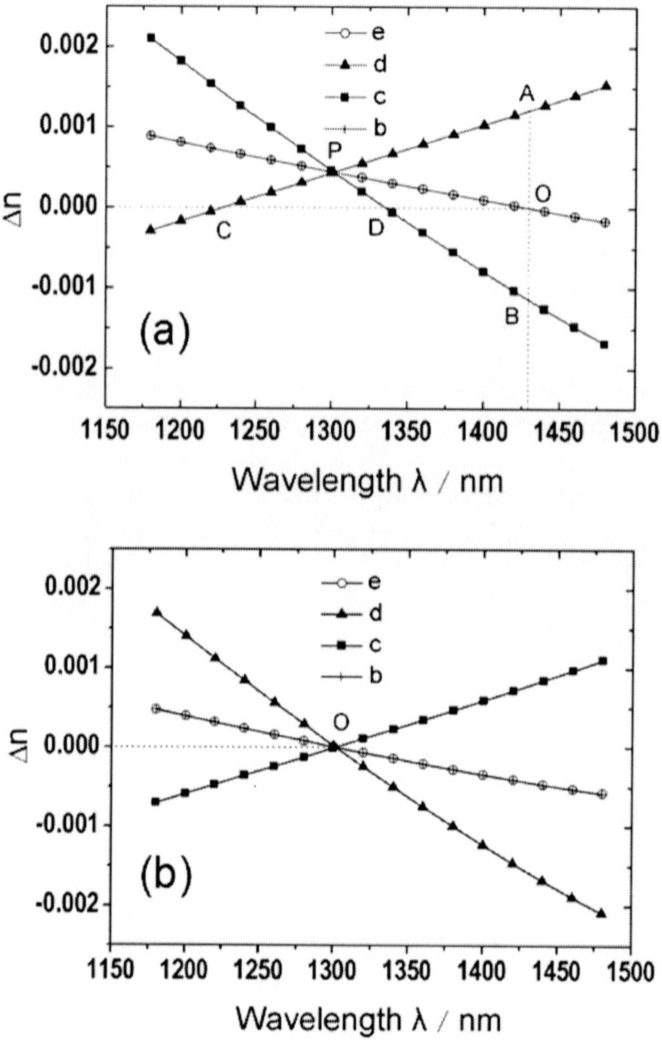

Figure 7.21. Plot of Δn versus λ of the EH_{15} cladding mode before and after considering fiber core and cladding dispersion: (a) $\Lambda = 330$ μm; (b) $\Lambda = 295$ μm.

In addition, the four curves in Figure 7.21a almost intersect at point P. According to Eq. (7.21), a decrease of grating period induces a decrease of detuning function Δn, thus the position of point P moves downward. When the grating period is appropriate, point P will move to the special position $\Delta n = 0$, where the resonance wavelength will not be influenced by fiber core and cladding dispersion. Figure 7.21b shows the curves for Δn versus λ of the EH_{15} cladding mode of a gold-coated LPFG with a grating period of 295 μm. The four curves in Figure 7.21b almost intersect at point O. The detuning corresponding to point O is zero, which indicates that there is almost no resonance wavelength shift, whether the fiber core and cladding dispersion are considered or not.

5.2.2. Metal Film Dispersion

In addition to fiber core and cladding dispersion, metal dispersion can also influence the resonance wavelength of a metal-coated LPFG. Figure 7.22 shows the relation between the resonance wavelength of the EH_{15} cladding mode and the grating period. Lines B, C, D, and E in Figure 7.22 represent the curves of the grating period and resonance wavelength with regard to no dispersion, core dispersion, cladding dispersion, and gold dispersion, respectively. The figure shows that, for a grating period of more than 295 μm, the resonance wavelength shifts toward a shorter wavelength upon considering gold dispersion. The larger the grating period, the more the resonance wavelength shifts. However, for a grating period of less than 295 μm, the resonance wavelength is immune to the influence of gold dispersion. So the grating period 295 μm can be called the de-dispersion grating period. When the grating period is less than the de-dispersion grating period, gold dispersion does not influence the resonance wavelength.

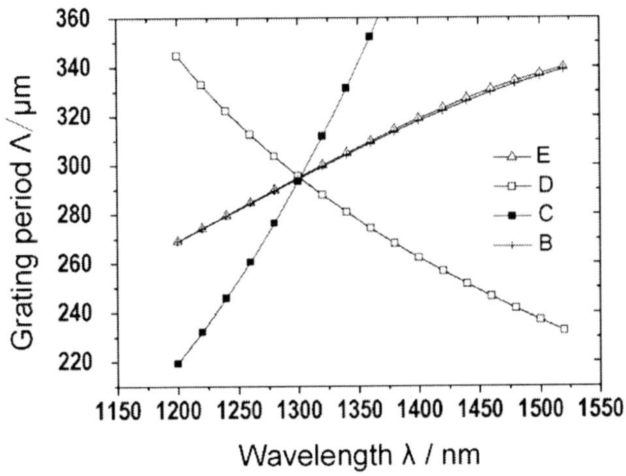

Figure 7.22. Relationships between grating period and resonance wavelength with consideration of metal film dispersion.

5.2.3. Sensitive-Film Dispersion

Taking TiO_2 as the sensitive film, the relation between the resonance wavelength for the EH_{15} cladding mode and the grating period are shown in Figure 7.23. The

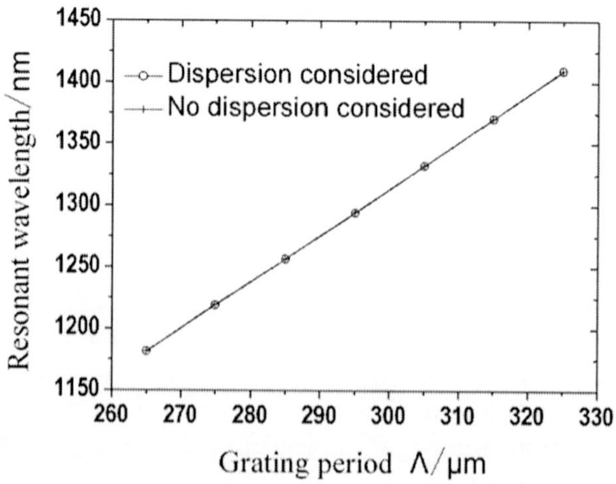

Figure 7.23. Relationships between resonance wavelength and grating period with consideration of sensitive film dispersion.

resonance wavelengths hardly change before and after considering sensitive-film dispersion for grating periods ranging from 265 to 325 μm, which indicates that the sensitive-film dispersion has little influence on the resonance wavelength. The reason is that the energy of the electromagnetic waves is mostly distributed in the fiber core and cladding, and the energy decreases sharply due to strong absorption of the metal film, which results in relatively weak energy in the sensitive film. Thus the influence of sensitive-film dispersion on the resonance wavelength can be neglected.

5.3. JUMP REGION IN RESPONSE OF RESONANCE WAVELENGTH

Nicholas et al. (2002) pointed out there exists a jump regions in the response of the resonance wavelength to the thin-film thickness, where the resonance wavelength disappears and there is an abnormal phenomenon in the transmission spectrum. The jump region is caused by the imaginary part of the film refractive index and the variation of the cross-coupling constant. In order to investigate the influence of material dispersion on the jump regions, the transmission spectra of the EH_{15} cladding mode before and after considering material dispersion are shown in Figure 7.24, in which the metal film thickness is 101 nm.

From Figure 7.24, the left transmission spectrum of the $EH_{1,10}$ cladding mode becomes abnormal when the material dispersion is taken into consideration. In order to investigate the influence of the material dispersion on the jump regions, Figure 7.25 shows the dependence of the dual resonance wavelengths on the metal film thickness before and after considering the material dispersion. This figure

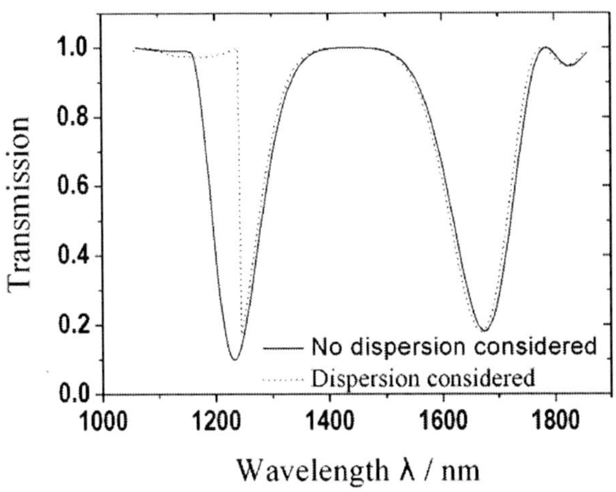

Figure 7.24. Transmission spectrum of $EH_{1,10}$ around the jump change regions.

shows two distinct jump regions, which are indicated by arrows. Without regard to the material dispersion, there exists a jump region in the metal film thickness at 105–167 nm. Considering material dispersion, the jump region is broadened and located in the range of 97–166 nm. Therefore, the metal film thickness of 101 nm is located in the jump region when considering the material dispersion,

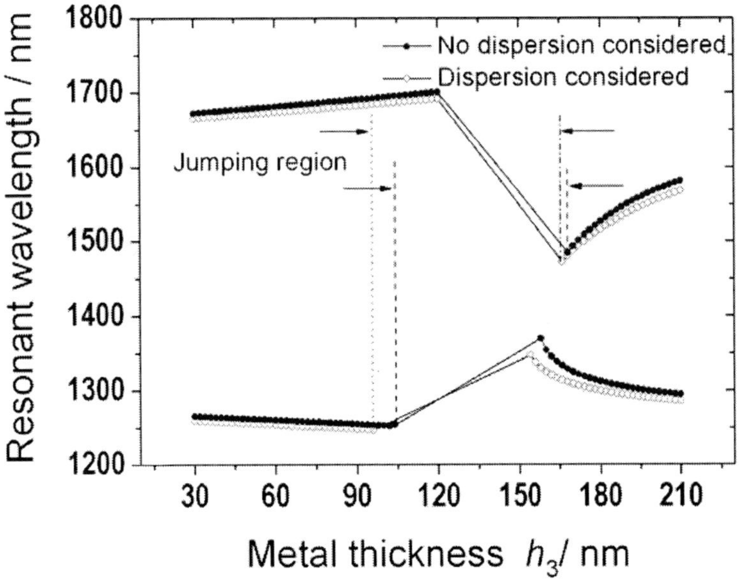

Figure 7.25. Relationship between resonance wavelength and metal film thickness with material dispersion considered and not considered.

which explains the abnormal phenomenon of the transmission spectrum of the $EH_{1,10}$ cladding mode. In practice, the material dispersion should be taken into consideration when coating metal film on the LPFG so that the dual resonance wavelengths are not located in the jump region.

5.4. OPTIMIZATION BASED ON CONSIDERATION OF DISPERSION

Because the influence of dispersion cannot be ignored, it is very important to consider the dispersion in designing a metal-coated LPFG sensor. In the following discussion, a metal-coated LPFG liquid sensor based on DPR and SPR is designed including consideration of the dispersion. The sensitivity S_n of this kind of sensor is expressed by the rate of change of the dual peak interval $\Delta\lambda$ relative to the refractive index (RI) of the surrounding liquid, n_4. Obviously, S_n is a function of the incident wavelength, the optical parameters of the metal film, and the grating parameters. In order to bring the variation of the dual-peak resonance wavelength interval into play to the utmost, those parameters must be optimized. By analyzing the response of the sensor to the RI of the surrounding liquid, the optimum parameters can be obtained. In this section, for a specified liquid analyte RI (n_4 ~ 1.35), an optimized design of a dual-peak LPFG sensor is implemented by optimizing the metal film thickness h_3 and the grating period Λ when the other grating parameters are given ($\sigma = 4 \times 10^{-4}$, $L = 1$ cm).

Figures 7.26a and 7.26b show the dependence of the sensitivity $S_{n|\max}$ on the silver film thickness h_3 (30–110 nm) and grating period Λ (178–180 μm) before and after considering material dispersion, respectively. These figures show explicitly that there is a great difference in sensitivity $S_{n|\max}$ for the same parameter combination of h_3 and Λ before and after considering the material dispersion. Without considering the material dispersion, the maximum sensitivity $S_{n|\max}$ is 6.6 when Λ is 178.6 μm and h_3 is 60 nm. With consideration of the material dispersion, the $S_{n|\max}$ is only 1.36 under the same parameters combination of Λ and h_3, as shown in Figure 7.26b. In addition, the maximum sensitivity in Figure 7.26b is 7.8, which is higher than the maximum sensitivity without considering the dispersion. Obviously, the material dispersion has a significant influence on the sensitivity of the dual-peak-based metal-coated LPFG sensor. In addition, the $S_{n|\max}$ of blank space is zero, which means that there is no dual resonance peak for given combinations of Λ and h_3 in the blank area, so these blank areas should be avoided when designing a dual resonance peak sensor.

For convenience in selecting the optimum metal film thickness h_3 and the optimum grating period Λ for sensor design, the contour lines of $S_{n|\max}$ are plotted. Figures 7.27a and 7.27b show the contour lines of $S_{n|\max} = 4.5$ before and after considering material dispersion. It is evident that those contour lines change greatly when material dispersion is considered. Only from the contour line

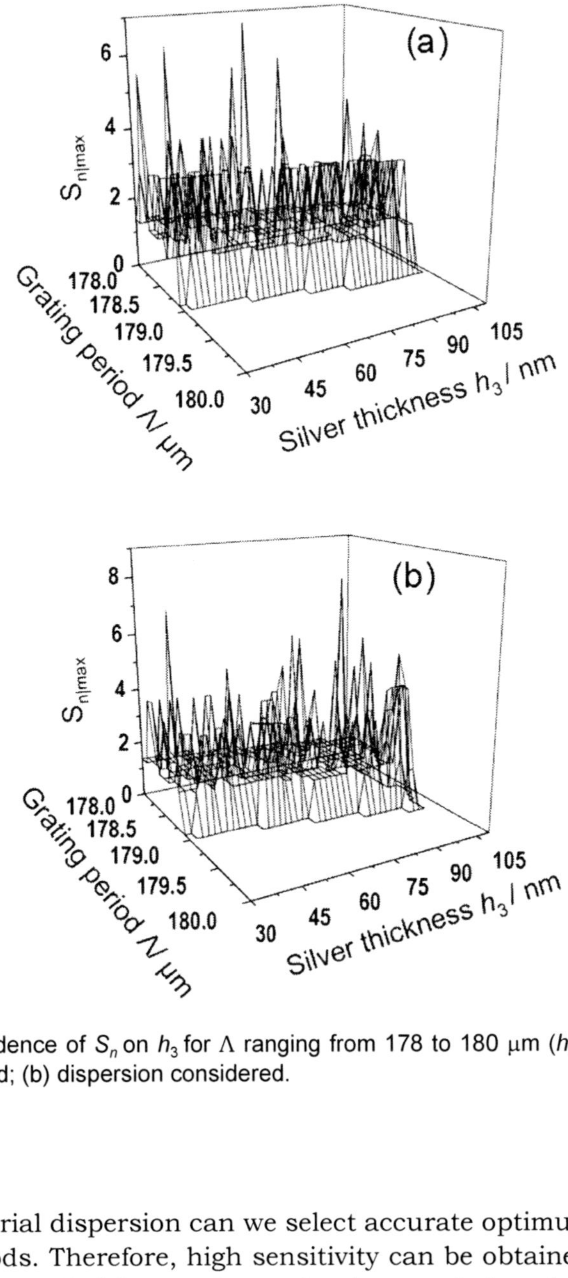

Figure 7.26. Dependence of S_n on h_3 for Λ ranging from 178 to 180 μm (h_3 = 30–110 nm): (a) no dispersion considered; (b) dispersion considered.

considering material dispersion can we select accurate optimum film thicknesses and grating periods. Therefore, high sensitivity can be obtained by adjusting the film thickness controlled by coating technology and choosing a grating period determined by a precise mobile platform. In a practical application, the designed grating structure parameters and film optical parameters must be feasible. For example, the grating length must not be too long (usually in the range of 1–5 cm) in consideration of fiber grating toughness.

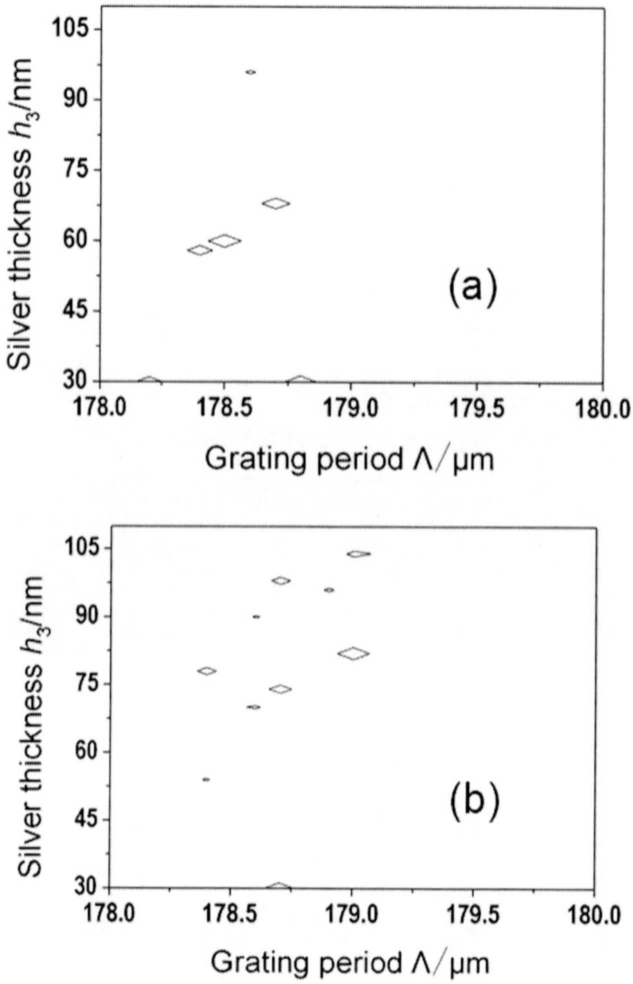

Figure 7.27. Contour line of $S_{n|max} = 4.5$: (a) no dispersion considered; (b) dispersion considered.

6. CONCLUSION

A novel LPFG sensor based on surface plasmon resonance and dual peak resonance has been presented for high-sensitivity online monitoring of the environment. The sensor is easily optimized for maximum sensitivity to the refractive index of the sensitive film, and the resolution of the refractive index is predicted to be 10^{-8}. So, this sensor can be used far more widely in different sensing applications, such as the detection of gas, liquid, humidity, biomaterials, and all surroundings which can affect the refractive index of sensitive films.

Dual peak resonance in a coated LPFG has been studied based on a rigorous coupled-mode theory. The dual peak resonance wavelengths are determined by resolving the eigenvalue equation of the coated LPFG. The results show that the smaller the grating period, the higher is the cladding mode order corresponding to the dual peak resonance. The dual resonance wavelengths shift in opposite directions with variation of the refractive index of the environment, and the change of wavelength intervals between the dual resonance peaks in the coated LPFG is much larger than that in the noncoated LPFG. So, the sensitivity of the coated LPFG sensor based on dual peak resonance is much higher than that of the noncoated LPFG sensor. Further, the film optical parameters and grating structural parameters exert different effects on the dual peak separation, which must be considered when optimizing the design of a coated LPFG sensor.

For a SPR-based LPFG with a metal coating, the complex characteristic equation of the LPFG is set up based on the coupled-mode theory. In solving the complex characteristic equation, numerical overflow is resolved through approximate deformation and scale contraction factors in the Bessel function. Further, the simulation indicates that the proportion of the even mode's intensity in the core is much larger than that of the odd mode's intensity. The resonance wavelength shifts toward shorter wavelengths with an increase of the metal film thickness, the sensitive film thickness, and the refractive index. In addition, there is a jump phenomenon in certain regions of the film parameters which must be avoided in actual applications of this LPFG sensor.

Further, a novel LPFG sensor based on SPR and DPR has been presented and studied. The sensor sensitivity, which indicates the shifting degree of the interval between dual resonance wavelengths changing with the refractive index of the thin film, is defined. By analyzing the relationship between the sensitivity and the thin-film optical parameters (thickness and refractive index) and the fiber grating parameters (grating period, index average modulation, and grating length), the optimal design parameters of the LPFG film sensor can be obtained. Data simulation shows that the resolution of the refractive index of this LPFG sensor based on SPR and DPR is predicted to be 10^{-8}, so this kind of LPFG sensor can be used as a refractive index sensor with high sensitivity.

In addition, the influence of the material dispersion of the core, cladding, metal film, and sensing films on the resonance characteristics of the presented LPFG sensor has been analyzed. The results show that the core and cladding material dispersion must be considered simultaneously so as to eliminate the dispersion influence only from the core or cladding. The dispersion from the sensitive films has little influence on the resonance characteristics of this LPFG sensor. The metal film dispersion influences the resonance peak position, so it is necessary to consider metal dispersion in designing a high-sensitivity sensor. By analyzing a typical liquid LPFG sensor based on DPR and SPR, the sensitivity of this LPFG sensor as a function of the regional distribution of the grating structural parameter and the metal-film optical parameters is discussed with consideration of material dispersion or not, and the contour line of the sensor sensitivity provides an accurate and valid means for optimal design of sensor sensitivity.

ACKNOWLEDGMENTS

This work was supported by the National Science Foundation of China (Grant No. 60777035), the Scientific Research Key Project Fund (Grant No. 208040), the Innovation Program of the Shanghai Municipal Education Commission (Grant No. 11ZZ131), and a Shanghai Leading Academic Discipline Project (Grant No. S30502).

REFERENCES

Bhatia V. and Vengsarkar A.M. (1996) Optical fiber long-period grating sensors. *Opt. Lett.* **21**, 692–694. DOI: 10.1364/OL.21.000692

Cao Z.X., Wu L.N., and Li D.Y. (2006) Distributed optical fiber surface plasmon resonance sensors. *Chin. Opt. Lett.* **4**(3), 160–163.

Chen Y., Huang H., Yu X., and Qi L. (2005) Chiral recognition of dextran sulfate with D- and L-cystine studied by multiwavelength surface plasmon resonance. *Carbohydr. Res.* **340**(12), 2024–2029. DOI: 10.1016/j.carres.2005.06.012

Chong J.H., Shum P., Haryono H., Yohana A., Rao M.K., Lu C., and Zhu Y.N. (2004) Measurements of refractive index sensitivity using long-period grating refractometer. *Opt. Commun.* **229**, 65–69. DOI: 10.1016/j.optcom.2003.10.044

Corres J.M., Del V.I., Matias I.R., and Arregui F.J. (2008) Two-layer nanocoatings in long-period fiber gratings for improved sensitivity of humidity sensors. *IEEE T. Nanotechnol.* **7**(4), 394–400. DOI: 10.1109/TNANO.2008.926337

Costantini D.M., Muller C.A.P., Vasiliev S.A., Limberger H.G., and Salathé R.P. (1999) Tunable loss filter based on metal-coated long period fiber grating. *IEEE Photon. Technol. Lett.* **11**(11), 1458–1460. DOI: 10.1109/68.803078

Cusano A., Iadicicco A., Pilla P., Contessa L., Campopiano S., and Cutolo A. (2006) Mode transition in high refractive index coated long period gratings. *Opt. Express* **14**(1), 19–34. DOI: 10.1364/OPEX.14.000019

Deng C.L., Gu Z.T., and Zhang J.T. (2009) Resonant characteristics of long-period fiber grating coated with double-layer films including metal film. *Acta Opt. Sin.* **29**(10), 2656–2664. DOI: 10.3788/AOS20092910.2656

Dostalek J., Ctyroky J., Homola J., Brynda E., Skalsky M., Nekvindova P., Spirkova J., Skvor J., and Schröfel J. (2001) Surface plasmon resonance biosensors based on integrated optical waveguides. *Sens. Actuators B* **76**, 8–12. DOI: 10.1016/S0925-4005(01)00559-7

Duhem O., Dascota A., Henninot J.F., and Douay M. (1999) Long period copper-coated grating as an electrically tunable wavelength-selective filter. *Electron. Lett.* **35**(12), 1014–1016. DOI: 10.1049/el:19990659

Erdogan T. (1997) Cladding-mode resonances in short- and long-period fiber grating filters. *J. Opt. Soc. Am. A* **14**(8), 1760–1773. DOI: 10.1364/JOSAA.14.001760

Flavio S.D., Rita C.S.L., and Lauro T.K. (2005) Study of poly(methylene blue) ultrathin films and its properties by electrochemical surface plasmon resonance. *J. Electroanal. Chem.* **581**, 231–240. DOI: 10.1016/j.jelechem.2005.04.021

Gloge D. (1971) Weakly guiding fibers. *Appl. Opt.* **10**(10), 2252–2258. DOI: 10.1364/AO.10.002252

Gu Z.T., Xu Y.P., and Gao K. (2006) Optical fiber long-period grating with solgel coating for gas sensor. *Opt. Lett.* **31**(16), 2405–2407. DOI: 10.1364/OL.31.002405

Gupta B.D. and Sharma A.K. (2005) Sensitivity evaluation of a multi-layered surface plasmon resonance-based fiber optic sensor: A theoretical study. *Sens. Actuators B* **107**(1), 40–46. DOI: 10.1016/j.snb.2004.08.030

He W., Li J.Z., and Mei J.C. (2005) Experimental study on photolytic index changes in germanosilicate fiber. *J. Inorg. Mater.* **20**, 210–214.

He Y.J., Lo Y.L., and Huang J.F. (2006) Optical-fiber surface-plasmon-resonance sensor employing long-period fiber gratings in multiplexing. *J. Opt. Soc. Am. B* **23**(5), 801–811. DOI: 10.1364/JOSAB.23.000801

Ho H.P., Wu S.Y., Yang M., and Cheung A.C. (2001) Application of white light-emitting diode to surface plasmon resonance sensors. *Sens. Actuators B* **80**(2), 89–94. DOI: 10.1016/S0925-4005(01)00881-4

Iadicicco A., Campopiano S., Paladino D., Cutolo A., Cusano A., and Bock W. (2010) Gold coated long period grating in single and multilayer configuration for sensing applications. *Sens. Microsyst.* **54**, 133–136. DOI: 10.1007/978-90-481-3606-3_23

Iga M., Seki A., and Watanabe K. (2005) Gold thickness dependence of SPR-based hetero-core structured optical fiber sensor. *Sens. Actuators B* **106**(1), 363–368. DOI: 10.1016/j.snb.2004.08.017

Ignacio D.V., Ignacio R.M., and Francisco J.A. (2006) Influence on cladding mode distribution of overlay deposition on long-period fiber gratings. *J. Opt. Soc. Am. A* **23**(3), 651–658. DOI: 10.1364/JOSAA.23.000651

Ignacio D.V., Ignacio R.M., Francisco J.A., and Miguel A. (2005) Nanodeposition of materials with complex refractive index in long-period fiber gratings. *J. Lightwave Technol.* **23**(12), 4192–4199. DOI: 10.1109/JLT.2005.858246

Ignacio D.V., Miguel A., Ignacio R.M., and Francisco J.A. (2005) Deposition of overlays by electrostatic self-assembly in long-period fiber gratings. *Opt. Lett.* **30**(7), 720–722. DOI: 10.1364/OL.30.000720

James S.W. and Tatam R.P. (2003) Optical fibre long-period grating sensors: Characteristics and applications. *Meas. Sci. Technol.* **15**, R49–R61. DOI: 10.1088/0957-0233/14/5/201

Jesus M.C., Ignacio D.V., Ignacio R.M., and Francisco J.A. (2007) Fiber-optic pH-sensors in long-period fiber gratings using electrostatic self-assembly. *Opt. Lett.* **32**(1), 29–31. DOI: 10.1364/OL.32.000029

Lee B.H., Liu Y., Lee S. B., Choi S.S., and Jang J. N. (1997) Displacements of the resonant peaks of a long-period fiber grating induced by a change of ambient refractive index. *Opt. Lett.* **22**, 1769–1771. DOI: 10.1364/OL.22.001769

Li M., Liao Y.B., and Shi C.Z. (2000) Dispersion characteristics of single mode As_2S_3 fiber. *Acta Photon. Sin.* **29**(2), 171–173.

Li Y.Q. and Cui M. (2002) *Optical Waveguide Theory and Technology.* Posts & Telecommunications Press, Peking.

Liang K. M. (2004) *Methods in Mathematical Physics.* Higher Education Press, Peking.

Lotierzo M., Henry O.Y.F., Piletsky S., Tothill I., Cullen D., Kania M., Hock B., and Turner A.P.F. (2004) Surface plasmon resonance sensor for domoic acid based on grafted imprinted polymer. *Biosens. Bioelectron.* **20**(2), 145–152. DOI: 10.1016/j.bios.2004.01.032

Mathews J.H. and Fink K.D. (2008) *Numerical Methods Using MATLAB,* 4th ed. Pearson Education, Singapore.

Nicholas D.R., Stephen W.J., and Ralph P.T. (2002) Optical fiber long-period gratings with Langmuir-Blodgett thin-film overlays. *Opt. Lett.* **27**(9), 686–688. DOI: 10.1364/OL.27.000686

Palik E.D. (1985) *Handbook of Optical Constants of Solids.* Academic Press, New York.

Patrick H.J., Kersey A.D., and Bucholtz F. (1998) Analysis of the response of long period fiber gratings to external index of refraction. *J. Lightwave Technol.* **16**, 1606–1612. DOI: 10.1109/50.712243

Pilla P., Iadicicco A., Contessa L., Campopiano S., Cutolo A., Giordano M., Guerra G., and Cusano A. (2005) Optical chemo-sensor based on long period gratings coated with δ form syndiotactic polystyrene. *IEEE Photon. Technol. Lett.* **17**(8), 1713–1715. DOI: 10.1109/LPT.2005.851979

Shin Y.B., Lee J.M., Park M.R., Kim M.G., Chung B.H., Pyo H.B., and Maeng S. (2007) Analysis of recombinant protein expression using localized surface plasmon resonance (LSPR). *Biosens. Bioelectron.* **22**(9–10), 2301–2307. DOI: 10.1016/j.bios.2006.12.028

Shu X.W. and Huang D.X. (1999) Highly sensitive chemical sensor based on the measurement of the separation of dual resonant peaks in a 100-μm-period fiber grating. *Opt. Commun.* **171**, 65–69. DOI: 10.1016/S0030-4018(99)00522-2

Shu X.W., Zhu X.M., Jiang S., Shi W., and Huang D.X. (1999) High sensitivity of dual resonant peaks of long-period fibre grating to surrounding refractive index changes. *Electron. Lett.* **35**(18), 1580–1581. DOI: 10.1049/el:19991040

Tang J.L., Cheng S.F., Hsu W.T., Chiang T.Y., and Chau L.K. (2006) Fiber optic biochemical sensing with a colloidal gold modified long period fiber grating. *Sens. Actuators B* **119**, 105–109. DOI: 10.1016/j.snb.2005.12.003

Tong Z., Wei H., Wang M.G., Wang Z., and Jian S.S. (2002) Effect of change of ambient refractive index on characteristics of long-period fiber gratings. *Acta Opt. Sin.* **22**(9), 1088–1091.

Tsao C. (1992) *Optical Fibre Waveguide Analysis.* Oxford University Press, New York.

Vengsarkar A.M., Lemaire P.J., Judkins J.B., Bhatia V., Erdogan T., and Sipe J.E. (1996) Long-period fiber gratings as band-rejection filters. *J. Lightwave Technol.* **14**, 58–65. DOI: 10.1109/50.476137

Vengsarkar A.M., Pedrazzani J.R., Judkins J.B., Lemaire P.J., Bergano N.S., and Davidson C.R. (1996) Long period fiber grating based gain equalizers. *Opt. Lett.* **21**, 336–338. DOI: 10.1364/OL.21.000336

Wang Y.J. and Knoll W. (2006) In situ electrochemical and surface plasmon resonance (SPR) studies of aniline-carboxylated aniline copolymers. *Anal. Chim. Acta* **558**, 150–157. DOI: 10.1016/j.aca.2005.11.006

Wang Z.Y., Heflin J.R., Stolen R.H., and Ramachandran S. (2005) Analysis of optical response of long period fiber gratings to nm-thick thin-film coatings. *Opt. Express* **13**(8), 2808–2813. DOI: 10.1364/OPEX.13.002808

Wei H., Tong Z., Wang M.G., and Jian S.S. (2003) Long-period fiber grating with metal coating. *Acta Opt. Sin.* **23**(8), 931–936.

Xu Y.P., Gu Z.T., and Chen J.B. (2005) Effect of surface film optical parameters on the characteristic of long-period fiber grating. *Chin. J. Lasers* **32**(11), 1519–1524.

Zhang Z.J. and Shi W.K. (2004) Investigation of the cladding modes of metal coated fibers. *J. Shanghai Jiao Tong Univ.* **38**(11), 1847–1850.

Zhang Z.J., Shi W.K., Gao K., Ji X.J., and Fang Z.J. (2004) Theoretical and experimental investigation of metal coated long-period fiber gratings. *Acta Opt. Sin.* **24**(7), 897–901.

CHAPTER 8

Analytical Approaches to Optimization of Gas Detection Using Fabry-Pérot Interferometers

E. Vargas-Rodriguez
H. N. Rutt
D. Claudio-Gonzalez
R. Rojas Laguna
J. A. Andrade-Lucio

1. INTRODUCTION

The plane Fabry-Pérot interferometer (FPI) is formed by two parallel and flat mirrors (Figure 8.1), which produces a set of transmission fringes (Born and Wolf 1970). The transmission spectrum it is described by Airy's formula, which can be written as

$$T_{\text{FPI}}(\nu, R, n, d, \Delta d, \theta) = \frac{1}{1 + \dfrac{4R}{(1-R)^2}\sin^2\left[\dfrac{4\pi\, n(d+\Delta d)\cos(\theta)\,\nu}{2}\right]} \tag{8.1}$$

DOI: 10.5643/9781606503201/ch8

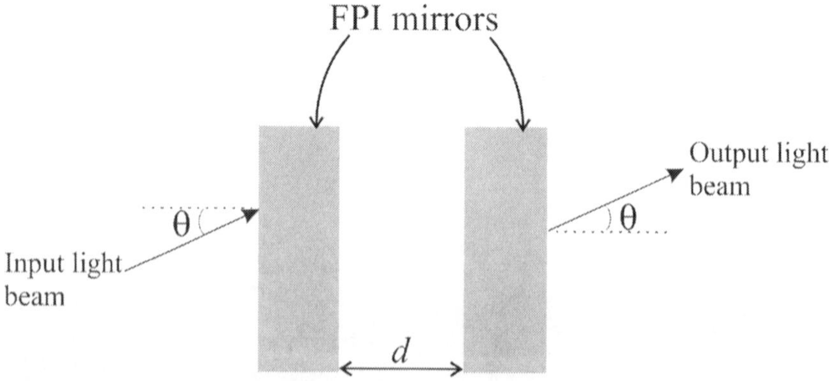

Figure 8.1. Plane Fabry-Pérot interferometer.

where R is the mirrors' reflectivity, n is the refractive index of the medium between the FPI mirrors, d is the cavity length, Δd represents the cavity length scan, θ is the incidence angle, and ν is the frequency given in wavenumbers.

The cavity length (d) is very important since it determines the free spectral range (FSR) of the spectral fringes, which is the spectral distance between two consecutive fringes. For instance, in Figure 8.2 we show three different FPI transmission spectra with different cavity lengths, where a dramatic change in the FSR can be observed since it is inversely proportional to the optical thickness, $\mathrm{FSR} = 1/(2nd)$.

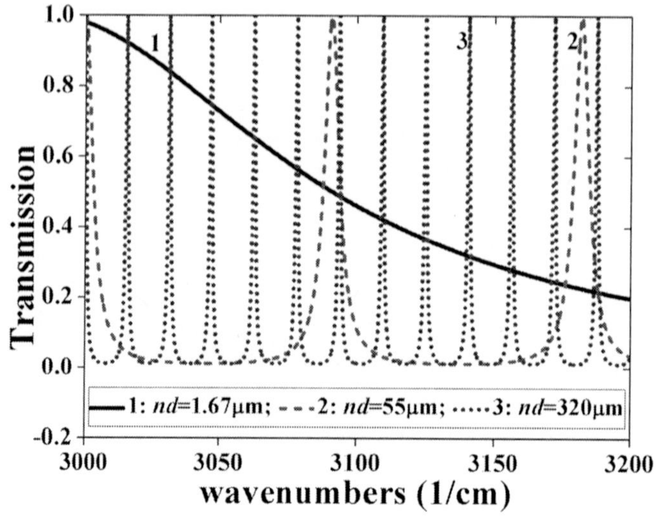

Figure 8.2. Simulated spectral response for a FPI with three different cavity lengths and reflectivity of 0.8.

1.1. GAS SENSOR DESIGN

A typical gas sensor setup, which uses a FPI as a modulator, is shown in Figure 8.3. This setup is relatively simple since it requires only a few basic components: a light source, a gas cell, the Fabry-Pérot interferometer, a band-pass filter, and an optical detector.

Mathematically it is possible to represent the sensor response as

$$I_D(C,l,\nu,n,d,\Delta d,\theta)=\int_0^{\infty} T_{\mathrm{FPI}}(\nu,n,d,\Delta d,\theta)T(\nu,l,C)\mathrm{Fil}(\nu)S(\nu)\,d\nu \tag{8.2}$$

where $I_D(C,l,\nu,n,d,\Delta d,\theta)$ is the power reaching the detector, $T_{\mathrm{FPI}}(\nu,n,d,\Delta d,\theta)$ is the spectral transmission profile of the Fabry-Pérot interferometer, $T(\nu,l,C)$ describes the spectral transmission through the gas cell with a path length l and a gas concentration C, the bandpass filter is given by Fil(ν), and $S(\nu)$ is the light source.

In this type of sensor the main idea is to use the FPI as a modulator; however, depending on the FPI optical thickness (nd), we can design sensors with very different operating principles. It is possible to classify these sensors by the optical thickness of the FPI into two groups: the narrow-gap and the wide-gap sensors. For both cases the overall sensor response is given by Eq. (8.2).

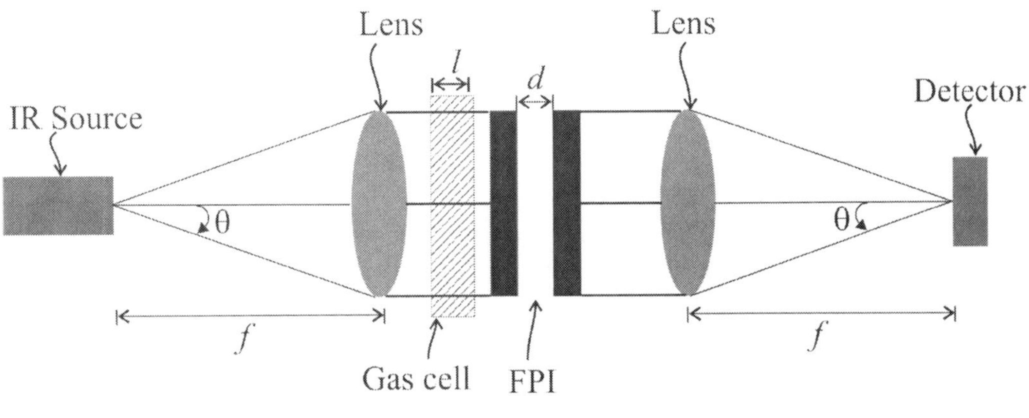

Figure 8.3. Simple configuration of a gas sensor with a FPI modulator.

2. THE NARROW-GAP FPI GAS SENSOR

In the narrow-gap type of sensor, the FPI's cavity length is adjusted to produce fringes with large free spectral range. Typically we have one or two FPI spectral transmission fringes over a wide spectral range to cover one absorption band of

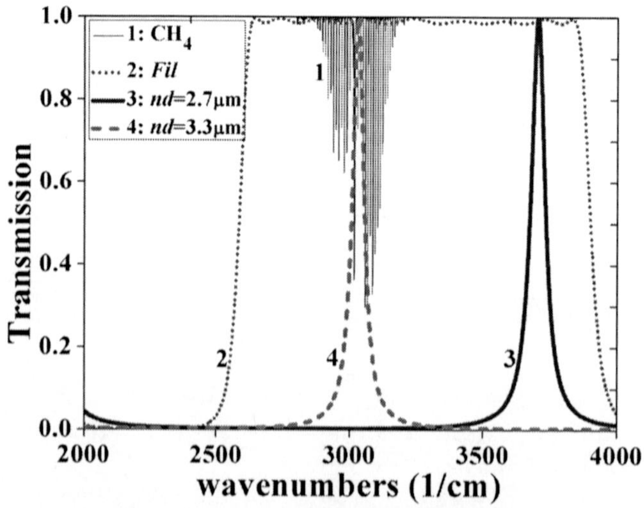

Figure 8.4. Simulated wide FPI fringe for a CH_4 narrow-gap sensor, R = 0.9.

a target molecule. To induce modulation, the FPI fringe is shifted along the wavenumber axis by scanning the FPI cavity length. Hence the FPI's fringe shifting means that at some point it will "select" some gas absorption lines and eventually will illuminate a spectral region without gas absorptions. Moreover, the band-pass filter is used to isolate a spectral region where the absorption band of the target molecule is in the reference region. As an example, let us to consider the simulated spectra shown in Figure 8.4, where the spectral reference region is described by the FPI fringe obtained when $nd = 2.70\ \mu m$ and the spectral measurement region is selected when $nd + \Delta d = 3.30\ \mu m$, and the CH_4 spectrum was simulated using the line positions and intensities provided by the HITRAN database (Rothman et al. 2005). The simulated sensor response, using Eq. (8.2) and considering different gas concentrations, is shown in Figure 8.5. Here two values $I_D(C,l,\nu,n,d,\Delta d,\theta)$ are obtained for two different values of the cavity length scan (Δd), one to shift the FPI fringe to the reference region and the second to shift the FPI fringe to the measurement spectral region. The change in $I_D(C,l,\nu,n,d,\Delta d,\theta)$ will be related directly to the gas concentration.

Narrow-gap sensors, physically, can be seen as *single-path, single-source* sensors since they require only one optical detector, one gas path, and one optical source. However, these sensors have the advantage that in practice they operate as *single-path, dual-wavelength* sensors since we have the reference and the measurement wavelengths, which are obtained by spectrally shifting the FPI fringe (Carlson 1997). This helps to reduce errors due to differential drift when two detectors are used (Alause et al. 1997), as in the typical dual-wavelength sensors. Gas sensors using this technique have been explored by several authors;

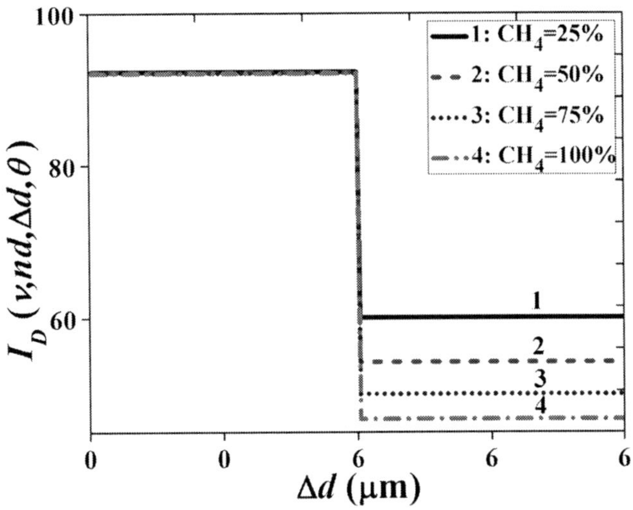

Figure 8.5. Simulated narrow-gap sensor response.

for instance, Carlson (1997) and Blomberg et al. (1997b) presented gas sensor designs to detect CO_2 using this principle of operation.

2.1. SENSOR RESPONSE AND THE CROSS-CORRELATION SPECTROSCOPY PRINCIPLE

The cross correlation spectroscopy principle was described by Goody in 1968. In that work Goody stated that the cross correlation between spectra of different gases is very small. Before analyzing this spectrometer, let us to recall that a cross correlation of two real-valued functions is defined by

$$f * g = \int_{-\infty}^{\infty} f(\tau) g(t + \tau) d\tau \tag{8.3}$$

In this operation the function $g(\tau)$ is shifted along the τ axis and therefore the integral must be evaluated for each t value. In Goody's hypothetical cross-correlation spectrometer the absorption spectra of the target gas must be shifted along the wavenumber axis. Mathematically this concept is easily understandable; in practice, however, it is impossible to shift the gas absorption spectrum as required by the cross-correlation function. However, Goody (1968) proposed a conceptual spectrometer to obtain an "experimental simulation" of a two-point spectral shift. This cross-correlation spectrometer was very simple since it consisted of a two-arm arrangement with two gas cells (the reference and the measurement cells). In this arrangement the light first passes by the measurement cell, which contains the target gas which can be mixed with others gases. After that the light alternately passes through the arms arrangement. In one of the two arms (arm 1) is placed the reference cell, which contains a pure target gas or a mixture of pure target gas with a nonabsorbing gas (e.g., N_2 for infrared). Finally, the light that exits from the arms is directed to an optical detector. The overall response for this system is given by

$$I(C_x, \Delta\nu) = \int_{\nu 1}^{\nu 2} S(\nu) A(\nu)\, T_1(C_x, \nu)\, T_2(C_r, \nu + \Delta\nu)\, d\nu \tag{8.4}$$

where $A(\nu)$ is the transmission spectrum of gases other than the target in the measurement cell, $T_1(C_x, \nu)$ is the transmission spectrum of the target gas within the measurement cell, with unknown concentration C_x, $T_2(C_r, \nu + \Delta\nu)$ is the transmission of the target gas within the reference cell, with known concentration C_r, and $\Delta\nu$ is the spectrum shift. Here losses due to optical windows are not considered. Therefore it is possible to write Eq. (8.4) for the two arms as

$$I(C_x, \Delta\nu_1) = \int_{\nu 1}^{\nu 2} S(\nu) A(\nu)\, T_1(C_x, \nu)\, T_2(C_r, \nu + \Delta\nu_1)\, d\nu \tag{8.5}$$

and

$$I(C_x, \Delta\nu_2) = \int_{\nu 1}^{\nu 2} S(\nu) A(\nu)\, T_1(C_x, \nu)\, T_2(C_r, \nu + \Delta\nu_2)\, d\nu \tag{8.6}$$

where $\Delta\nu_2$ indicates the spectral shifts in the absorption spectrum when light follows arm 2 and $\Delta\nu_1$ is the spectral shift when light follows arm 1. Hence the spectral shifting of the reference gas spectrum is obtained by alternately passing the light by arm 1, with the reference gas cell, having a ($\Delta\nu_1 = 0$), and later by the arm without gas cells, where we have no reference gas absorptions, which thus can be considered as if the spectrum was spectrally shifted ($\Delta\nu_2$) far away from the reference spectral region.

Now, according to Goody (1968), if the source profile is flat over the band-pass filter region and if the absorptions of gases other than the target do not change, it is possible to rewrite Eqs. (8.5) and (8.6) as

$$I(C_x, \Delta\nu_1) = \int_{\nu 1}^{\nu 2} S(\nu)\, d\nu \int_{\nu 1}^{\nu 2} A(\nu)\, d\nu \int_{\nu 1}^{\nu 2} T_1(C_x, \nu) T_2(C_r, \nu + \Delta\nu_1)\, d\nu \tag{8.7}$$

and

$$I(C_x, \Delta\nu_2) = \int_{\nu 1}^{\nu 2} S(\nu)\, d\nu \int_{\nu 1}^{\nu 2} A(\nu)\, d\nu \int_{\nu 1}^{\nu 2} T_1(C_x, \nu) T_2(C_r, \nu + \Delta\nu_2)\, d\nu \tag{8.8}$$

respectively.

At this point it is possible to evaluate the ratio between $I(C_x, \Delta\nu_1)$ and $I(C_x, \Delta\nu_2)$:

$$I(C_x) = \frac{I(C_x, \Delta\nu_1)}{I(C_x, \Delta\nu_2)} \cong \frac{\int_{\nu 1}^{\nu 2} T_1(C_x, \nu) T_2(C_r, \nu + \Delta\nu_1)\, d\nu}{\int_{\nu 1}^{\nu 2} T_1(C_x, \nu) T_2(C_r, \nu + \Delta\nu_2)\, d\nu} \tag{8.9}$$

The result of this approach is very important since it shows that if we perform the cross correlation between the spectra of the target gas and the spectra of an unknown mixture, the depth of modulation $I(C_x)$ will be independent of absorptions due to other gases.

Another alternative to determine the depth of modulation is by evaluating the average difference between $I(C_x, \Delta\nu_1)$ and $I(C_x, \Delta\nu_2)$, which can be expressed as

$$\begin{aligned} D_m &= \frac{2\left[I(C_x, \Delta\nu_2) - I(C_x, \Delta\nu_1)\right]}{I(C_x, \Delta\nu_2) + I(C_x, \Delta\nu_1)} \\ &\cong \frac{2\left[\int_{\nu 1}^{\nu 2} T_1(C_x, \nu) T_2(C_r, \nu + \Delta\nu_2)\, d\nu - \int_{\nu 1}^{\nu 2} T_1(C_x, \nu) T_2(C_r, \nu + \Delta\nu_1)\, d\nu\right]}{\int_{\nu 1}^{\nu 2} T_1(C_x, \nu) T_2(C_r, \nu + \Delta\nu_2)\, d\nu + \int_{\nu 1}^{\nu 2} T_1(C_x, \nu) T_2(C_r, \nu + \Delta\nu_1)\, d\nu} \end{aligned} \tag{8.10}$$

This last equation is another option to evaluate the depth of modulation (Goody 1968). Indeed, other authors have proposed equations similar to (8.9) and (8.10) to determine the overall sensor signal based on cross-correlation spectroscopy. The final form of the depth-of-modulation equation can vary depending on the optical setup. Cross-correlation spectroscopy gas sensors can be implemented in very different ways, without the need of a FPI modulator; for instance, Johnston (1992), Dakin et al. (1995, 2003), and Chambers et al. (2004) have proposed different gas sensor setups based on this principle.

The cross-correlation principle can be applied to the narrow-gap sensors since we can have two values of the sensor response, one for the reference and one for the measurement. The reference and the measurement channels are obtained by spectrally shifting the FPI fringe. Hence, following the cross-correlation principle for the narrow-gap sensors, the depth of modulation $I_D(C)$ can be expressed as a ratio between the measurement and the reference responses (Alause et al. 1997; Blomberg et al. 1997b; Carlson 1997). It can be expressed as

$$I_D(C)=\frac{I_D(C,l,\nu,nd,\Delta d_x,\theta)}{I_D(C,l,\nu,nd,\Delta d_r,\theta)}$$
$$=\frac{\int_0^{\infty} T_{\mathrm{FPI}}(\nu,nd,\Delta d_x,\theta)T_2(\nu,l.C)\,A(\nu)\mathrm{Fil}(\nu)\,S(\nu)\,d\nu}{\int_{\nu 1}^{\nu 2} T_{\mathrm{FPI}}(\nu,nd,\Delta d_r,\theta)T_2(\nu,l.C)\,A(\nu)\mathrm{Fil}(\nu)\,S(\nu)\,d\nu}$$
$$=\frac{\int_0^{\infty} T_{\mathrm{FPI}}(\nu,nd,\Delta d_x,\theta)\,d\nu\int_0^{\infty} T_2(\nu,l.C)d\nu\int_0^{\infty} A(\nu)\,d\nu\int_0^{\infty}\mathrm{Fil}(\nu)\,d\nu\int_0^{\infty} S(\nu)d\nu}{\int_0^{\infty} T_{\mathrm{FPI}}(\nu,nd,\Delta d_r,\theta)\,d\nu\int_0^{\infty} T_2(\nu,l.C)d\nu\int_0^{\infty} A(\nu)\,d\nu\int_0^{\infty}\mathrm{Fil}(\nu)\,d\nu\int_0^{\infty} S(\nu)d\nu}$$
$$=\frac{\int_0^{\infty} T_{\mathrm{FPI}}(\nu,nd,\Delta d_x,\theta)\,d\nu\int_0^{\infty} T_2(\nu,l.C)d\nu}{\int_0^{\infty} T_{\mathrm{FPI}}(\nu,nd,\Delta d_r,\theta)\,d\nu\int_0^{\infty} T_2(\nu,l.C)d\nu} \tag{8.11}$$

where $I_D(C,l,\nu,nd,\Delta d_x,\theta)$ is the signal obtained when the FPI fringe is covering the absorption band of the target gas and $I_D(C,l,\nu,nd,\Delta d_r,\theta)$ is the sensor response for the reference channel. To obtain Eq. (8.11) we assume that absorptions due to molecules $A(\nu)$ different from our target gas do not change. Thus the depth of modulation based on the cross-correlation principle, Eq. (8.11), helps to reduce cross-sensitivity errors of narrow-gap sensors. Finally, to evaluate the depth of modulation numerically, a processing stage is required to perform the cross-correlation operation.

2.2. OPTIMAL FPI MIRROR REFLECTIVITY

In the narrow-gap sensors an important parameter requiring optimization is the FPI mirror reflectivity (R). Depending on the reflectivity, the depth of modulation can be affected; but it is also necessary to consider that fabrication of FPI mirrors with high reflectivity is more complex. Hence it is necessary to determine the optimal value of this parameter which permits the best compromise between the depth-of-modulation factor and the cost of the system. Hence, considering as an example the gas sensor described in Section 2, whose response, using Eq. (8.2), is shown in Figure 8.5, we get the depth of modulation, by using Eq. (8.11), presented in Figure 8.6. Here we considered a FPI with different R values and a constant cavity length. Also, when FPI mirrors with low reflectivity are assumed, the modulation depth is not strongly affected by changes in the gas concentration, compromising the sensor sensitivity. In contrast, when FPI mirrors with higher reflectivity are considered, the depth of modulation is more sensitive to concentration changes. Therefore, ideally, a high reflectivity is a better choice for this

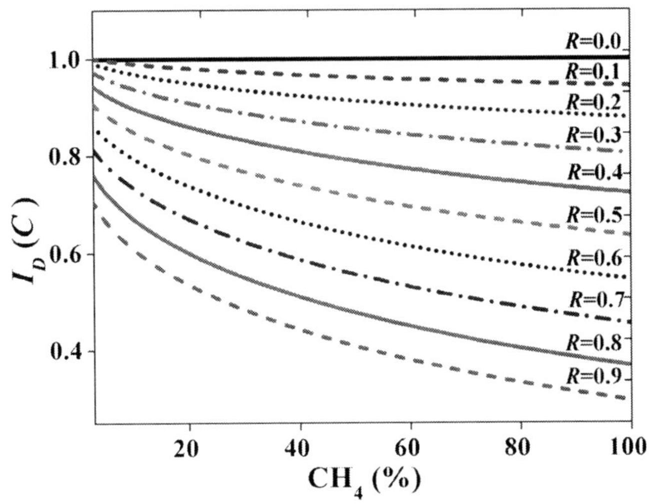

Figure 8.6. Modulation depth for a narrow-gap sensor as a function of gas concentration and for different reflectivity values.

sensor. However, we need to determine the best compromise between the depth of modulation and the reflectivity, because above some level of reflectivity, the modulation depth will not improve significantly. For instance, in Figure 8.6 the slope of the depth of modulation for $R = 0.9$ is similar to that obtained for $R = 0.8$. Therefore we can use this reflectivity for the design, as we can expect a minimal loss in the overall sensor response. Furthermore, even at $R = 0.7$, the reduction in the slope of the depth of modulation will not be dramatic. The optimum values obtained are low compared to those generally intuitively expected, and very low compared to those used in analytical Fabry-Pérot interferometers. Fortunately, such low-reflectivity mirrors are cheap and simple to fabricate and have a high tolerance to manufacturing errors, rendering the system more practical in use.

2.3. MULTIPLE GAS SENSORS BASED ON THE NARROW-GAP DESIGN

The narrow-gap design (Figure 8.3) can be also used to detect two or more gases without the need to add more optical components. This can be achieved by fixing the optical cavity length in such a way that a FPI spectrum with a wide FSR is produced, allowing us to cover the spectral region where two or more absorption bands of different molecules occur. Hence, by scanning the cavity length, the FPI transmission fringe will be alternately covering different absorption bands (Figure 8.7). Moreover, in this spectral region, ideally, the FPI fringe must cover a spectral region where there are no gas absorptions, which is used as the reference

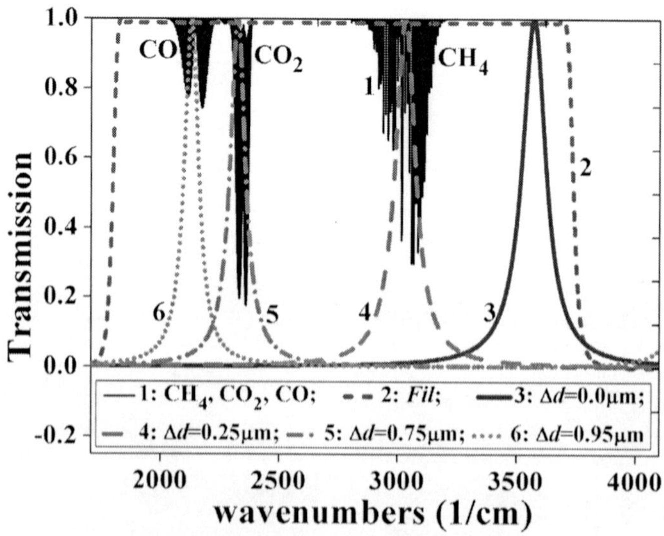

Figure 8.7. Simulation of a typical FPI transmission spectrum for a dual-gas sensor. In this simulation, nd = 1.4 μm and R = 0.9 were considered.

region. For instance, Calaza et al. (2004) presented a gas sensor design to detect CH_4 and CO in which the FPI fringe can be shifted from 3.30 μm, where the fundamental CH_4 absorption occurs, to 3.95 μm, which is used as a reference, and to 4.70 μm, where the fundamental absorption band of CO occurs. Similar designs using this principle of operation can be found in the literature as well. For example, Alause et al. (1997) presented a sensor capable of detecting CH_4 and CO, Noro et al. (2002) proposed a sensor to detect H_2O and CO_2, and Calaza et al. (2003) presented a sensor able to detect three gases (CH_4, CO_2, and CO). Other authors have presented FPI filters that can be used to implement sensors capable of detecting more than two different gases (Blomberg et al. 1997a; Neumann et al. 2007).

An important issue that needs to be considered when designing a narrow-gap multiple-gas sensor using only one optical detector is that the sensor response will be composed of the response for each target gas (Alause et al. 1997). Hence the sensor response signal must be temporally separated to determine the response for each particular gas and for the reference. As an example, let us to consider that we would like to design a gas sensor to detect CO, CO_2, and CH_4, and for this we use an FPI with nd = 1.4 μm and R = 0.9, a flat source $S(\nu)$ = 1, a gas cell path length l = 5 cm, and the band-pass filter whose transmission spectrum is shown in Figure 8.7. In this case, to shift the FPI fringe from the reference position to reach the spectral region, where the fundamental ro-vibrational absorption band of CO lies (4.70 μm), it is necessary to scan the cavity length $0 \leq \Delta d \leq 0.95\,\mu m$. In this way, when $\Delta d \cong 0.25\,\mu m$, the sensor will produce the response for CH_4; when $\Delta d \cong 0.75\,\mu m$, the response will correspond to CO_2; and

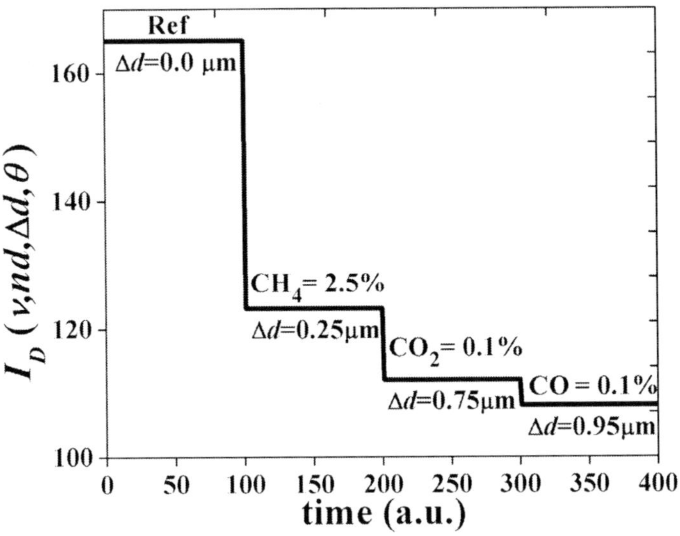

Figure 8.8. Overall response of a multiple-gas sensor based on the narrow-gap design.

finally, when $\Delta d \cong 0.95\,\mu\text{m}$, the sensor response will be for CO. Therefore the signal response will be temporally segmented and in each interval of time will be for a specific gas (Figure 8.8). Hence it will be necessary to add appropriate signal processing to determine the values of $I_D\left(C, l, \nu, nd, \Delta d_0, \theta\right)$ for each target gas and later perform the cross-correlation operation, given by Eq. (8.11), to determine the concentration of each gas.

2.4. NARROW-GAP SENSOR DESIGN WITH A NARROW BAND-PASS FILTER

In narrow-gap sensors it is usually necessary to have a reference region and to perform the cross-correlation spectroscopy operation given by Eq. (8.11) in order to reduce drift and cross-sensitivity errors. However, it is also possible to reduce the cross-sensitivity errors by using a narrow band-pass filter (NBPF) instead of a relatively wide band-pass filter. This is because the NBPF can isolate only the spectral region where the absorption lines of the target gas occurs or even isolate only a few absorption lines. Additionally, the NBPF can help us to reduce the sensor response due to general background light.

In order to clarify this point, let us to consider as an example a CH_4 sensor that uses a NBPF, whose spectrum is shown in Figure 8.9, and with the following parameters: gas path length l = 5 cm, a FPI with mirror reflectivity R = 0.8, nd = 2.90 μm, $0 \leq \Delta d \leq 0.7$ μm, $\theta = 0$, and a broad-band flat source $S(\nu)$ = 1. The simulated response for this gas sensor is presented in Figure 8.10, where can be observed that when the FPI's spectral fringe is shifted so that it covers the

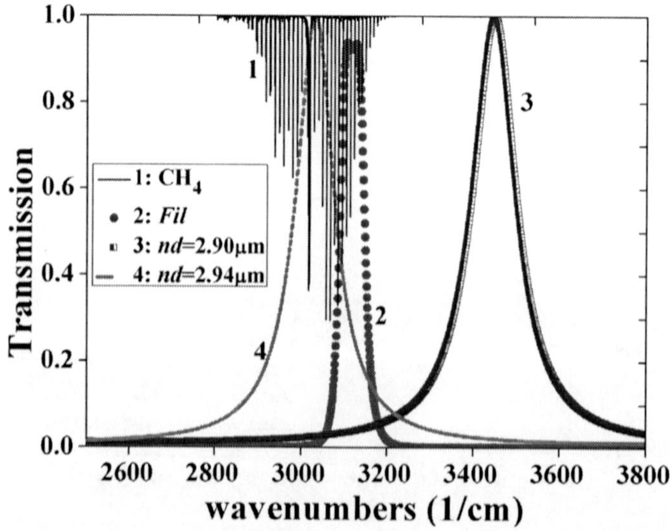

Figure 8.9. Simulated spectra of a narrow band-pass filter and a FPI for a narrow-gap sensor design.

transmission spectral region of the NBPF, there is a maximum in $I_D(C,l,\nu,nd,\Delta d,\theta)$; and as the FPI fringe moves away from the central wavelength of the NBPF, it decays to a minimum. Moreover, the amplitude of the function $I_D(C,l,\nu,nd,\Delta d,\theta)$ depends on the gas concentration (Figure 8.10). In these sensors, as the gas concentration increases, the $I_D(C,l,\nu,nd,\Delta d,\theta)$ amplitude decreases. Finally, to obtain the modulation, the FPI mirror needs to be returned to its initial position. In our

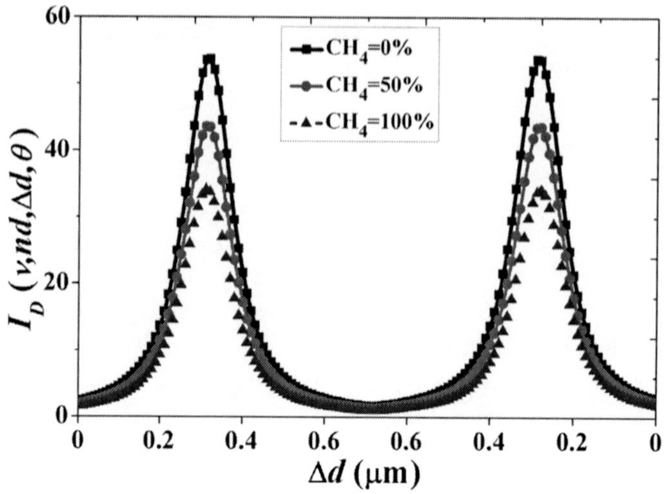

Figure 8.10. Simulated narrow-gap sensor response as a function of cavity length scan and gas concentration.

example the FPI mirror is moved forward from 0 up to 0.7 μm, and after it returns from 0.7 μm to 0 (Figure 8.10). Now, as the function $I_D(C,l,\nu,nd,\Delta d,\theta)$ will be periodic, it can be useful to relate its RMS value to the gas concentration instead of its amplitude. This design has the advantage that it is not necessary to shift the FPI fringe to a reference spectral region, and additionally, the NBPF can help to reduce the cross sensitivity since we isolate the spectral region of the target absorption band.

2.5. NARROW-GAP SENSOR DESIGN AND MEMS

Technology used for fabrication of micro-electro-mechanical-systems (MEMS) devices have been developing very quickly in the last two decades, and it is now possible to integrate micro-mechanical, micro-optical, and micro-electronic elements within the same device. The MEMS tunable FPI (MFPI) has gained attention, and several works aiming at the integration of MFPIs can be found in the literature (see Table 8.1). For instance, Alause et al. (1997), Carlson (1997), and Calaza et al. (2004) all presented gas sensors with an electrically tuned MFPI. Moreover, some of these MFPIs integrate an optical detector, such as the CO_2 sensor design of Mayrwöger et al. (2011), which includes a bolometer. Furthermore, some MFPIs can be used to detect multiple gases (Noro et al. 2002; Neumann et al. 2008).

Most MFPIs that can be found in the literature have a narrow gap, where the cavity length of the FPI is just a few micrometers. Also, in most of them the reflectivity of the FPI mirrors is generally high ($R > 0.9$). In gas sensing, the narrow-gap sensors based on MFPIs are very reliable and help to reduce considerably the size and power consumption of the sensor.

Table 8.1. Examples of MFPI devices

Reference	R	nd (μm)	Integrated optical detector	Target gas
Alause et al. 1997	0.3	2	No	CH_4 and CO
Carlson 1997	n/a	n/a	No	CO_2
Calaza et al. 2004	n/a	2.4	No	CH_4, CO_2, and CO
Noro et al. 2002	n/a	3.4	No	CO_2 and H_2O
Neumann et al. 2007	>0.9	2.5	Yes	CH_4, CO_2, and CO
Mayrwöger et al. 2011	0.36	4.3	Yes	CO_2
Saito et al. 2008	0.30	6	No	n/a
Srivastava et al. 2004	>0.9	4.07	No	n/a
Crocombe 2004	>0.9	≈1	No	C_2H_2
Domash 2005	n/a	n/a	No	CO

n/a: data not available.

3. THE WIDE-GAP FPI GAS SENSOR

In narrow-gap sensors the FPI produces one or two relatively wide spectral fringes to illuminate the spectral region where the absorption band of the target gas occurs. In these sensors the structure of the absorption band, such as the shape and the separation of absorption lines, is not so important. In narrow-gap sensors the most important issue is to select a proper cavity length and shift the FPI fringes to produce the measurement and reference spectral regions. However there are gases that present absorption bands with very well-defined shape and almost equidistant absorption lines, such as some diatomic and linear molecules—for example, CO, CO_2, N_2O, etc.—and some specific absorptions bands of symmetric top and spherical top molecules such as NH_3, CH_4 at 1 atm pressure (Herzberg 1945). It is possible to design gas sensors that can take advantage of this physical property, such as, for instance, the wide-gap FPI sensors. These sensors have essentially the same design and components as narrow-gap sensors (Figure 8.3), discussed previously. The only difference is the FPI optical thickness (*nd*), which must be considerably longer to reduce the free spectral range of the FPI transmission fringe pattern. Therefore, by selecting a proper value of the cavity length, it is possible to match the FPI transmission fringes with several ro-vibrational absorption lines of the target molecule. After the FPI fringes are matched with absorption lines, these can be shifted along the wavenumber axis by moving one of the FPI mirrors, and therefore the cavity length is varied by $d + \Delta d$ (Figure 8.11), where

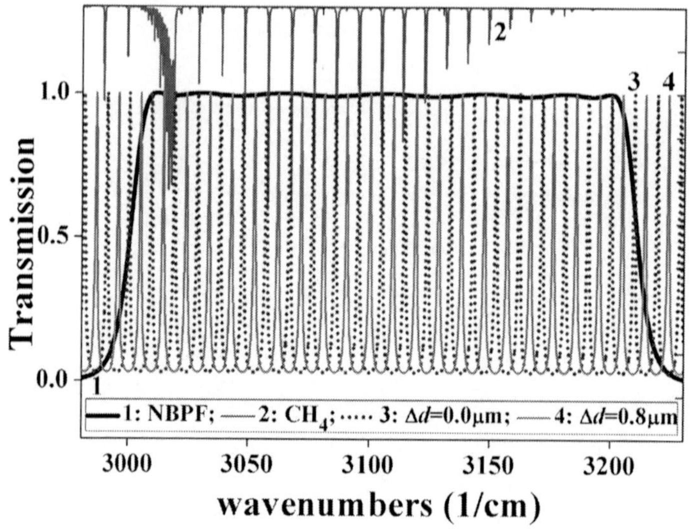

Figure 8.11. Simulated FPI transmission fringes matched with CH_4 rotational lines (Δd = 0 μm) and FPI fringes shifted by scanning the cavity length (Δd = 0.8 μm). The CH_4 spectrum is shifted up 0.3 unit for clarity. The optical thickness of the FPI was *nd* = 525 μm.

$0 \le \Delta d \le \lambda_0/2$ (Jacquinot 1960); λ_0 is the reference wavelength. Hence, as the cavity length is scanned, a change in the amount of transmitted light will be observed. Consequently, the optical detector will observe a change in power, which will be related directly to the gas concentration. Optical gas sensors using the FPI as a modulator have been studied for a long time, and several authors have presented gas sensors using the technique of matching the FPI fringes with the ro-vibrational absorption lines (Batchellor and Dakin 1987; Dakin et al. 1995; Jin et al. 1993; Mohebati and King 1998). In most of these works, the target gas was methane CH_4.

Wide-gap sensors present high selectivity to the target molecule (Mohebati and King 1998), since the FPI spectral transmission fringes are matched with the absorption lines of the target molecule. Hence, ideally, the sensor response will be minimally affected by the presence of absorption lines of different gases, which do not have the same spectral separation as the absorptions lines of the target molecule. Therefore it is important to determine the optimal FPI parameters such as the cavity length and reflectivity in order to guarantee the best performance of a wide-gap gas sensor. The sensor response for the wide-gap sensor is given by Eq. (8.2), and therefore, by changing the parameters of this equation systematically, one can determine the optimal parameter values for the sensor, such as the FPI cavity length and reflectivity, as well as the optimum source and detector characteristics which guarantee maximum sensor response as a function of the target gas concentration.

The simplest optimization procedure consists of evaluating Eq. (8.2) for all the possible combinations of values of the parameters R, d, n, Δd, and $\mathrm{Fil}(\nu)$. However, this procedure is time-consuming, since it takes a long time to compute all the possible combinations of R (from 0 to 1), nd (from 0 to ∞), and Δd (from 0 to $\lambda_0/2$), since each point involves integration over a rapidly varying function of frequency. Furthermore, the same procedure should be carried out for different filter shapes $\mathrm{Fil}(\nu)$. Moreover, this procedure does not give information about what *kind* of filter will produce the lowest background sensor response, that is, when the target gas concentration is 0. Hence, after the optimization procedure is carried out taking into account different filters, the background amplitude modulation values are compared to determine which filter produces the lowest background modulation value. Therefore, although the direct numerical method (DNM) is very simple to implement, it is nevertheless a very slow procedure and gives little insight into the sensor design.

Here, to avoid direct numerical evaluation of Eq. (8.2), the convolution method (Vargas-Rodríguez and Rutt 2007), a simple method based on the Fourier transform, can be used to determine the optimal FPI parameters. This method permits evaluation of $I_D(C, l, \nu, nd, \Delta d, \theta)$ in a much more efficient form than the direct numerical method. Moreover, the optimum optical thickness nd can be determined easily from the Fourier transforms involved in the method. Furthermore, it will be shown that the optimum R can be determined more efficiently (faster). Additionally, the convolution method gives guidance on choosing the best band-pass filter to reduce the background amplitude modulation. The results of these optimizations are to a considerable extent counterintuitive, and sensor design without their guidance will lead to poor performance.

3.1. THE CONVOLUTION METHOD

Wide-gap sensors depend on the interaction of the spectral response of the FPI, the absorption band of the target gas, the spectral response of the NBPF, and the source spectral profile, as indicated by Eq. (8.2). In these sensors, to induce modulation the FPI cavity length is scanned by $\lambda_0/2$, producing a shift of the FPI fringes along the wavenumber axis which changes the total transmission. However while the transmission FPI fringes are shifted by scanning the cavity length, the filter response Fil(ν), the source profile $S(\nu)$, and the absorptions due to the molecule $T(\nu, l, C)$, are not affected. Consequently, the product of these three functions will remain constant regardless of the cavity length scan. Thus it is possible to consider that, in general, the sensor response depends on two spectral functions, and therefore Eq. (8.2) can be rewritten as

$$I_D(C,l,\nu,nd,\Delta d,\theta) = \int_0^\infty T_{\mathrm{FPI}}(\nu,R,nd,\Delta d,\theta)\, G(\nu,l,C)\, d\nu \tag{8.12}$$

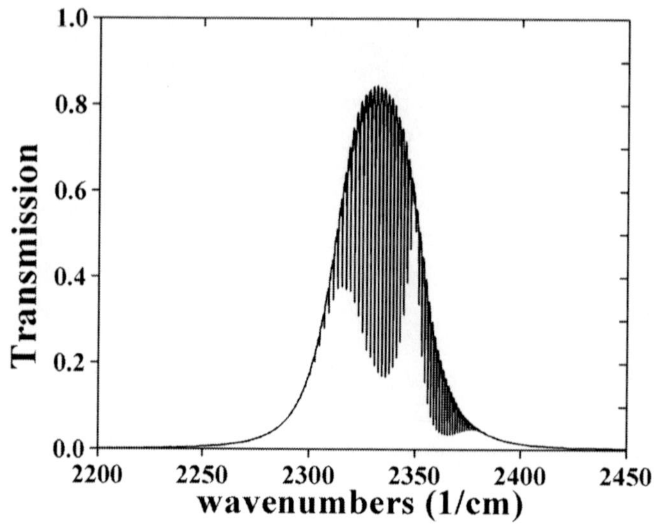

Figure 8.12. Transmission profile $G(\nu, l, C)$.

where the function $G(\nu,l,C)$ is the product of fixed spectra, given by

$$G(\nu,l,C) = T(\nu,l,C)\,\mathrm{Fil}(\nu)\,S(\nu) \tag{8.13}$$

and $T_{\mathrm{FPI}}(\nu,R,n,d,\Delta d,\theta)$ is the FPI transmission profile described by Eq. (8.1). An example of the transmission profile of $G(\nu,l,C)$ is shown in Figure 8.12; this simulation was generated using a CO_2 concentration of 5000 ppm, a path length of 5 cm, and $S(\nu)$ = 1. The overall shape of the function profile $G(\nu,l,C)$ is described by the filter band-pass shape, while the absorption lines observed in the $G(\nu,l,C)$ profile are produced by the ro-vibrational absorption lines of CO_2. In this example, we use as Fil(ν) an experimental measurement of a real narrow band-pass filter (Dennis 2004).

3.1.1. System Response Described as a Convolution

A small shift in the FPI's fringes along the wavenumber axis ($\Delta\nu$) is produced for each small change in the cavity length Δd (Figure 8.11). It is important to recall that the cavity length scan Δd is given in centimeters, while the fringes shift ($\Delta\nu$) is expressed in wavenumber units. Therefore, here it is useful to make a change of variable in Eq. (8.12) to get

$$I_D(C,l,\nu,nd,\theta,\Delta\nu) = \int_0^{\infty} T_{\mathrm{FPI}}(\nu,R,nd,\Delta\nu,\theta)\,G(\nu,l,C)\,d\nu \tag{8.14}$$

Now, considering that the width of the FPI's fringes and their symmetry do not vary when they are shifted over one FSR, Eq. (8.14) can be rewritten as

$$I_D(C,l,\nu,nd,\theta,\Delta\nu) = \int_0^{\infty} T_{\mathrm{FPI}}(\nu-\Delta\nu,R,nd,\theta)\,G(\nu,l,C)\,d\nu \tag{8.15}$$

Hence, the system response described by Eq. (8.2) is reduced to Eq. (8.15), which has the form

$$h(t) = \int_{-\infty}^{\infty} f(\tau - t)\,k(\tau)\,d\tau \tag{8.16}$$

which is the definition of the convolution and can be easily solved by using the Fourier transform (**FT**). It is important to recall that the Fourier transform of a convolution (*) is given by

$$\mathbf{FT}\{f(t) * k(t)\} = H(\omega) = F(\omega)K(\omega) \tag{8.17}$$

Therefore, to solve Eq. (8.15) it is only necessary to evaluate the Fourier transforms of two functions, $T_{\text{FPI}}(\nu, R, nd, \Delta d = 0, \theta)$ and $G(\nu, l, C)$, involved in Eq. (8.12). All this can be expressed as

$$\mathbf{FT}\{I_D(C, l, \nu, nd, \theta, \Delta\nu)\} = I_D(C, l, nd, \theta, \xi) = T_{\text{FPI}}(R, n, d, \theta, \xi)\, G(\xi, l.C) \qquad (8.18)$$

where

$$T_{\text{FPI}}(R, n, d, \theta, \xi) = \mathbf{FT}\{T_{\text{FPI}}(R, n, d, \theta, \nu)\} \qquad (8.19)$$

and

$$G(\xi, l, C) = \mathbf{FT}\{G(\nu, l, C)\} \qquad (8.20)$$

Equation (8.18) describes the Fourier transform of the systems response.

Now, to obtain the system response in terms of the FPI's fringes shift $\Delta\nu$, it is necessary to apply the inverse Fourier transform (**IFT**) to the product given by Eq. (8.18).

Here it is important to mention that in our work the units in the Fourier domain (ξ) are in terms of centimeters. This can be very confusing, since generally signals are in the time domain (expressed in seconds), so when the Fourier transform is applied the Fourier domain is in units of hertz (1/s). However, ours is a special case since our signal is already in the frequency domain with wavenumber units (cm^{-1}), so when we apply the Fourier transform the units that we get for a particular Fourier domain are in centimeters (cm). Therefore, in this chapter, when we mention frequency represented by ξ, it will be in units of centimeters.

After the inverse Fourier transform is applied to Eq. (8.18), $\mathbf{IFT}\{T_{\text{FPI}}(R, n, d, \theta, \xi)\, G(\xi, l.C)\}$, the system response will be expressed in terms of $\Delta\nu$ rather than in terms of the cavity length scan Δd (cm). To get the sensor response in terms of Δd it is necessary to extract a segment from $\mathbf{IFT}\{T_{\text{FPI}}(R, n, d, \theta, \xi)\, G(\xi, l.C)\}$. Here, a cavity length scan $\Delta d_{max} = \lambda_0/2$ shifts the FPI fringes by one FSR along the wavenumber axis, which is given by

$$\Delta\nu_{max} = \text{FSR} = \frac{1}{2nd} \qquad (8.21)$$

Therefore, from $I_D(R, n, d, \Delta\nu)$ we just need to extract the segment from 0 to 1 FSR, which is equivalent to a cavity length scan from 0 to $\lambda_0/2$. This segment is equivalent to $I_D(R, n, d, \Delta d)$.

Up to this point we have shown that the convolution method can help us to determine the sensor response in a more efficient way than direct numerical evaluation. Now we need to optimize the design for a band-pass filter shape, source, and the ro-vibrational absorption lines of the target gas. Here we can use Fourier transform characteristics of the $G(\nu, l.C)$ and $T_{\mathrm{FPI}}(\nu - \Delta\nu, R, n, d, \theta)$ functions of the convolution method to determine the optimal system parameters (Vargas-Rodríguez and Rutt 2007; Vargas-Rodriguez 2007).

3.1.2. *Fourier Characterization of the FPI Transmission Spectrum*

For simplicity we will start by analyzing the Fourier transform of the Fabry-Pérot transmission fringes given by the Airy formula, Eq. (8.1), which can be expanded as a cosine series (Cooper 1971; Vaughan 1989). This representation is quite convenient since the Fourier transform of this series can be easily evaluated. The Airy formula expanded as a cosine series is given by (Cooper 1971; Vaughan 1989)

$$T_{\mathrm{FPI}}(\nu, R, n, d, \theta) = \frac{(1-R)}{(1+R)}\left\{1 + 2\sum_{m=1}^{\infty} R^m \cos\left[m4\pi n d \nu \cos(\theta)\right]\right\} \qquad (8.22)$$

Here we can observe that the Airy function is given by a sum of cosine functions of "frequency" $\xi = m2nd$. These "frequencies" are due to the period of the fringes which is defined by the free spectral range, FSR = $1/2nd$, which is the separation between two consecutive fringes peaks or valleys. Thus, the first harmonic (fundamental) of the Airy function has a "frequency" of $2nd$, the second harmonic has a "frequency" of $4nd$, the third harmonic of $6nd$, and so on. The amplitudes of these cosine functions are given by

$$A_m = \frac{(1-R)}{(1+R)} 2R^m \qquad m = 1, 2, 3, \ldots \qquad (8.23)$$

Now the Fourier transform of the Airy function can be expressed as

$$\mathbf{FT}\left\{T_{\mathrm{FPI}}(\nu, R, n, d, \theta)\right\} = \mathbf{FT}\left\{\frac{1-R}{1+R}\right\} + \sum_{m=1}^{\infty} A_m\, \mathbf{FT}\left\{\cos\left[4m\pi n d \nu \cos(\theta)\right]\right\} \qquad (8.24)$$

This expression simplifies the problem, since now we only need to calculate the Fourier transform of the cosine functions, and after that sum their respective transforms. Here it is useful to recall that the Fourier transform of a cosine function is given by a pair of impulses which can be defined by

$$\left|\mathbf{FT}\{A\cos(2\pi\omega_0 t)\}\right| = \begin{cases} \dfrac{A}{2} & |\omega| = \omega_0 \\ 0 & |\omega| \neq \omega_0 \end{cases} \tag{8.25}$$

Substituting (8.25) into (8.24), we find that the Fourier transform of the FPI fringes is given by (Cooper 1971; Vargas-Rodriguez 2007)

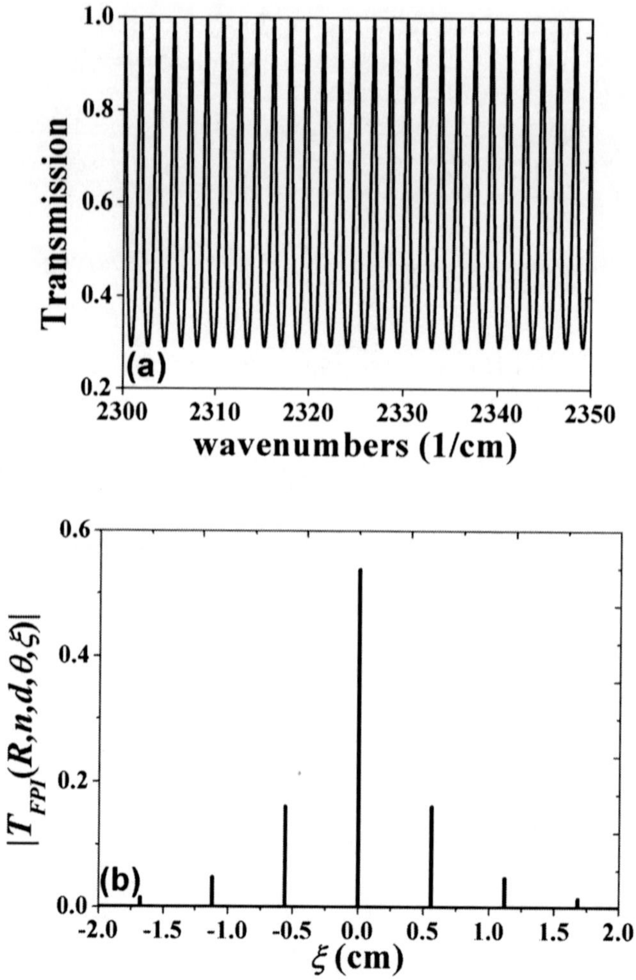

Figure 8.13. Simulated profiles of (a) FPI transmission spectrum and (b) Fourier transform magnitude of the FPI transmission spectrum. (Reprinted with permission from Vargas-Rodriguez and Rutt 2007. Copyright 2007 Optical Society of America.)

$$T_{\mathrm{FPI}}(R,n,d,\theta,\xi) = \frac{1-R}{1+R}\delta(0) + \sum_{j=1}^{m}\left\{\frac{A_j}{2}[\delta(\xi - m\xi_1) + \delta(\xi + m\xi_1)]\right\} \quad (8.26)$$

where $\xi_1 = 2nd$ (cm) and the impulse magnitudes are $A_m/2$.

As examples, the FPI transmission spectrum for $R = 0.3$ and $nd = 0.28$ cm is presented in Figure 8.13a, and the magnitude of the Fourier transform of the FPI spectrum is shown in Figure 8.13b. This latter plot consists of a very well defined set of equidistant impulses. In this figure it is clear that the impulse related to the first harmonic "frequency" is centered at $\xi = 2nd = 0.56$ cm, the impulse due to the second harmonic is found at $\xi = 4nd = 1.12$ cm, the third at $\xi = 6nd = 1.68$ cm, and so on. It is important to point out that the Fourier transform magnitude of the impulses depends only on the reflectivity value as defined by Eq. (8.23). For instance, $|T_{\mathrm{FPI}}(R,n,d,\theta,\xi)|$ as a function of the reflectivity is shown in Figure 8.14, where we considered an optical thickness $nd = 0.28$ cm. It is important to stress that the optical thickness does not affect the magnitude of impulses; it just determines their positions in the Fourier domain, which is given by $\xi_m = m2nd$, where $m = 0,1,2,\ldots$ Moreover, in Figure 8.14 it can be appreciated that for low reflectivity values the DC component and the first harmonic are considerably stronger than higher harmonics. In contrast, for higher reflectivity values, higher harmonics get stronger with similar magnitudes to first harmonics.

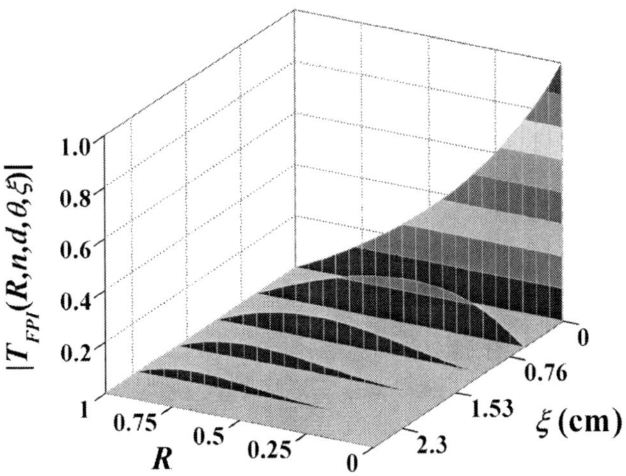

Figure 8.14. Magnitude of the Fourier transform of the FPI transmission spectrum as a function of reflectivity and optical thickness.

3.1.3. Fourier Characterization of the $G(\nu,l,C)$ Function

The $G(\nu,l,C)$ function was defined in Eq. (8.13) and is given as the product of the source $S(\nu)$, the filter transmission function $Fil(\nu)$, and the transmission through the gas path length $T(\nu)$. Therefore the Fourier transform of the $G(\nu,l,C)$ profile will contain information related to the separation of the absorption ro-vibrational lines of the target molecule and also will contain side lobes which are related to the filter function and source profile. For instance, the $|G(\xi,l,C)|$ function corresponding to the CO_2 sensor system presented previously (see Figure 8.12) is shown in Figure 8.15. Additionally, in this figure we present different $|G(\xi,l,C)|$ functions corresponding to different CO_2 gas concentrations. Side lobes very close to the DC component ($|\xi| < 0.05$ cm) are more related to the filter's shape rather than to the target molecule, since their magnitudes do not change appreciably as the concentration of the target molecule increases. In contrast, side lobes at frequencies $|\xi| \approx 0.56$ cm and $|\xi| \approx 1.12$ cm are strongly related to the target molecule, since clearly their magnitude increases with the molecule concentration. Therefore, these side lobes give information about where the impulses due to the Fourier transform of the FPI fringes must be placed.

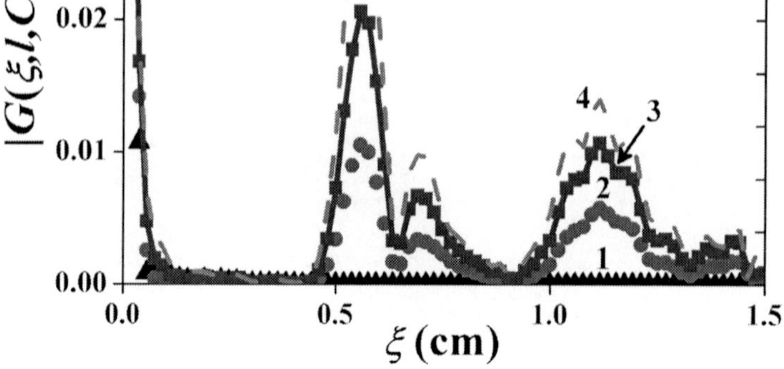

Figure 8.15. Fourier transform magnitude of the $G(\nu,l,C)$ function.

3.1.4. Fourier Characterization of the System Response

The Fourier transform of the system response can be expressed as

$$\begin{aligned} I_D(C,l,R,n,d,\theta,\xi) &= \frac{1-R}{1+R}G(0)\delta(0) + \frac{A_1G(\xi_1)}{2}\delta(\xi-\xi_1) + \frac{A_1G(\xi_1)}{2}\delta(\xi+\xi_1) \\ &+ \frac{A_2G(2\xi_1)}{2}\delta(\xi-2\xi_1) + \frac{A_2G(2\xi_1)}{2}\delta(\xi+2\xi_1) \\ &+ \frac{A_3G(3\xi_1)}{2}\delta(\xi-3\xi_1) + \frac{A_3G(3\xi_1)}{2}\delta(\xi+3\xi_1) \\ &+ \cdots + \frac{A_mG(m\xi_1)}{2}\delta(\xi-m\xi_m) + \frac{A_mG(m\xi_1)}{2}\delta(\xi+m\xi_1) \\ &= \frac{1-R}{1+R}G(0)\delta(0) + \sum_{j=1}^{m}\frac{A_jG(j\xi_1)}{2}\left[\delta(\xi-j\xi_1)+\delta(\xi+j\xi_1)\right] \end{aligned} \quad (8.27)$$

Equation (8.27) shows that the Fourier transform of the system response is given by a sum of impulses; here, for simplicity, we intentionally use $G(\nu)$, since we assumed that l and C are constant. Furthermore, the position of these impulses depends only on the cavity length of the FPI, which is described by $\delta(\xi \pm m\xi_1)$. Moreover, their amplitudes are related to the FPI reflectivity and the magnitude of the $G(m\xi_1, l, C)$ function. Finally, to obtain the system response in terms of the wavenumber shift, it is just necessary to evaluate the inverse Fourier transform of Eq. (8.27), which will lead to a sum of sinusoidal functions of frequencies $m\xi_1 = m2nd$ whose amplitudes are given by $A_mG(m\xi_1, l, C)/2$.

In Figure 8.16 we present the sensor response as a function of the cavity length scan. Here we assumed the $G(\nu, l, C)$ shown in Figure 8.12 and the FPI whose transmission spectrum is presented in Figure 8.13a. Moreover, the depth of modulation for this sensor can be determined by evaluating the RMS of the function $I_D(C,l,R,n,d,\theta,\xi)$ as shown in Figure 8.17. Additionally, in this figure we show that results obtained with the direct numerical method (DNM) and the convolution method (CON) are practically identical. It is important to mention that here we assuming the ideal case, where the FPI transmission fringes match the gas absorption lines before we start to scan the optical cavity length. However, some changes in the form of the sensor response can be observed, depending on the initial position of the fringes. Therefore a calibration procedure is required to reduce errors and also to determine how to scan the cavity length to improve the overall sensor signal (Vargas-Rodriguez et al. 2008). For instance, if we are moving the FPI mirror by using a piezoelectric transducer, we can use a triangular or a sawtooth waveform, but we need to analyze which is better for the application.

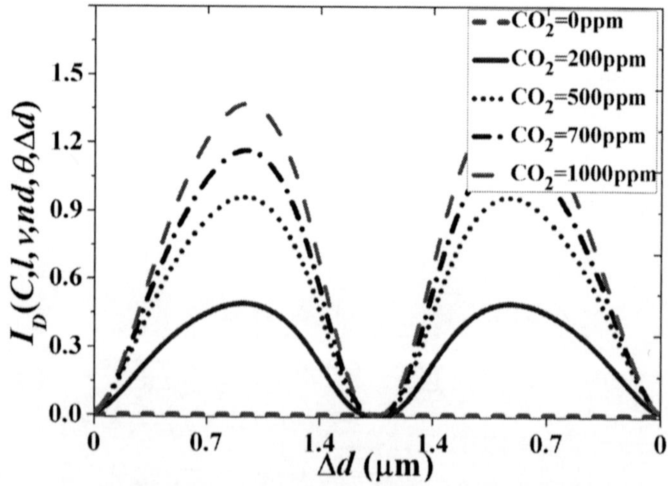

Figure 8.16. Typical wide-gap sensor response as a function of cavity length scan.

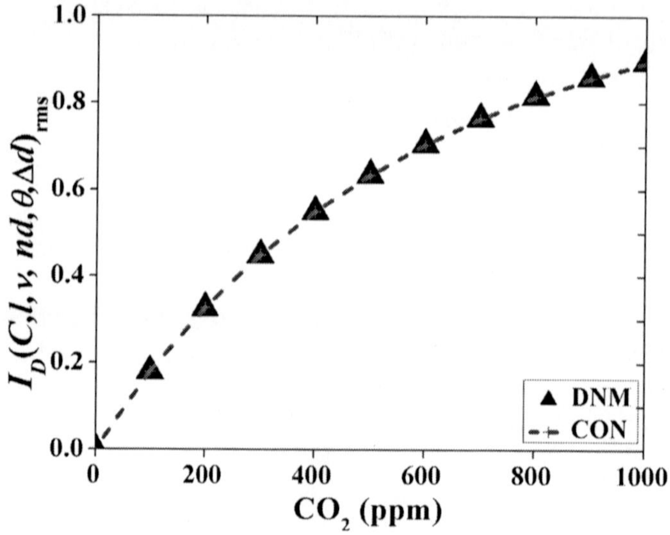

Figure 8.17. Simulated depth of modulation for a wide-gap sensor obtained by the direct numerical method and the convolution method.

3.1.5. Parameter Optimization Using the Convolution Method

Previously we showed that the Fourier transform of the system response can be expressed as a sum of impulses; see Eq. (8.27). The position of these impulses is related directly to the FPI's cavity length ($\xi_1 = 2nd$), and the magnitude of these impulses depends on the reflectivity of the FPI's mirrors. From Eq. (8.27) we can infer that in order to have the maximum amplitude for the first harmonic of the system response $I_D(C,l,R,n,d,\theta,\xi)$, we need to place the first Fourier impulse due to the Fourier transform spectrum, $T_{FPI}(R,n,d,\theta,\xi)$, where the peak of the strongest side lobe of $G(\xi,l,C)$ occurs, which is labeled as $\xi_{max} = \xi_1$ (Figure 8.15). Therefore, if we know ξ_{max}, we can determine the optimal cavity length value of the FPI directly, as $d = \xi_{max}/2n$. The main advantage of this approach is that the optimal FPI cavity length can be calculated by direct observation of the Fourier transform magnitude of $G(\xi,l,C)$.

As was mentioned previously, the system response $I_D(C,l,R,n,d,\theta,\Delta d)$ consists of a series of cosine functions. The amplitudes of these cosine functions are described in the Fourier domain by the product $T_{FPI}(R,n,d,\theta,\xi)\,G(\xi,l,C)$. The magnitude of the side lobes of $G(\xi,l,C)$ will remain fixed unless there is a change in the gas concentration. However, the magnitude of impulses of the $T_{FPI}(R,n,d,\theta,\xi)$ function can be manipulated by changing the reflectivity of the FPI mirrors (Figure 8.13b), which consequently modifies the amplitudes of the system response harmonics. Therefore, from Eq. (8.27) we can say that in order to have the maximum amplitude in the first harmonic of the system response, it is necessary to find the value of R for which the magnitude of the first impulse of $T_{FPI}(R,n,d,\theta,\xi)$ is a maximum. The magnitude of the FPI impulses as a function of the reflectivity is given by

$$|A_m| = \frac{(1-R)}{(1+R)}R^m \qquad m = 1,2,3,\ldots \tag{8.28}$$

Hence, the magnitude of the first impulse due to the first harmonic ($m = 1$) is given by

$$|A_1| = \frac{1-R}{1+R}R \tag{8.29}$$

From differentiation of Eq. (8.29) it is found that the maximum of $|A_1|$ occurs at $R \approx 0.41$. This is very useful result since it shows that if we use a phase sensitivity detector (PSD) to recover the first harmonic of the sensor system response $I_D(C,l,R,n,d,\theta,\Delta d)$, the optimal value of R is 0.41 regardless the cavity length. This result is surprisingly low compared to "intuitive" designs.

In this way, for the CO_2 sensor design we have been using as an example and for which we assumed the $G(\nu,l,C)$ shown in Figure 8.12, we obtain directly from the Fourier transform characteristics of the function that the optimal optical thickness must be $nd = 0.28$ cm and $R = 0.41$. In order to probe these values we performed direct numerical evaluation and used the convolution method for different values of nd and R, and the results obtained are shown in Figure 8.18. With

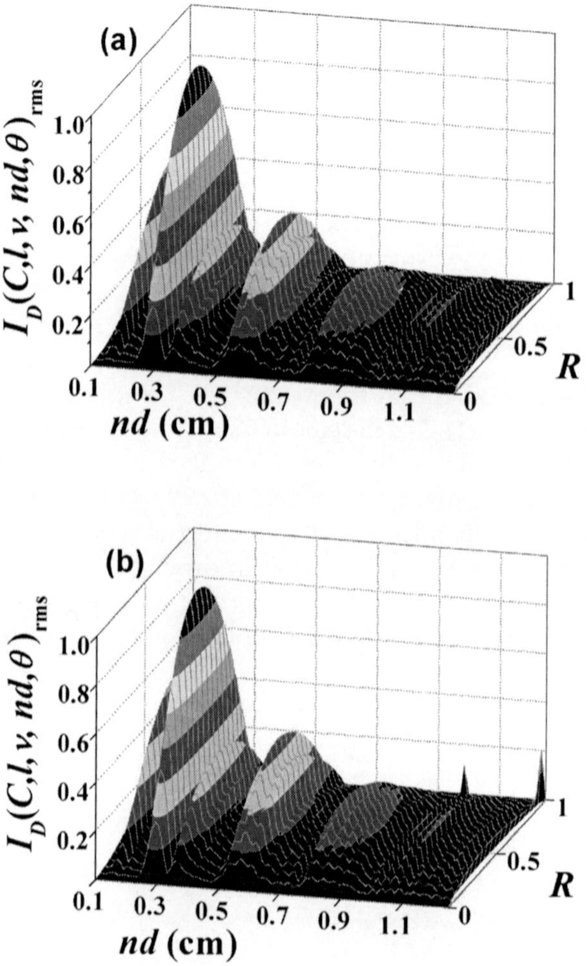

Figure 8.18. Sensor response as a function of optical thickness and reflectivity evaluated (a) by using the direct numerical method and (b) the convolution method.

both methods we can conclude that the maximum of $I_D(C,l,R,n,d,\theta,\Delta d)_{rms}$ occurs when $nd = 0.28$ cm and $R = 0.42$. For these simulations we assumed a CO_2 concentration of 1000 ppm. Moreover, if we extract two slices of $I_D(C,l,R,n,d,\theta,\Delta d)_{rms}$ (Figure 8.18) corresponding to the optimal optical thickness and reflectivity, we see that both methods produce the same result (Figure 8.19). The optimal parameters found by numerical evaluation using the convolution method and the direct numerical method are in agreement with values obtained directly from the Fourier transform characteristics of $G(\nu,l,C)$ and $T_{FPI}(R,n,d,\theta,\xi)$ analysis.

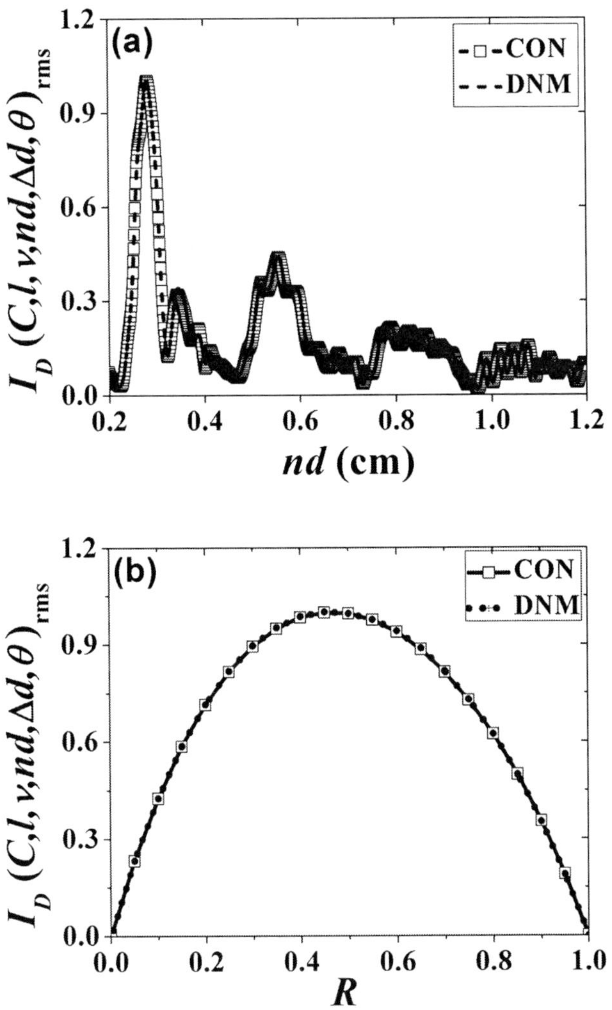

Figure 8.19. Sensor response $I_D(C,l,R,n,d,\theta,\Delta d)$ as a function of (a) optical thickness and (b) reflectivity.

3.1.6. Filter Selection Approach

Initially we mentioned that an optimization procedure should look for the parameters that guarantee the maximum amplitude modulation produced by the sensor when the target gas is present; this part is completed using the optimization methods based on the convolution method. However, it is also necessary to guarantee that our gas sensor produces a background depth of modulation as close to zero as possible. The background case is when there is a zero concentration of the target gas, $T(\nu) = 1$. Therefore we need to determine what factors help us to minimize the background depth of modulation; otherwise our sensor will lose its sensitivity to very small gas concentrations.

The background modulation depth is the particular case when $T(\nu) = 1$. Therefore it can be useful to review the characteristics of the $G(\nu, l, C)$ function for this case (Vargas-Rodriguez and Rutt 2005), since we can obtain that:

$$G(\nu,l,C) = \mathrm{Fil}(\nu)S(\nu)T(\nu,l,0) \Rightarrow G_{\mathrm{BK}}(\nu) = \mathrm{Fil}(\nu)S(\nu) \tag{8.30}$$

Here the function $G_{\mathrm{BK}}(\nu)$ depends just on the filter performance $\mathrm{Fil}(\nu)$ and the source profile $S(\nu)$, since the $T(\nu, l, 0) = 1$. As was stated by Eq. (8.27), the Fourier transform of the sensor response $I_D(C, l, R, n, d, \theta, \Delta d)$ is related to $G(\xi, l, C)$ and $T_{\mathrm{FPI}}(R, n, d, \theta, \xi)$. Therefore the background Fourier transform $I_D(C, l, R, n, d, \theta, \Delta d)$ can be evaluated using Eq. (8.27), just replacing $G(\xi)$ with $G_{\mathrm{BK}}(\xi)$ as follows:

$$I_D(R,n,d,\xi)_{\mathrm{BK}} = \frac{1-R}{1+R}G(0)_{\mathrm{BK}}\,\delta(0) + \sum_{j=1}^{m}\frac{A_j G_{\mathrm{BK}}(j\xi_1)}{2}\left[\delta(\xi - j\xi_1) + \delta(\xi + j\xi_1)\right] \tag{8.31}$$

where A_m represents the magnitude due to the FPI fringes. Therefore, using the last equation, we can say that if we require our sensor to produce a signal equal to zero for the background case, it is necessary that $G_{\mathrm{BK}}(m\xi_1) = 0$. Here, considering that the cavity length and the reflectivity are constant, $|T_{\mathrm{FPI}}(R, n, d, \theta, \xi)|$ will remain constant. However, $|G_{\mathrm{BK}}(\xi)|$ will present some side lobes that do not depend on the target gas, but do depend directly on the band-pass filter and the source profile, so these will "control" the background modulation depth. Consequently, for this case the problem consists of determining which $G_{\mathrm{BK}}(\nu)$ function produces the minimum background depth of modulation for a given cavity length and reflectivity. Furthermore, if we consider a thermal broad-band infrared source, it is reasonable to set $S(\nu) = 1$, which leads to

$$G_{\mathrm{BK}}(\nu) = \mathrm{Fil}(\nu) \tag{8.32}$$

This result is very helpful, since it shows that the background modulation produced by our sensor will depend directly on the band-pass filter. Therefore the background amplitude modulation can be minimized by selecting a band-pass filter whose Fourier transform satisfies the condition $G_{\mathrm{BK}}(m\xi_1) \approx 0$. For the CO_2 sensor design that we have been considering as an example, we find that the optimal optical thickness is $nd = 0.28$ cm and $R = 0.41$. Now, to determine which band-pass filter is the best choice, we can select a set of filters (Figure 8.20a) and determine their Fourier transform as shown in Figure 8.20b.

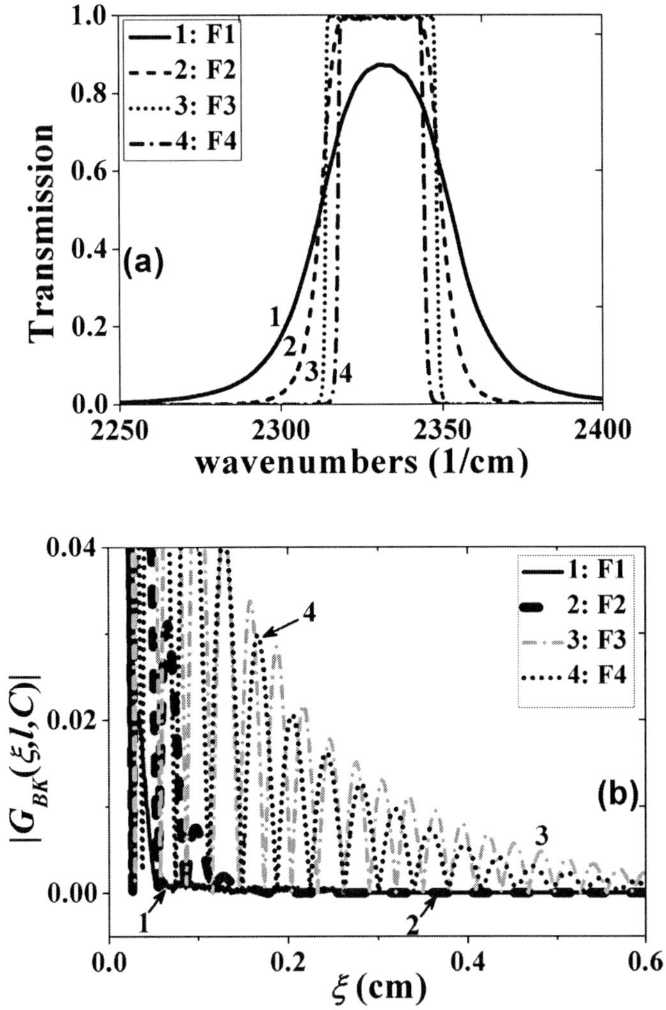

Figure 8.20. Set of narrow band-pass filters for a CO_2 sensor design: (a)transmission spectra; (b) magnitude of their Fourier transforms.

Here we need to check which filter has the lower magnitude in the region near $\xi = \xi_1 = 2nd = 0.56$ cm, since we can expect, based on Eqs. (8.27) and (8.31), that it will produce the lower background depth of modulation $I_D(C = 0, l, R, n, d, \theta, \Delta d)_{\text{rms}}$. Hence for our example we can observe that the filters F1 and F2 have lower magnitudes in the region near ξ_1, while filters F3 and F4 have the stronger magnitudes in this region. Consequently, we can predict that filter F1 or F2 will be the best choice for our sensor, and we can expect that F3 will produce the worst background modulation depth.

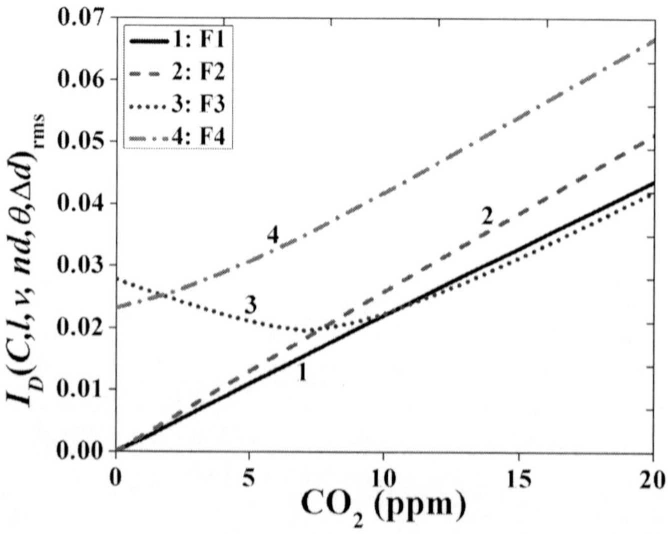

Figure 8.21. Sensor response for low gas concentration for different band-pass filters.

This point is supported by the results obtained of the evaluation by the sensor response $I_D(C,l,R,n,d,\theta,\Delta d)_{\text{rms}}$ considering low gas concentrations and optimal parameters previously established (Figure 8.21). These results show that the worst case of background depth of modulation, the highest value of $I_D(C,l,R,n,d,\theta,\Delta d)_{\text{rms}}$ when $C = 0$, is produced by filter F3. In contrast, the lowest background depth of modulation is produced by filters F1 and F2. Moreover, the depth of modulation for filters F1 and F2 is very similar, so we can use whichever one has the lower cost. It is fair to say that, intuitively, a designer would be more likely to choose filter F3 or F4. These filters isolate the target spectral region much better. However, the steep transition to the stop band results in strong side lobes in $G_{\text{BK}}(\xi)$ which result in large zero gas concentration signals, which is a major disadvantage. Furthermore, the F3 and F4 designs are typically more complex, expensive, and also more subject to manufacturing error, temperature drift, and alignment or system F/# effects. This result is very interesting since the counter-intuitive optimum design is actually the cheap and easy filter.

4. INTERNAL REFLECTION EFFECTS

In a real FPI, additional spectral FPI fringe patterns can be present due to internal reflections occurring in the FPI mirrors (Figure 8.22); we call these "spectral parasitic reflection fringes" (SPRF). These SPRF can produce a sort of "interference" with the main FPI fringe pattern. In practice, our FPI is formed by two plane parallel mirrors (Figure 8.1), so we have two main sources of SPRF. As stated before,

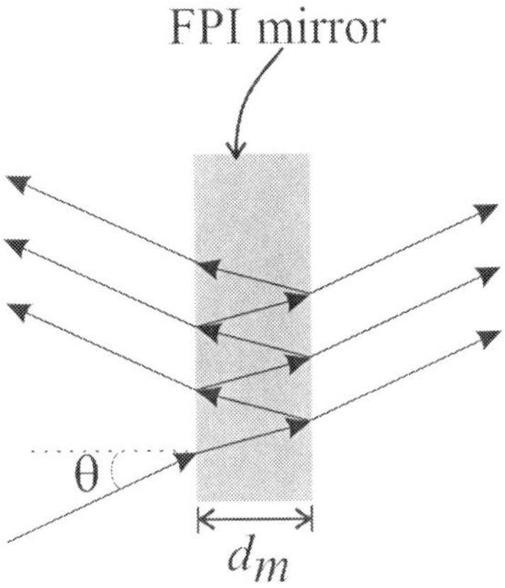

Figure 8.22. Multiple beam fringes on the FPI mirrors.

SPRF are basically Fabry-Pérot fringes (Born and Wolf 1970), and therefore these are again described by the Airy formula given in Eq. (8.1). For clarity, let us re-write this equation for the SPRF as

$$I_{mul}(R_m, n_m, d_m) = \frac{1}{1 + \left(\frac{1}{1+4R_m}\right)^2 \sin^2(2\pi n_m d_m \nu \cos\theta)} \tag{8.33}$$

where R_m is the reflectivity, n_m is the refractive index, and d_m is the thickness of the plate.

SPRF can introduce several undesirable effects on the final FPI fringe pattern, especially losses in fringe contrast. SPRF can be reduced and sometimes practically eliminated by using an antireflection coating on the incident and the output surfaces of FPI plates (Kuhn 1951; Macleod 2001) or by wedging the plate. The wedge can be conceived as a type of defect in the parallelism of the plate, but it is useful in some cases and therefore the plates can be deliberately fabricated with a wedge angle (Mansfield and Rutt 1999). Further, Saito et al. (2008) presented an interesting tunable FPI with silicon wedged mirrors to reduce SPRF. Therefore it is quite important to take into account SPRF during the design procedure of either narrow-gap or wide-gap sensors, in order to minimize them adequately. They can represent a significant threat to system performance, but are easily eliminated.

5. CONCLUSIONS

An overview of the basic types of gas sensors that use a Fabry-Pérot interferometer as a modulator has been presented. It has been shown that, using the same gas sensor setup, it is possible to design two different types of sensors simply by modifying the optical thickness of the FPI. We have called these narrow- and wide-gap sensors, and they have very different principles of operation. In narrow-gap sensors, only one or even two spectral transmission FPI fringes interact with the absorption band of the target gas or gases. In wide-gap sensors, several FPI fringes interact directly with the absorption lines of the target gas. Moreover, each type has advantages and disadvantages; for instance, the FPI for a narrow-gap sensor can be integrated within a MEMS device, while, on the other hand, high selectivity to the target gas is the main advantage of the wide-gap sensor design. The selectivity of the narrow-gap sensor can be improved by using the cross-correlation principle, which also can help to improve the gas selectivity and to reduce drift errors. Furthermore, based on this mathematical model, we have shown that, using the convolution method, it is possible to determine some optimal parameters for the wide-gap sensor such as FPI optical thickness and reflectivity. Additionally, this method gives us guidance to select the best narrow band-pass filter for the application.

REFERENCES

Alause H., Grasdepot F., Malzac J.P., Knap W., and Hermann J. (1997) Micromachined optical tunable filter for domestic gas sensors. *Sens. Actuators B* **43**(1–3), 18–23. DOI: 10.1016/S0925-4005(97)00139-1

Batchellor C.R. and Dakin J.P. (1987) Wavelength Scanning Optical Sensor. UK Patent Application GB 2181536A.

Blomberg M., Rusanen O., Keranen K., and Lehto A. (1997a) A silicon microsystem - miniaturised infrared spectrometer. In: *Proceedings of the TRANSDUCERS '97 1997 International Conference on Solid-State Sensors and Actuators*, Chicago, June 16–19, 1257–1258. DOI: 10.1109/SENSOR.1997.635463

Blomberg M., Torkkeli A., Lehto A., Helenelund C., and Viitasalo M. (1997b) Electrically tuneable micromachined Fabry-Perot interferometer in gas analysis. *Phys. Scripta* **1997**(T69), 119–-121. DOI: 10.1088/0031-8949/1997/T69/018

Born M. and Wolf E. (1970) *Principles of Optics*. Pergamon Press, Oxford, UK.

Calaza C., Fonseca L., Moreno M., Marco S., Cané C., and Gracia I. (2004) A surface micromachining process for the development of a medium-infrared tuneable Fabry-Perot interferometer. *Sens. Actuators A* **113**(1), 39–47. DOI: 10.1016/j.sna.2004.01.047

Calaza C., Meca E., Marco S., Moreno M., Samitier J., Fonseca L., Gracia I., and Cane C. (2003) Assessment of the final metrological characteristics of a MOEMS-based NDIR spectrometer through system modeling and data processing. *IEEE Sensors J.* **3**(5), 587–594. DOI: 10.1109/JSEN.2003.817672

Carlson H. (1997) CARBOCAP—New carbon dioxide sensor electrically tuneable interferometer for infrared gas analysis. *Sensor Rev.* **17**(4), 304–306. DOI: 10.1108/02602289710185126

Cooper V.G. (1971) Analysis of Fabry-Perot interferograms by means of their Fourier transforms. *Appl. Optics* **10**(3), 525–530. DOI: 10.1364/AO.10.000525

Crocombe R.A. (2004) MEMS technology moves process spectroscopy into a new dimension. *Spectroscopy Europe* **16**(3), 16–19.

Chambers P., Austin E.A.D., and Dakin J.P. (2004) Theoretical modelling studies of gas sensing systems using correlation spectroscopy. *Proc. SPIE* **5272**, 123–138. DOI: 10.1117/12.536962

Dakin J.P., Edwards H.O., and Weigl B.H. (1995) Progress with optical gas sensors using correlation spectroscopy. *Sens. Actuators B* **29**(1–3), 87–93. DOI: 10.1016/0925-4005(95)01667-8

Dakin J.P., Gunning M.J., Chambers P., and Xin Z.J. (2003) Detection of gases by correlation spectroscopy. *Sens. Actuators B* **90**(1–3), 124–131. DOI: 10.1016/S0925-4005(03)00043-1

Dennis R. (2004) Private communication.

Domash L.H. (2005) Tunable thin films as sensors. *Sensors Magazine Online*, 1–5.

Goody R. (1968) Cross-correlating spectrometer. *J. Opt. Soc. Am. B* **58**(7), 900–908. DOI: 10.1364/JOSA.58.000900

Herzberg G. (1945) *Molecular Spectra and Molecular Structure, Vol II, Infrared and Raman Spectra of Polyatomic Molecules*. Van Nostrand Reinhold Company International. Princeton, NJ.

Jacquinot P. (1960) New developments in interference spectroscopy. *Rep. Prog. Phys.* **23**(1), 267–312. DOI: 10.1088/0034-4885/23/1/305

Jin W., Stewart G., Culshaw B., Murray S., and Pinchbeck D. (1993) Absorption measurement of methane gas with a broadband light and interferometric signal processing. *Opt. Lett.* **18**(16), 1364–1366. DOI: 10.1364/OL.18.001364

Johnston S.F. (1992) Gas monitors employing infrared LEDs. *Meas. Sci. Technol.* **3**(2), 191–195. DOI: 10.1088/0957-0233/3/2/008

Kuhn H. (1951) New techniques in optical interferometry. *Rep. Prog. Phys.* **14**, 64–94. DOI: 10.1088/0034-4885/14/1/303

Macleod H.A. (2001) *Thin-Films Optical Filters*. Institute of Physics Publishing, Bristol, UK.

Mansfield C.D. and Rutt H.N. (1999) Evaluation of multiple beam interference effects in infrared gas spectroscopy. *Meas. Sci. Technol.* **10**(3), 206–210. DOI: 10.1088/0957-0233/10/3/016

Mayrwöger J., Reichl W., Hauer P., Krutzler C., and Jakoby B. (2011) CO_2 monitoring using a simple Fabry-Perot-based germanium bolometer. *Sens. Actuators B* **154**(2), 245–250. DOI: 10.1016/j.snb.2009.12.038

Mohebati A. and King T.A. (1998) Remote detection of gases by diode laser spectroscopy. *J. Mod. Opt.* **35**(3), 319–324. DOI: 10.1080/09500348814550361

Neumann N., Ebermann M., Hiller K., and Kurth S. (2007) Tunable infrared detector with integrated micromachined Fabry-Perot filter. *Proc. SPIE* **6466**, 646606. DOI: 10.1117/12.698871

Neumann N., Ebermann M., Kurth S., and Hiller K. (2008) Tunable infrared detector with integrated micromachined Fabry-Perot filter. *J. Micro/Nanolith. MEMS MOEMS* **7**(2), 021004. DOI: 10.1117/1.2909206

Noro M., Suzuki K., Kishi N., Hara H., Watanabe T., and Iwaoka H. (2002). CO_2/H_2O gas sensor with a MEMS Fabry-Perot filter. In: *Proceedings of the 2002 IEEE/LEOS International Conference on Optical MEMS*, Lugano, Switzerland, August 20–23, 27–28. DOI: 10.1109/OMEMS.2002.1031427

Rothman L.S., Jacquemart D., Barbe A., Chris Benner D., Birk M., Brown L.R., Carleer M.R., Chackerian J.C., Chance K., Coudert L.H., Dana V., Devi V.M., Flaud J.M., Gamache R.R., Goldman A., Hartmann J.M., Jucks K.W., Maki A.G., Mandin J.Y., Massie S.T., Orphal J., Perrin A., Rinsland C.P., Smith M.A.H., Tennyson J., Tolchenov R.N., Toth R.A., Vander Auwera J., Varanasi P., and Wagner G. (2005) The HITRAN 2004 molecular spectroscopic database. *J. Quant. Spectrosc. Radiat.Transfer* **96**(2), 139–204. DOI: 10.1016/j.jqsrt.2004.10.008

Saito M., Yamada I., Mizuta T., Matsuura A., Kawashima H., Sakaki A., Otsuka N., Shindo Y., and Kuwabara T. (2008) Tunable infrared filter made of thin silicon wedges. *Infrared Phys. Techn.* **51**(3), 236–241. DOI: 10.1016/j.infrared.2007.07.009

Srivastava R., Shenoy U.G., Forrest S.R., Chinnayelka S., Shaikh M.J., Besser R.S., and McShane M.J. (2004) Bulk micromachining of a MEMS tunable Fabry-Perot interferometer: Effect of residual silicon on device performance. *J. Microlith. Microfab. Microsyst.* **3**(4), 579–588. DOI: 10.1117/1.1794707

Vargas-Rodriguez E. (2007). Optical gas sensors based on correlation spectroscopy using a Fabry-Perot interferometer. PhD thesis, Optoelectronics Research Centre, University of Southampton, Southampton, UK.

Vargas-Rodriguez E. and Rutt H.N. (2005) Method to minimize spurious background signals in gas detectors based on correlation spectroscopy using a Fabry-Perot bandpass filter shape optimization. *Opt. Eng.* **44**(10), 103002-1. DOI: 10.1117/1.2065808

Vargas-Rodríguez E. and Rutt H.N. (2007) An analytical method to find the optimal parameters for gas detectors based on correlation spectroscopy using a Fabry-Perot interferometer. *Appl. Opt.* **46**(21), 4625–4632. DOI: 10.1364/AO.46.004625

Vargas-Rodriguez E., Rutt H.N., Rojas-Laguna R., and Alvarado-Mendez E. (2008) Calibration of a gas sensor based on cross-correlation spectroscopy. *J. Opt. A: Pure Appl. Opt.* **10**(10), 104018. DOI: 10.1088/1464-4258/10/10/104018

Vaughan J.M. (1989) *The Fabry-Perot Interferometer: History, Theory, Practice and Applications.* Adam Hilger, Bristol, UK.

CHAPTER 9

Spectroscopic Modeling of Mid-Infrared Chemical Sensors

A. Wilk
B. Mizaikoff

1. INTRODUCTION

Simulations have been applied as the connecting link between theory and experiment during a wide variety of scenarios including, e.g., architectural, structural, and chemical engineering for process simulations of chemical plants (Crowe 1971). Automotive industries use simulations from stress analysis of chassis (solid mechanics) to testing the ride and handling characteristics of entire passenger vehicles on virtual platforms (virtual prototyping), and for studying processes in internal combustion engines (Han and Reitz 1995; Dai et al. 1996; Tseng et al. 1998; Ottosson 2002; Bourlioux and Gander 2002; Rauh 2003). Even in designing computer architectures, computers (emulators) simulate novel computer designs in order to verify hardware functionality or to debug software (Edwards et al. 1997). According to Bell and Waag (1998), simulation of technology not only supports design and performance optimization, it also serves as a tool for developing practical skills during training of simulated real-world scenarios (e.g., flight simulators). Consequently, computer-based simulations provide significant support during the development of innovative products in automotive, machine, and industrial design (Ottosson 2002), in semiconductor device manufacturing, and in nanotechnology (Smith 1997). Furthermore, complex natural systems such

DOI: 10.5643/9781606503201/ch9

as climate systems or traffic flows have been simulated by Fritzsche (1994) to study the impact of certain parameters that are difficult to address or deliberately change in the real-world environment. Last but not least, national economies and mathematical finance are nowadays routinely simulated (Bourlioux and Gander 2002), as are social interactions within, e.g., game theory to understand the mechanisms and reasons for social learning (Gilbert 1995). While this list is far from being complete, it is evident that simulations are credible exploratory procedures for studying the design and dynamics of complex systems, and usually attain to closely resemble the real-world environment.

In the context of chemistry, (theoretical) physicists and (physical) chemists use computer simulations to study both steady-state and dynamic processes by applying Monte Carlo (MC) and molecular dynamics simulation methods, respectively. In chemical or physical modeling, frequently the same fundamental descriptions are applied, e.g., for describing the physical phenomenon of the interaction of high-energy particles appearing in fusion devices, as for studying, e.g., thin-film growth dominated by chemical effects (Evans et al. 1990). The aim is either to predict the macroscopic behavior of the system and to compute thermodynamic properties from the microscopic motion of individual molecules (Hoover 1986; Haile 1997; Van Gunsteren and Berendsen 1990), or to describe kinetic processes (Barshop et al. 1983; Heermann 1986) based on analyzing how position, velocity, orientation, conformation, and structure changes with time. Global low-free-energy regions of a molecular system in thermal equilibrium can be determined via MC methods (Sadus 1999; Binder and Heermann 2002). The computational power available today allows—besides solids and gases—modeling molecular systems comprising many particles without restrictions on complexity, including polymers, fluids, or other fluidlike systems with many degrees of freedom (Allen and Tildesley 1987; Smith 1997; Sadus 1999; Kotelyanskii and Theodorou 2004; Gould et al. 2006). This includes modeling in biological/biochemical applications (Van Gunsteren and Berendsen 1990) involving folding and kinetic simulations of large macromolecules such as proteins (Sadus 1999), or even studying the behavior of complex molecular assemblies such as viruses (Freddolino et al. 2006), just to name a few.

Specific to the topic of this chapter, in the past decades substantial progress has been made in the development and refinement of chemical sensors and components thereof, and some of them have been supported or became feasible only because of the successful application of appropriate simulations. In turn, only a minor fraction of those sensors are devices operating in the mid-infrared (MIR; 2–25 μm) spectral regime; in the following, this chapter will provide a detailed overview on simulating IR sensing devices for designing advanced optical chemical sensing systems.

In particular, with the development of advanced spectrometer technology in the 1940s, mid-infrared absorption spectroscopy was rapidly accepted as a useful analytical tool in a wide range of applications due to its inherent molecular selectivity. Among these applications, absorption spectroscopy is of particular interest for chemical sensing. In general, MIR absorption techniques rely on the excitation of vibrational-rotational modes (hence, the name "vibrational spectroscopy") resulting from the absorption of MIR radiation interacting with the sample (e.g., a gas, liquid, or solid) at resonant frequencies. Especially, Fourier-transform infrared (FT-IR) spectroscopy is a reliable analytical tool for the simultaneous label-free identification, quantification, and online monitoring of multiple species present in complex mixtures, because radiation modulation via an interferometer eliminates the need for sequential wavelength selection, which is inherent to dispersive optical instruments. The molecular selectivity provided, along with rapid response times, predestinated FT-IR spectroscopy for the development of multicomponent sensing systems operating in the MIR regime, where most molecules are characterized by distinct fundamental vibrations and corresponding spectra.

In contrast to sensor optimization based entirely on experimentation, simulations have proven to be a time-saving and less expensive alternative during the development of optical sensors, if the established model closely resembles the real-world performance of the sensing system. Hence, virtual optical chemical sensors should provide a sufficiently detailed computer-based representation of the real-world optical device to enable studying its functionality. More specifically, this virtual model should also attempt to estimate or ideally even predict the behavior or key characteristics of a system, including, e.g., the obtainable sensitivity and limit of detection. In the rare case of a system that may be exactly represented by a theoretical mathematical model, simulations try to establish closed-form analytical solutions. For more complex scenarios for which direct mathematical models do not exist, simulation is inevitably constrained by simplifying approximations. Consequently, the simulations obtained have to be tested for the accuracy of the established model, and have to be validated by corresponding experiments. Furthermore, as a general principle, experimental data must be used as the source of information for establishing the fundamental simulation parameters. Thus, "numerical experiments" conducted with the established models, as well as the insights they provide, have much in common with real-world experiments and ideally resemble them as closely as possible. Consequently, the interrelation among theory, simulation, and experiment has led to establishing simulation as a third discipline in between theory and experiment (Winsberg 2001; Binder and Heermann 2002; Winsberg 2003). The following examples will show that the concepts of simulation can be successfully extended to model not only key components of chemical sensors operating in the MIR, but even to entire optical chemical sensor devices and their response function with possibly close resemblance to the real-world experiment.

Given the limitations of a book chapter, details of the optics involved and the simulation methods have to be omitted, and it has to be assumed that the reader is already familiar with the fundamental concepts of optical modeling, simulations, and electromagnetism. However, the most relevant relations essential for understanding this chapter will be provided. For a more expanded discussion the reader is referred to pertinent literature, which is summarized in the reference section of this chapter. Particularly useful reading in preparation for the methods discussed herein are governing optical principles (Born and Wolf 1999; Hecht 2002; Winston et al. 2005; Saleh and Teich 2007) and geometric optics (Loshin 1991; Mouroulis and Macdonald 1997; Keating 2002), Monte Carlo methods (Landau and Binder 2000; Gentle 2003; Rubinstein and Kroese 2008; Robert and Casella 2010; Kroese et al. 2011), and the principles of chemical sensing (Harvey 2000; Patnaik 2004; Skoog et al. 2007; Rouessac and Rouessac 2007; Fraden 2010); furthermore, the fundamentals of FT-IR spectroscopy (Günzler and Gremlich 2002; Griffiths and de Haseth 2007; Smith 2011; Larkin 2011).

2. EXAMPLE #1: ACCESSING THE "ACTIVE" SENSING REGIONS OF AN ATR ELEMENT VIA RAY TRACING

The law of reflection states that if light is incident at an interface (i.e., the surface that two phases have in common), the angle of incidence (θ_i) in consequence equals the angle of reflection (θ_r) with respect to the surface normal (*specular reflection*):

$$\theta_i = \theta_r \tag{9.1}$$

However, any wave *crossing* an interface with different indices of refraction alters its phase velocity v as well as the direction of propagation upon entering the other medium (index $m = 2$). The equation describing the direction of the refracted ray is commonly referred to as Snell's law:

$$\frac{v_1}{v_2} = \frac{\hat{n}_2}{\hat{n}_1} = \frac{\sin(\theta_i)}{\sin(\theta_r)} \qquad \hat{n}_m = n_m + i\kappa_m \tag{9.2}$$

where θ_i denotes the angle of the incoming ray relative to the surface normal (incident angle); θ_r is the angle of the outgoing (refracted) ray; $\hat{n}_1$ and $\hat{n}_2$ are the complex refractive indices of the two media; n is the refractive index, and κ is the extinction coefficient. Note that the law of refraction only states the direction into which an incoming wave is refracted.

Augustin-Jean Fresnel deduced the equations which allow calculating the fractions of the reflected and transmitted field amplitudes and their relation to the initial electric field for electromagnetic radiation incident at an interface. With the complex refractive index $\hat{n} = n + i\kappa$ [Eq. (9.2)] and under the assumption that the permeabilities μ_m of both media are $\mu_1 = \mu_2 = 1$, the Fresnel equations are written as follows:

$$\hat{r}_{\mathrm{TE}} = \frac{\hat{n}_1 \cos(\theta_i) - \hat{n}_2 \cos(\theta_r)}{\hat{n}_1 \cos(\theta_i) + \hat{n}_2 \cos(\theta_r)} \tag{9.3}$$

$$\hat{r}_{\mathrm{TM}} = \frac{\hat{n}_2 \cos(\theta_i) - \hat{n}_1 \cos(\theta_r)}{\hat{n}_2 \cos(\theta_i) + \hat{n}_1 \cos(\theta_r)} \tag{9.4}$$

$$\hat{t}_{\mathrm{TE}} = \frac{2\hat{n}_1 \cos(\theta_i)}{\hat{n}_1 \cos(\theta_i) + \hat{n}_2 \cos(\theta_r)} \tag{9.5}$$

$$\hat{t}_{\mathrm{TM}} = \frac{2\hat{n}_1 \cos(\theta_i)}{\hat{n}_2 \cos(\theta_i) + \hat{n}_1 \cos(\theta_r)} \tag{9.6}$$

with θ_r = angle of reflection and θ_i = angle of incidence. Thereby, the index TE stands for "transverse electric field," i.e., the electric field vector **E** is polarized perpendicular to the plane of incidence. TM ("transverse magnetic field") stands for **E** being polarized in the plane of incidence. θ_i and θ_r are related via Snell's law [Eq. (9.2)]. Squaring Eq. (9.2) and using the identity $\cos^2(\xi) \equiv 1 - \sin^2(\xi)$ allows us to express $\cos(\theta_r)$ as

$$\cos(\theta_r) = \frac{\sqrt{\hat{n}_2^2 - \hat{n}_1^2 \sin^2(\theta_i)}}{\hat{n}_2} \tag{9.7}$$

Reflectance $\hat{\mathcal{R}}$ and transmittance $\hat{\mathcal{T}}$ refer to the fraction of incident power that is reflected/transmitted at an interface. These may be calculated applying

$$\hat{\mathcal{R}}_i = \hat{r}_i \cdot \overline{\hat{r}_i} \tag{9.8}$$

and

$$\hat{\mathcal{T}}_i = \left| \frac{\hat{n}_2 \cos(\theta_r)}{\hat{n}_1 \cos(\theta_i)} \right| \hat{t}_i \cdot \overline{\hat{t}_i} \tag{9.9}$$

wherein $\overline{\hat{r}_i}$ and $\overline{\hat{t}_i}$ denote the complex conjugates of $\hat{r}_i$ and $\hat{t}_i$. The index i is either TE or TM. For dielectrics (extinction coefficient $\kappa_i = 0$),

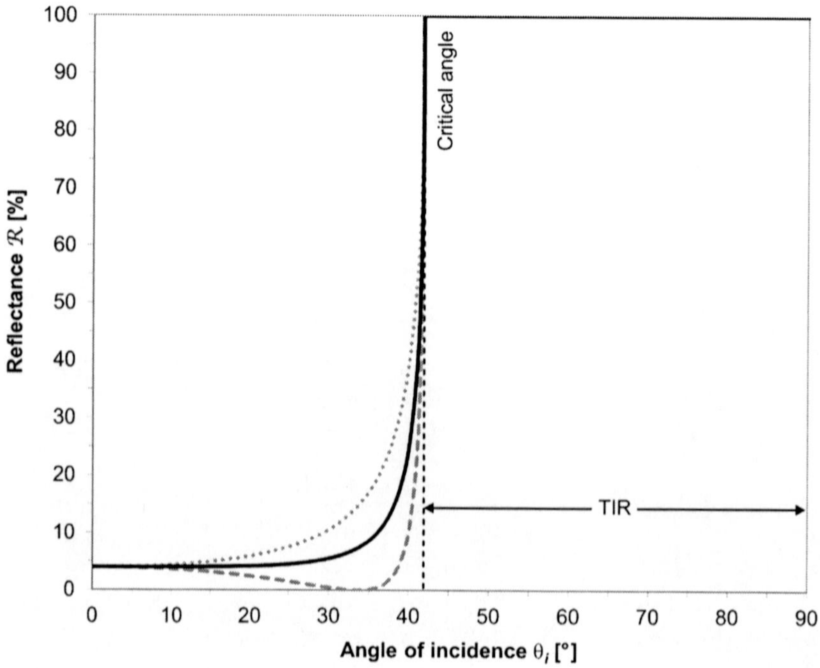

Figure 9.1. Total internal reflectance—light incident from denser medium. Reflectance versus angle of incidence for light incident from a dielectric ($n_1 = 1.5$) into air ($n_2 = 1.0$), calculated with Eqs. (9.3), (9.4), and (9.12). Dotted line = $\mathcal{R}_{TE}$, dashed line = $\mathcal{R}_{TM}$, solid line = $\mathcal{R}$.

$$\mathcal{T}_{TE} + \mathcal{R}_{TE} = 1 \tag{9.10}$$

$$\mathcal{T}_{TM} + \mathcal{R}_{TM} = 1 \tag{9.11}$$

holds. Moreover, if the incident radiation is unpolarized,

$$\mathcal{R} = \frac{\mathcal{R}_{TE} + \mathcal{R}_{TM}}{2} \tag{9.12}$$

and

$$\mathcal{T} + \mathcal{R} = 1 \tag{9.13}$$

As Eqs. (9.3), (9.4), and (9.8) show, $\mathcal{R}$ depends on the angle of incidence. When light enters the optically rare from the optically denser medium ($n_1 > n_2$), *total internal reflection* (TIR) is observed if the critical angle of incidence θ_c becomes larger than

$$\theta_C = \arcsin\left(\frac{n_2}{n_1}\right) \qquad \text{with } n_1 > n_2 \tag{9.14}$$

Above θ_c the light is lossless reflected ($\mathcal{R} = 1$), following Eq. (9.1), as illustrated in Figure 9.1.

Although the incident radiation is totally internally reflected, the wave function trespasses perpendicular across the boundary (i.e., in the z direction) as an *evanescent wave* with an electric field intensity $I(z)$ exhibiting an exponential decay in intensity according to

$$I(z) = I_0 e^{-z/d} \tag{9.15}$$

where

$$d = \frac{\lambda_0}{4\pi}\left[n_1^2 \sin^2(\theta_i) - n_2^2\right]^{-1/2} \tag{9.16}$$

is the characteristic penetration depth at which the amplitude of the evanescent field has decreased to $1/e$. λ_0 is the vacuum wavelength of the incident light, and the field intensity I_0 is proportional to $|\mathbf{E}|^2$ at $z = 0$; different proportionality factors apply for parallel and perpendicular polarizations. The existence of an evanescent field follows from continuity requirements of the wave function at the interface. A comprehensive theoretical description of the evanescent field is given by Axelrod et al. (1984).

If fractions of the evanescent field are absorbed, it is considered *attenuated TIR* (or ATR), i.e., the net energy flux across the boundary surface is no longer 0. ATR sensing techniques take advantage of this particular optical phenomenon: Exposing the area of an evanescent field resulting from internally reflected radiation to any sample featuring absorptions in the MIR induces measurable changes in the intensity of the MIR spectrum, as the sample absorbs characteristic fractions of the probing radiation via the evanescent field, thereby enabling both qualitative measurements due to the characteristic absorption features of different chemical species and quantitative measurements. For the latter we introduce

$$\mathcal{T}_\lambda = \frac{I_\lambda}{I_{0,\lambda}} = 10^{-\mathcal{A}_\lambda} = e^{-\alpha_\lambda \cdot l} \tag{9.17}$$

with I being the fraction of light reaching the detector (or, alternatively, being outcoupled of the ATR sensing element), I_0 the amount of light emitted by the source (or, incoupled into the ATR element respectively), and $\mathcal{T}$ the ratio of the two (i.e., the *transmission*). This should not be confused with the different usage of $\mathcal{T}$ in the previous context, i.e., with the definition of the amount of light "transmitted" through the ATR unit via total internal *reflection*. $\mathcal{A}$ in Eq. (9.17) is the absorbance, l the interaction path length, and α the absorptivity, which is further related to the molar absorptivity ε_λ, the concentration c, the wavenumber $\tilde{\nu}$, and the extinction coefficient κ via

$$\alpha_\lambda = \ln(10) \cdot \varepsilon_\lambda \cdot c = \frac{4\pi \cdot \kappa_\lambda}{\lambda} = 4\pi \cdot \kappa_\lambda \cdot \tilde{\nu} \tag{9.18}$$

Accordingly, the absorbance may be written as

$$\mathcal{A}_\lambda = -\log_{10}(\mathcal{T}_\lambda) = -\log_{10}\left(\frac{I_\lambda}{I_{0,\lambda}}\right) = \varepsilon_\lambda \cdot c \cdot l = \frac{1}{\ln(10)} \cdot \alpha_\lambda \cdot l \tag{9.19}$$

The index λ in Eqs. (9.17)–(9.19) is a reminder that all values carrying that index are dependent on the frequency (or, alternatively, the wavelength or wavenumber) of the incident radiation. The fact that $\mathcal{A}$ is linearly dependent on the concentration c [$\mathcal{A} \propto c$; Eq. (9.19)] is called the *Beer-Lambert law*, also known as *Beer's law*. For more than one analyte it is further written as (Laqua et al. 1988)

$$\mathcal{A}_\lambda = l \sum_{i=1}^{N} \varepsilon_{i,\lambda} c_i \tag{9.20}$$

Moreover, it is important to note that *transmission and absorption spectra* are derived by dividing a set of transmittance values (i.e., a *sample single-channel spectrum*, SSC) by a second set of transmittance values, the *background* (or *reference*) *single-channel spectrum* (RSC). Accordingly, the variable I_0 has a slightly different meaning when speaking in terms of absorption (or transmission). I_0 is then (as well as I) representing a *transmitted fraction*, in particular that of the RSC (or the SSC, respectively). To be accurate, I_0 and I in the context of Eqs. (9.17) and (9.19) are better envisaged as among the many possible outcomes of a random experiment. Occasionally, confusion may arise because the real numbers of inserted rays in a ray-tracing experiment (i.e., the number of rays emitted by the light source), I_0 and I_0' (one for the RSC and one for the SSC) do not show up in the absorbance $\mathcal{A}$.

Evanescent field absorption measurements satisfy a pseudo-Beer's law relationship according to

$$\mathcal{A}_\lambda^* = (\varepsilon_\lambda \cdot c \cdot l)\rho = (\varepsilon_\lambda \cdot c \cdot l)\frac{P_{e,\lambda}}{P_{t,\lambda}} \tag{9.21}$$

where ρ is the *evanescent wave ratio*, i.e., the fraction of power guided outside the denser medium (P_e) divided by the total power of guided light (P_t).

While solid samples are in general best analyzed using, e.g., single reflection (also called "single-bounce") ATR systems, horizontal ATR (HATR) accessories with multiple internal reflections are ideal implementations for analyzing liquids or low sample concentrations. The optically denser medium is usually a trapezoidal crystal (ATR element or waveguide). In particular and with respect to our example, Dobbs and Mizaikoff (2006) used polycrystalline ZnSe with a length of 72 mm (i.e., the major axis), a thickness of 6 mm, and 45° beveled radiation coupling facets (cf. Figure 9.2 for a CAD rendering of the HATR crystal). According to this geometry, the radiation exhibits 6 TIRs along the measurement surface boundaries before the infrared beam is directed onto the detector of a FT-IR spectrometer. Using $n_1 = 2.43$ at a wavelength of $\lambda = 5$ μm as the refractive index of ZnSe in Eq. (9.14) (compare Figure 9.1)—in the absence of an absorber—all radiation incident at an angle larger than 24.3° will be totally internal reflected.

Each reflection is the origin of an evanescent wave extending a few micrometers beyond the surface [typically 0.1–2 μm for MIR frequencies; Eq. (9.16)] into the adjacent medium, thus creating discrete "active" sensing regions along the measurement surface. However, due to the nonuniform spatial distribution of the probing radiation, quantitative analysis is enabled in practice only if the sample coverage is homogeneous along the entire crystal, which is usually accomplished by covering the entire waveguide surface with liquid or solid sample. However, if the sample is not distributed homogeneously at the surface of the HATR element but is instead deposited in, e.g., small droplets along the major axis of the crystal, a variation in the analytical response [i.e., signal generation according to Eq. (9.21)] depending on the position of droplet position relative to the incoupling facet is experimentally observable [compare Dobbs and Mizaikoff (2006) for a more detailed description of these findings].

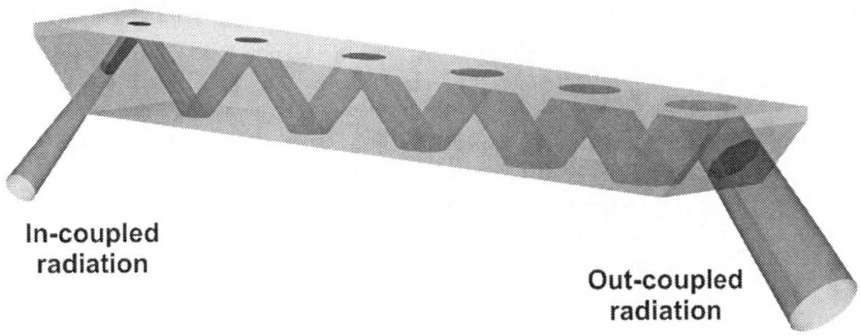

Figure 9.2. Schematic 3-D view of MIR radiation confined within an HATR crystal, illustrating increasing probing spot sizes. (Adapted with permission from Dobbs 2006. Copyright 2006 Society for Applied Spectroscopy.)

Although a qualitative interpretation is straightforward (in fact, this is already provided by stating that the potential for signal generation is related to the electric field strength perpendicular to the boundary [Eq. (9.21)], an accurate assessment of the experimental results is not readily available. Thus, complementary simulations may be used in addition to confirm the experimental values, assist in explaining them quantitatively, aid the visualization of the "active" sensing regions or to provide insight on parameters that are experimentally not (easily) accessible. A straightforward and convenient way to localize the regions of such interactions at a ZnSe ATR crystal is a ray-tracing approach, as discussed in the following.

In the context of optics, a ray is an instrumentalized abstraction of a classic electromagnetic wave traveling orthogonal to its electric and magnetic field. Therefore, radiation may be conceptualized as a symbolized beam of light rays with zero extent perpendicular to their direction of propagation. In optical ray tracing, these rays are propagated through a modeled optical system, calculating their paths under the rules provided by geometric optics. Although geometric optics are the most stringent simplification of Maxwell's equations, frequently considerations derived from geometric optics provide sufficiently accurate results, if the dimensions of the objects and structures within the established model are significantly larger (or smaller) than the wavelength of the modeled radiation.

In terms of probability theory, a simulation experiment based on ray tracing is an example for a random experiment, which is an experiment whose outcome cannot be determined in advance (Rubinstein and Kroese 2008). Simulations that show this particular behavior are commonly summarized within a concept called Monte Carlo (MC) methods. MC simulations approximate real-world behavior statistically, rather than exactly describing the real-world environment (Robert and Casella 2010). Consequently, MC simulations are particularly applicable to systems that show no intrinsic dynamics, which in consequence would require alternative modeling techniques such as, e.g., molecular dynamics simulations or ab-initio quantum chemical models.

Generally speaking, in order to accurately simulate an optical system via ray tracing, a full description of all components that exert influence (or are in principle able to influence) on the cross section, radiation density distribution, direction, angular spread, and/or absolute intensity of the radiation propagating through the setup until the detector is reached, must be included in the computer model. That is, starting from the radiant light source, the (relative) 3-D location, size, (relative) orientation, surface textures, and material properties of all mirrors, lenses, windows, apertures, beam splitters, etc., must precisely match the experimental setup. Moreover, the geometry and directional characteristics of the source itself, and—if the simulation should encompass different wavelengths—its spectral characteristics must be matched. Needless to say, in order to satisfy all requirements, all variables must be known, which leads us to the finding that simulations are always dependent on experimental results for obtaining and adjusting the first set of simulation parameters. The conditions set to the light source are most often the most difficult to access experimentally, and all aforementioned parameters rely on experimental measurements.

At this stage, one should note an aspect that is fundamental to any simulation except ab-initio simulations: The quality of the raw data defining these parameters is essential for the quality of the derived model. This is also the point where one has to consider appropriate model simplifications in order to come up with results within a reasonable calculation time. Many of the aforementioned parameters may not even be known, and frequently they do not even have to be entirely known, if they are only of minor influence on the simulation result or do not limit correct conclusions.

The example illustrated herein has been shown by Dobbs and Mizaikoff (2006), and has used the commercial software package SPRAY (W. Theiss Hard- and Software, Aachen, Germany) (Theiss 2008a). SPRAY is a software package providing nonsequential ray tracing in a 3-D space for complex optical components and systems at MIR frequencies. Section 4.5 provides more information on the available software options.

To realize the optical configuration of this particular example, i.e., the simulation of the location of the "active" sensing regions, the simulation model must consist of a minimum of three "objects," which are specified by geometric parameters (location, size, orientation) and assigned material properties (e.g., permeability, permittivity, susceptibility, refractive index) or functions (i.e., some special object types, e.g., source, screen, detector) describing the optical behavior of the objects (e.g., gases, solids, mirrors, absorbers, etc.) within the virtual sensor scenery. As with any optical configuration, this scenery is illuminated by a light source (LS), in this case with planar circular geometry, which emits discrete rays that are finally propagated through the optical system following the laws of geometric optics. Moreover, the model comprises an exact geometric representation of the HATR crystal.

As the third and last component, the virtual setup features a pixelated rectangular screen (the screen resolution is user-defined), which is responsible for counting the number of incident rays (intensity I) at each pixel location.

In order to render the ray-tracing approach successful, the screen must be placed parallel to the top surface of the HATR element but still *inside* the crystal, e.g., 0.1 μm below the surface. This requirement is a consequence of the ray optics approach: The evanescent field itself cannot be simulated with geometric optics. As a consequence, every ray striking the top surface must hit the screen twice, once on its way toward the boundary and once on its way back toward the lower crystal surface. To ensure that not every ray is counted twice, the screen can easily be defined to only count rays incident from one predefined direction (in this example, it is of minor relevance which direction is selected, as both of them yield very similar results).

The optical (response) characteristics of a material itself are defined by its frequency-dependent optical properties. In this case, the dielectric function model of ZnSe from the SPRAY materials database may be used directly and assigned to the virtual representation of the HATR element. Alternatively, one could take values from a table (e.g., published literature), or the dielectric function model might be based on experimental measurements of the optical response spectrum of the material of interest in the frequency range of interest. For comparison, Example #2 offers a scenario under which the dielectric functions were extracted from measured MIR spectra rather than from a database or literature values.

Once the virtual model is established, the ray-tracing procedure may be executed. Here, one should observe the number of in-coupled rays for the ray-tracing procedure: Using a greater number of rays per simulation, the simulation results naturally converge, as predicted by the law of large numbers. Yet, for practical considerations one usually starts with computing a small number of rays and continues adding more until the desired accuracy is reached, given the calculation time required to consecutively execute the required calculations per ray.

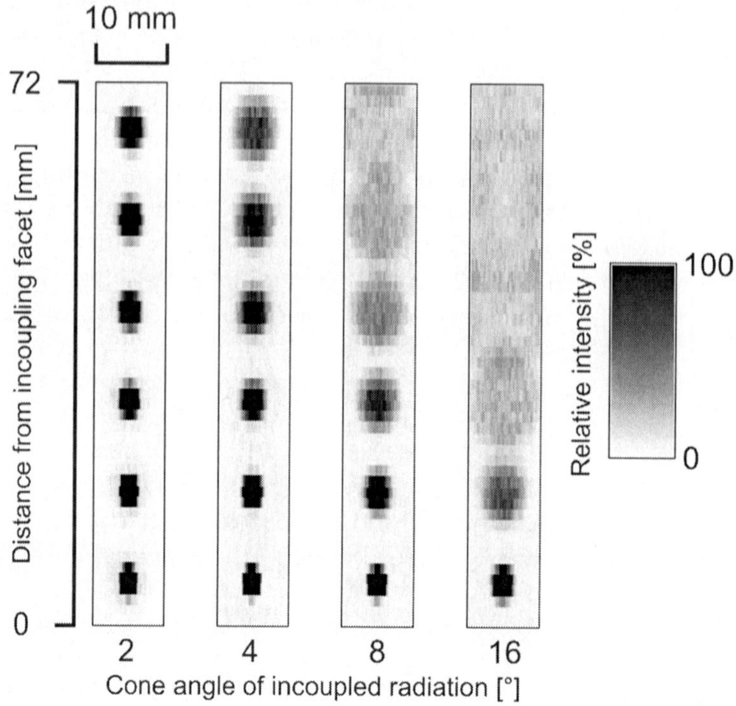

Figure 9.3. Visualization of internal reflection regions at an ATR crystal's top surface via simulated radiation density maps. (Adapted with permission from Dobbs 2006. Copyright 2006 Society for Applied Spectroscopy.)

Compared to the general all-encompassing modeling concept mentioned above, the described "three-object model" of Dobbs and Mizaikoff (2006) is simplified in such a way that a circular light source (LS) is inserted at the beveled in-coupling facet of the HATR crystal itself, in order to replace the real-world focal point of the in-coupling radiation. Accordingly, its size resembles the diameter and its location the position of the focal point; as a consequence, all other optical (coupling) elements can be omitted. Furthermore, the radiation density and angular distribution of the light source are considered homogeneous in a simplified assumption providing a constant solid cone angle. Homogeneity in this context means that the Monte Carlo process randomly selects emission points that are statistically uniformly distributed across the surface. It should be noted that this is an example of a rather stringent simplification, as the true focal point is not circular, and it is characterized by a nonhomogeneous radiation density distribution and a nonuniform divergence angle.

For brevity, this example is limited to showing the influence of increasing the radiation cone angles on the spatial distribution of radiation at the sensing regions along the measurement surface of the simulated HATR element (Figure 9.3). The screen pixel brightness indicates the number of rays that have hit this location at the screen and is proportional to the absolute radiation density in this region, thus providing surface maps of the radiation distributions according to the different cone angles of the LS. Darkest spots are generally located closest to the in-coupling facet (i.e., nearest the focusing point, that is, the LS) and successively decreasing at each of the six individual reflection regions with respect to the beam divergence.

While all simulation runs shown in Figure 9.3 were generated with a source radius of 0.1278 cm, another simulation series (not shown here) was performed with varying source radii while maintaining a constant cone angle of 2° in order to estimate the diameter of the in-coupled radiation cone, which is experimentally only indirectly observable. A direct comparison between experimentally generated data (compare Dobbs and Mizaikoff 2006) and ray optic values revealed that both become quantitatively equivalent if applying 0.1278 cm as LS diameter and a 2° diverging angle of the in-coupled radiation cone, as shown in Figure 9.3. Whenever such excellent agreement between simulation and experiment is achieved, it is obvious that the virtual setup provides an appropriate representation of the experimental setup, thereby validating the model at hand. Hence, despite the simplifications introduced, it may be concluded that the outlined model approximates the actual dimensions of "active" sensing regions along the measurement surface of an ATR waveguide with reasonable accuracy, thereby rendering the simplifications legitimate.

The fact that this ray-tracing approach successfully relates simulated radiation densities with real-world evanescent field interactions—which one may believe would in fact necessitate physical optics—is not immediately intuitive.

Indeed, the radiation density map (as shown, e.g., in Figure 9.3) is independent of the angle of incidence (i.e., screens treat rays equally, independent of their angle of incidence on the screen; they only *count* them), but the absorbance [Eq. (9.21)] is not, as the penetration depth [Eqs. (9.15) and (9.16)] contains θ_i as a variable. The reason why proportionality between simulated intensity and measured absorbance is maintained is due simply to the narrow entrance cone of only 2°, which effectively forces the HATR sensor to obey a pseudo-Beer-Lambert law within the tolerances of the measurement.

However, ray theory fails to appropriately describe phenomena such as interference, diffraction, and polarization, as well as the phase of a wave. Moreover, a ray-tracing approach is not applicable if the size of the objects falls within the size range of the wavelength. If that is the case, simulation approaches based on wave optics are needed, which are covered in Chapter 10 of this volume. Finally, describing, e.g., ultrashort optical pulses or the mechanical forces of light on matter requires simulation using quantum optics.

3. EXAMPLE #2: SENSOR RESPONSE SIMULATION—TOWARD VIRTUAL CALIBRATIONS

Prior to the simulation era, development progress depended mainly on the experience and experimental skills of the researcher. Whether a novel optical concept will work or yield an analytical benefit is frequently not immediately evident, as a sensor or component thereof may react quite sensitively to some critical design parameters. Accordingly, sensor enhancements frequently rely on minor continuous modifications of already-existing devices. However, radically new ideas are frequently limited by experimental resources and trials. Moreover, in the field of research, the fabrication of many devices with only small iterations/modifications is frequently time-consuming and cost-extensive if, e.g., clean-room techniques such as, e.g., epitaxial growth, reactive ion etching, or focused ion beam milling—to name just a few—are required. Hence, it is advantageous to determine in advance (i.e., prior to, e.g., fabrication of a modified device) whether the alternative yields an overall enhancement of the device—here, an optical chemical sensor—figures of merit. The next example illustrates the benefits of modeling entire sensing devices to advance the overall sensor concept prior to experimentation.

The careful reader may have noticed already that the previous example was executed at a particular MIR wavelength. As a consequence, this example was—strictly speaking—not an approach illustrating *spectrally resolved* modeling. However, one may consider this example as the most stringent simplification of spectral modeling: namely, narrowing the spectral band to a single wavelength. As discussed, the previous example may have been executed for multiple wavelengths as well. Furthermore, the first example introduced fundamental principles that are also applied in Example #2, which is to our best knowledge the only demonstration of the spectrally resolved simulation of an entire optical chemical sensor operating in the MIR reported to date. While the first example has shown simulations for liquid-phase samples, the second example models a gas-phase sensing system, where the achievable limit of detection depends mainly on the optical interaction path length within the gas sample (i.e., usually the interaction path length provided by the sample cell) as dictated by the Beer-Lambert law [Eq. (9.19)]. If the gases of interest are characterized by low absorptivity and for trace-gas analysis with ppb-level detection limits, a sufficiently long interaction path length is thus required.

Hollow cylindrical waveguides (HWGs) are a potential solution for gas-sensing applications to replace conventional multipath absorption gas cells. These HWGs serve as both capillary flow cell and efficient light guide in the MIR, with effective optical path lengths up to several meters (Kozodoy et al. 1996; Worrell and Gallen 1997). Tightly constricted radiation within the hollow waveguide core ensures intimate interaction with the molecules present within the same confined volume. Consequently, HWGs have several advantages over multipass absorption cells, including their small volumes, short associated response times, a flexible structure, and relatively low fabrication costs (Kozodoy et al. 1996; Haan and Harrington 1999; Fetzer et al. 2002; Harrington 2004). The HWG used by Wilk et al. (2009) is shown schematically in Figure 9.4. It comprises a metallic reflector made from a layer of silver coated with a dielectric film to enhance the reflection, both of which are deposited on the inside of a hollow silica capillary.

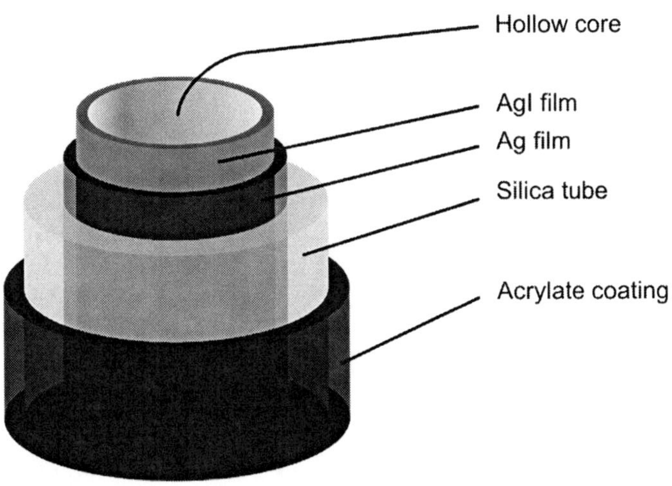

Figure 9.4. Schematic layout of a dielectric coated glass HWG. (Reprinted with kind permission from Springer Science+Business Media: *Analytical and Bioanalytical Chemistry,* An approach to the spectral simulation of infrared hollow waveguide gas sensors, 395, 2009, 1661–1671, Wilk A., Kim S-S., and Mizaikoff B., Copyright 2009.)

In the following, the previously discussed basic concepts will be expanded to establish computational modeling techniques to simulate MIR spectroscopy inside the hollow core of a dielectrically coated waveguide utilizing spectral ray-tracing techniques for selected analytes, as shown by Wilk et al. (2009).

Any successful sensor simulation comprises main objectives:

1. Establish a 3-D simulation model of the sensor that resembles the real-world optical configuration closely. As a first step, this task includes developing dielectric function models for the optical components as well as for the analytes based on experimental data.
2. Demonstrate the utility of the developed model (validation step). This may be performed by synthetically generating spectrally resolved responses of individual analytes and mixtures of analytes at various concentrations, and comparing thus-obtained virtual results to experimental data, thereby confirming or disproving the model.
3. If—and only if—the validation step was successful, the model may be used for predictions of novel sensor concepts and advanced device design.

3.1. EXPERIMENTAL SPECTRA ACQUISITION

Spectra simulation studies may theoretically be performed using reference spectra derived from databases. However, as spectra from such collections are frequently measured using dispersive instruments, it may not necessarily be expected that the obtained simulation results coincide with measurements obtained with a particular sensor, e.g., the developed FT-IR/HWG sensor of Wilk et al. (2009) discussed herein, which generates data at entirely different optical and chemical environments inside the hollow core of the waveguide. Moreover, database spectra are frequently not suited for obtaining quantitative simulations due to missing information on necessary experimental conditions such as, e.g., temperature and pressure conditions. As the quality of the raw data is essential for the quality of the derived model, it must be concluded that superior quantitative models rely on experimental data determined with the system to be simulated rather than using database spectra, which are usually applied for the identification of unknown constituents rather than for quantification. Consequently, for the present study, IR spectra of the analytes of interest were recorded at controlled conditions and with certified standards for establishing a sound modeling basis. A gas cell was designed and certified gas standards of the target analytes carbon dioxide, methane, butane, isobutylene, and cyclopropane were measured within that gas cell using a Bruker Equinox 55 FT-IR spectrometer (Bruker Optics, Billerica, MA, USA). Figure 9.5 shows the background transmittance spectrum of the setup. This spectrum summarizes the combined response function of all involved optical

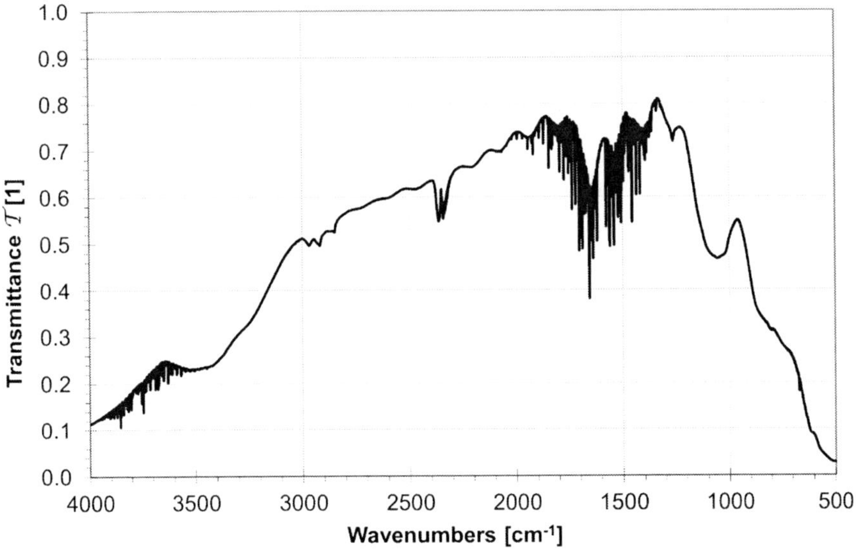

Figure 9.5. Background transmittance spectrum. Bruker Equinox 55 FT-IR spectrometer including the gas cell in the optical path; 1/cm resolution, 1000 averaged spectral scans.

elements without absorbing sample inside the gas cell. It includes the optical characteristics of the LS (SiC filament), the coatings of all mirrors in the optical path, the ambient atmospheric conditions, the KBr windows sealing the sample compartment, the ZnSe windows of the gas cell, and of the detector. This background spectrum was later used for modeling the "combined system response" of the simulated spectrometer, as will be discussed.

Gas concentrations of 100 ppm were provided directly from reference gas cylinders. Since the absorbances of those standards were comparatively low, the average of 1000 sample scans was used for spectrum acquisition, yielding acceptable signal-to-noise ratios [S/N; compare Eq. (9.28)]. Thus obtained reference spectra of gas-phase samples provided a proper basis for the modeling and simulation studies, and were used as raw data for the parameter fit of modeling the dielectric function.

3.2. DIELECTRIC FUNCTION MODELING

The fact that IR sensors may provide the capability of detecting multiple constituents as a result of the inherent molecular selectivity of vibrational and rotational spectra suggests utilizing wavelength-resolved ray-tracing approaches to simulate the sensor response across selected frequency bands. Extending this concept, spectral ray tracing consequently implies that the rays used in Monte

Carlo simulations carry frequency information. In consequence, the software package SPRAY, version 2.35, object generation 2.9100 [W. Theiss Hard- and Software, Aachen, Germany (Theiss 2008a)], which provides this feature, was used for ray tracing.

Regarding Example #1, it has already been stated that the dielectric response of a material may be derived from its optical response spectrum. As the spectra of the materials (e.g., analytes, HWG, etc.) of interest were obtained prior to the modeling efforts, the dielectric functions of the involved materials could be simulated. SCOUT 98, object generation 2.8400 [W. Theiss Hard- and Software, Aachen, Germany (Theiss 2008b); SCOUT], formerly used predominantly for thin-film studies (Hilbrich et al. 1997; Bjerke et al. 1999; Hasegawa et al. 2001; Venkataraj and Geurts 2001; Zhao et al. 2004; Zhao and Wäckelgård 2006), was applied for dielectric function modeling. Thus-obtained dielectric function models simulate possibly closely the real-world spectral response of a material, and may be directly assigned to 2-D and 3-D surfaces/objects in SPRAY. Together, the two software packages provide a sound modeling platform enabling spectral ray tracing of complex optical systems, and even of the chemical response of IR sensors.

For spectrum analysis via computer simulation, the correct optical constant model has to be selected. Physically, the resonance frequencies for the excitation of vibrational modes in the IR depend on the vibrating masses of the functional groups and the strengths of the chemical bonds. When the atomic nuclei are displaced from their equilibrium positions, the chemical bonds act as restoring forces according to their bond strengths. The type of model that best describes a system experiencing restoring forces in analogy to Hooke's law is an oscillator model. In this particular case, the susceptibilities of the materials involved were modeled with harmonic oscillators, which are flexible models defined by parameters for the peak position (resonance—i.e., oscillator frequency—position), peak height (oscillator strength), peak width (oscillator damping), and peak shape (Gauss-Lorentz switch).

In order to execute a parameter fit, the spectrum of the material of interest—database spectrum or experimentally obtained spectrum—is loaded as a reference into SCOUT, and the experimental conditions for obtaining the real-world spectrum are defined in the software. The above-mentioned parameters are manually adjusted in such a way that simulated and measured spectra match reasonably well. Then, an automatic fitting routine implemented in SCOUT is executed to optimize all oscillator parameters, and account for the mutual influence of overlapping oscillators. If an IR spectrum comprises many oscillators (20 or more), and dependent on the computational power, one has to smartly group the parameters selected for fitting, in order to obtain an adequate fit of the resulting dielectric function in a reasonable time. As a rule of thumb, strongly overlapping oscillators have to be treated as a collective, whereas nonoverlapping oscillators may be treated separately as individual oscillators. Furthermore, grouping for, e.g., only

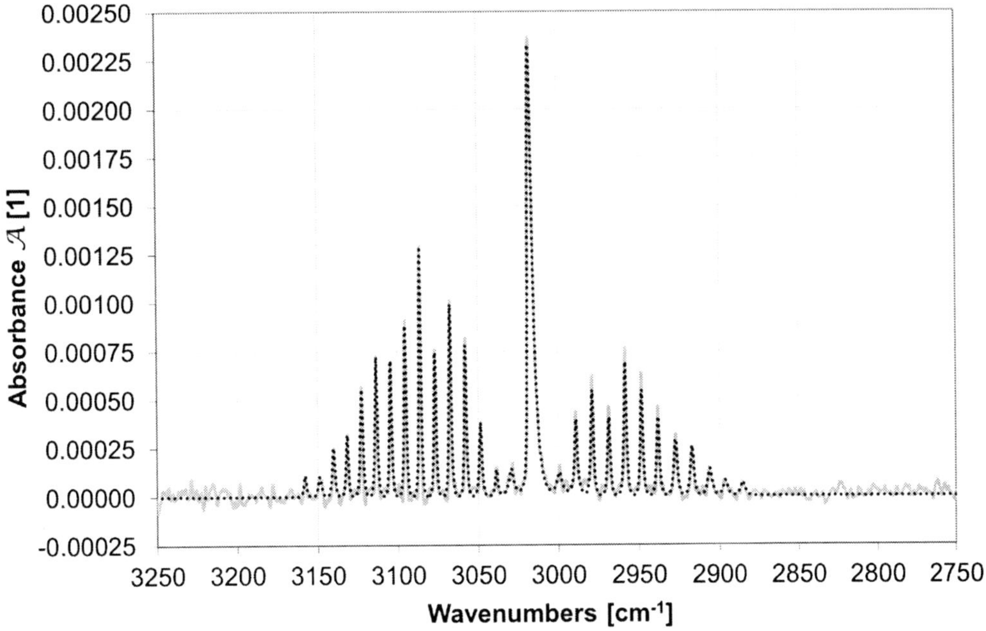

Figure 9.6. Methane, SCOUT oscillator fit. *c* = 100 ppm ± 2%. Dotted black line = SCOUT fit with 30 Kim oscillators; continuous gray line = original data (1000 scans, 1/cm resolution). (Reprinted with kind permission from Springer Science+Business Media: *Analytical and Bioanalytical Chemistry,* An approach to the spectral simulation of infrared hollow waveguide gas sensors, 395, 2009, 1661–1671, Wilk A., Kim S-S., and Mizaikoff B., Copyright 2009.)

peak positions or dampings is possible. Finally, the main absorption features should be modeled with particular attention, as their, e.g., heights or areas may potentially be used for substance quantification.

Figure 9.6 shows an example of an accomplished fit to an experimentally determined methane IR spectrum. Here, so-called Kim oscillators (Kim et al. 1992) were applied, as they enabled an improved adjustment to the original data. The solid gray curve represents the original data, and the dotted black line the associated parameter fit. Like all other models reported herein, this model is in excellent agreement with the experimental spectra. The quality of the fit may readily be quantified by testing the residuals against the null hypothesis, i.e., that they result from a normally distributed sample. One such test statistic is the probability plot correlation coefficient (PBCC), which was used by Wilk et al. (2009) at a 1% significance level. A formula for executing the necessary wavenumber rescaling in order to match real and simulated absorption frequencies is given in that article. The modeled methane response served as "virtual analyte" by assigning it to the material, which filled the core of the virtual HWG during simulation of the entire IR sensor. Similar parameter fits may be executed for all other target analytes of interest.

3.3. MODEL ESTABLISHMENT AND VALIDATION

The next step is to assign the obtained dielectric responses to the individual components of the virtual sensor established in SPRAY, and to perform ray-tracing simulations for retrieving virtual spectra generated by the sensor model. Beforehand, a virtual representation of the entire FT-IR spectrometer had to be realized.

The virtual spectrometer was modeled essentially following the procedure described for Example #1. The model is an exact clone of the geometric dimensions of the Bruker Equinox 55 FT-IR spectrometer, except for the Michelson interferometer, which is not required for the ray optics simulation. Figure 9.7 shows a schematic layout of the FT-IR spectrometer and the position of the crucial components including off-axis parabolic mirrors (OAPMs) that direct the IR beam into the sample chamber, as well as elliptic mirrors that refocus the beam onto the

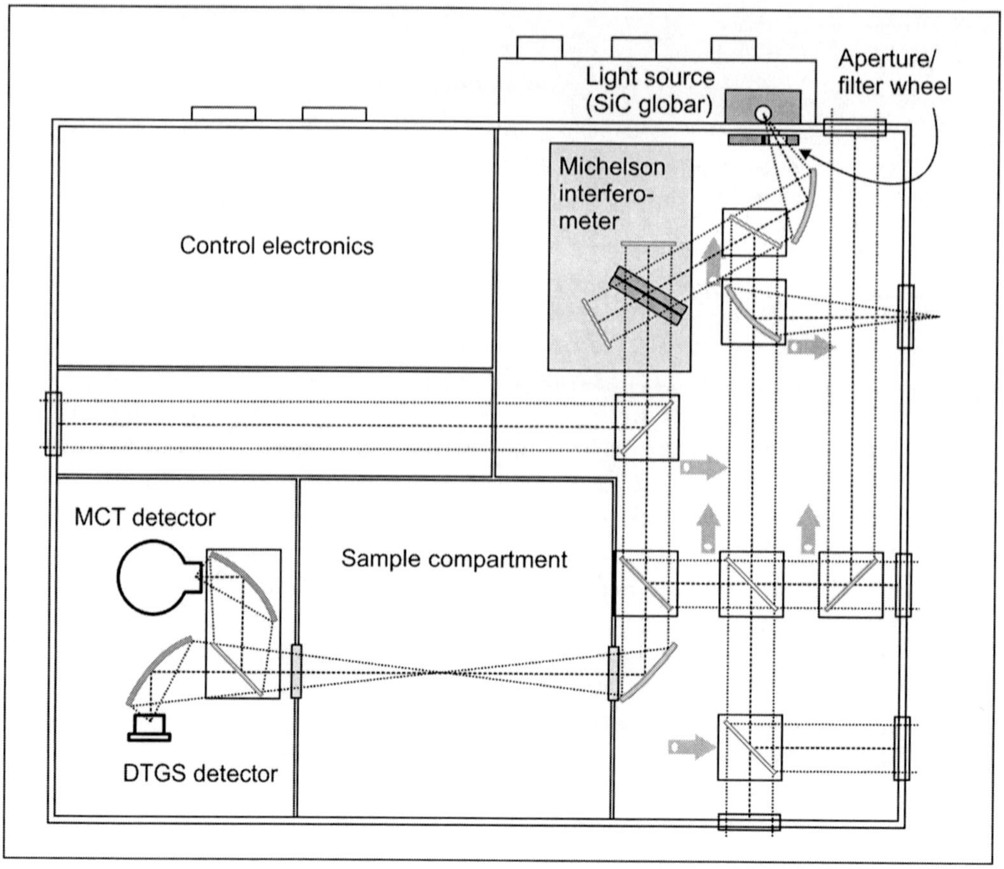

Figure 9.7. FTIR spectrometer—Top view. Optical path and mirror configuration as provided by Bruker Optics (Bruker Optik GmbH, Ettlingen, Germany).

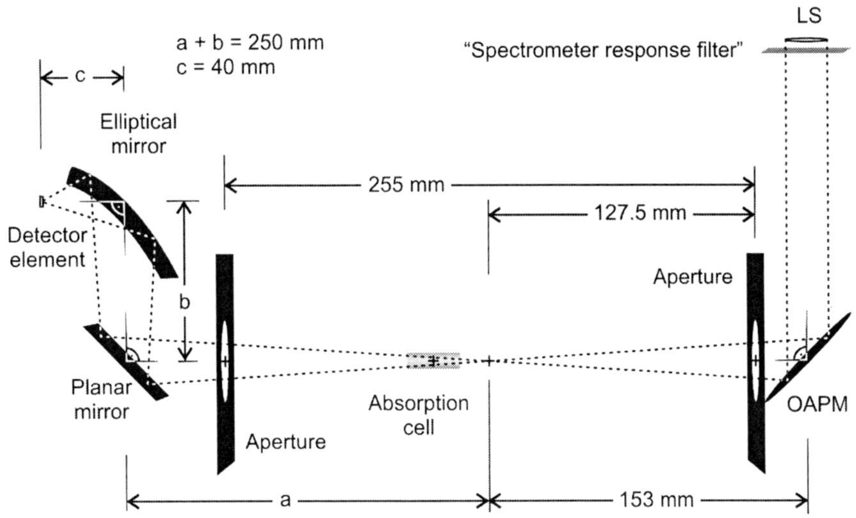

Figure 9.8. SPRAY scenery—FT-IR spectrometer clone. (Reprinted with kind permission from Springer Science+Business Media: *Analytical and Bioanalytical Chemistry,* An approach to the spectral simulation of infrared hollow waveguide gas sensors, 395, 2009, 1661–1671, Wilk A., Kim S-S., and Mizaikoff B., Copyright 2009.)

mercury cadmium telluride (MCT) detector element. The completed virtual replica of the spectrometer is given in Figure 9.8.

The MCT detector element is represented by a square with an edge length of 2 mm. As opposed to a "screen object," a "detector" provides information on neither the intensity nor the position of a ray, but does record the spectral information (wavenumber $\tilde{\nu}$) of the ray. All objects defined in SPRAY are surrounded by the material that is assigned to the light source, i.e., in this example dry CO_2-free air. As the LS sends a predefined and constant number of rays into the scenery for each spectral data point (s.p.) independent of the wavenumber, a "spectrometer response filter" was set right in front of the LS. This filter is defined as partially transparent and partially absorbing within the model. By assigning the single-channel reference spectrum (Figure 9.5) to this filter, it is ensured that the filter statistically retains exactly that segment of the IR radiation that is used to obtain the transmittance characteristics of the real-world sensor. As that response already represents a combined response function of the entire sensor setup, all mirrors as well as the wall of the gas cell (which is represented by an open cylinder) could be defined as "ideal mirror" interfaces. Note that that the KBr and ZnSe windows sealing the sample compartment of the spectrometer and the gas cell are absent. Their implementation can be dispensed with, as the response filter already contains their contributions to the background spectrum. Finally, sample spectra are generated by assigning the modeled optical constants of one of the target analytes (or a virtual mixture of several analytes) to the inside of the absorption cell (i.e., an object of the type "closed cylinder").

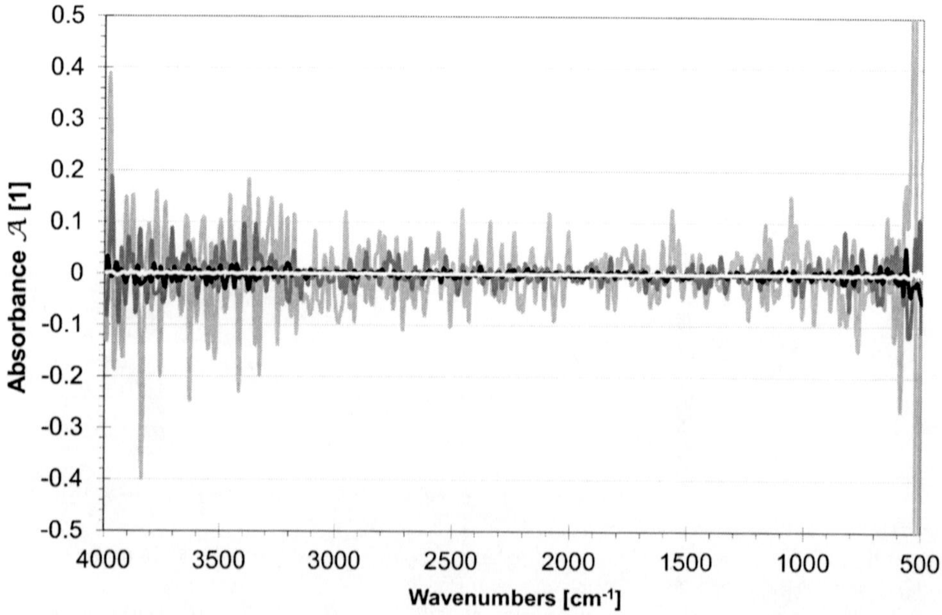

Figure 9.9. Simulated background absorbance spectra. SPRAY, gas cell simulation, different number of inserted rays I_0 per s.p.; 10/cm resolution. I_0: light gray = 100; dark gray = 10^3; black = 10^4; white = 10^6.

Now that all material and geometric parameters are set, performing the ray-tracing simulation leads to the blank spectra depicted in Figure 9.9 for a varying numbers of inserted rays. Those spectra were calculated using Eq. (9.19) by calculating the negative base-10 logarithm of the ratio of two independent simulation runs [i.e., the background single-channel spectrum $I_0(\lambda)$, and the sample single-channel spectrum $I(\lambda)$] in the absence of analytes.

In agreement with experimentally obtained blank spectra (where the same relation is true for the number of averaged scans), it can be found that the standard deviation of the noise, s_n', is inversely proportional to the square root of the number of inserted rays:

$$s_n' = \Xi \cdot \frac{1}{\sqrt{I_0}} \tag{9.22}$$

Comparing the noise levels in Figure 9.9 and the low absorbtion values of methane in the real-world spectrum (solid line in Figure 9.6), it could be estimated that in order to achieve the same noise level (or, alternatively, the limit of detection) in the simulated spectra, roughly more than a quarter-billion rays per spectral data point (I_0) would be necessary. Such large numbers by far exceed the computational power of such simulations.

However, there is strong evidence that a simulated detector behaves in a linear fashion for almost all concentrations (i.e., that the Beer-Lambert law is valid). If so, the absorbance may instead be simulated for concentrations which are a factor Λ higher than the ones of interest or experimentally measured. Hence, by multiplying the complex refractive indices [compare Eq. (9.2)] representing the target analytes in the model by Λ, and dividing all absorbance signals by the same factor Λ after the calculation is completed, computations may be accelerated. The outlined procedure has the advantage that the noise level will drop to $1/\Lambda$. However, without this modification/assumption, $\Lambda^2 I_0$ rays would be needed in the simulation to achieve the same s_n' [Eq. (9. 22)]. The obtained benefit in calculation time is thus $\propto \Lambda^2$.

Downscaling by Λ works, as long as the simulated detector does not show saturation effects at higher concentrations Λc; i.e., the absorbance should be restricted to values < 2. This means that at least a mean of one-hundredth of I_0 (i.e., one of 100 inserted rays) is present at the detector. Otherwise, deviations from the linear behavior of the detector have to be anticipated. While those could be eliminated by increasing I_0, this would be counterproductive, as again the calculation time would increase with an increasing number of inserted rays.

The spectra of methane in Figure 9.10, showing methane at various concentrations and at a resolution of 1 cm^{-1}, were calculated using $\Lambda = 100$ and I_0 = 50,000, in order to simulate spectra in the 100- to 20-ppm concentration range.

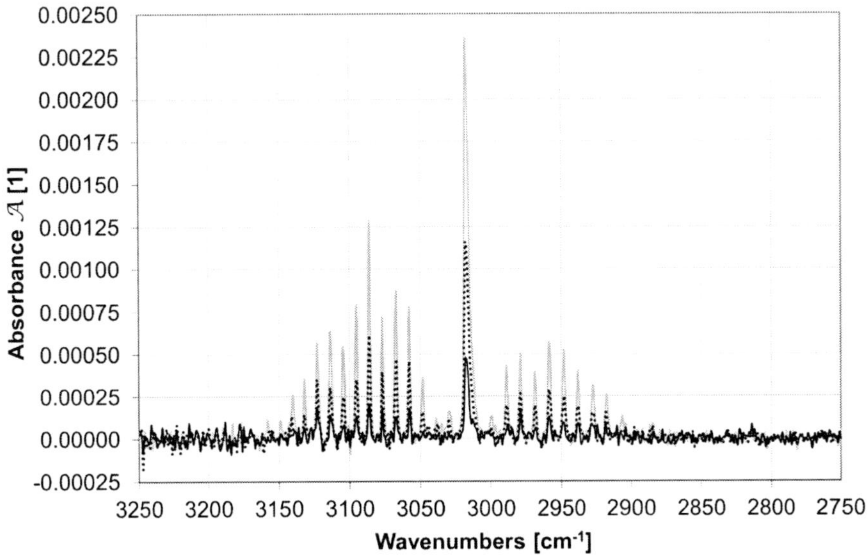

Figure 9.10. Methane spectra, different concentrations: SPRAY simulation experiment. 1/cm resolution, 5×10^4 rays/s.p., Λ = 100; c_0 = 100 ppm ± 2%. Dilution: gray solid line = c_0, black dotted line = $\frac{1}{2}c_0$, black solid line = $\frac{1}{5}c_0$. (The $\frac{1}{5}c_0$ curve was adapted with kind permission from Springer Science+Business Media: *Analytical and Bioanalytical Chemistry,* An approach to the spectral simulation of infrared hollow waveguide gas sensors, 395, 2009, 1661–1671, Wilk A., Kim S-S., and Mizaikoff B., Copyright 2009.)

Simulating each single spectrum required approx. 8 h of calculation time using an Intel Conroe processor (Intel Corp., Santa Clara, CA, USA). Without the outlined procedure the simulation would have taken approx. 9 years to complete (500 billion rays in total) on that same computer.

At this point it is important to note that the simulated 50- and 20-ppm methane spectra in Figure 9.10 are visually fully consistent if compared with the (independently obtained) corresponding real-world spectra for the same concentrations (data not shown; compare Wilk et al. 2009). Mathematical ways of comparing spectra with each other are deduced by Wilk et al. (2009), allowing quantification of the similarity of the spectrum obtained. As shown, the statement that absorbance spectra obtained with real spectrometers may be reproduced by virtual spectrometers in superb agreement with the experimental data even holds true for quantitative data. In consequence, this implies that the virtual spectrometer (Figure 9.8) is indeed an eligible model of the real FT-IR spectrometer, thereby enabling predictions of spectra at different concentrations (e.g., 35 ppm) or of artificial mixtures (compare Section 3.5 in this chapter). Such a validation step (i.e., the outlined comparison of real and simulated spectra) is required, as this verifies the modeling strategy and the model simplifications with the discussed assumptions and limitations. Once the model is verified, in a next step the dielectric

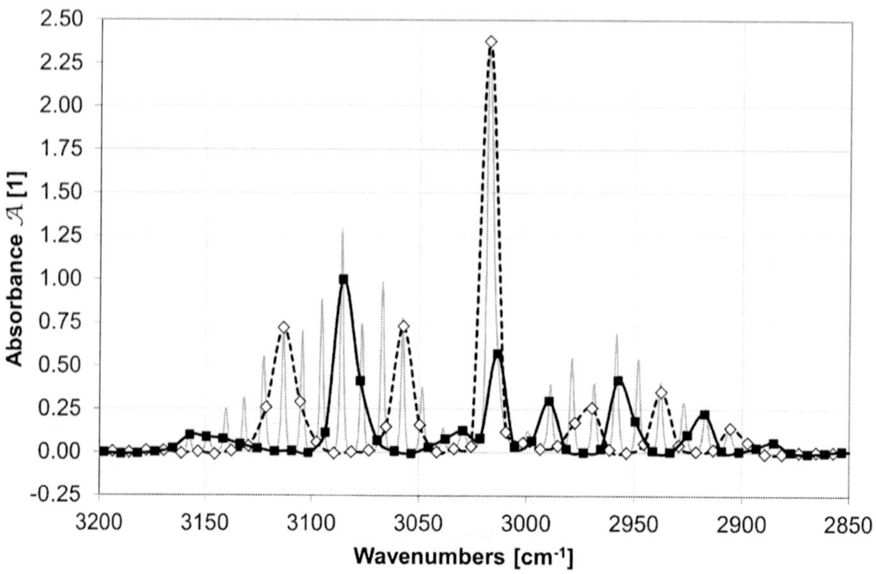

Figure 9.11. Spectra simulation experiment with reduced resolution. Methane, 8/cm resolution; curves shifted 4 cm^{-1}; 5000 rays/s.p.; concentration c = 10% ± 2%. Black dashed line = data point at 3017.6 cm^{-1}, black solid line = no data point at 3017.6 cm^{-1}, gray solid line = absorption spectrum calculated from the dielectric function model of methane (Figure 9.6).

function models are transferable to another sensor platform, thus enabling the prediction of the spectral response of, e.g., novel sensor configurations.

Before this is shown, one final remark concerning the resolution of simulated spectra is provided: a decrease in resolution (shown here: $R = 8\ \text{cm}^{-1}$) will strongly impact the peak shape of the signal, as shown in Figure 9.11. Here the gray curve is the calculated absorbance derived from the SCOUT oscillator fit, and all ray-traced data points are—unsurprisingly—located on this line. However, the characteristic band structure of methane is lost, as the ray-tracing calculation pays no attention to proximate oscillators. Even worse than the dashed curve is the black solid curve, as there is no data point at the resonance frequency of methane. Compared to a real-world experiment, one has to trade off losing the recognition ability for peaks entirely by reducing the spectral resolution. Thus, trading off spectral resulution is not a suitable option to speed up computation time. Consequently, in order to obtain comparable results to real spectra at lower resolution, the SPRAY simulation has to be performed with a higher resolution, and the results have to be smoothed by subsequent averaging across neighboring data points.

3.4. SENSOR RESPONSE SIMULATION

Experimentally, the sensor to be simulated comprised a compact FT-IR spectrometer coupled with a HWG, and a MCT detector cooled with liquid nitrogen. Both ends of the HWG were sealed with ZnSe windows, allowing for an enclosed gas cell. An OAPM coupled broad-band MIR radiation emitted from the spectrometer source onto the proximal HWG entrance facet, and two identical OAPMs were used to focus output radiation at the distal end of the HWG onto the detector. An exponential dilution experiment was executed for the analytes CO_2 and CH_4. The virtual representation of this sensor configuration is shown in Figure 9.12. Once again, the realized virtual sensor duplicates the experimental sensor 1:1, except that the (uncoated) ZnSe windows sealing the HWG were omitted for simplicity.

Based on the underlying material properties of ZnSe, this change appears negligible. ZnSe offers a transmission window from 0.6 to 20 μm, therefore rendering it the most commonly used transparent optical material for MIR applications. Its fairly high and relatively low dispersed refractive index (2.45–2.3 @ 2–20 μm; Li 1984) causes reflection losses of roughly 15% per interface within the entire MIR spectral band. While these windows may be implemented in the simulation, their effect on the results of the simulations are negligible, as reflection losses—at least in this case; a counterexample is given in Section 4.1—may only influence the underlying noise levels, but do not cause a fundamental failure of the model.

The dielectric function of gold was assigned to the objects representing the OAPMs using the Drude Au model included in the software database. For the

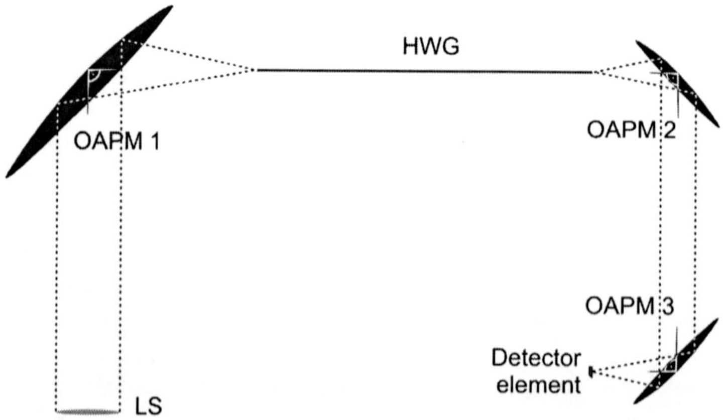

Figure 9.12. SPRAY representation of the experimental single-pass HWG sensor setup. (Reprinted with kind permission from Springer Science+Business Media: *Analytical and Bioanalytical Chemistry,* An approach to the spectral simulation of infrared hollow waveguide gas sensors, 395, 2009, 1661–1671, Wilk A., Kim S-S., and Mizaikoff B., Copyright 2009.)

simulation of this particular chemical sensor setup and for future modifications thereof, an appropriate representation of the complex structure of the HWG (Figure 9.4) is essential, but less straightforward. The challenge in modeling HWGs—as in any simulation—is to make the correct assumptions and simplifications to establish a workable model that provides a sufficiently accurate representation of the real-world device.

Light guiding inside HWGs has been described in simulation experiments from the viewpoint of ray, wave, and quantum optics. Encompassing the analysis of the waveguide mode structure (Snitzer 1961; Ibanescu et al. 2003), laser propagation (Marcatili and Schmeltzer 1964; Davies and Mendonça 2000), and optical transmission theory (Karbowiak 1955; Marcatili 1966; Snyder and Love 1975; Garmire et al. 1976; Yeh and Yariv 1978; Miyagi et al. 1983; Marksteiner et al. 1994; Lines and Klocek 1998; Jackson 1999), accurate insights into HWG transmission have been achieved. In particular, the investigation of (hollow) optical fibers with ray optical methods has been pursued for decades, and dates back nearly as long as the history of MIR fiber optics themselves. Already in 1974, leaky modes of optical fibers had been studied via ray-tracing techniques (Snyder and Mitchell 1974). A couple of years later, this method was expanded toward the treatment of multimode step index waveguides (Love et al. 1978). Loss evaluations of HWGs were reported soon thereafter (Miyagi and Kawakami 1983; Miyagi 1985). Based on ray models, it was proved in 1987 that both straight and bent tubes with metal and metal-dielectric inner coatings transmit IR radiation (Croitoru et al. 1987). The calculation of transmission as a function of the geometric dimensions of a bent waveguide was published by Mendlovic et al. (1989) using a ray model. Ray-tracing calculations of multiply bent HWGs followed soon thereafter (Morhaim et al. 1991). One year before, it was demonstrated that the losses in bent HWGs may be estimated based exclusively on geometric optics (Saito et al. 1990). For the analysis of signal propagation in highly multimode optical waveguides with rough surfaces, a hybrid ray-tracing technique presented by Bierhoff et al. (2001) was determined to be suitable. Recently, the beam shape distribution of guided laser light was analyzed using this technique (Gannot et al. 2002). Furthermore, ray tracing was used to support the experimental development of HWGs. Matsuura et al. (1997) designed a launching coupler consisting of a lens and a tapered HWG using a ray-tracing method for optimization. Iwai et al. (2002) proposed substituting the refractive indices of two-layered HWG structures by a virtually monolayered material with a single "effective refractive index" to speed up ray-tracing calculations.

In the following, a different approach will be described, following Wilk et al. (2009): By assigning the spectral loss function of a leaky HWG to its hollow core, a convenient and easily implementable virtual representation of HWGs is achieved, which emulates the given HWG-based IR optical gas sensor with sufficient accuracy for fundamental studies. This virtual HWG is shown in Figure 9.13. In particular, a representation of the straight HWG was constructed in SPRAY by defining a solid cylinder (core), and a hollow cylinder (hull) with slightly larger diameter. Figure 9.13 is not drawn to scale; in the actual representation the diameter of the mirror is only negligibly (0.001%) larger than the core. The hull is assigned a perfect mirror interface that is a 100% reflecting surface. The fact that a ray bouncing through the HWG has to penetrate through the core interface ("3" in Figure 9.13) is insignificant, as $n_{\text{core}} \cong n_{\text{air}}$ are both very close to 1.

In this study (Wilk et al. 2009), the dielectric response of the HWG was modeled based on digitized data provided by the manufacturer of the HWG (Polymicro Technologies, Phoenix, AZ, USA). The model consists of 14 Kim oscillators and simulates the original data with reasonable accuracy (Figure 9.14).

Subsequently, the dielectric function model of the observed HWG dielectric response was assigned to the core volume. Consequently, κ_{HWG} was treated as an

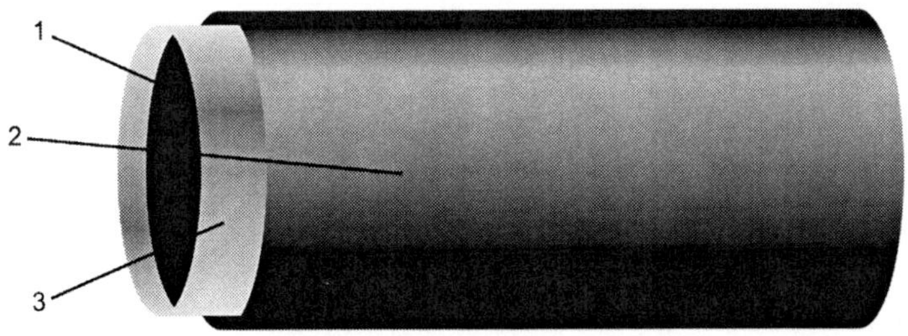

Figure 9.13. Straight HWG—schematic. 1, core filling; 2, hull interface; 3, core interface. (Reprinted with kind permission from Springer Science+Business Media: *Analytical and Bioanalytical Chemistry*, An approach to the spectral simulation of infrared hollow waveguide gas sensors, 395, 2009, 1661–1671, Wilk A., Kim S-S., and Mizaikoff B., Copyright 2009.)

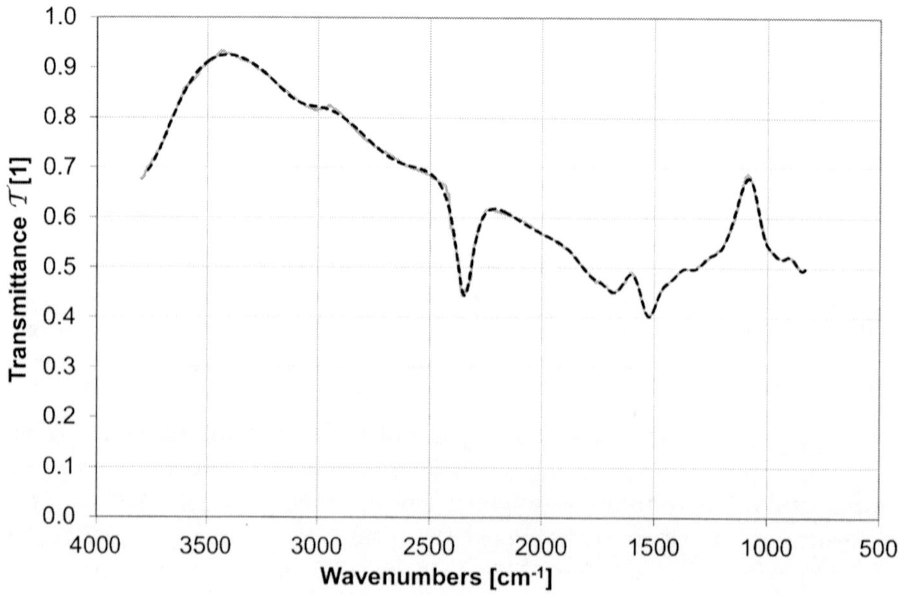

Figure 9.14. Response of a HWG, 1-m length—SCOUT oscillator fit. Black dashed line = SCOUT fit with 14 Kim oscillators; gray solid line = original data HWEA7501200. (Reprinted with kind permission from Springer Science+Business Media: *Analytical and Bioanalytical Chemistry,* An approach to the spectral simulation of infrared hollow waveguide gas sensors, 395, 2009, 1661–1671, Wilk A., Kim S-S., and Mizaikoff B., Copyright 2009.)

analyte, and set as an offset value to the extinction coefficient of the core [compare Eqs. 2 and 10 of Wilk et al. (2009)]:

$$\kappa'_{core} = \kappa_{core} + \kappa_{HWG} \tag{9.23}$$

Although this definition of a HWG is a major systematic difference from the construction of a real HWG (see Figure 9.4), the obtained results confirm that this is a solution approximating the real-world scenario with sufficient accuracy. However, this description may not be sufficient to model more complex HWG geometries (e.g., the coiled HWG, as introduced in Section 4.5).

One should be aware that assigning κ_{HWG} to the volume [Eq. (9.23)] renders the HWG response (i.e., transmission) independent of the number of reflections inside the HWG, but dependent on the optical path length (OPL). In consequence, representing the HWG by a volume surrounded by an ideal mirror fails if the effect of various hollow-core diameters (or in-coupling angles) of the HWG are investigated.

Figure 9.15 illustrates the effect that doubling the hollow-core diameter of a HWG has on the number of interface reflections. As evident, a doubled diameter decreases the number of reflections by one-half, while the OPL remains unaltered (the same physical length of the HWG implied). Consequently, with a less than 100% reflecting surface, the attenuation loss is expected to decrease with decreasing number of internal reflections. The theory of wave optics for HWG transmission predicts for either metallic or dielectric HWGs that (Marcatili and Schmeltzer 1964)

$$L \propto \frac{1}{\mathrm{ID}^3} \tag{9.24}$$

where L is the attenuation coefficient and ID is the internal diameter of the HWG. The current representation of the HWG is not able to describe this expected behavior; the detected fraction will not show an observable increase in radiation throughput. However, the calculation time will be reduced by one-half for the wider HWG, as a reduced number of interface contacts must be processed (compare Figure 9.15). In reality, however, the energy throughput is dependent on the number of surface reflections. This correct description is lost if κ_{HWG} becomes a volume property, which restricts this simulation method to alternating lengths of the HWG.

The presented model was established in a typical top-down approach used for rapidly explaining a selected measurement situation. For future studies, the latter findings may motivate fundamental changes in the strategy, leading to a more generic bottom-up approach with the HWG model featuring the inner-layer stack structure comprising Ag-AgI. Thereby, the loss function becomes dependent not only on the number of interactions with the interface but also on the angle of incidence at the layer stack. However, to establish this model accurately, further data is required from the manufacturer, such as how exactly the attenuation of the HWG was measured, and which in-coupling conditions were used. The major drawback of this more generic (refined) model of a HWG would be significantly more computation time. In addition, the length of the HWG could no longer be reduced (as for the current simulations), and unrealistically small light sources could not be used without having a major impact on the HWG transmission characteristics.

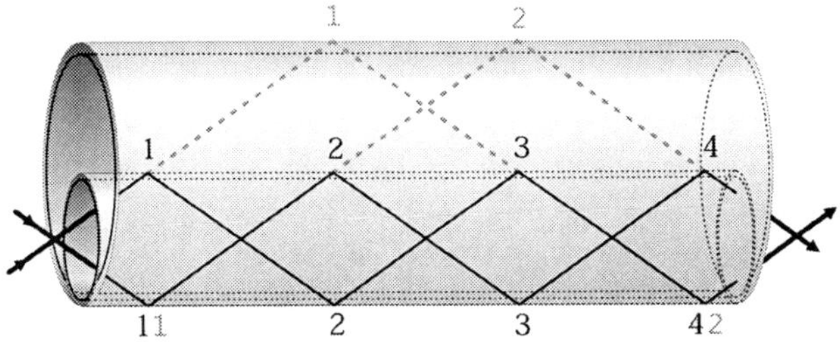

Figure 9.15. Effect of doubling the bore diameter of the HWG. Black digits = number of interface contacts for rays in the small-bore cylinder; gray digits = corresponding number for the wide-bore cylinder.

Accordingly, it was assumed for the current model that the manufacturer must have used in-coupling conditions similar to the ones in the simulated sensor, which in consequence means that both OPLs are comparable, and the results obtained are sufficiently close to the experimental data. Moreover, only absorbance spectra were studied herein. Hence, the resulting error should be within the noise level, and should be significantly smaller than the analyte signals; presuming a constant optical path length (OPL), the absorbance $\mathcal{A}$ is independent of experimental conditions, which renders it superior to single-channel spectra that are not universally comparable.

Finally, the dielectric function of methane was assigned to the core of the established HWG, and different concentrations were modeled. In order to save computation time, the simulations were restricted to the MIR band between 2750 and 3250 cm^{-1}, where the most prominent CH_4 absorptions occur. The data were simulated using a collimated beam and a LS diameter of 100 μm, to further speed up the computation. This simplification dramatically minimized the number of interface contacts, and consequently reduced the computational time by a factor of 7.3 compared to a LS with wider diameter of 2 cm. One has to be aware, though, that this simplification leads to a reduced influence of the HWG transmittance characteristics, as the interaction length is longer describing a ray traveling in a "zigzag" pattern through the HWG compared to a ray which is passing straight through the HWG. Nonetheless, this appears to be an arguable simplification, as it may be calculated from the geometrics of the in-coupling OAPM that a ray entering 1 cm off the central axis is reflected with an in-coupling angle $\theta = 12.5°$ for the side of the OAPM, which is closer to the HWG (and 10.2° for the more distant side). A brief estimation of the maximum error that results assuming the worst-case scenario of all rays entering at an in-coupling angle of 12.5° yields an increased influence by 2.4% compared to the simplification mentioned above. Consequently, both the signal and the HWG transmittance should drop at maximum by a factor of 2.4%. In the simulation, the averaged transmittance indeed dropped by less than 2% (from 60% to 58%) in the simulated spectral region (i.e., 3250–2750 cm^{-1}). A last simplification to further reduce the number of interface interactions was the down-scaling of the HWG length to one-tenth of its original value, and simultaneously increasing its response and the analyte concentrations by a factor of 10.

Figure 9.16a shows a simulation of the sensor response with a methane concentration of 116.8 ppm (corresponding to the dilution after 15 min), both with the narrow LS (gray solid curve) and the wider LS (black dashed curve). In fact, the main absorption peak at 3017.6 cm^{-1} increased by roughly 10%. Nonetheless, this difference appears less dramatic because this increase lies well within the ambient noise level (Figure 9.16b is the difference spectrum between the two calculations).

It should be noted that all simulated IR spectra were simulated with 1-cm^{-1} resolution and smoothed over 5 data points by applying

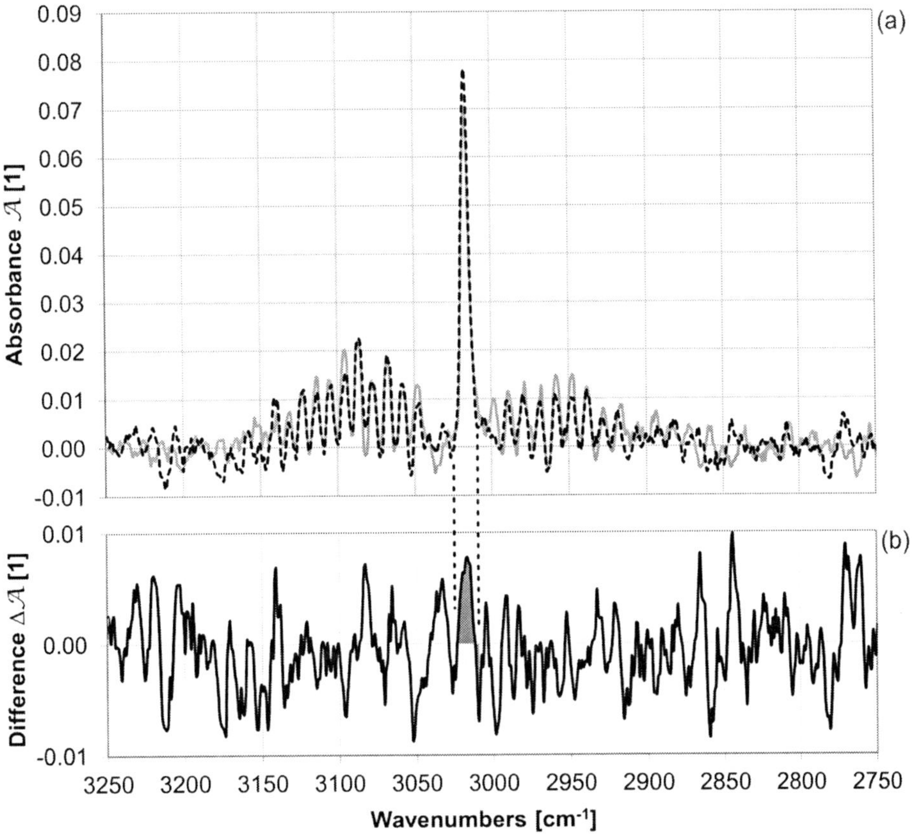

Figure 9.16. Error estimation. (a) solid gray line = spectra obtained for narrow light source (LS), dashed black line = wide LS. (b) Difference spectrum, wide LS minus narrow LS.

$$Z_0' = \frac{1}{5}\sum_{j=-2}^{2} Z_j \tag{9.25}$$

where $Z_{j=0}$ is the original value and Z_0' is the smoothed value, in order to match the experimental data, which was recorded at a resolution of 4 cm^{-1} (Kim et al. 2009).

For this example, it shall be sufficient to exemplarily overlay the simulated response spectrum with an experimentally obtained spectrum after 9 min of dilution time had elapsed. Both correspond to a concentration of 835.8 ppm in Figure 9.17; this picture was also included in the original experimental paper. When compared to the experimental spectrum, the spectral profile of the simulated spectrum still looks strikingly similar (especially the Q branch), albeit the absorptions in the R band are significantly higher in the simulation, and there are considerable mismatches evident in the obtained noise level, especially at the

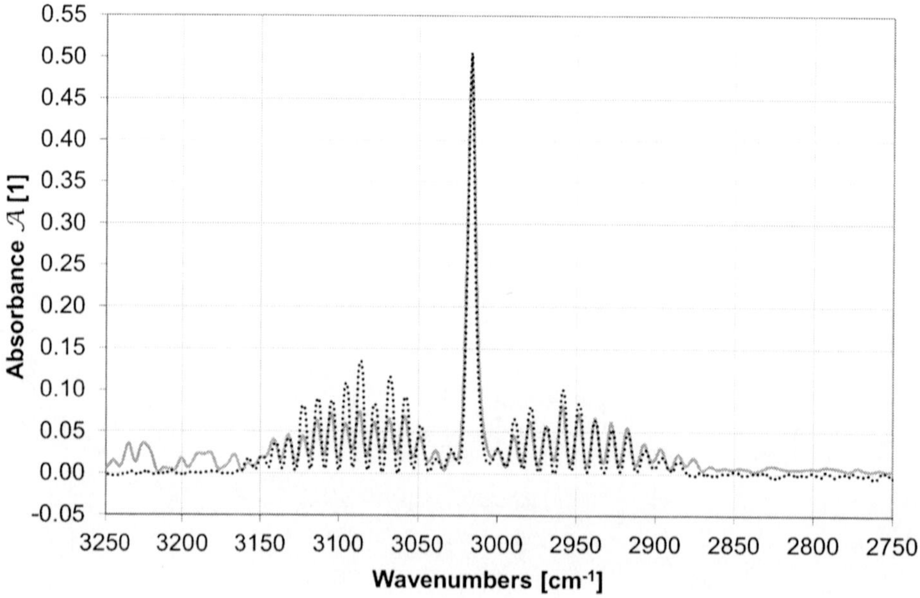

Figure 9.17. Overlay of simulated and real methane spectrum, 9 min after injection (835.8 ppmv), smoothed over 5 data points. Solid gray line = experiment; dotted black line = simulation. (Reprinted with kind permission from Springer Science+Business Media: *Analytical and Bioanalytical Chemistry*, An approach to the spectral simulation of infrared hollow waveguide gas sensors, 395, 2009, 1661–1671, Wilk A., Kim S-S., and Mizaikoff B., Copyright 2009.)

higher-wavenumber end of the spectrum. Although some possible explanations are discussed by Wilk et al. (2009), future studies will show whether this experimentally observed phenomenon is intrinsic or due to unknown effects that yet have to be clarified.

3.5. SPECTRUM PREDICTION

Linear combinations of previously modeled dielectric functions may be used to simulate the spectral response of any compound mixture without access to real-world mixture spectra. Ultimately, a single spectrum of an individual analyte measured at one concentration is sufficient to simulate any other concentration, or a spectral mixture of this component with other constituents.

The predicted mixture spectra of two and three components, as obtained with the virtual FT-IR spectrometer (Figure 9.8), are shown in Figures 9.18 and 9.19, respectively. As previously discussed, the initial concentrations were simulated with a factor $\Lambda = 100$ higher concentrations, reducing the number of inserted rays to an acceptable value for generating high-quality spectral simulations at the sub-100-ppm concentration range. The absorbance values were divided by the same Λ afterwards in order to establish the simulated spectra shown.

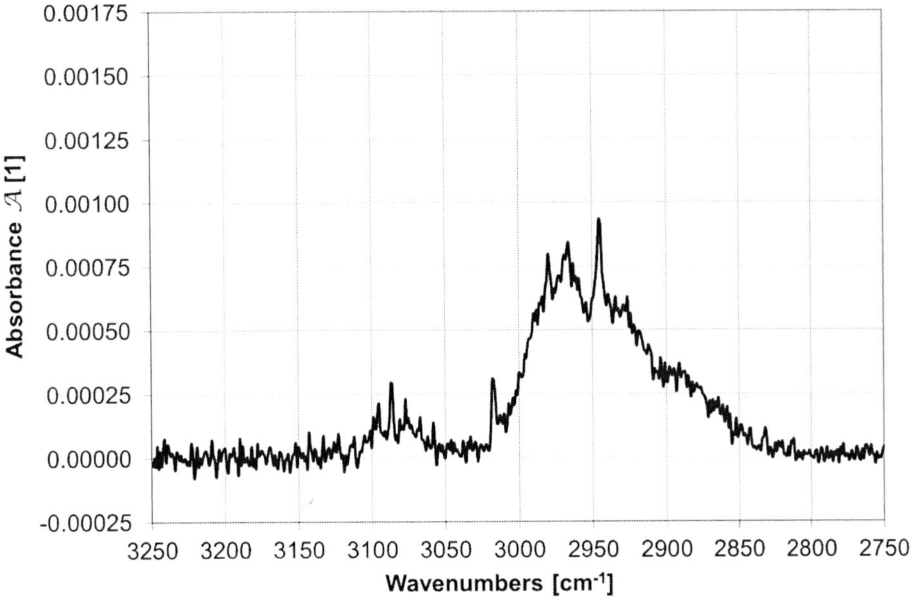

Figure 9.18. Virtual spectrum of 10-ppm methane, 20-ppm butane, and 80-ppm isobutylene generated by mixing the dielectric functions. 1/cm resolution, 5×10^4 rays per s.p. Simulated with 1000, 2000, and 8000 ppm, respectively (Λ = 100).

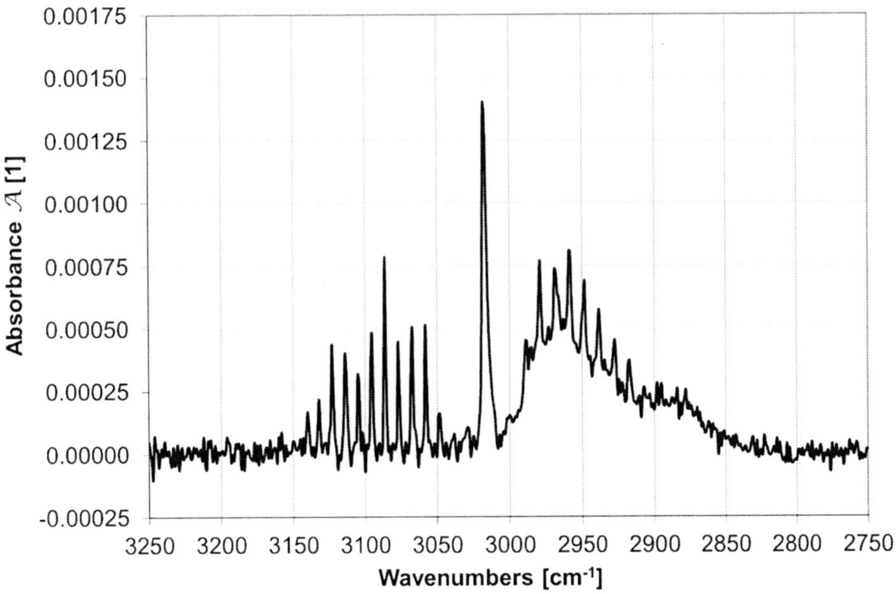

Figure 9.19. Virtual spectrum of 60-ppm methane and 30-ppm butane generated by dielectric function mixing. 1/cm resolution, 5×10^4 rays per s.p. Simulated with 6000 and 3000 ppm, respectively (Λ = 100).

The methane concentration in Figure 9.18 was intentionally selected to be below the limit of quantification (LOQ) of the real-world sensor [i.e., the FT-IR spectrometer (Figure 9.7) combined with the 1-in. OPL gas cell mentioned in Section 3.1], assuming the same noise in the simulation. The determined limit of detection (LOD) of the real experiment is 4.14 ppm for 1000 scans. Future studies may establish refined virtual mixture spectra close to the real-world concentrations. Such spectra may provide high-quality raw data that might be useful for testing, optimizing, and selecting appropriate multivariate data analysis routines and chemometric data evaluation strategies (Miller and Miller 2005; Mark and Workman 2007; Brereton 2007; Otto 2007; Varmuza and Filzmoser 2009). Moreover, we anticipate that virtual calibrations may potentially provide a useful addition to real-world data sets to increase calibration robustness, if upcoming studies verify the possibility of creating calibrations of actual sensors by using simulated spectra.

4. EXAMPLE #3: EXTENDED APPLICATIONS

So far, the modeled examples have been based on sensing systems that have already been realized experimentally. However, a particular strength of the proposed simulations is to either explain experimentally observed phenomena that are not readily accessible from experiment, such as deviations from Beer's law, and predictive modeling in order to forecast the functionality and behavior of novel IR sensor geometries. Moreover, simulations should enable evaluating their robustness (and potential vulnerability) as well as their analytical performance. Accordingly, it may be concluded that simulations are valuable tools for the development of novel and improved sensing concepts, which may in part replace complex experiments or at least reduce the number of experiments to a minimum.

4.1. EXPLORATION OF THE SOURCES OF DEVIATIONS IN BEER'S LAW PLOTS

Substantial negative deviations from Beer's law behavior are usually a manifestation of the effects of a nonlinear detector response. However, if the highest absorbance values included in the Beer's law plot are below 0.5, such effects are usually very small (Richardson et al. 1998a).

Unexpectedly, a particular double-pass IR sensor configuration utilizing hollow waveguides has revealed considerable nonlinearity independent of the analyte concentration and absorbance strength. Figure 9.20 shows the corresponding experimental results for the example analytes isobutylene, methane, and cyclopropane (labeled "measured" values). The double-pass MIR gas-sensing assembly

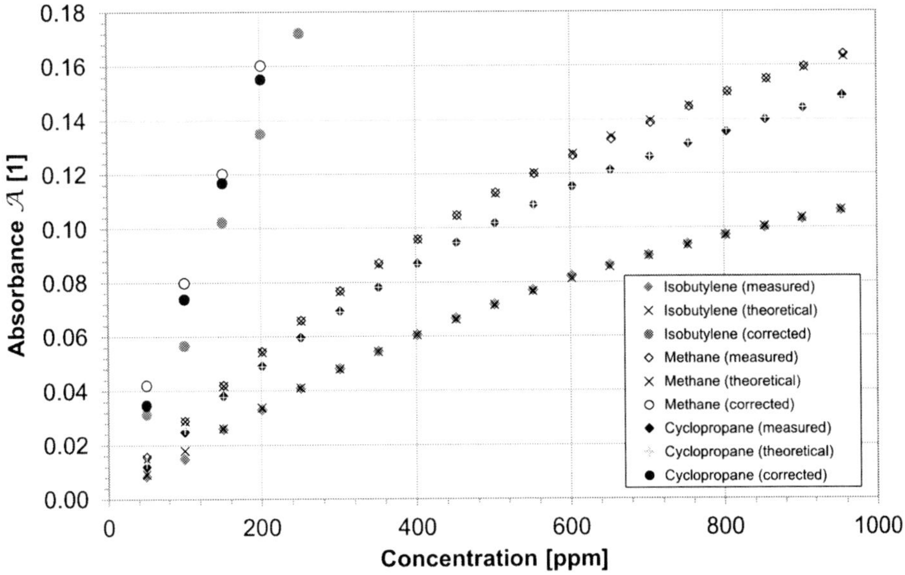

Figure 9.20. Beer's law plot at the wavenumbers of maximum absorption for the analytes isobutylene, methane, and cyclopropane measured on the double-pass HWG sensor at concentrations ranging from 50 to 950 ppm. In particular, the 890.0 cm^{-1} band was used for isobutylene, the 3017.7 cm^{-1} band for methane, and the 3101.6 cm^{-1} band for cyclopropane.

comprised a FT-IR spectrometer, a 50:50 beam splitter, two gold-coated OAPMs, a HWG with a length of 40 cm (hollow-core diameter = 750 μm, Ag/AgI dielectric layered), uncoated ZnSe windows for sealing the HWG end facets, a gold-coated spherical mirror, and a MCT detector.

The SPRAY scenery depicted in Figure 9.21 provides a precise virtual model of the aforementioned sensor elements and their geometry in real dimensions. The beam splitter diverts the collimated beam emanating from the spectrometer, and the in-coupling OAPM focuses the radiation into the HWG. The spherical mirror at the distal end of the HWG retro-reflects the IR radiation into the HWG in order to achieve a double pass. During the return pass, the radiation is again collected by the in-coupling OAPM, and guided through the beam splitter. A second OAPM focuses the radiation onto the detector element. It should be noted that in this configuration, 50% of the incident radiation is lost during the first pass and another 50% during the second pass through the beam splitter.

A straightforward ray-tracing simulation similar to those discussed in the previous examples with 5×10^6 rays @ 2000 cm^{-1} has qualitatively revealed that the ZnSe window sealing the proximal end of the HWG (i.e., the front facet) is the source of the observed nonlinear response inherent to this specific sensor configuration. Due to the reflectance of this window, a major fraction of the radiation

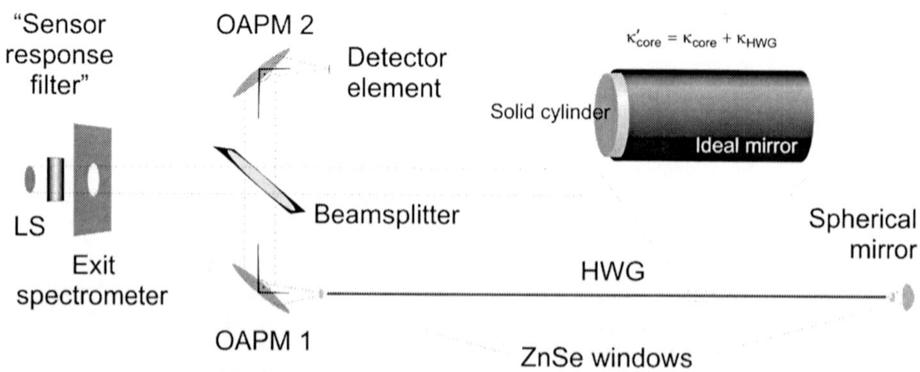

Figure 9.21. SPRAY representation of the FTIR-HWG double-pass sensor.

is reflected onto the detector without having been guided into the HWG and probing the containing gas sample (i.e., this part of the radiation does not contribute to the analytical signal). As Figure 9.22 shows, both the directly reflected radiation and the radiation carrying analytical signal are strongly overlapping. Consequently any usual detector alignment procedure aimed toward maximizing the ADC counts will cause stray light contributions of the order of up to 90%, thus giving rise to the observed nonlinearity (Figure 9.23).

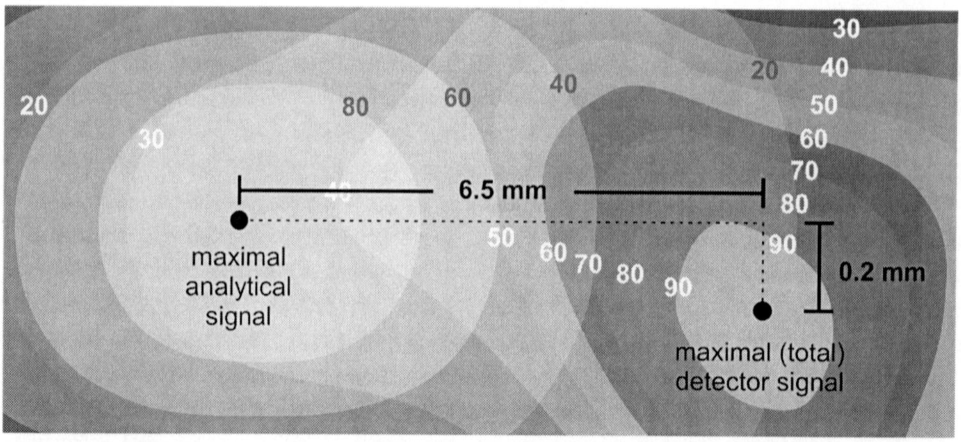

Figure 9.22. Height profile map showing the simulated convolution of the probing radiation (i.e., the radiation carrying analytical information) incident at the detector element with reflected radiation. The numbers are percentage values normalized to the height of the maximal analytical signal, and the maximal detector signal.

From Figure 9.24 an analytical function could be derived which enabled predicting the absorbance quantitatively for any arbitrarily chosen concentration:

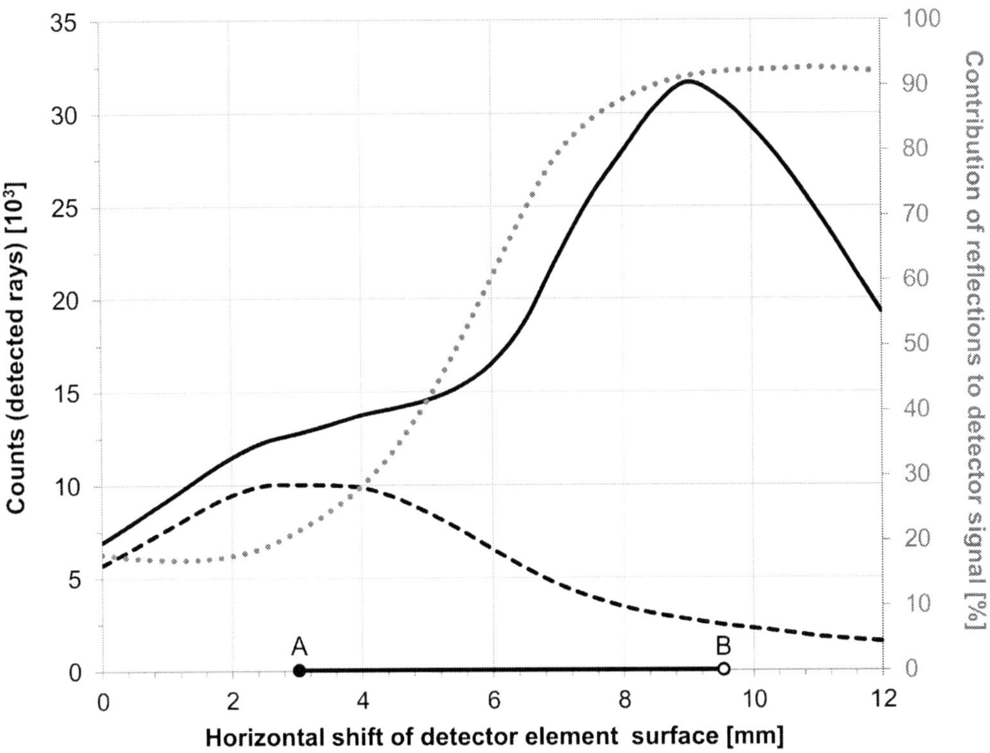

Figure 9.23. Cross section along the dashed horizontal line indicated in Figure 9.22, showing the percentaged contributions (dotted) of the probing radiation (dashed) to the total detector signal (solid). (A) location of maximal analytical signal and (B) projection of the maximal detector signal (B) on that horizontal line.

$$\mathcal{A} = -\log_{10}\frac{I(\nu)}{I_0(\nu)} = -\log_{10}\left\{\frac{1+\xi\cdot 10^{-\varepsilon(\nu)cl-\alpha(\nu)l}\cdot\left[\dfrac{1+\mathcal{R}(\nu)}{2\mathcal{R}(\nu)f(\nu)}-1\right]}{1+\xi\cdot 10^{-\alpha(\nu)l}\cdot\left[\dfrac{1+\mathcal{R}(\nu)}{2\mathcal{R}(\nu)f(\nu)}-1\right]}\right\} \tag{9.26}$$

The meaning of the symbols in Eq. (9.26) is explained in Figure 9.24. With 19 observables per analyte (i.e., the different concentrations), an overdetermined equation system arises (57 equations in total) with only 8 independent variables: 3 attenuation coefficients $\alpha(\nu)$ of the HWG at the 3 wavelengths where the analytes were evaluated, 3 molar absorption coefficients $\varepsilon_i(\nu)$ (1 per analyte i), 1 geometric alignment parameter ξ (which is used to describe the (unknown) detector position and the relative location and orientation of all components to each other in the existing setup), and 1 dimensionless constant m used for linear scaling of the single-channel background intensity spectrum. Hence, an optimization using least-

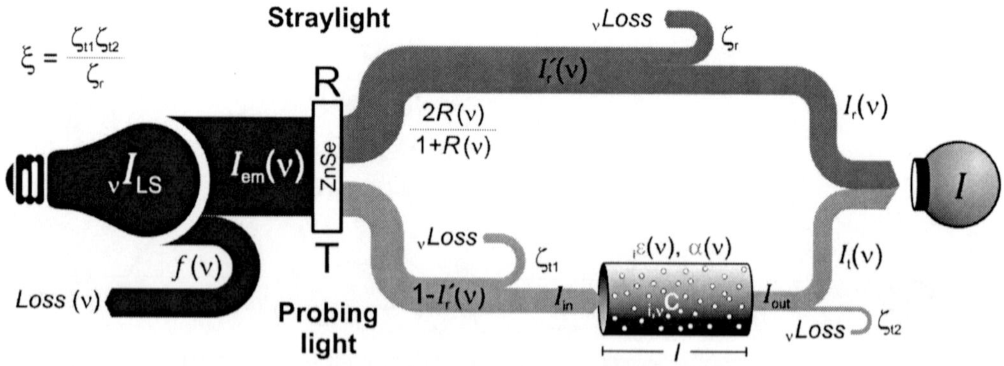

ξ	geometric alignment parameter (dependent on detector position and incoupled fraction)
ν	frequency
${}_{i}\varepsilon$	molar absorption coefficient of analyte i
${}_{i,\nu}c$	concentration of analyte i
l	length of HWG
α	attenuation coefficient of HWG
R	reflectance of ZnSe resulting from a dispersion equation(*)
f	sensor response function which is a dependent variable: $f(\nu) = f\{m \cdot P(\nu), \xi, \alpha(\nu), R(\nu), l\}$
P	single-channel background
m	constant.

Figure 9.24. Quantitative description of the radiation flux through the double-pass sensor from source to detector. A leading subscript ν indicates that this particular variable is frequency-dependent. *The reflectance of ZnSe was calculated using the dispersion equation given by Li (1984).

squares regression is feasible. Using those 8 optimized values, the nonlinearity may be restored ("theoretical" values in Figure 9.20) with remarkably close resemblance to the measured values, thus proving the simulation result (i.e., the prediction of stray light contributions) correct on a theoretical basis. Hence, stray light correction is not only legitimate, but certainly advisable. In analogy to the stray light correction for dispersive spectrometers (Ingle and Crouch 1988), Richardson et al. (1998b) suggested calculating a corrected transmittance $\mathcal{T}_c$ by using

$$\mathcal{T}_c = \frac{\mathcal{T}_m - \tau}{\mathcal{T} - \tau} \tag{9.27}$$

where $\mathcal{T}_m$ is the measured transmittance and τ is a correction factor. The corrected absorbance, $\mathcal{A}_c$, is then calculated using Eq. (9.19). The optimum values of τ for each analyte, i.e., the values resulting in the highest Pearson's correlation coefficients in the Beer's law plot, were estimated using all "measured" data points in Figure 9.20, and (as determined via the Microsoft Excel Solver add-in) resulting in $\tau \cong 62.5\%$ and $R^2 = 0.99993$ for methane and isobutylene, and $\tau \cong 72.1\%$ ($R^2 = 0.99990$) for cyclopropane. Those different τ values are explained as the sensor had been realigned in between those measurement series. The stray light–corrected values are plotted in Figure 9.20 as well and labeled as "corrected." The slope of those data points indicates the high sensitivity that would be achievable with such a sensor if the proximal ZnSe window had been, e.g., broad-band antireflection-coated or tilted to minimize the stray light contributions to the analytical signal.

4.2. OPTICAL TOLERANCE STUDY I: ASSESSING THE IN-COUPLING RATIO—SHARPNESS OF THE FOCAL POINT

Without proof, any spectrometer is usually expected or at least assumed to provide a perfectly collimated MIR beam. Simulations have corroborated the significance of that assumption by demonstrating how crucial the collimation of the beam in fact is for obtaining highly efficient in-coupling.

Figure 9.25 shows via visualization of the ray counts on a virtual screen placed right in front of the HWG in-coupling facet the focusing performance of the in-coupling OAPM in Figure 9.21 ("OAPM 1"). OAPMs are particularly sensitive to aberrations of the incident beam, and are characterized by rapidly decreasing focusing abilities with increasing incident beam divergence. A diverging beam angle θ_D (note that θ_D is twice the cone angle θ_{max} of the LS) of only 1° leads to a drop of the in-coupling ratio by one-third (Figure 9.25b). Assuming a divergent angle of 4° already results in an in-coupling ratio that drops below 5% (Figure 9.25d). The blurred focus not only results in a considerable decrease in energy throughput of the sensing system, it also prohibits the precise optical alignment of the sensor components, which is important for maximizing the radiation throughput.

4.3. OPTICAL TOLERANCE STUDY II: MIRROR MISALIGNMENT

A crucial component in the introduced double-pass sensor setup is the back-reflecting spherical mirror located at the distal end of the HWG. Figure 9.26 magnifies this particular mirror (q.v., the overview given with Figure 9.21). The depicted mirror is perfectly aligned along the horizontal axis, but tilted by 0.5° along the vertical axis. This minute misalignment results in already more than one-third (gray) of the 20 test rays (black) missing the HWG entrance facet. Furthermore, additional simulations yielded an intensity drop of 10% for an angular rotation of this mirror of only 0.1°. It is thus clear that the current sensor system responds

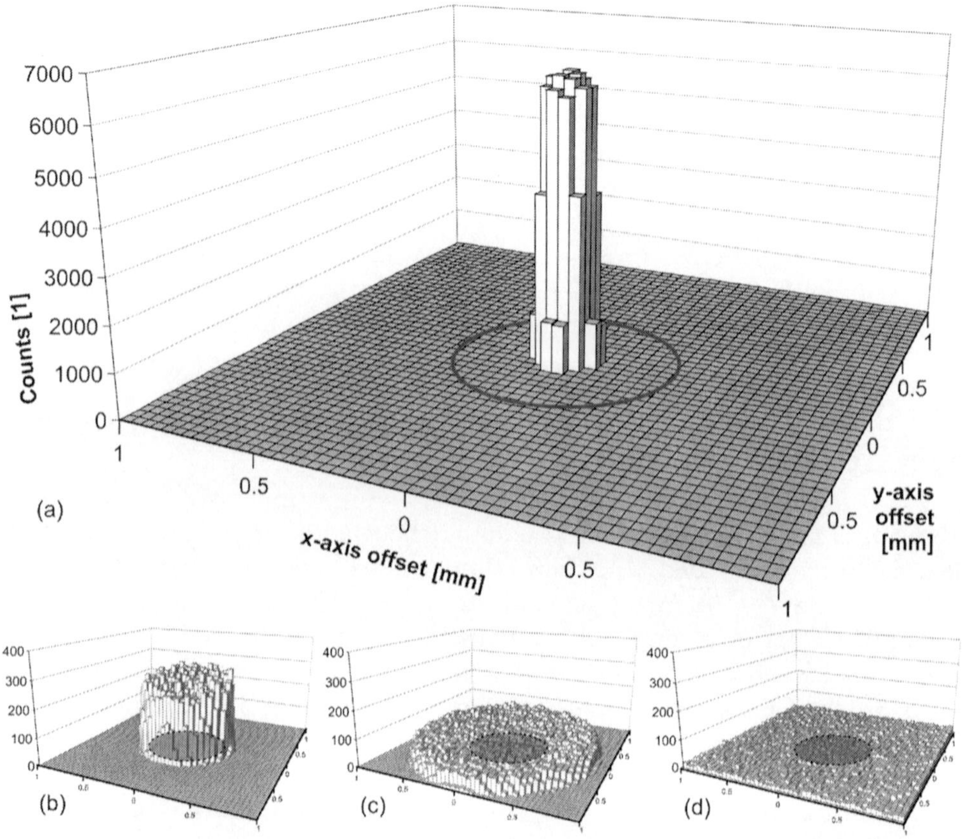

Figure 9.25. Focusing performance of the in-coupling OAPM ("OAPM 1" in Figure 9.21) with varying diverging angle of the LS. (a) 0.2°; (b) 1.0°; (c) 2.0°; (d) 4.0°. Simulated with a LS diameter of 1 cm. The ring in (a) and the circles (b–d) mark the internal diameter of this particular HWG (750 μm).

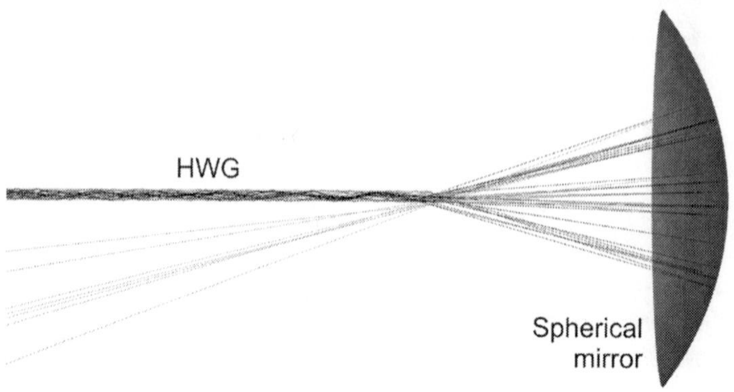

Figure 9.26. Misalignment of the spherical mirror installed in the double-pass sensor (compare Figure 9.21). Mirror tilted 0.5° downwards.

in a highly sensitive way to misalignment of this retro-reflecting mirror. While an intensity drop does not necessarily affect the finally obtained absorbance signal strength, any optical misalignment that reduces the total energy throughput of the sensor system increases the noise in the resulting absorbance spectra.

The signal-to-noise ratio (S/N) of an absoprtion measurement is given by

$$S/N = \frac{\bar{\mathcal{A}}}{\sigma_A} \tag{9.28}$$

where $\bar{\mathcal{A}}$ is the signal mean (or expected value) of the analyte absorption and σ_A is the standard deviation of the instrument noise or an estimate thereof. Misalignments affect the S/N [Eq. (9.28)], and increase the LOD, because S/N and LOD are related to each other; for details, refer to "The Orange Book" (Inczédy et al. 1998, chap. 18, subsec. 4.3.7). This fundamentally demonstrates how components of an optical sensor system may be individually optimized and evaluated for their qualitative impact on the radiation throughput or LOD of the entire sensor using such virtual models. In particular, tolerances in misalignment or novel geometries of selected components may be qualitatively tested prior to experiments.

4.4. SHINING LIGHT ON EXPERIMENTALLY CHALLENGING DOMAINS: EFFECTS OF DISPLACING THE IN-COUPLING OAPM

One of the main advantages of simulations is the fact that any scenery may be evaluated with an infinite number of virtual screens and detectors visualizing the ray distribution at any location of interest throughout the simulated optical setup. "Virtual" screens or detectors do not influence the propagation of rays (like the screen that was used in Example #1). Therefore, simultaneous information on any desired observable is provided, even at locations that are difficult or impossible to access to obtain experimental data (e.g., energy/intensity distributions inside an optical element). An example is the determination of the intensity distribution of radiation within a HWG, which is experimentally laborious in that it requires elaborate methods (Dahan et al. 1995) and is in fact impossible without affecting the further propagation of the radiation. On the other hand, MIR cameras providing the desired resolution are expensive devices and are usually not readily available in routine sensing laboratories for optimizing the sensor configuration.

Figure 9.27 displays the application of virtual screens to study the radiation intensity distribution inside HWGs at any location. Strategically positioned along the waveguide, virtual screens visualize the intensity distribution at any desired location. Off-center coupling via the OAPM unavoidably results in asymmetric intensity distributions within the core of the HWG, as radiation propagates to the

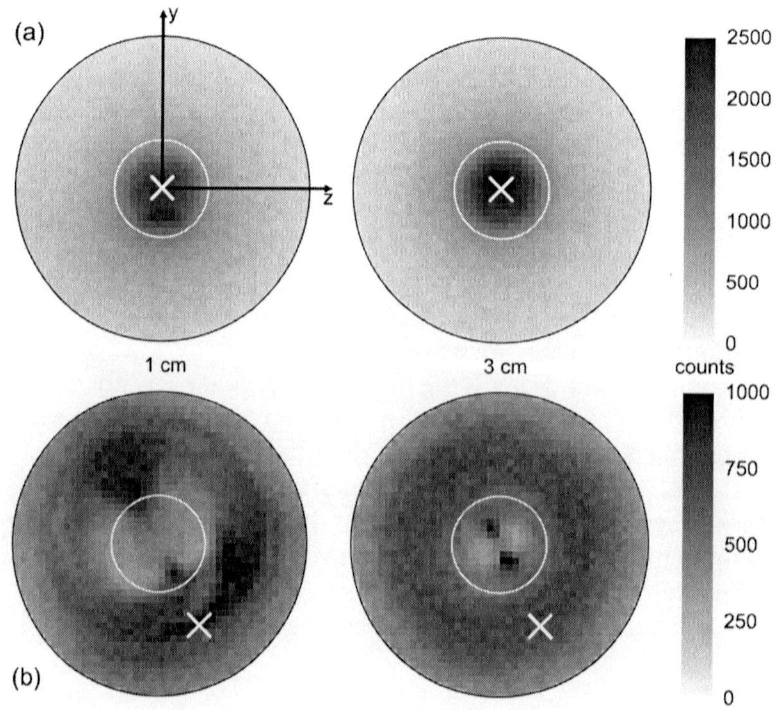

Figure 9.27. Intensity distribution inside a HWG. Collimated beam, 2-cm LS diameter, simulated with 10^6 rays in total. HWG internal diameter = 750 μm, ideal mirror interface. Images taken 1 cm and 3 cm behind the entrance. (a) HWG and OAPM perfectly aligned; (b) in-coupling OAPM shifted 200 μm in −*y* direction and 100 μm in +*z* direction. "counts" = ray density per 256 μm^2. X = focal point of the in-coupling OAPM, circle = centered 10% of the area of the HWG cross section.

distal end (Figure 9.27b). The drawn circles display 10% of the entire HWG cross-sectional area. Perfectly aligned components (Figure 9.27a) cause more than 30% of the overall radiation to be located within this area, while this fraction drops to roughly 11% with a displaced in-coupling OAPM. Note that, for the selected simulation parameters, the focal point is sharp enough not to affect the in-coupling ratio, which is still 100% (q.v. Figure 9.25a).

The fact that the radiation intensity inside the HWG is significantly less constricted to the center of the HWG in the case of misaligned components is not only accountable for a decreased throughput. It may become an even more important factor during the construction of "holey" HWGs for, e.g., passively (i.e., via diffusion) controlled gas-sensing applications, as the radiation-guiding capability is expected to decrease, the more the areas close to the wall are illuminated. A perforation of the HWG would permit gases to rapidly diffuse into the HWG, thereby allowing for online surveillance of, e.g., industrial gas lines, enabling the detection of leaks which would be observable as a change of the environmental air composition. Virtual screens could aid in finding the fewest illuminated positions which constitute the best locations to perforate a HWG.

4.5. EXTENDED SENSOR DESIGN EVALUATION: SIMULATION OF COMPLEX OBJECTS

Despite the advantages of HWGs, when applied as straight waveguides they remain space-consuming, which contradicts the objective of miniaturized IR sensing systems. Hence, in recent years research was focused on establishing more compact configurations by, e.g., coiling the waveguides. However, experiments have shown that coiling significantly increases the propagation losses, as documented by various papers discussing bending losses in curved HWG structures (Marhic 1981; Miyagi 1981; Saito et al. 1990; Haan et al. 1998; Haan and Harrington 1999; Young et al. 2007).

As SPRAY enables importing CAD data for establishing "complex objects" within a virtual optical setup, testing a wide variety of optical sensor configurations by modeling is facilitated, and may provide directions for future studies. To demonstrate this opportunity, Figure 9.28 shows—beside other objects—a coiled

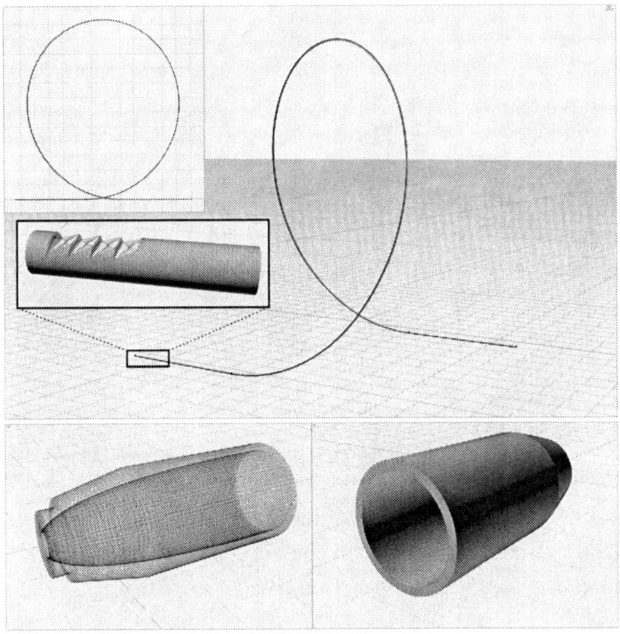

Figure 9.28. Various potentially interesting optical sensor elements rendered with MoI 3D. Top: coiled HWG; upper left corner: side view. The zoomed section shows a fiber grating in-coupling (assume the HWG would be a fiber instead). Bottom: Winston cone (rear and front views).

HWG with a total length of 50 cm, straight ends of 5 cm at each side, and an internal diameter of 750 μm. The coiled segment has a lateral offset of 10 cm, resulting in a bending radius of 6.16 cm.

However, coiled HWGs are only one of many object shapes that could be useful for improving optical IR sensors. As shown previously, it is experimentally delicate to focus on the input facet of HWGs using OAPMs. Fiber couplers and lenses are alternative solutions for increasing the alignment tolerance (Matsuura et al. 1997). Further—albeit more difficult to realize—concepts include tapered HWGs with widened ends (e.g., "horn-shaped" coupling antennas, Dossou et al. 2006), and Winston cones (the latter is the object depicted to the bottom of Figure 9.28). Speaking abstractly, a Winston cone is a cut-out segment of the outer surface generated by a parabola revolved around an axis that is not parallel to its symmetry axis (Winston and Hinterberger 1975). Practically, it is a nonimaging light-funneling device. Implemented in a chemical sensor setup, a Winston cone is used to collect all the MIR radiation incident on the large entrance aperture, and to concentrate the radiation onto the smaller exit aperture situated right in front of a HWG entrance facet. As such high-order surfaces are difficult to manufacture and therefore quite expensive, simulations allow one to (1) explore to what extent a Winston cone is capable of increasing both in-coupling ratio and alignment tolerances, and (2) to establish appropriate dimensions.

In the field of optical chemical sensors in general, solid-core fibers and planar waveguides are of increasing relevance as transducer elements, as they enable tailoring the molecule–photon interface to the needs of the application. Small waveguide structures guide only a few discrete wavemodes. In order to excite these modes, MIR radiation may illuminate the cleaved waveguide ends directly. However, efficient butt coupling between LS and waveguide relies crucially on the surface quality of the cleavages. Alternatively, the in- and out-couplings of a guided mode to these transducer elements may be achieved by diffraction gratings (Figure 9.28, zoomed section) (Waldhäusl et al. 1997; Hecht 2002). Such structures may not only bypass the problem discussed above, they may allow for miniaturization as well by replacing classic lenses and mirrors. Moreover, gratings serve as highly selective coupling filters for specific wavelengths at a given angle, and allow for selecting discrete polarization states.

All the objects in Figure 9.28 were constructed utilizing the shareware "Moment of Inspiration," Version 1.0 (MoI; Triple Squid Software Design, Seattle, WA, USA) during the software trial period. Three-dimensional shapes generated with MoI are natively stored in the 3dm-file format, which uses NURBS (Non-Uniform Rational B-Splines) for representation. However, complex objects in SPRAY demand a triangular mesh (stl-file format), which requires that the 3-D object be composed of molded triangles. A model that represents a curved object with sufficient quality therefore demands a possibly small mesh angle, and consequently many triangles. "Sufficient quality" for optical simulations translates as a surface smooth enough to obtain realistic results upon the incidence of rays during the ray-tracing simulation. In the following, the minimum number of triangles was determined to yield a sufficiently fine mesh for the representation of the coiled HWG mentioned above. This estimation is crucially important, as simply using more triangles than essential will lead to increased computational effort.

Figure 9.29 depicts the overlayed end cross sections of cylinders meshed with n triangles. It is evident that every single distance between two reflections of a ray emerging from the center is shorter than the diameter of the circumscribing circle. Consequently, this ray needs more reflections to exit. On the other hand, with an absorbing inner surface, the (coiled) HWG throughput is expected to increase with a reduced number of reflections.

That said, a number of straight HWGs of 50-cm length, 750-μm internal diameter, and 99% reflecting surface were generated using mesh angles of 60°, 30°, 10°, 2°, and 1°. A SPRAY simulation was executed with a pointy LS (radius = 1 μm) located centered at a distance of 100 μm in front of the in-coupling facet of the HWG coil. As a next step, a constant 99% reflecting interface [i.e., $\alpha = 0.01$; compare Eqs. (9.17)–(9.19)] was assigned to the inner surface of the HWG. Varying the divergent angle θ_D of the LS, the radiative power output at the distal end of the coil was compared to the theoretical transmittance $\mathcal{T}$ of a straight HWG (i.e., a perfect cylindrical shape) with the same optical properties and length, which is given by

$$\mathsf{T} = \frac{I}{I_0} = \frac{1}{1-\cos(\theta_{\max})}\int_0^{\theta_{\max}} (1-\alpha)^{l\tan(\theta)/d} \sin(\theta)\, d\theta \tag{9.29}$$

where I is power out, I_0 is power in, $\theta_{\max}$ is the cone angle of the LS, α is the absorption coefficient, l is the length of the HWG, and d is the internal diameter of the HWG. It should be noted that θ_D equals twice the cone angle of the LS.

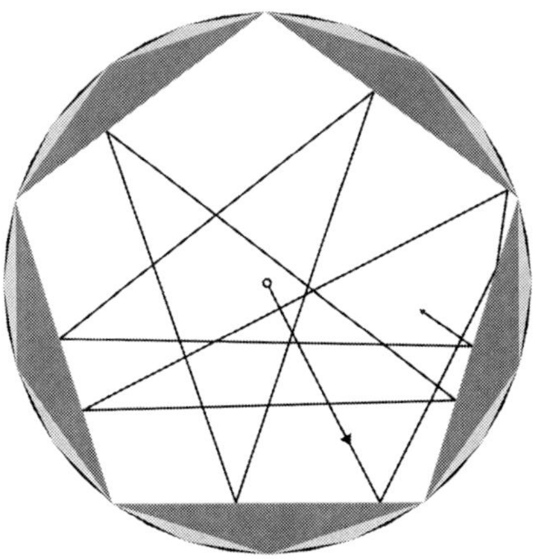

Figure 9.29. Mesh quality. First 10 reflections of a random ray emerging from the center. Mesh angle/number of triangles a cylinder consists of: white = 72°/10, dark gray = 36°/20, light gray = 18°/40, black = 9°/80.

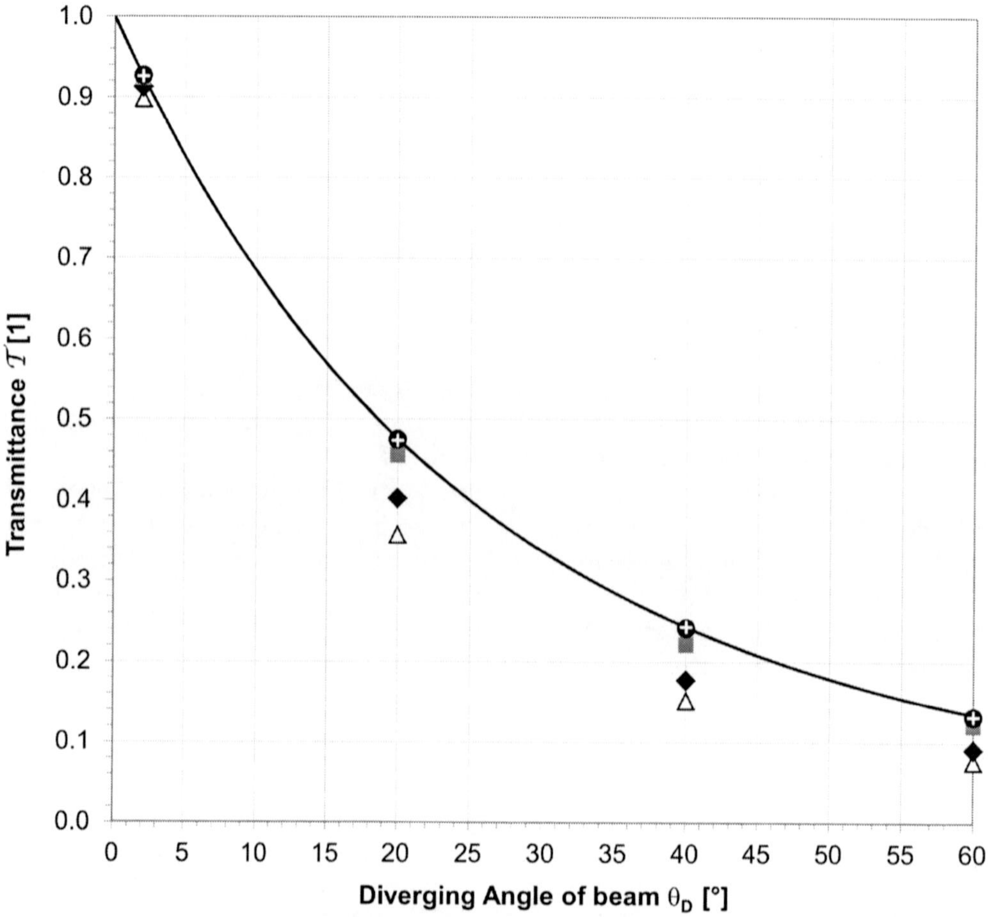

Figure 9.30. Smoothing options of a straight HWG. 50-cm length, 750-μm internal diameter; simulated with a 99% reflecting surface, 1-μm LS diameter and 10^5 inserted rays per data point. Mesh angle: white triangle = 60° (14 polygons), black rhomb = 30° (26 polygons), gray square = 10° (76 polygons), black circle = 2° (640 polygons), white cross = 1° (1280 polygons), solid line = result of Eq. (9.29).

Equation (9.29) can be solved numerically. This marks the theoretical optimum (upper limit) of the transmitted fraction. The result is shown in Figure 9.30. It is not surprising that the simulated throughput converges to the theoretical value for all angles. The surface smoothness of the mesh can be estimated by comparison with the simulation results in Figure 9.30. Consequently, a mesh, angle of 2° seems to be "practicably close" for defining a straight HWG, and is expected to be sufficient for the coiled HWG as well.

Accordingly, the coiled HWG was exported from MoI using a mesh angle of 2°, which still results in 206,080 polygons. Figure 9.31 illustrates the findings of the ray tracing with the coiled HWG. The well-known negative effect of bending the HWG as described in pertinent literature (Marhic 1981; Miyagi 1981; Saito et al. 1990; Haan et al. 1998; Haan and Harrington 1999; Young et al. 2007) is clearly evident, as the transmittance stays below the values of a straight HWG meshed with a 2° mesh angle (compare Figure 9.30) for all cone angles that were simulated. Moreover, the bending loss appears to be a function of the in-coupling conditions as well, and depends on the divergent angle, which itself is an approximation for different focal lengths (FLs) of the focusing OAPM, usually positioned prior to the HWG in-coupling facet. The results shown in Figure 9.31 provide strong evidence that there may be an optimum FL for a given bending radius, as the additional loss declines significantly at 20°, where the throughput is still 90% of a straight HWG.

It should be noted that the present model may be too simple, as it neglects the frequency dependence of the reflectivity of both the metal substrate and the dielectric layer. However, due to the generally straightforward approach for constructing even complex virtual objects, a wide variety of different materials/layers

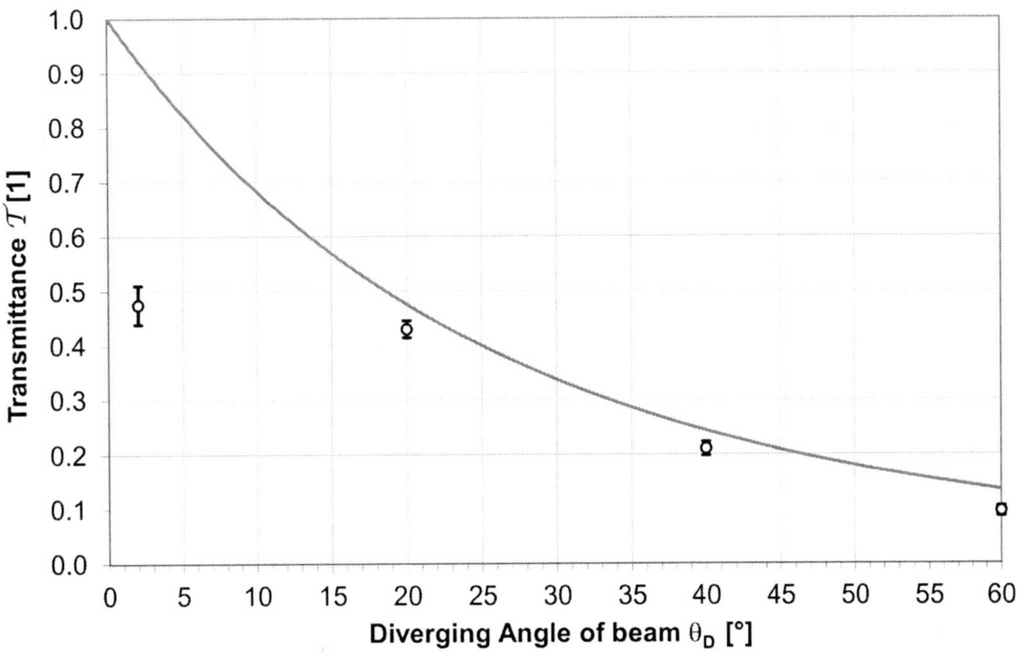

Figure 9.31. Throughput drop in a coiled HWG with increasing divergence of the in-coupled radiation. Individual data points = transmittance of the coiled HWG, gray curve = theoretical transmittance of the corresponding straight HWG [Eq. (9.29)]. The error bars indicate the remaining simulation uncertainty (i.e., standard deviation of the corresponding binomial distribution) due to the number of inserted rays.

may be assigned to these structures, thereby providing a promising path toward developing entirely novel sensor concepts during simulation prior to usually time- and cost-consuming experiments. Concerning optical geometries such as, e.g., coiled HWGs, for experimental studies an optimum between compactness of the active transducer (i.e., HWG) and optical performance has to be established. To our knowledge, no systematic simulation studies have been published that pinpoint optimized optical parameters (length, diameter of the HWG, coil diameter, etc.). Here, we have shown the feasibility in principle of implementing coiled HWGs, which should enable addressing more complex HWG systems during future studies using the same approach.

Only four data points were simulated in this preliminary study (Figure 9.31), as the computational effort is very high due to the large number of polygons simulating the coiled HWG surface, resulting in approx. 23 h to simulate each data point. It should be noted that already-implemented calculation-accelerating simplifications like the ones used in Example #3 cannot be adopted for scenerios with unreducable complexity (e.g., a coiled HWG cannot be shortened). The only possible solution to accelerate simulations for such scenarios is an increase in calculation power, which may be achieved, e.g., by using the ray-tracing software in a distributed computing mode.

The reason for the increased computation times is that the ray-tracing algorithm of SPRAY itself is based on object-orientated programming. This requires that *all* objects have to decide for each emitted ray individually, whether they get hit by that ray or not, prior to selecting the closest hit point (Theiss 2008a). This is believed to be the key to the main limiting factor in calculation speed of the present algorithm. Indeed, the majority of individual polygons building a complex object (like the coiled HWG, Figure 9.28) could be excluded from the decision, as it is completely unlikely for them to get hit *first*. The same is true for all triangles within sections of the HWG through which the ray has already traveled. Consequently, optimizing the calculation speed may require implementation of a prescreening algorithm for the established virtual sensor scenery, and preferably reducing the number of interfacial calculations to rays that ultimately deliver analytical information to the virtual detector to maximize computational efficiency. It should be noted that SPRAY is of limited suitability for executing this simulation, as it is not designed for processing millions of rays through scenarios containing complex objects composed of hundreds of thousands of polygons. It may be anticipated that using mainstream commercial forward ray tracers such as ASAP (Breault Research Organization, Tucson, AZ, USA), LightTools (Synopsys Inc., Mountain View, CA, USA), TracePro (Lambda Research Corp., Littleton, MA, USA), or ZEMAX (Radiant ZEMAX LLC, Bellevue, WA, USA), to name just a few, could enable simulations of MIR chemical sensors with a high level of complexity, especially if they can handle the import of spline curve–based CAD file exchange formats. On the other hand, all those ray-tracing engines may reveal their own downsides and limitations as well (e.g., spectrally resolved simulations might not be standard applications, hence difficult to implement). Hence, one must be aware that an optimal solution to all problems hardly exists; the usability of a specific illumination software package depends on the purpose.

Finally, it may be concluded that any simulation may be further refined to more closely match a particular experiment. However, more experimental parameters would be required, including, e.g., the spectral radiation density distribution of the MIR source (a SiC filament in the case of a FTIR spectrometer), as well as its precise angular intensity distribution. Once these characteristics are known, the right filters may be applied to simulate the LS more correctly. Likewise, any nonlinearity of the detector may be modeled if it is established on a sound experimental basis. The quality of the established models is currently not limited by the ideas of refinement, but current limitations on the experimental knowledge of experimental parameters and efficient use of available computational power limit the possibilities for improvement of the models. With the ongoing progress of computer development, future simulations may include refined models of HWGs and other optical components, thereby providing highly detailed models of entire sensor setups with reduced deviation from experimental results.

5. CONCLUSIONS AND OUTLOOK

Following the description of basic ray-tracing principles, this chapter discussed universal model generation and model validation techniques as illustrated by visualizing the active sensing regions responsible for signal generation on conventional horizontal attenuated total reflection crystals. In addition, an optical ray-tracing routine for simulating broad-band IR absorbance spectra of individual gases and constituent mixtures generated for spectrally resolved modeling of entire hollow waveguide–based sensing devices was established.

Virtual analytes were created by deriving the complex refractive indices of corresponding mid-infrared spectroscopic data based on a refined harmonic oscillator model, and subsequent assignment of these dielectric responses to virtual analytes inside a virtual 3-D replica of the optical sensing module. It was shown that absorbance spectra reproduced by virtual spectrometers can be in excellent agreement with experimental data obtained with corresponding real-world sensing systems. Moreover, we demonstrated that modeling a single spectrum per analyte is sufficient to generate a virtual spectrum of any desired concentration of this molecule, and any arbitrarily selected mixture with other analytes, which paves the way toward generating modeled mixture spectra for virtual multivariate calibrations.

By assigning the spectral loss function of a leaky HWG to its hollow core, a convenient and easily implementable virtual representation of HWGs was achieved, which emulates the given HWG-based IR optical gas sensor with sufficient accuracy for fundamental design and analytical response studies. Accompanied by a critical in-depth review of potential simplifications and limitations of the current model, solutions to refine the described model have been derived.

The chapter was completed by highlighting illustrative application examples guiding the development of future sensing concepts. For example, simulations exposing stray light effects as the origin of deviations from Beer's law in a double-pass sensor configuration demonstrate the utility of 3-D spectroscopic simulations. Furthermore, the analytical robustness of MIR sensors was studied for a variety of optical configurations. Evidently, the collimation quality of the IR beam is critical for efficient radiation transfer into the active sensing module along with the effects of angular and/or directional misalignments.

Finally, complementary ray-tracing studies on a variety of optical components with complex geometries illustrate the utility of optical simulations for sensing systems where both measurable properties and parameters that are experimentally difficult to address may be computed and visualized. In summary, it has been shown that for detailed optimization of optical sensors and their spectral response, the physical dimensions and the optical response functions of the components composing the sensor system, including the analytes of interest, are required for virtual sensor modeling complementary to experimental sensor design. However, it has also been shown that the quality of the simulation is crucially dependent on the quality of and access to detailed experimental parameters, and on validation of thus-derived models via accurate experiments.

6. FUTURE TRENDS

The current achievements discussed in this chapter indicate that simplified models of HWG-based sensors may be transferable to even more complex sensing scenarios, as the quality of the established models is currently limited only by computational resources. Consequently, the development of improved models of HWGs and other optical components is in progress to minimize the deviation from corresponding experimental results. It is anticipated that future applications of optical sensor simulations including their spectroscopic response will provide direct access to a system response function providing directions for optimization or alternative sensing configurations prior and complementary to experiments. As even complex virtual objects may be defined, a wide variety of different materials or layered coatings may be assigned to these structures, thereby potentially enabling predictive modeling of entirely novel sensor concepts via simulation, thus substituting for time- and cost-consuming experiments.

Although currently applied ray optics–based models are not yet exhausted in their capabilities, simulating sensor setups of similar complexity using wave optics models are of significant interest. Furthermore, 3-D simulations may also aid in gas-phase sensor design by incorporating single-mode radiation propagation and photonic band-gap materials, which have lately also been implemented as HWG materials. Finally, future studies may focus on refined virtual mixture spectra in order to generate scenarios close to complex real-world sample matrices for testing, optimizing, and selecting appropriate multivariate data analysis routines. Consequently, it is anticipated that virtual calibrations may provide a useful addition to real-world datasets for increasing the robustness of the calibration, if the studies currently in progress verify that the calibration sets of actual sensors may usefully be supplemented by simulated spectra.

ACKNOWLEDGMENTS

The authors would like to express thanks to Dr. Wolfgang Theiss (W. Theiss Hard- and Software, Aachen, Germany) for his assistance with SPRAY and SCOUT. Especially, the authors would like to thank Dr. Chance Carter at Lawrence Livermore National Laboratory for his unremitting support, his ideas, and all the fruitful discussions.

This work was performed in part under the auspices of the U.S. Department of Energy by the University of California, Lawrence Livermore National Laboratory, under contract No. W-7405-Eng-48. This project was funded by the Laboratory Directed Research and Development Program at LLNL under sub-contract Nos. B565491, B590992, and B598643.

REFERENCES

Allen M.P. and Tildesley D.J. (1987) *Computer Simulation of Liquids.* Oxford University Press, Oxford, UK.

Axelrod D., Burghardt T.P., and Thompson N.L. (1984) Total internal reflection fluorescence. *Ann. Rev. Biophys. Bioeng.* **13**, 247–268. DOI: 10.1146/annurev.bb.13.060184.001335

Barshop B.A., Wrenn R.F., and Frieden C. (1983) Analysis of numerical methods for computer simulation of kinetic processes: Development of KINSIM—A flexible, portable system. *Anal. Biochem.* **130**(1), 134–145. DOI: 10.1016/0003-2697(83)90660-7

Bell H.H. and Waag W.L. (1998) Evaluating the effectiveness of flight simulators for training combat skills: A review. *Int. J. Aviat. Psychol.* **8**(3), 223–242. DOI: 10.1207/s15327108ijap0803_4

Bierhoff T., Wallrabenstein A., Himmler A., Griese E., and Mrozynski G. (2001) Ray tracing technique and its verification for the analysis of highly multimode optical waveguides with rough surfaces. *IEEE Trans. Magnet.* **37**(5–1), 3307–3310. DOI: 10.1109/20.952601

Binder K. and Heermann D.W. (2002) *Monte Carlo Simulation in Statistical Physics,* 4th ed. Springer-Verlag, Berlin. DOI: 10.1007/978-3-642-03163-2

Bjerke A.E., Griffiths P.R., and Theiss, W. (1999) Surface-enhanced infrared absorption of Co on platinized platinum. *Anal. Chem.* **71**(10), 1967–1974. DOI: 10.1021/ac981093u

Born M. and Wolf E. (1999) *Principles of Optics: Electromagnetic Theory of Propagation, Interference and Diffraction of Light,* 7th ed. Cambridge University Press, Cambridge, UK.

Bourlioux A. and Gander M.J. (eds.) (2002) *Modern Methods in Scientific Computing and Applications.* Kluwer, Dordrecht, The Netherlands.

Brereton R.G. (2007) *Applied Chemometrics for Scientists.* John Wiley, Chichester, UK. DOI: 10.1002/9780470057780

Croitoru N., Dror J., Goldenberg E., Mendlovic D., and Ruschin S. (1987) Use of metallic and dielectric films for hollow fibers. *Fiber Int. Opt.* **6**(4), 347–361. DOI: 10.1080/01468038708222974

Crowe C.M. (1971) *Chemical Plant Simulation: An Introduction to Computer-Aided Steady-State Process Analysis.* Prentice-Hall, Englewood Cliffs, NJ.

Dahan R., Dror J., Inberg A., and Croitoru N. (1995) Nondestructive method for attenuation measurements in optical hollow waveguides. *Opt. Lett.* **20**(14), 1536–1538. DOI: 10.1364/OL.20.001536

Dai F., Felger W., Frühauf T., Göbel M., Reiners D., and Zachmann G. (1996) Virtual prototyping examples for automotive industries. *Proc. Virt. Real. World 1996.*

Davies J.R. and Mendonça J.T. (2000) Basic physics of laser propagation in hollow waveguides. *Phys. Rev. E* **62**(5), 7168–7180. DOI: 10.1103/PhysRevE.62.7168

Dobbs G.T. and Mizaikoff B. (2006) Shining new light on old principles: Localization of evanescent field inter-actions at infrared-attenuated total reflection sensing interfaces. *Appl. Spectrosc.* **60**(6), 573–583. DOI: 10.1366/000370206777670585

Dossou K., Botten L.C., de Sterke C.M., McPhedran R.C., Asatryan A.A., Chen S., and Brnovic J. (2006) Efficient couplers for photonic crystal waveguides. *Opt. Commun.* **265**(1), 207–219. DOI: 10.1016/j.optcom.2006.02.052

Edwards S., Lavagno L., Lee E.A., and Sangiovanni-Vincentelli A. (1997) Design of embedded systems: Formal models, validation, and synthesis. *Proc. IEEE* **85**(3), 366–390. DOI: 10.1109/5.558710

Evans J.W., Sanders D.E., Thiel P.A., and DePristo A.E. (1990) Low-temperature epitaxial growth of thin metal films. *Phys. Rev. B* **41**(8), 5410–5413. DOI: 10.1103/PhysRevB.41.5410

Fetzer G.J., Pittner A.S., Ryder W.L., and Brown D.A. (2002) Tunable diode laser absorption spectroscopy in coiled hollow optical waveguides. *Appl. Opt.* **41**(18), 3613–3621. DOI: 10.1364/AO.41.003613

Fraden J. (2010) *Handbook of Modern Sensors: Physics, Designs, and Applications,* 4th ed. Springer-Verlag, New York. DOI: 10.1007/978-1-4419-6466-3

Freddolino P., Arkhipov A., Larson S.B., McPherson A., and Schulten K. (2006) Molecular dynamics simulations of the complete satellite tobacco mosaic virus. *Structure* **14**(3), 437–449. DOI: 10.1016/j.str.2005.11.014

Fritzsche H.-T. (1994) A model for traffic simulation. *Traffic Eng. Control* **35**(5), 317–321.

Gannot I., Ben-David M., Inberg A., Croitoru N., and Waynant R.E. (2002) Beam shape analysis of waveguide delivered infrared lasers. *Opt. Eng.* **41**(1), 244–250. DOI: 10.1117/1.1420191

Garmire E., McMahon T., and Bass M. (1976) Propagation of infrared light in flexible hollow waveguides. *Appl. Opt.* **15**(1), 145–150. DOI: 10.1364/AO.15.000145

Gentle J.E. (2003) *Random Number Generation and Monte Carlo Methods*, 2nd ed. Springer-Verlag, New York. DOI: 10.1007/b97336

Gilbert N. (1995) *Artificial Societies: The Computer Simulation of Social Life*. UCL Press, London.

Gould H., Tobochnik J., and Christian W. (2006) *An Introduction to Computer Simulation Methods: Applications to Physical Systems*, 3rd ed. Addison-Wesley, San Francisco.

Griffiths P. and De Haseth J.A. (2007) *Fourier Transform Infrared Spectrometry*, 2nd ed. John Wiley, Hoboken, NJ. DOI: 10.1002/047010631X

Günzler H. and Gremlich H.-U. (2002) *IR Spectroscopy: An Introduction*. Wiley-VCH, Weinheim, Germany.

Haan D.J., Gibson D.J., Rabii C.D., and Harrington J.A. (1998) Coiled hollow waveguides for gas sensing. *Proc. SPIE* **3262**, 125--129. DOI: 10.1117/12.309465

Haan D.J. and Harrington J.A. (1999) Hollow waveguides for gas sensing and near-IR applications. *Proc. SPIE* **3596**, 43–49. DOI: 10.1117/12.346725

Haile J.M. (1997) *Molecular Dynamics Simulation: Elementary Methods*. John Wiley, Chichester, UK.

Han Z. and Reitz R.D. (1995) Turbulence modeling of internal combustion engines using RNG κ–ε models. *Combustion Sci. Technol.* **106**(4–6), 267–295. DOI: 10.1080/00102209508907782

Harrington J.A. (2004) *Infrared Fibers and Their Applications*. SPIE, Bellingham, WA. DOI: 10.1117/3.540899

Harvey D. (2000) *Modern Analytical Chemistry*. McGraw-Hill, Boston.

Hasegawa T., Nishijo J., Umemura J., and Theiss W. (2001) Simultaneous evaluation of molecular-orientation and optical parameters in ultrathin films by oscillators—Model simulation and infrared external-reflection spectrometry. *J. Phys. Chem. B* **105**(45), 11178–11185. DOI: 10.1021/jp011739d

Hecht E. (2002) *Optics*, 4th ed. Addison-Wesley, San Francisco.

Heermann D.W. (1986) *Computer Simulation Methods in Theoretical Physics*. Springer-Verlag, Berlin.

Hilbrich S., Arens-Fischer R., Küpper L., Theiß W., Berger M.G., Krüger M., and Thönissen M. (1997) The application of porous silicon interference filters in optical sensors. *Thin Solid Films* **297**(1–2), 250–253. DOI: 10.1016/S0040-6090(96)09364-9

Hoover W.G. (1986) *Molecular Dynamics*. Springer-Verlag, Berlin.

Ibanescu M., Johnson S.G., Soljačić M., Joannopoulos J.D., Fink Y., Weisberg O., Engeness T.D., Jacobs S.A., and Skorobogatiy M. (2003) Analysis of mode structure in hollow dielectric waveguide fibers. *Phys. Rev. E* **67**(4), 046608/1–046608/7. DOI: 10.1103/PhysRevE.67.046608

Inczédy J., Lengyel T., and Ure A.M. (eds.) (1998) *IUPAC Compendium of Analytical Nomenclature (Definitive Rules 1997)*, 3rd ed. Blackwell Science, Oxford, UK.

Ingle J.D. and Crouch S.R. (1988) *Spectrochemical Analysis*. Prentice Hall, Englewood Cliffs, NJ.

Iwai K., Abe Y., Matsuura Y., and Miyagi M. (2002) Equivalent complex refractive index for ray-tracing evaluation of dielectric-coated hollow waveguides. *Opt. Eng.* **41**(7), 1471–1472. DOI: 10.1117/1.1485090

Jackson J.D. (1999) *Classical Electrodynamics*, 3rd ed. John Wiley, New York.

Karbowiak A.E. (1955) Theory of imperfect waveguides: The effect of wall impedance. *Proc. IEE B* **102**, 698–708.

Keating M.P. (2002) *Geometric, Physical, and Visual Optics*, 2nd ed. Butterworth-Heinemann, Woburn, MA.

Kim C.C., Garland J.W., Abad H., and Raccah P.M. (1992) Modeling the optical dielectric function of semiconductors: Extension of the critical-point parabolic-band approximation. *Phys. Rev. B* **45**(20), 11749–11767. DOI: 10.1103/PhysRevB.45.11749

Kim S.-S., Menegazzo N., Young C., Chan J., Carter C., and Mizaikoff B. (2009) Mid-infrared trace gas analysis with single-pass FT-IR hollow waveguide gas sensors. *Appl. Spectrosc.* **63**(3), 331–337. DOI: 10.1366/000370209787598924

Kotelyanskii M. and Theodorou D.N. (eds.) (2004) *Simulation Methods for Polymers.* Marcel Dekker, New York.

Kozodoy R.L., Micheels R.H., and Harrington J.A. (1996) Small-bore hollow waveguide infrared absorption cells for gas sensing. *Appl. Spectrosc.* **50**(3), 414–417. DOI: 10.1366/0003702963906393

Kroese D.P., Taimre T., and Botev Z.I. (2011) *Handbook of Monte Carlo Methods.* John Wiley, Hoboken, NJ. DOI: 10.1002/9781118014967

Landau D.P. and Binder K. (2000) *Monte Carlo Simulations in Statistical Physics,* 2nd ed. Cambridge University Press, New York.

Laqua K., Melhuish W.H., and Zander M. (1988) IUPAC Publications on Nomenclature, Symbols, Units and Their Usage in Spectrochemical Analysis. Part VII: Molecular Absorption Spectroscopy, Ultraviolet and Visible (UV/VIS). *Pure Appl. Chem.* **60**(9), 1449–1460. DOI: 10.1351/pac198860091449

Larkin, P.J. (2011) *IR and Raman Spectroscopy; Principles and Spectral Interpretation.* Elsevier, Amsterdam.

Li H.H. (1984) Refractive index of ZnS, ZnSe, and ZnTe and its wavelength and temperature derivatives. *J. Phys. Chem. Ref. Data* **13**(1), 103–150. DOI: 10.1063/1.555705

Lines M.E. and Klocek P. (1998) Optical transmission theory. In: Sanghera J.S. and Aggarwal I.D. (eds.), *Infrared Fiber Optics.* CRC Press, New York, 1–30.

Loshin D.S. (1991) *The Geometrical Optics Workbook.* Butterworth-Heinemann, Newton, MA.

Love J.D., Winkler C., Sammut R., and Barrell K. (1978) Leaky modes and rays on multi-mode step index waveguides—A comparison. *Electron. Lett.* **14**(15), 489–490. DOI: 10.1049/el:19780328

Marcatili E.A.J. and Schmeltzer R.A. (1964) Hollow metallic and dielectric waveguides for long distance optical transmission and lasers. *Bell Syst. Tech. J.* **43**, 1783–1809.

Marcatili E.A.J. (1966) Light transmission in a multiple dielectric (gaseous and solid) guide. *Bell Syst. Tech. J.* **45**(1), 97–103.

Marhic M.E. (1981) Mode-coupling analysis of bending losses in IR metallic waveguides. *Appl. Opt.* **20**(19), 3436–3441. DOI: 10.1364/AO.20.003436

Mark H. and Workman J. Jr. (2007) *Chemometrics in Spectroscopy.* Academic Press, Amsterdam.

Marksteiner S., Savage C.M., Zoller P., and Rolston S.L. (1994) Coherent atomic waveguides from hollow optical fibers: Quantized atomic motion. *Phys. Rev. A* **50**(3), 2680–2690. DOI: 10.1103/PhysRevA.50.2680

Matsuura Y., Hiraga H., Wang Y., Kato Y., Miyagi M., Abe S., and Onodera S. (1997) Lensed-taper launching coupler for small-bore, infrared hollow fibers. *Appl. Opt.* **36**(30), 7818–7821. DOI: 10.1364/AO.36.007818

Mendlovic D., Goldenberg E., Ruschin S., Dror J., and Croitoru N. (1989) Ray model for transmission of metallic dielectric hollow bent cylindrical waveguides. *Appl. Opt.* **28**(4), 708–712. DOI: 10.1364/AO.28.000708

Miller J.N. and Miller J.C. (2005) *Statistics and Chemometrics for Analytical Chemistry*, 5th ed. Prentice Hall, Harlow, UK.

Miyagi M. (1981) Bending losses in hollow and dielectric tube leaky waveguides. *Appl. Opt.* **20**(7), 1221–1229. DOI: 10.1364/AO.20.001221

Miyagi M. (1985) Waveguide-loss evaluation in circular hollow waveguides and its ray-optical treatment. *J. Lightwave Technol.* **3**(2), 303–307. DOI: 10.1109/JLT.1985.1074177

Miyagi M., Hongo A., and Kawakami S. (1983) Transmission characteristics of dielectric-coated metallic waveguides for infrared transmission: Slab waveguide model. *IEEE J. Quantum Electron.* **19**, 136–145. DOI: 10.1109/JQE.1983.1071837

Miyagi M. and Kawakami S. (1983) Waveguide loss evaluation by the ray-optics method. *J. Opt. Soc. Am.* **73**(4), 486–489. DOI: 10.1364/JOSA.73.000486

Morhaim O., Mendlovic D., Gannot I., Dror J., and Croitoru N. (1991) Ray model for transmission of infrared radiation through multibent cylindrical waveguides. *Opt. Eng.* **30**(12), 1886–1891. DOI: 10.1117/12.56016

Mouroulis P. and Macdonald J. (1997) *Geometrical Optics and Optical Design.* Oxford University Press, New York.

Otto, M. (2007) *Chemometrics: Statistics and Computer Application in Analytical Chemistry*, 2nd ed. Wiley-VCH, Weinheim, Germany.

Ottosson S. (2002) Virtual reality in the product development process. *J. Eng. Design* **13**(2), 159–172. DOI: 10.1080/09544820210129823

Patnaik P. (2004) *Dean's Analytical Chemistry Handbook*, 2nd ed. McGraw-Hill, New York.

Rauh J. (2003) Virtual development of ride and handling characteristics for advanced passenger cars. *Vehicle Syst. Dynam.* **40**(1–3), 135–155. DOI: 10.1076/vesd.40.1.135.15876

Richardson R.L. Jr., Yang H., and Griffiths P.R (1998a) Effects of detector nonlinearity on spectra measured on three commercial FT-IR spectrometers. *Appl. Spectrosc.* **52**(4), 572–578. DOI: 10.1366/0003702981943897

Richardson R.L. Jr., Yang H., and Griffiths P.R (1998b) Evaluation of a correction for photometric errors in FT-IR spectrometry introduced by a nonlinear detector response. *Appl. Spectrosc.* **52**(4), 565–571. DOI: 10.1366/0003702981943888

Robert C.P. and Casella G. (2010) *Monte Carlo Statistical Methods.* Springer-Verlag, New York.

Rouessac F. and Rouessac A. (2007) *Chemical Analysis: Modern Instrumentation Methods and Techniques*, 2nd ed. John Wiley, Chichester, UK.

Rubinstein R.Y. and Kroese D.P. (2008) *Simulation and the Monte Carlo Method*, 2nd ed. John Wiley, Hoboken, NJ.

Sadus R.J. (1999) *Molecular Simulation of Fluids.* Elsevier Science, Amsterdam.

Saito M., Matsuura Y., Kawamura M., and Miyagi M. (1990) Bending losses of incoherent light in circular hollow waveguides. *J. Opt. Soc. Am. A* **7**(11), 2063–2068. DOI: 10.1364/JOSAA.7.002063

Saleh B.E.A. and Teich M.C. (2007) *Fundamentals of Photonics,* 2nd ed. John Wiley, Hoboken, NJ.

Skoog D.A., Holler F.J., and Crouch S.R. (2007) *Principles of Instrumental Analysis,* 6th ed. Thomson Brooks/Cole, Belmont, MA.

Smith R. (ed.) (1997) *Atomic and Ion Collisions in Solids and at Surfaces: Theory, Simulation and Applications.* Cambridge University Press, Cambridge, UK.

Smith B.C. (2011) *Fundamentals of Fourier Transform Infrared Spectroscopy,* 2nd ed. CRC Press, Boca Raton, FL. DOI: 10.1201/b10777

Snitzer E. (1961) Cylindrical dielectric waveguide mode. *J. Opt. Soc. Am.* **51**(5), 491–498. DOI: 10.1364/JOSA.51.000491

Snyder A.W. and Love J.D. (1975) Reflection at a curved dielectric interface—Electromagnetic tunneling. *IEEE Trans. Microwave Theory Technol.* **23**, 134–141. DOI: 10.1109/TMTT.1975.1128514

Snyder A.W. and Mitchell D.J. (1974) Leaky rays on circular optical fibers. *J. Opt. Soc. Am.* **64**(5), 599–607. DOI: 10.1364/JOSA.64.000599

Theiss W. (2008a) *SPRAY: Spectral Ray Tracing.* www.wtheiss.com/docs/spray/index.html.

Theiss W. (2008b) *SCOUT: Spectrum Interpretation by Simulation.* www.wtheiss.com/docs/scout2/index.html.

Tseng M.M., Jiao J., and Su C.-J. (1998) Virtual prototyping for customized product development. *Integr. Manuf. Syst.* **9**(6), 334–343. DOI: 10.1108/09576069810238682

Van Gunsteren W.E. and Berendsen H.J.C. (1990) Computer simulation of molecular dynamics: Methodology, applications, and perspectives in chemistry. *Angew. Chem. Int. Ed. Engl.* **29**, 992–1023. DOI: 10.1002/anie.199009921

Varmuza K. and Filzmoser P. (2009) *Introduction to Multivariate Statistical Analysis in Chemometrics.* CRC Press, Boca Raton, FL. DOI: 10.1201/9781420059496

Venkataraj S. and Geurts J. (2001) Structural and optical properties of thin lead oxide films produced by reactive direct current magnetron sputtering. *J. Vac. Sci. Technol. A* **19**(6), 2870–2878. DOI: 10.1116/1.1410948

Waldhäusl R., Schnabel B., Dannberg P., Kley E.-B., Bräuer A., and Karthe W. (1997) Efficient coupling into polymer waveguides by gratings. *Appl. Opt.* **36**(36), 9383–9390. DOI: 10.1364/AO.36.009383

Wilk A., Kim S-S., and Mizaikoff B. (2009) An approach to the spectral simulation of infrared hollow waveguide gas sensors. *Anal. Bioanal. Chem.* **395**(6), 1661–1671. DOI: 10.1007/s00216-009-3104-2

Winsberg E. (2001) Simulations, models and theories: Complex physical systems and their representations. *Phil. Sci.* **68**(3), S442–S454. DOI: 10.1086/392927

Winsberg, E. (2003) Simulated experiments: Methodology for a virtual world. *Phil. Sci.* **70**(1), 105–125. DOI: 10.1086/367872

Winston R. and Hinterberger H. (1975) Principles of cylindrical concentrators for solar energy. *Solar Energy* **17**(4), 255–258. DOI: 10.1016/0038-092X(75)90007-9

Winston R., Miñano J.C., and Benítez P.G. (2005) *Nonimaging Optics.* Elsevier, Burlington, VT.

Worrell C.A. and Gallen N.A. (1997) Trace-level detection of gases and vapours with mid-infra-red hollow waveguides. *J. Phys. D: Appl. Phys.* **30**, 1984–1995. DOI: 10.1088/0022-3727/30/14/005

Yeh P. and Yariv A. (1978) Theory of Bragg fiber. *J. Opt. Soc. Am.* **68**(9), 1196–1201. DOI: 10.1364/JOSA.68.001196

Young C., Hartwig S., Lambrecht A., Kim, S.-S., and Mizaikoff B. (2007) Optimizing gas sensors based on quantum cascade lasers and photonic bandgap hollow waveguides. *2007 IEEE Sens.*, 1345–1348. DOI: 10.1109/ICSENS.2007.4388660

Zhao Z.W., Tay B.K., Sun C.Q., and Ligatchev V. (2004) Oxygen lone-pair states near the valence band edge of aluminum oxide thin films. *J. Appl. Phys.* **95**(8), 4147–4150. DOI: 10.1063/1.1686905

Zhao S. and Wäckelgård E. (2006) The optical properties of sputtered composite of Al-AlN. *Solar Energy Mater. Solar Cells* **90**(13), 1861–1874. DOI: 10.1016/j.solmat.2005.11.006

CHAPTER 10

Finite-Element Modeling of Infrared Sensors

X. Wang

S.-S. Kim

A. Wilk

B. Mizaikoff

1. INTRODUCTION

Most optical sensor systems are composed of optical components with macroscopic dimensions, and are analyzed and designed based on geometric optics. Geometric optics treat optical rays as lines with zero width (Jenkins and White 1976). While this is a quite bold approximation to represent the nature of light, it has been proved quite successful when dealing with practical optical system with macroscopically sized optical components. Hence, also, most mid-infrared optical sensors are initially designed with the assumptions and limitations of geometric optics. Such optical sensors usually turn out to be very effective devices in practical applications where the size of the optical sensing system or the amount of analyte available to the analysis are no major concerns for obtaining a useful analytical result. Motivated by the widespread applicability of chem/bio sensors, branching out into new areas of application inherently implies novel demands including, e.g., miniaturization, portability, biocompatibility, or extremely small quantities of analyte available to the measurement. With the introduction

DOI: 10.5643/9781606503201/ch10

of nanophotonics, the obtainable scale may satisfy such challenging demands, thereby enabling novel applications for optical sensors, e.g., in the field of nanomedicine (Kim et al. 2008). Evidently, integrating photonic devices operating in the mid-infrared spectral range provides substantial benefits enabling more widespread access and use of optical sensing technologies based on molecular vibrations. While fabrication of photonic sensors at the nanoscale has already matured due to advances in nanotechnology and microfabrication of silicon electronics, the design and simulation of such devices based on geometric optics fails if the dimensions of the building blocks within the optical system approach the wavelength of light propagated within these structures. Furthermore, geometric optics do not provide any information on interference or diffraction phenomena of the involved photons. Therefore, it is inevitable to introduce modeling approaches based on wave optics for analyzing and designing nanophotonic sensors and their key components such as, e.g., waveguides.

This chapter will start by summarizing basic knowledge required for understanding commonly used simulation methods, i.e., finite-element modeling (FEM) theory, and will highlight its utility using exemplary applications in near-infrared sensor design. Trends in mid-infrared sensor technologies are discussed thereafter, along with simulation approaches that may contribute to advancing mid-infrared optical sensor technology toward next-generation mid-infrared chem/bio sensor design with optimized sensor performance.

2. INTRODUCTION TO FEM MODELING THEORY

2.1. ELECTROMAGNETIC WAVE EQUATION

Classical electromagnetics are considered among the most elegant theories in physics. The following four Maxwell's equations and additional two constitutive relationships are sufficient to interpret most classical electromagnetic phenomena observed in nature. The summary of Maxwell's equations in their differential forms (Jaskson 1999) can be written as

$$\nabla \cdot \vec{D} = \rho \tag{10.1}$$

$$\nabla \cdot \vec{B} = 0 \tag{10.2}$$

$$\nabla \times \vec{E} = -\frac{\partial \vec{B}}{\partial t} \tag{10.3}$$

$$\nabla \times \vec{H} = \frac{\partial \vec{D}}{\partial t} + \vec{J} \tag{10.4}$$

Basically, Maxwell's equations as written above define the causality between sources and effects. $\vec{E}$, $\vec{D}$, $\vec{B}$, and $\vec{H}$ represent effects, i.e., electric field, electric displacement field, magnetic flux density, and magnetic field, respectively. ρ and $\vec{J}$ are sources for such effects, i.e., charge density and current density.

$$\vec{D} = \varepsilon \vec{E} \tag{10.5}$$

$$\vec{B} = \mu \vec{H} \tag{10.6}$$

The two equations above are constitutive relationships, which define the properties of electric and magnetic fields in media with electric polarization and magnetization. ε and μ are called electric permittivity and magnetic permeability, respectively. Electromagnetic wave equations are derived in two steps, i.e., curl operation on both sides of Eq. (10.3) and substitution of $\vec{B}$ with Eq. (10.4). After rearranging the terms, we obtain the wave equation (10.7) for an electric field from Maxwell's equations.

$$\nabla^2 \vec{E}(\vec{r},t) = \mu_0 \varepsilon_0 \varepsilon_r \frac{\partial^2 \vec{E}(\vec{r},t)}{\partial t^2} \tag{10.7}$$

Similarly, the wave equation for $\vec{H}$ may be obtained.

$$\nabla^2 \vec{H}(\vec{r},t) = \mu_0 \varepsilon_0 \varepsilon_r \frac{\partial^2 \vec{H}(\vec{r},t)}{\partial t^2} \tag{10.8}$$

In the following, the general monochromatic phase expressions for $\vec{E}$, $\vec{D}$, $\vec{B}$, and $\vec{H}$ are given. ε_0 and μ_0 again represent the electric permittivity and magnetic permeability in vacuum, and ε_r is the relative permittivity.

$$\vec{E}(\vec{r},t) = \mathrm{Re}\left\{\vec{E}(\vec{r}) * \exp(i\omega t)\right\} \tag{10.9}$$

$$\vec{D}(\vec{r},t) = \mathrm{Re}\left\{\vec{D}(\vec{r}) * \exp(i\omega t)\right\} \tag{10.10}$$

$$\vec{H}(\vec{r},t) = \mathrm{Re}\left\{\vec{H}(\vec{r}) * \exp(i\omega t)\right\} \tag{10.11}$$

$$\vec{B}(\vec{r},t) = \mathrm{Re}\left\{\vec{B}(\vec{r}) * \exp(i\omega t)\right\} \tag{10.12}$$

By replacing $\vec{E}(\vec{r},t)$ and $\vec{H}(\vec{r},t)$ with the expressions above, we may reduce the wave equation to the time-independent Helmholtz equation.

$$\nabla^2\vec{E}(\vec{r})+k^2\vec{E}(\vec{r})=0 \tag{10.13}$$

$$k=k_0 n=k_0\sqrt{\varepsilon_r}=\omega\sqrt{\varepsilon_0\varepsilon_r\mu_0} \tag{10.14}$$

k and k_0 are wave numbers in a medium and in vacuum, respectively. n is the refractive index of the medium. The expression above represents a general three-dimensional case. However, if we are interested in supporting a light mode within a cross section of an optical waveguide such as, e.g., an optical fiber or channel waveguide, a two-dimensional analysis is usually sufficient. The three-dimensional Helmholtz equation may thus further be reduced to the two-dimensional case after assuming that the electromagnetic wave propagates along the z axis.

$$\nabla_\perp^2\vec{E}(x,y)+k_0^2\left(\varepsilon_r-n_{\text{eff}}^2\right)\vec{E}(x,y)=0 \tag{10.15}$$

$$\nabla_\perp^2\equiv\frac{\partial^2}{\partial x^2}+\frac{\partial^2}{\partial y^2} \tag{10.16}$$

$$n_{\text{eff}}\equiv\frac{\beta}{k_0} \tag{10.17}$$

β is defined as propagation constant, which is a component of wave number k along the wave propagation direction. The effective index n_{eff} is the ratio of the propagation constant in the medium to the wave number in a vacuum.

2.2. BACKGROUND ON FEM

Most natural phenomena may be expressed by partial differential equations. Therefore, finding solutions for such partial differential equations is very important to predict the outcome prior to an experimental trial in science and engineering. Exact analytical solutions are obtainable in only a limited number of cases, particularly systems with a high degree of symmetry. Perturbation theory allows determining approximate solutions for systems with "broken" symmetry, but a general procedure to determine analytical solutions for partial differential equations is not well established to date. In the early 1940s, A. Hrennikoff and R. Courant suggested a substantially different way to obtain approximate solutions for partial differential equations for almost any system regardless of its complexity (Hrennikoff 1941; Pelosi 2007). Their suggested strategy was to "first divide a continuous region of space into many smaller meshed elements, and sum up the approximate solutions of each individual meshed element." This method was increasingly adopted to solve engineering problems during the late 1950s. This strategy was particularly suitable to work with emerging computer technology that was introduced around the same time. Finite-difference methods (FDM), which are an alternative to finite-element methods (FEM) for solving partial differential equations, also depend on the strategy of dividing a continuous space into many smaller elements (Kawano and Kitoh 2001). While FDM is another numerical way of solving differential equations by introducing differential equations that are considered an approximation of the original differential equation, FEM tries to obtain an approximation of the exact solution via integral equations. Basically, there are two different subschemes to solve partial differential equations numerically in FEM, i.e., the *variational method* and the *weighted residual method.* In the *variational method,* the solution for the partial differential equation is obtained by solving an integral equation imposing stationary conditions onto the differential equations of interest, which is equivalent to solving the original differential equation. Alternatively, the *weighted residual method* takes the error residual of the trial solution multiplied by some weight function, which is then minimized to find an approximate solution for the original differential equation. In both cases, it is essential to divide the space of interest into small elements (i.e., finite elements), which is called a *mesh.*

2.2.1. *Variational Method*

The governing equations of many phenomena in nature, including the electromagnetic wave equation, have the mathematical form of the Helmholtz equation.

First, we introduce a functional Helmholtz equation.

$$\nabla^2\varphi + k^2\varphi = 0 \tag{10.18}$$

After multiplying $\delta\varphi$ on both sides of the equation, we integrate over this area.

$$\int_S \left(\nabla^2\varphi + k^2\varphi\right)\delta\varphi \, dS = 0 \tag{10.19}$$

$$\int_S \left(\nabla^2\varphi\,\delta\varphi + k^2\varphi\,\delta\varphi\right) dS = 0 \tag{10.20}$$

$$\int_S \left[\nabla \cdot (\delta\varphi\,\nabla\varphi) - \nabla(\delta\varphi) \cdot \nabla\varphi\right] dS + \int_S k^2\varphi\,\delta\varphi \, dS = 0 \tag{10.21}$$

Then we apply Green's theorem to Eq. (10.21) such that some part of the integration area is converted to a contour line integration, which readily accepts boundary conditions defined at a later stage.

$$\int_S \nabla \cdot (\delta\varphi \nabla\varphi)\, dS = \oint (\delta\varphi \nabla\varphi) \cdot \hat{n}\, dl = \oint \delta\varphi \frac{\partial\varphi}{\partial n} dl \tag{10.22}$$

$$-\int_S \nabla(\delta\varphi) \cdot \nabla\varphi\, dS + \int_S \frac{1}{2} k^2 \delta(\varphi^2)\, dS + \oint \delta\varphi \frac{\partial\varphi}{\partial n} dl = 0 \tag{10.23}$$

This equation is further reduced by applying appropriate boundary conditions, and by the following vector identity.

$$\begin{aligned} \delta(\nabla\varphi \cdot \nabla\varphi) &= \delta(\nabla\varphi) \cdot \nabla\varphi + \nabla\varphi \cdot \delta(\nabla\varphi) \\ &= 2\left[\delta(\nabla\delta) \cdot \nabla\varphi\right] \\ &= 2\left[\nabla(\delta\varphi) \cdot \nabla\varphi\right] \end{aligned} \tag{10.24}$$

$$-\int_S \delta\left[\frac{1}{2}(\nabla\varphi \cdot \nabla\varphi)\right] dS + \int_S \delta\left(\frac{1}{2} k^2 \varphi^2\right) dS + \oint \delta\left(\varphi \frac{\partial\varphi}{\partial n}\right) dl = 0 \tag{10.25}$$

$$\delta\left(\frac{1}{2}\int_S \nabla\varphi \cdot \nabla\varphi - k^2\varphi^2\, dS - \oint \varphi \frac{\partial\varphi}{\partial n} dl\right) = 0 \tag{10.26}$$

Now, we introduce functional I as follows.

$$I = \frac{1}{2}\int_S \nabla\varphi \cdot \nabla\varphi - k^2\varphi^2\, dS - \oint \varphi \frac{\partial\varphi}{\partial n} dl \tag{10.27}$$

In the two-dimensional case, we may more explicitly express the equation by introduction of the following terms.

$$\varphi = \varphi(x, y) \tag{10.28}$$

$$k^2 = k_0^2\left(\varepsilon_r - n_{\text{eff}}^2\right) \tag{10.29}$$

$$I = \frac{1}{2}\iint_S \left[\left(\frac{\partial \varphi}{\partial x}\right)^2 + \left(\frac{\partial \varphi}{\partial y}\right)^2 - k_0^2\left(\varepsilon_r - n_{\text{eff}}^2\right)\varphi^2\right] dx\, dy - \oint \varphi \frac{\partial \varphi}{\partial n} dl \tag{10.30}$$

We may now impose stationary conditions on the obtained functional I.

$$\delta I = 0 \tag{10.31}$$

$$\delta\varphi \iint_S \left[\frac{\partial^2 \varphi}{\partial x^2} + \frac{\partial^2 \varphi}{\partial y^2} - k_0^2\left(\varepsilon_r - n_{\text{eff}}^2\right)\varphi\right] dx\, dy = 0 \tag{10.32}$$

This equation is now valid for any arbitrary variation $\delta\varphi$. It should be noticed that the integrand has exactly the same functional form as the electromagnetic wave equation. Therefore, The electromagnetic wave equation may be obtained by applying stationary conditions for the functional I. In other words, solving the electromagnetic wave equation is equivalent to imposing stationary conditions onto the obtained functional I, thereby setting the variation of functional I to zero. Using this approach to solve partial differential equations is called the *variational method.* The general function φ satisfying such conditions will be equivalent to the solution $E(x,y)$ of the time-independent electromagnetic wave equation.

The Rayleigh-Ritz method is one of more popular schemes to implement discretization of areas of interests. Using this method, an unknown solution function of the partial differential equation is expanded by a linear combination of known basis functions. Furthermore, appropriate boundary conditions and variational principles are applied based on the linear combination of basis functions. First, the area of interests is divided into small regions, which are called *meshed elements.* Then, the function of the entire area of interests may be expressed as the sum of functions for each element e.

$$I = \sum_e I_e \tag{10.33}$$

By imposing stationary conditions, one obtains

$$\delta I = \sum_e \delta I_e = 0 \tag{10.34}$$

The functional I of a certain element e is expressed as follows.

$$I = \frac{1}{2}\iint_{Se} \left[\left(\frac{\partial \varphi_e}{\partial x}\right)^2 + \left(\frac{\partial \varphi_e}{\partial y}\right)^2 - k_0^2\left(\varepsilon_r - n_{\text{eff}}^2\right)\varphi_e^2\right] dx\, dy - \oint_e \varphi_e \frac{\partial \varphi_e}{\partial n} dl \tag{10.35}$$

In reply to:

$$\varphi_e = \sum_{i=1}^{n} N_{ei}\varphi_{ei} = [N_e][\varphi_e]^T \tag{10.36}$$

In Eq. (10.36), the wave equation within an element is expanded with respect to the field values at the nodes within the element. N_{ei} and φ_{ei} are the base functions, and the expansion coefficient of the *i*th node in the element *e*. In other words, φ_{ei} corresponds for the field value at the *i*th node within the element *e*. N_{ei} is also called the shape function, and corresponds to an interpolation function to obtain an approximation to the real function. *n* in Eq. (10.36) is the number of nodes within the element *e*.

$$[N_e] \equiv [N_{e1} \quad N_{e2} \quad . \quad . \quad . \quad N_{en}] \tag{10.37}$$

$$[\varphi_e] \equiv [\varphi_{e1} \quad \varphi_{e2} \quad . \quad . \quad . \quad \varphi_{en}] \tag{10.38}$$

$$\begin{aligned}\left(\frac{\partial\varphi_e}{\partial x}\right)^2 &= \left\{\frac{\partial\varphi_e}{\partial x}\right\}^T \frac{\partial\varphi_e}{\partial x} \\ &= \frac{\partial}{\partial x}\left\{[\varphi_e][N_e]^T\right\}\frac{\partial}{\partial x}\left\{[N_e][\varphi_e]^T\right\} \\ &= [\varphi_e]\frac{\partial[N_e]^T}{\partial x}\frac{\partial[N_e]}{\partial x}[\varphi_e]^T\end{aligned} \tag{10.39}$$

Similarly, we write

$$\left(\frac{\partial\varphi_e}{\partial y}\right)^2 = [\varphi_e]\frac{\partial[N_e]^T}{\partial y}\frac{\partial[N_e]}{\partial y}[\varphi_e]^T \tag{10.40}$$

$$\varphi_e^2 = \{\varphi_e\}^T \varphi_e = [\varphi_e][N_e]^T[N_e][\varphi_e]^T \tag{10.41}$$

$$\begin{aligned}I_e &= \frac{1}{2}[\varphi_e]\iint_{Se}\left\{\frac{\partial[N_e]^T}{\partial x}\frac{\partial[N_e]}{\partial x} + \frac{\partial[N_e]^T}{\partial y}\frac{\partial[N_e]}{\partial y} - k_0^2\left(\varepsilon_r - n_{\text{eff}}^2\right)[N_e]^T[N_e]\right\}dx\,dy[\varphi_e]^T \\ &\quad -\oint_e \varphi_e\frac{\partial\varphi_e}{\partial n}dl \\ &= \frac{1}{2}[\varphi_e]\{[\alpha_e] - \lambda^2[\beta_e]\}[\varphi_e]^T - \oint_e \varphi_e\frac{\partial\varphi_e}{\partial n}dl\end{aligned} \tag{10.42}$$

$$[\alpha_e] \equiv \iint_{Se} \frac{\partial [N_e]^T}{\partial x}\frac{\partial [N_e]}{\partial x} + \frac{\partial [N_e]^T}{\partial y}\frac{\partial [N_e]}{\partial y} dx\,dy \tag{10.43}$$

$$[\beta_e] \equiv \iint_{Se} [N_e]^T [N_e] dx\,dy \tag{10.44}$$

$$\lambda^2 \equiv k_0^2 \left(\varepsilon_r - n_{\text{eff}}^2\right) \tag{10.45}$$

$$\begin{aligned} I &= \sum_e I_e \\ &= \frac{1}{2}\sum_e [\varphi_e]\{[\alpha_e] - \lambda^2 [\beta_e]\}[\varphi_e]^T - \sum_e \oint_e \varphi_e \frac{\partial \varphi_e}{\partial n} dl \\ &= \frac{1}{2}[\varphi]\{[\alpha] - \lambda^2 [\beta]\}[\varphi]^T - \oint \varphi \frac{\partial \varphi}{\partial n} dl \end{aligned} \tag{10.46}$$

$$[\varphi] = \sum_e [\varphi_e] \tag{10.47}$$

$$[\alpha] = \sum_e [\alpha_e] \tag{10.48}$$

$$[\beta] = \sum_e [\beta_e] \tag{10.49}$$

Now, we are ready to apply the boundary conditions to function *I*. We impose the following two conditions onto the boundary of the region of analysis.

1. Dirichlet condition: $\varphi = 0$
2. Neumann condition: $\frac{\partial \varphi}{\partial n} = 0$

Thus, the function is further reduced.

$$I = \frac{1}{2}[\varphi]\{[\alpha] - \lambda^2 [\beta]\}[\varphi]^T \tag{10.50}$$

Assuming that there are three nodes within an element, then

$$[\varphi]=[\varphi_1 \quad \varphi_2 \quad \varphi_3] \tag{10.51}$$

$$[S]\equiv[\alpha]-\lambda^2[\beta]=\begin{pmatrix} S_{11} & S_{12} & S_{13} \\ S_{12} & S_{22} & S_{23} \\ S_{31} & S_{23} & S_{33} \end{pmatrix} \tag{10.52}$$

$$\begin{aligned} I &= \frac{1}{2}[\varphi][S][\varphi]^T \\ &= \frac{1}{2}(\varphi_1 \quad \varphi_2 \quad \varphi_3)\begin{pmatrix} S_{11} & S_{12} & S_{13} \\ S_{12} & S_{22} & S_{23} \\ S_{31} & S_{23} & S_{33} \end{pmatrix}\begin{pmatrix} \varphi_1 \\ \varphi_2 \\ \varphi_3 \end{pmatrix} \\ &= \frac{1}{2}\left(S_{11}\varphi_1^2+S_{22}\varphi_2^2+S_{33}\varphi_3^2+2S_{12}\varphi_1\varphi_2+2S_{23}\varphi_2\varphi_3+2S_{13}\varphi_1\varphi_3\right) \end{aligned} \tag{10.53}$$

$$\frac{\partial I}{\partial \varphi_1}=S_{11}\varphi_1+S_{12}\varphi_2+S_{13}\varphi_3 \tag{10.54}$$

$$\frac{\partial I}{\partial \varphi_2}=S_{22}\varphi_2+S_{12}\varphi_1+S_{23}\varphi_3 \tag{10.55}$$

$$\frac{\partial I}{\partial \varphi_3}=S_{33}\varphi_3+S_{13}\varphi_1+S_{23}\varphi_2 \tag{10.56}$$

$$\frac{\partial I}{\partial \begin{pmatrix} \varphi_1 \\ \varphi_2 \\ \varphi_3 \end{pmatrix}}=\begin{pmatrix} S_{11} & S_{12} & S_{13} \\ S_{12} & S_{22} & S_{23} \\ S_{31} & S_{23} & S_{33} \end{pmatrix}\begin{pmatrix} \varphi_1 \\ \varphi_2 \\ \varphi_3 \end{pmatrix} \tag{10.57}$$

$$\frac{\partial I}{\partial[\varphi]^T}=\{[\alpha]-\lambda^2[\beta]\}[\varphi]^T \tag{10.58}$$

Now, we apply the variational principle to Eq. (10.58) by imposing stationary conditions, which becomes equivalent to the electromagnetic wave equation.

$$\frac{\partial I}{\partial[\varphi]^T}=0 \tag{10.59}$$

$$\{[\alpha]-\lambda^2[\beta]\}[\varphi]^T=0 \tag{10.60}$$

Thus, it is proved that the variational method combined with the Rayleigh-Ritz method results in an equivalent of the electromagnetic wave equation in the form of an eigenvalue matrix equation.

2.2.2. First-Order Triangular Elements

Although the field value at each node point of every triangular element is calculated by eigenmatrix manipulation in FEM, it is almost impossible to obtain real values at any arbitrary point inside the triangular elements. Therefore, an approximation is required. Typical approximation schemes are the first-order triangular element and the second-order triangular element. The first-order triangular element has three node points at the vortex of the triangle. The second-order triangular element has a total of six nodes; three of them are at the vortex of the triangle, and the others are at the midpoint of each side of the triangle. The wave function at an arbitrary point inside the first-order triangular element is expressed by a linear function.

$$\varphi_e(x,y)=\sum_{i=1}^{3}N_{ei}\varphi_{ei}=N_{e1}\varphi_{e1}+N_{e2}\varphi_{e2}+N_{e3}\varphi_{e3} \tag{10.61}$$

The first-order triangular elements use linear interpolations based on field values at the three node points. Since this is a more than one-dimensional linear interpolation, it is reasonable to perform the linear interpolation based on an area ratio.

$$\varphi_e(x,y)=[N_e][\varphi_e]^T \tag{10.62}$$

$$\varphi_e(x,y)=\frac{A_1}{A}\varphi_{e1}+\frac{A_2}{A}\varphi_{e2}+\frac{A_3}{A}\varphi_{e3} \tag{10.63}$$

Here A is the area of the element and is not a function of coordinates, but a constant. A_1, A_2, and A_3 are the areas of the smaller triangles shown in Figure 10.1. It should be noted that A_1, A_2, and A_3 are functions of the coordinates of the point $P(x,y)$.

$$A=A_1+A_2+A_3 \tag{10.64}$$

$$N_{ei} = \frac{A_i}{A_1 + A_2 + A_3} \tag{10.65}$$

Such interpolated wave functions should also provide field values, if the arbitrary point becomes one of the three nodes. If $P(x,y)$ is coincident with node point 1, the wave function should be identical to the field value of the node point 1.

$$\varphi_e(x_1,y_1) = \frac{A}{A}\varphi_{e1} + \frac{0}{A}\varphi_{e2} + \frac{0}{A}\varphi_{e3} = \varphi_{e1} \tag{10.66}$$

$$N_{e1}(x_1,y_1) = 1 \tag{10.67}$$

$$N_{e2}(x_1,y_1) = 0 \tag{10.68}$$

$$N_{e3}(x_1,y_1) = 0 \tag{10.69}$$

The areas A, A_1, A_2, A_3 are then expressed explicitly as follows.

$$A = \frac{1}{2}\begin{vmatrix} x_2 - x_1 & y_2 - y_1 \\ x_3 - x_1 & y_3 - y_1 \end{vmatrix} = \frac{1}{2}\begin{vmatrix} x_1 & y_1 & 1 \\ x_2 & y_2 & 1 \\ x_3 & y_3 & 1 \end{vmatrix} = \frac{1}{2}\begin{vmatrix} x_1 & x_2 & x_3 \\ y_1 & y_2 & y_3 \\ 1 & 1 & 1 \end{vmatrix} \tag{10.70}$$

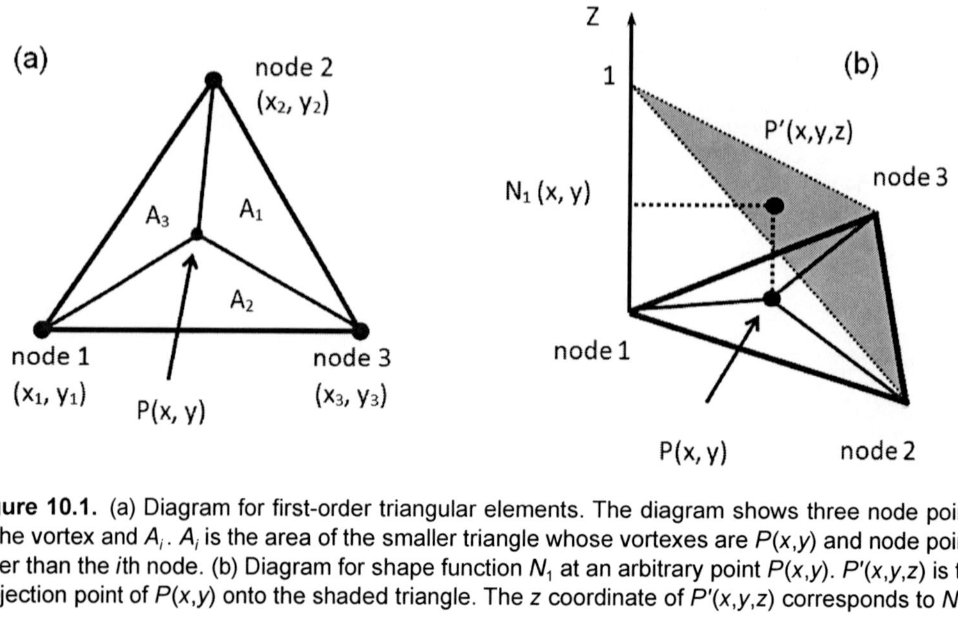

Figure 10.1. (a) Diagram for first-order triangular elements. The diagram shows three node points at the vortex and A_i. A_i is the area of the smaller triangle whose vortexes are $P(x,y)$ and node points other than the ith node. (b) Diagram for shape function N_1 at an arbitrary point $P(x,y)$. $P'(x,y,z)$ is the projection point of $P(x,y)$ onto the shaded triangle. The z coordinate of $P'(x,y,z)$ corresponds to N_1.

$$A_1 = \frac{1}{2}\begin{vmatrix} x_2 - x & y_2 - y \\ x_3 - x & y_3 - y \end{vmatrix} = \frac{1}{2}\begin{vmatrix} x & y & 1 \\ x_2 & y_2 & 1 \\ x_3 & y_3 & 1 \end{vmatrix} = \frac{1}{2}\begin{vmatrix} x & x_2 & x_3 \\ y & y_2 & y_3 \\ 1 & 1 & 1 \end{vmatrix} \tag{10.71}$$

$$A_2 = \frac{1}{2}\begin{vmatrix} x_3 - x & y_3 - y \\ x_1 - x & y_1 - y \end{vmatrix} = \frac{1}{2}\begin{vmatrix} x_1 & y_1 & 1 \\ x & y & 1 \\ x_3 & y_3 & 1 \end{vmatrix} = \frac{1}{2}\begin{vmatrix} x_1 & x & x_3 \\ y_1 & y & y_3 \\ 1 & 1 & 1 \end{vmatrix} \tag{10.72}$$

$$A_3 = \frac{1}{2}\begin{vmatrix} x_2 - x & y_2 - y \\ x_1 - x & y_3 - y \end{vmatrix} = \frac{1}{2}\begin{vmatrix} x_1 & y_1 & 1 \\ x_2 & y_2 & 1 \\ x & y & 1 \end{vmatrix} = \frac{1}{2}\begin{vmatrix} x_1 & x_2 & x \\ y_1 & y_2 & y \\ 1 & 1 & 1 \end{vmatrix} \tag{10.73}$$

3. NEAR-INFRARED EVANESCENT FIELD SENSORS: FROM SIMULATION TO APPLICATION

Modern near-infrared (NIR) sensors interface readily with microfluidics and may be integrated toward lab-on-a-chip device technology. A variety of transduction methods have been used in optical sensing applications, including refraction, absorbance, surface plasmon resonance, fluorescence, etc. Among these, near- and mid-infrared (MIR) sensing systems take particular advantage of absorption-based sensing principles due to the inherent potential of label-free detection, and for qualitative and/or quantitative analysis (Yang et al. 2007). Miniaturized MIR sensing systems frequently have to deal with a reduced optical path length, which limits the photon–sample interaction length, and thereby leads a decrease in sensitivity for absorption-based sensors. Consequently, novel photonic components have been introduced to mitigate this problem, including alternative waveguide geometries and photonic crystals (Hawkins and Schmidt 2008; Mortensen et al. 2008), which may enhance the interaction between the guided light and the target sample via a locally enhanced evanescent field or an extended effective absorption path length. For example, optical microresonators are attracting significant attention for IR evanescent field sensing, and have been proposed as powerful optical structures approaching label-free detection of only a few molecules. However, the development of such sophisticated sensing schemes requires fundamental understanding and complementary modeling of photonics and waveguide optics. Appropriate computational simulation of critical photonic components will contribute not only to optimized design of such sensing devices, but also to theoretically predicting the achievable sensitivity levels prior to constructing such devices. In particular, FEM simulations are among the most suitable methods for advanced mode analysis and mode propagation within optical waveguides owing to complicated waveguide structures/geometries and arbitrary refractive-index profiles. Moreover, FEM simulations are also applicable to thermal or stress analysis of optical waveguides. This section will highlight examples relevant to near-infrared evanescent field sensors, and will illustrate how FEM contributes to the design and development of NIR sensing system.

3.1. EXAMPLE 1

Recently, the group of Vahala has reported substantial progress on label-free, single-molecule detection using optical microcavities (Armani et al. 2007). By confining photons in resonant recirculation within optical cavities, the probability for photons to interact with target molecules is significantly increased. Figure 10.2a shows the setup of a microtoroid resonator combined with an optical fiber. The tapered fiber couples IR laser radiation (a single-mode, tunable cavity laser with an emission wavelength centered at 681.5 nm) into the whispering-gallery mode of the silica microtoroid. The resonantly confined mode generates an evanescent field penetrating into the surrounding liquid environment—here, water. Moreover, radiation within the microtoroid will generate periodic resonance peaks with an intensity correlated to the frequency. Such resonance peak shifts are highly sensitive to environmental changes, and may be used as a readout mechanism during a sensing process. For example, in the present experiments the silica surface was sensitized with antibodies for capturing target antigens. The binding behavior was monitored in real time by following the shift of the resonance peak. Using Comsol Multiphysics (COMSOL, Burlington, MA, USA), finite-element modeling was applied to study the intensity distribution of the electromagnetic field within the microtroid resonator. Simulation results shown in Figure 10.2b show that the majority of the optical field is confined within the toroid, with the portion of the evanescent field leaking into the ambient environment indicated by a white arrow.

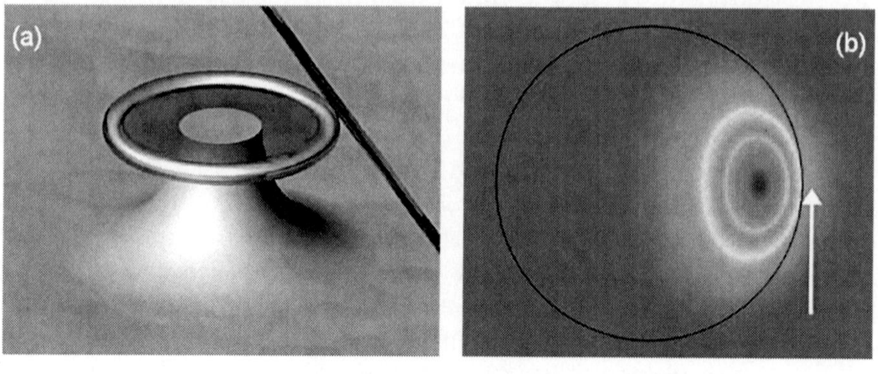

Figure 10.2. (a) A rendering of the microtoroid resonator coupled to a tapered optical fiber with highlighted whispering-gallery mode within the microtoroid resonator. (b) FEM modeling of a 4-μm (minor diameter) microtoroid resonator immersed in water. (Reprinted with permission from Armani et al. 2007. Copyright 2007 American Association for the Advancement of Science.)

Moreover, the intensity profile of the evanescent tail is asymmetrically distributed across the surface of the toroid, which suggests a nonuniform system response if target molecules bind to regions of variable evanescent field intensity. This prediction was experimentally verified by studying a distribution of response signals according to localized binding events during real-time monitoring.

The other key contribution of FEM simulations is the potential optimization of the toroid geometry, i.e., the major (D) and minor (d) toroid diameters, as indicated in Figure 10.3a. The intrinsic thermal-optical properties of the microtoroid result in a shift of the resonance frequency, which is also the key parameter for sensing

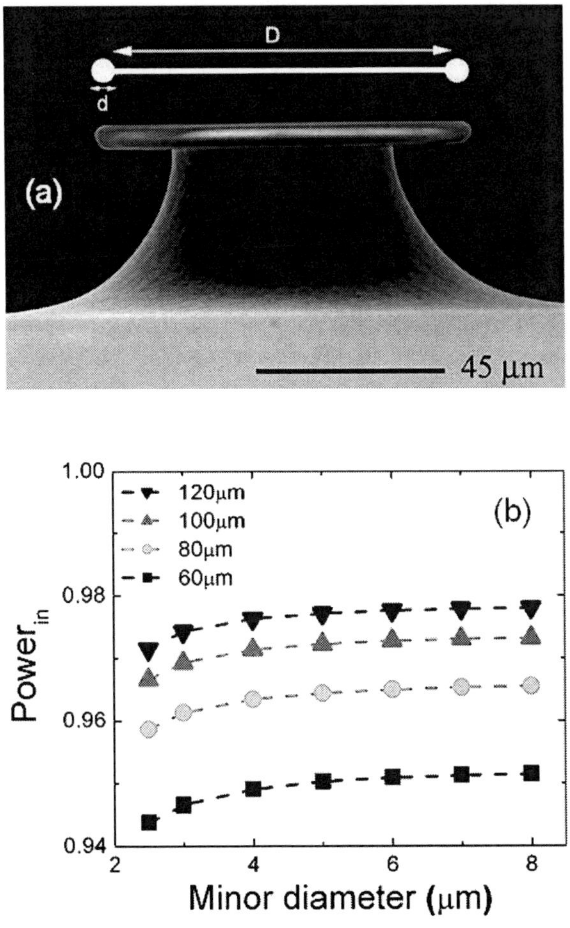

Figure 10.3. (a) SEM image of the microtoroid indicating the major (*D*) and minor (*d*) diameters. (b) FEM modeling of the optical mode within the toroid at a series of major and minor diameters. Modeling was performed at discrete microtoroid major/minor diameters. Lines were added to indicate the overall trend based on the modeling results. The plot shows the power overlap with the dielectric microcavity versus the minor diameter. (Reprinted with permission from Armani et al. 2007. Copyright 2007 American Association for the Advancement of Science.)

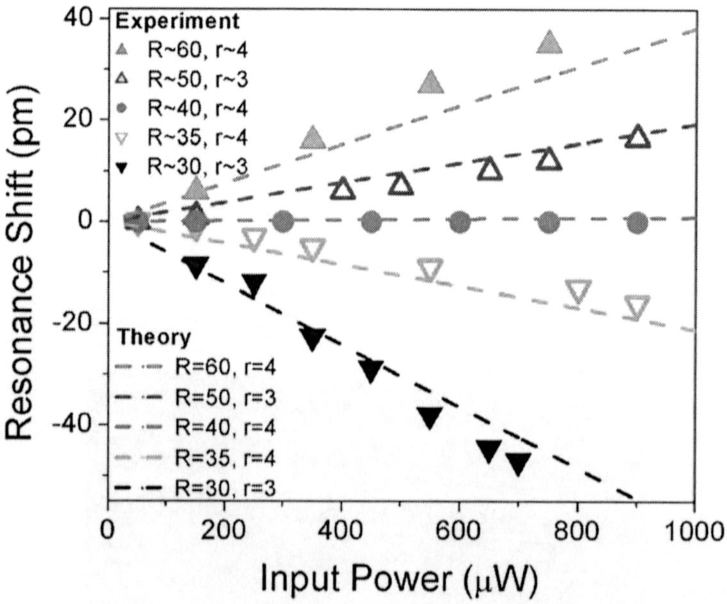

Figure 10.4. Tuning of the whispering-gallery mode versus in-coupled optical power. Each data set corresponds to the indicated major and minor radii (in micrometers). Larger radii result in a red shift, while smaller radii produce a blue shift with increasing in-coupled optical power. (Reprinted with permission from Armani et al. 2007. Copyright 2007 American Association for the Advancement of Science.)

binding of the target molecules. However, the ambient aqueous environment counteracts the shift due to the positive value of its thermal-optical coefficients. Hence, it is necessary to neutralize the overall thermal shift contributed by the silica and the surrounding water matrix such that the detection of the binding interaction is not affected by the external shift. This condition was computed via the spatial overlap of the whispering-gallery mode with water using FEM simulations provided by the Chemical Engineering Module within COMSOL Multiphysics.

While the details of how the model was developed are not provided, a series of results is illustrated in Figure 10.3b. The portion of optical power confined within the toroid is determined by the combination of major and minor diameters. For large diameters, a higher percentage of optical power is confined within the toroid. Therefore, the silica properties will dominate the shift of the resonance peak—a red shift, as predicted by theory. Accordingly, with decreasing dimensions, a steadily enhanced evanescent field overlaps with the surrounding water matrix, and the shift will be countertuned in the opposite direction. A close-to-neutral shift condition is evident at D = 80 μm and d = 8 μm, as shown in Figure 10.4. The theoretical prediction based on the FEM simulation is indicated as dashed lines, while the corresponding experimental data are marked as discrete symbols. Excellent agreement between the FEM simulation and the experimental data is achieved across a variety of toroid dimensions and optical input powers.

3.2. EXAMPLE 2

Another example for a NIR evanescent field sensor was reported by the group of Lipson and collaborators (Nitkowski et al. 2008). They have demonstrated both integrated and broad-band NIR sensing with a combination of microring resonators and microfluidic channels. As shown in Figure 10.5a, the device comprises a silicon microring resonator covered by a fluidic channel. Similar to the previously discussed microtoroid resonant cavity, incoupled light propagates within the inlet strip waveguide and couples evanescently to the ring resonator. At resonance,

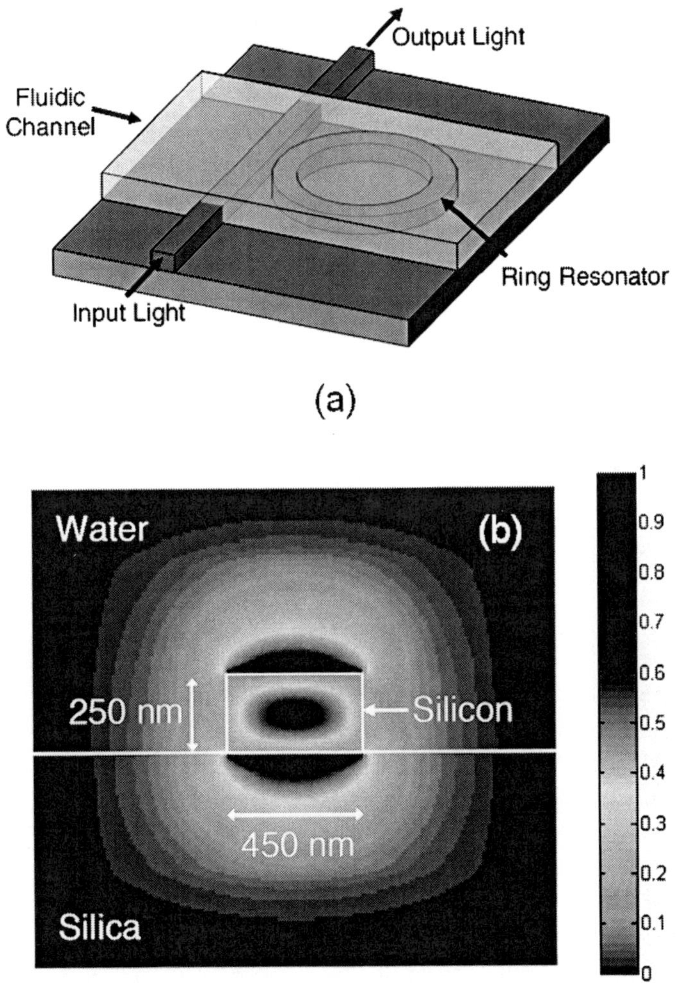

Figure 10.5. (a) Illustration of the photonic device comprising a straight waveguide and a ring resonator covered by a microfluidic channel. (b) Mode analysis of quasi-TM single-mode intensity profile at cross section of a silicon waveguide. (Reprinted with permission from Nitkowski et al. 2008. Copyright 2008 Optical Society of America.)

light circulates within the ring structure and interacts with the ambient environment several times until the mode is extinguished. Target analytes present within the evanescent field may interact with the electromagnetic field, which leads to damping of the radiation in a transmission-absorption experiment.

FEM simulations were applied aiding the design of the optical waveguide in order to maximize the interaction with the cladding fluid, i.e., to achieve maximum photon energy within the evanescent field versus the intensity of guided light. Mode analysis with finite-element modeling at a cross section of the ring waveguide is shown in Figure 10.4b. Silicon was selected as the core material of the waveguide, with a cladding layer made from silica. Water was selected as upper cladding layer to mimic the aqueous environment within the microfluidic channel. The waveguide is designed for single-mode TM polarized light. The E_y mode profile is indicated as a well-defined single-mode distribution of the electromagnetic field within the silicon core and the evanescent field distributed within water and silica. Considering the fact that the evanescent field within silica does not contribute to actively sensing analyte molecules, the integrated intensity of the evanescent tail within the water matrix should be maximized to enhance the analytical sensitivity of the device. For a NIR light source emitting at 1.5 μm, optimization of the waveguide dimension was applied according to the width and height of the silicon structure. Waveguides with dimensions of 450 nm versus 250 nm were computed as the final parameters for subsequent microfabrication. Experimental analysis revealed outstanding performance of such integrated on-chip NIR sensors, enabling the determination of less than 2 nL of fluid using a sensing footprint of $< 0.03\ mm^2$.

4. FROM NEAR-INFRARED TO MID-INFRARED SENSING: POTENTIAL AND CHALLENGES

Mid-infrared (3- to 20-μm) spectroscopy is based on the fundamental vibrational and rotational modes associated with most organic and inorganic molecules interacting with MIR photons, and provides inherent molecular selectivity, which is particularly attractive for sensing applications (Mizaikoff 2003). Despite the potential for optical sensing, early applications of MIR spectroscopy were mostly confined to laboratory environments, due to the dimensions of conventional spectroscopic equipment (Rotermund et al. 2000). Considering the already-realized miniaturization/integration of UV-Vis and near-infrared (NIR) optical devices, the opportunities for miniaturized MIR sensors are evident, in particular for in-situ and on-site sensing, as well as biomedical applications. However, to develop portable sensing systems, appropriate technologies enabling miniaturizion/integration of each optical building block are required to facilitate on-chip MIR sensor technology (Kim et al. 2008).

4.1. QUANTUM CASCADE LASERS—A REVOLUTION IN MIR SENSOR TECHNOLOGY

For simplicity, MIR sensing system may be considered as comprising four major components: (1) a light source, (2) waveguides, (3) a wavelength selection device, and (4) a detector. Similar to NIR devices, photons emitted from the light source are guided via a radiation delivery system comprising optical elements (lenses, mirrors, waveguides, etc.) and are coupled to an active sensing interface/compartment, where signal-generating interaction of electromagnetic radiation with the target analyte(s) occurs. The generated signal—in most cases an optical signal—is further guided through a wavelength-selection device (grating, filter, etc.) if broad-band radiation is utilized, and finally collected onto an appropriate MIR-sensitive photonic detector.

Among the most commonly applied MIR radiation sources are broad-band–emitting black-body (e.g., SiC) filaments, which provide incoherent radiation throughout the entire MIR spectrum. However, black-body radiation sources generally provide limited radiation intensity. While most absorption frequencies characteristic for, e.g., organic analytes are addressable within relatively narrow frequency bands, the radiation energy of such light sources is distributed across the entire spectral regime, thus providing comparatively low energy at the wavelengths of interest for detecting low analyte concentrations. Additionally, broad-band light sources usually require wavelength modulation or wavelength-selection devices to obtain wavelength-resolved spectral information, which translates into a sizable footprint for the sensing system.

After gas lasers (e.g., CO_2 laser), which are usually bulky as well, and in contrast to broad-band light sources, narrow-band-gap semiconductor laser diodes (e.g., PbS, PbSe, PbTe, SnSe, SnTe, or CdS) were developed as alternative MIR light sources. However, until recently their application was limited by the available lasing wavelength, the output power, potential cooling requirements, and the lifetime. Quantum cascade lasers (QCLs), which were initially conceptualized by Kazarinov and Suris (1971) and demonstrated experimentally for the first time by Faist et al. (1994), are nowadays widely accepted as the most advanced light-source technology available for MIR sensors. QCLs are semiconductor lasers generating light emission via intersubband transitions of electrons in the conduction band within a series of quantum heterostructures, rather than electron–hole recombinations characteristic of conventional lasing (Saleh and Teich 2007). In contrast to conventional laser diodes, the lasing wavelength of QCLs is controlled by the thickness of the quantum well structure and not by the composition, thus providing access to the entire MIR spectral range for band structure engineering, while lead-salt lasers are limited in emission wavelength by the energy band gap defined by their material composition. Furthermore, QCLs provide superior output power within narrow spectral bands, which facilitates the detection of low concentrations and provides for phase-related detection schemes owing to the coherent nature of the emitted photons. Additionally, QCL provide compact dimensions (i.e., a few hundreds of micrometers), extended lifetimes, robustness, stability of the emitted wavelength, and tunability—sometimes exceeding 200 cm^{-1} if coupled to an external cavity—across almost the entire MIR spectral range (Capasso et al. 2002; Young et al. 2009). The utility of QCLs in waveguide-based evanescent field absorption sensing has recently been demonstrated (Chen et al. 2005; Charlton et al. 2005b, 2006). Those studies indicate that the combination of high optical output power within a narrow spectral band and the compact size of QCLs along with wavelength-matched waveguide technology provides a promising research trend for miniaturized MIR sensing platforms addressing gas-phase and condensed-phase samples and promising unsurpassed sensitivity with inherent molecular selectivity.

4.2. TRENDS IN MINIATURIZATION

The introduction of QCLs has contributed significantly to the miniaturization of MIR sensing systems. Besides appropriate radiation sources, efficient interaction of MIR radiation with the target analyte is required and may be achieved by scaling the remaining optical components as well. It should be noted that merely decreasing the dimensions of, e.g., the waveguides which are commonly applied as active transducers, will probably lead to a decrease of interaction path length, and consequently, reduce the achievable sensitivity. Hence, the following paragraphs will briefly summarize efforts toward “small but smart” component scaling for miniaturizing MIR sensing systems—potentially without sacrificing sensitivity—categorized into waveguides, detectors, and spectrometers.

In gas-phase sensing, conventional multipass gas cells are commonly applied to extend the optical path length; these are composed of a set of mirrors that folding the IR beam to achieve an extended optical path length within an acceptable device footprint, and thus provide exquisite sensitivity. However, these gas cells (White cell, Herriott cell, etc.) do not fit the requirement of miniaturization, as their physical size requires comparatively large sample gas volumes (i.e., at least several hundreds of milliliters). Hollow waveguides (HWGs) are alternatives to conventional gas cells with a significantly smaller volume, and ensure highly efficient photon–analyte interaction within the hollow waveguide core (Charlton et al. 2003). The fact that HWGs behave as both waveguides and absorbers makes them ideal candidates for miniaturized IR gas sensors. Recently, novel HWG concepts based on photonic structures have been introduced, which are termed photonic band-gap hollow waveguides (PBG-HWGs). These waveguides are composed of alternating dielectric layers that provide a deliberately designed refractive-index contrast, thereby establishing an optical stop band, rather than a metallic reflection coating, which is commonly used to coat the inside of conventional HWGs.

These PBG-HWGs minimize propagation losses within a designed narrow optical frequency region, and achieve even smaller gas-sensing volumes paired with the sensitivity that characterizes HWGs. Recently, the first combination of PBG-HWGs with QCLs was demonstrated (Charlton et al. 2005c), establishing a milestone on the way toward miniaturized MIR gas-sensing techniques.

As this chapter focuses on liquid-phase sensing schemes, solid-state waveguide design will be predominantly discussed, omitting new trends in hollow-core waveguide technology. Attenuated total reflection (ATR) crystals are commonly used as MIR waveguides for probing liquid-phase samples, taking advantage of the evanescent field generated via internal total reflection for absorption-based sensing applications. If the surface of the ATR waveguide is chemically modified (e.g., with a polymer coating), (1) target analyte molecules may additionally be enriched for enhanced sensitivity, (2) water may be excluded from interaction with the evanescent field if hydrophobic membranes with a thickness exceeding the penetration depth of the evanescent field are selected, and (3) the waveguide may be protected, e.g., against degradation in harsh sensing environments. Alternatively, fiber optic waveguides such as, e.g., solid-core fibers made from polycrystalline silver halide (AgBr-AgCl, usually abbreviated as AgX) or chalcogenide glasses (As-Se-Te) have been used for evanescent field absorption sensing in aqueous solutions. For example, the research group of Gmachl has reported progress in liquid-phase sensing by combining QCLs with silver halide fiber segments (Chen et al. 2005).

Significant progress has been contributed by the research group of Mizaikoff and collaborators, focused on developing an entire new generation of MIR waveguides based on semiconductor technology. For example, they have shown the first combination of QCLs with thin-film planar silver halide waveguides, and the first planar single-mode MIR GaAs/AlGaAs waveguides fabricated entirely on-chip. Very recently, a new generation of strip waveguides has been microfabricated from GaAs/AlGaAs semiconductor wafers, demonstrating outstanding analytical performance probing minute sample amounts with unsurpassed sensitivity (Wang et al. 2012).

In addition to detectors, entire IR spectrometers are also increasingly miniaturized and integrated, facilitating, e.g., the development of compact FT-IR spectrometers for deep-sea applications (Mizaikoff 1999; Kraft et al. 2003). Furthermore, quantum well engineering, which has enabled QCLs, has also recently triggered the development of quantum cascade photodetectors (QCDs) (Gendron et al. 2005; Graf et al. 2006), which share similar materials and active structures with QCLs.

Based on the development of these key components, MIR sensing systems are increasingly revealing their potential for integration toward monolithic on-chip-level photonic devices. As already evident with the development of integrated UV/Vis and near-IR optical sensors, systemic design and fabrication of individual components requires considerable experimental efforts that may be substantially aided by modeling and simulation of MIR photonic structures.

5. SIMULATION OF MID-INFRARED EVANESCENT FIELD WAVEGUIDES

Without simulations, significant enhancements of already-existing photonic devices rarely emerge as radical new ideas. Given the required experimental efforts, usually only a few interations of a novel optical design are verified, particularly if sophisticated fabrication techniques are required, including, e.g., epitaxial growth, reactive ion etching, or focused ion beam milling, to name just a few. Hence, cost- and time-efficient development of novel photonic sensing approaches should take advantage of scalable simulations that provide insight into the governing parameters for design, device, and performance optimization with reduced experimental effort.

5.1. PLANAR WAVEGUIDES

Due to the need for portable systems, system miniaturization and integration are of paramount importance for in-situ and on-site monitoring applications. However, most conventional MIR spectroscopy equipment is still limited to operation under laboratory conditions because of its dimensions.

Condensed-phase sensing in the MIR region frequently takes advantage of total internal refection (i.e., evanescent field absorption) spectroscopy propagating photons within attenuated total reflection (ATR) crystals or infrared-transmitting optical fibers. IR-absorbing analytes interact with the evanescent field generated at the waveguide–sample interface, resulting in attenuation of the propagating wave at analyte-specific absorption frequencies. Thereby, an ATR absorption spectrum is generated. In contrast to conventional transmission absorption spectroscopy, which illuminates the sample within a cuvette, the evanescent field penetrates exponentially, decaying into the adjacent medium for only a few micrometers, and thereby enabling probing IR-opaque media such as, e.g., water, within a quasi-thin-film cell extending from the surface of the waveguide. The absoption within the evanescent field follows a pseudo-Lambert-Beer relationship:

$$A = (\varepsilon c l) r \tag{10.74}$$

where ε is the molar absorptivity, c is the concentration of the analyte, l is the equivalent optical path length, and r is the fraction of power residing outside the waveguide core. Consequently, enhancing the intensity of the evanescent field directly affects the obtainable signal-to-noise ratio (SNR) of the absorption measurement, and thus the achievable sensitivity. The intensity of the evanescent field is directly dependent on the dielectric constant at the waveguide–sample interface, and on the cross-sectional dimensions of the waveguide, which affect the mode propagation. As the dielectric constant is determined by the materials involved, the most significant sensitivity gain is thus expected from tailoring and optimizing the waveguide structure and geometry.

Condensed-phase MIR waveguides are usually fabricated from a limited set of available IR-transparent materials, predominantly polycrystalline silver halides and amorphous chalcogenide glasses. For example, the research group of Mizaikoff and collaborators have recently demonstrated the first MIR evanescent field sensing application taking advantage of a thin-film, planar silver halide waveguide combined with an InGaAs/AlInAs/InP distributed-feedback (DFB) quantum cascade laser (QCL) (Charlton et al. 2005b). In their research, the collaborating research group of Katzir fabricated planar silver halide waveguides (PWGs) by press-tapering a cylindrical fiber with a diameter of 700 μm at elevated temperatures (170°C). The PWG, with a final thickness of 300 μm, was tested with acetic anhydride as an analyte, due to the CH_3–C absorption feature, which overlaps with the emission frequency of a 974-cm^{-1} QCL available to this study. Droplets of acetic anhydride containing a series of concentrations diluted in acetonitrile were deposited onto the planar waveguide surface. Using the 3σ criterion (i.e., defining a detectable signal as an analytical signature with a magnitude at least 3 times the standard deviation), acetic anhydride was detected with a limit of detection (LOD) of 0.01 μL (10.8 μg). These results were compared with a setup using a Fourier transform infrared (FT-IR) spectrometer as the light source for the same PWG. PWGs have been proved to enhance the evanescent field at the waveguide surface compared to nontapered waveguide structures. Taking a closer look at the tested silver halide waveguide with a thickness around 100 μm, compared to conventional ATR crystal structures, radiation still propagates within the waveguide via total internal reflection, as shown schematically in Figure 10.6. A waveguide with geometric parameters significantly larger than the wavelength of the propagating photons will result in multimode distribution of propagated light. In order to match the single-mode emission characteristics provided by QCLs and to achieve well-defined single-mode propagation within a planar waveguide, the thickness of the waveguide has to be decreased to the order of magnitude of the propagated wavelength.

Simulation studies by Charlton et al. (2005a) have used beam propagation methods (BPMs) to determine an optimum waveguide thickness for QCL sensing applications using the Olympios software by C2V (Enschede, Netherlands) to simulate waveguide design based on mode optics calculations. In this design, a free-standing waveguide with an analyte layer at the waveguide surface was constructed, as shown in Figure 10.7. A radiation source couples light into the

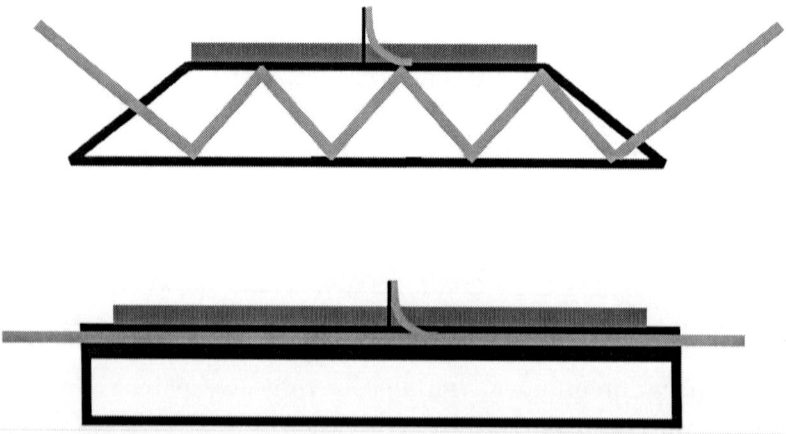

Figure 10.6. Illustration of radiation propagation within a conventionalATR crystal (top) and a single-mode thin-film waveguide (bottom). Multireflection within ATR crystals usually results in discrete active sensing spots at the waveguide surface, while single-mode thin-film waveguides support a well-defined evanescent field uniformly distributed along the waveguide surface.

waveguide from the left with a Gaussian profile, [TM (0,0) mode, λ = 10 μm]. In order to avoid coupling loss, the light source is simulated inside the waveguide material. By changing the waveguide thickness, the beam propagation may be displayed as a two-dimensional radiation intensity profile.

Figure 10.8 shows example simulation results, which reveal single-mode propagating evident by the Gaussian profile of the intensity decreasing from the core of waveguide to the edges. Additionally, the fact that interference patterns resulting from higher-order modes are not detected in this case indicates that only one mode is supported by the waveguide.

The intensity of the evanescent field is an important parameter that is closely related to the obtainable analytical sensing performance via the signal-to-noise

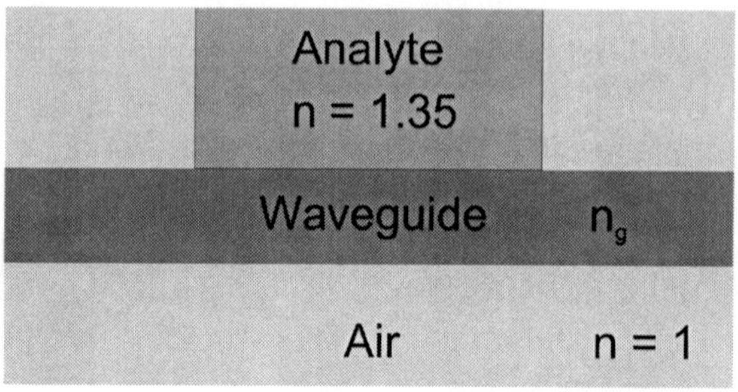

Figure 10.7. Cross section of a free-standing thin-film waveguide with corresponding refractive-index profile. (Reprinted with permission from Charlton 2005.)

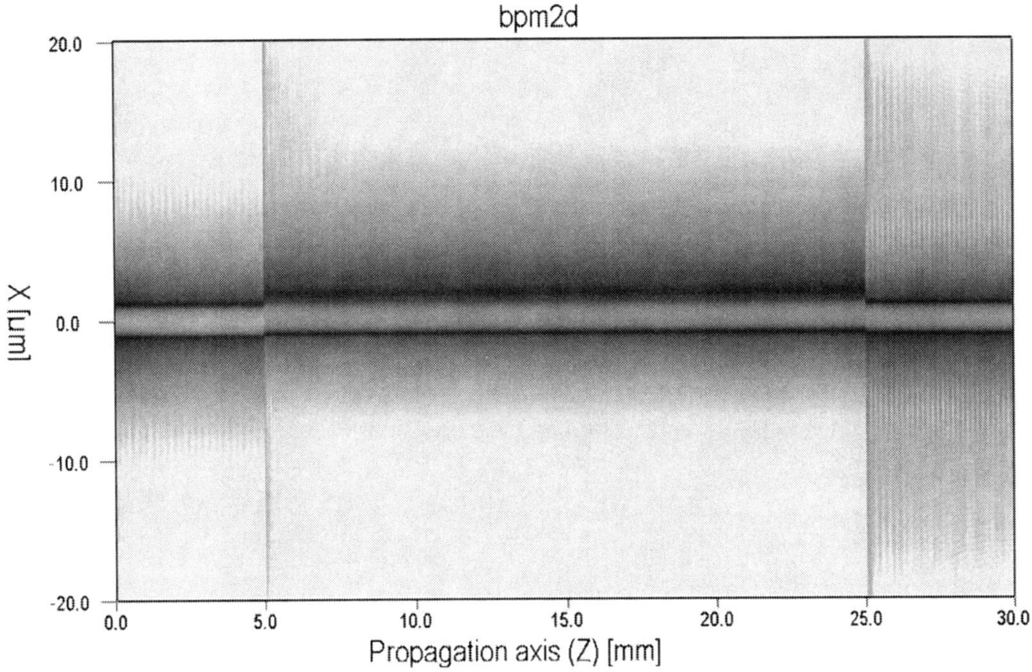

Figure 10.8. Simulation example revealing that only one mode is supported by the waveguide. Direction of propagation is along the *Z* axis, and the region within *Z* (5.0, 25.0) defines the section of the PWG covered by an analyte layer. (Reprinted with permission from Charlton 2005.)

ratio, which may be accessed by integrating the intensity of the evanescent field outside the waveguide core and within the analyte layer. System parameters of planar waveguide that may affect the evanescent field intensity include the waveguide thickness, the refractive index, the in-coupling angle, and wavelength of light. While omitting the optical details, only crucial parameters affecting the evanescent field intensity, and thus the achievable sensitivity, are pinpointed herein.

The intensity of the evanescent field depends strongly on the geometry of waveguide, as discussed in previous studies (Simhony et al. 1988; Messica et al. 1996). Decreasing the thickness of a planar waveguide results in an increase of the evanescent field prior to reaching an optimum value at a certain wavelength, as shown in Figure 10.9. The optimum thickness for maximum evanescent field intensity scales linearly with the wavelength of light. In conclusion, to maximize the evanescent field intensity for sensing purposes, the waveguide thickness should exhibit at least comparable dimensions to the wavelength of the photons that will be propagated, or potentially even smaller dimensions.

Based on the argument above, a AgX planar waveguide with a thickness of several micrometers should provide ideal waveguide structures combined with QCL light sources. However, further tapering of such waveguides is difficult to

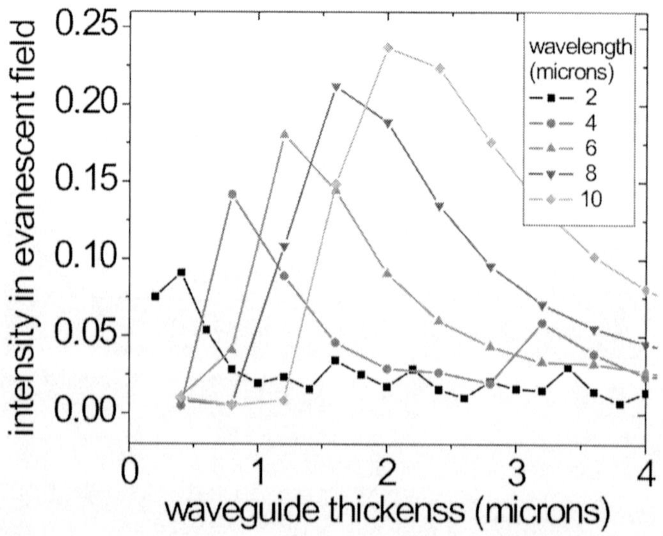

Figure 10.9. Intensity of the evanescent field as a function of waveguide thickness for various wavelengths of light. (Reprinted with permission from Charlton 2005.)

achieve with silver halides because of the intrinsic material property that results in inhomogeneous thicknesses and mechanical instability of the waveguide when approaching a thickness of only few micrometers. Hence, for practical handling of thin-film waveguides, it appears essential that the waveguiding layer be deposited at an appropriately refractive-index–matched substrate such as, e.g., a wafer surface providing a structurally rigid supporting.

Using advanced deposition methods such as, e.g., physical vapor deposition (PVD), chemical vapor deposition (CVD), or molecular beam epitaxy (MBE), thin-film planar waveguides may be fabricated with well-defined thicknesses and compositions. Accordingly, Mizaikoff and collaborators have reported a new generation of semiconductor (GaAs/AlGaAs) thin-film IR waveguides grown via MBE and later via metal-organic vapor-phase epitaxy (MOVPE), which provide a useful spectral window up to 13 μm at a waveguide thickness of 6 μm (Charlton et al. 2006). As shown in Figure 10.10 (top), a Si-doped GaAs wafer was used as substrate, and a 6-μm $Al_{0.2}Ga_{0.8}As$ cladding layer was epitaxially grown on top, followed by a 6-μm GaAs waveguide core layer to complete the epitaxial growth. A 290-nm Si_3N_4 was optionally grown via plasma-enhanced chemical vapor deposition (PECVD) to potentially passivate the waveguide surface.

An optical mode profile for TM polarized light was computed and is shown in Figure10.10 (bottom), which also indicates the corresponding refractive index of the waveguide. The mode is predominantly confined by the GaAs core layer, and partially guided by $Al_{0.2}Ga_{0.8}As$. The evanescent field shares 0.23% of the intensity with a gaseous species (set at n = 1); this value drops to 0.16% with the Si_3N_4 layer on top. Analytes with a higher refractive index therefore result in an increase of that overlap value. It should be noted, though, that the geometric parameters in Figure 10.10 are not optimized to achieve maximum evanescent field intensity. The corresponding waveguide absorption is calculated to be only 0.27 cm^{-1}, which enables single-mode propagation along even a few centimeters of waveguide length at yet acceptable losses.

However, an improvement in sensitivity is not the only goal of advanced waveguide design. Additionally, it is of interest to improve control of the evanescent

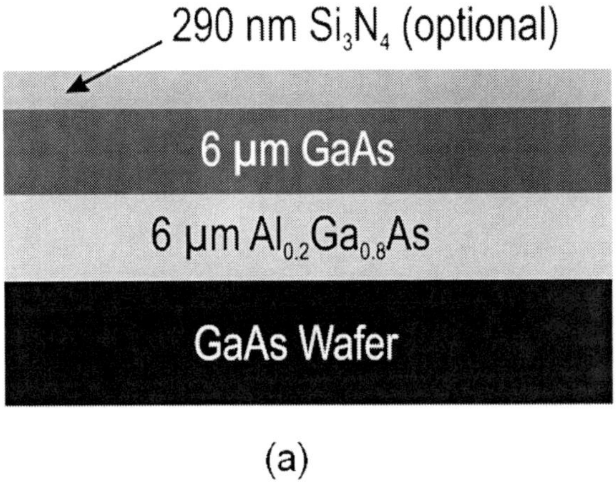

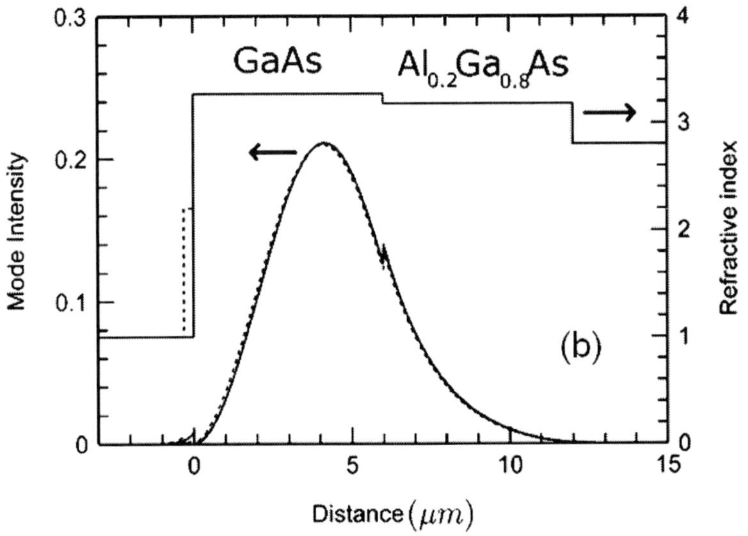

Figure 10.10. (a) Schematic cross section of the GaAs/AlGaAs waveguide structure with epitaxially grown AlGaAs layer (cladding) and GaAs waveguide layer (core). (b) Calculated mode intensity profile of thin-film GaAs/AlGaAs waveguides without silicone nitride coating. (Reprinted with permission from Charlton et al. 2006. Copyright 2006 American Chemical Society.)

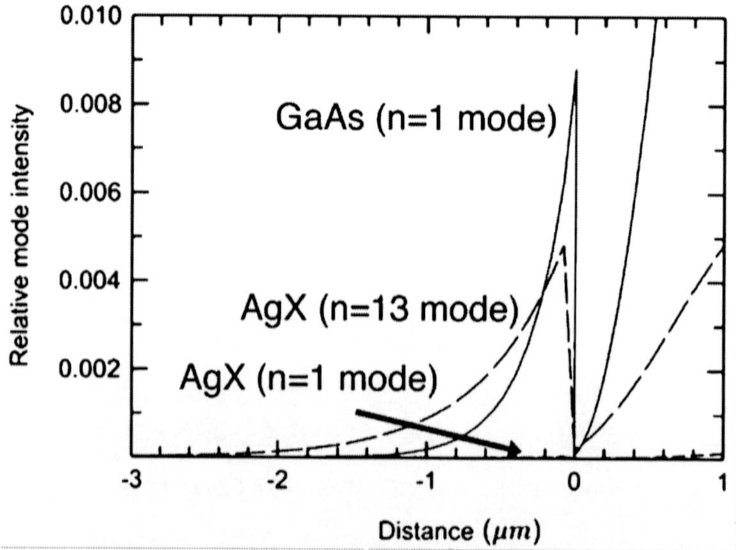

Figure 10.11. Mode profiles of the GaAs fundamental mode (solid) compared to selected AgX high-order modes (long dashed) and the AgX fundamental mode (short dashed). (Reprinted with permission from Charlton et al. 2006. Copyright 2006 American Chemical Society.)

field in order to facilitate quantitative measurements. For light propagating within a multimode waveguide, each guided mode has its individual evanescent field profile. Furthermore, each mode intensity profile relies on the coupling conditions, which in turn indicates that improving mode control should yield improved reproducibility of quantitative measurements. Accordingly, single-mode waveguides designed and fabricated to support a selected mode only provide a well-defined evanescent field, thereby ensuring optimally robust sensing conditions and resulting in an analytical response that is independent of the coupling parameters.

Figure 10.11 shows the relative mode intensity along the direction of exponential evanescent field decay perpendicular to the propagation direction of a TM polarized light beam. The electromagnetic field of the evanescent wave extends in the negative direction from the zero point at the air–core interface, where a discontinuity of the electromagnetic wave exists. The single-mode profile of a GaAs planar waveguide is shown in comparison to a multimode AgX waveguide with a low-order mode (n = 1) and a high-order mode (n = 13). In this example calculation, the fundamental mode of AgX shares only a minute fraction of the guided beam and thus generates a significantly weaker evanescent field compared to that of the 13th-order mode. On the other hand, a GaAs planar waveguide supports only the fundamental mode (solid line, n = 1) with a well-defined intensity profile. The characteristic intensity distribution of this single mode close to the surface is substantially higher than that of any mode supported by a silver halide waveguide. Hence, these calculations predict a substantially enhanced sensitivity for detecting analyte molecules present at or near the surface of the waveguide.

Experimental studies have been performed using single-mode GaAs/$Al_{0.2}Ga_{0.8}As$ waveguides using laser radiation sources such as, e.g., distributed-feedback quantum cascade lasers (DFB-QCLs). The QCL (Alpes Lasers, Neuchatel, Switzerland) applied in these studies emits TM_{00} polarized light at a wavelength of 10.3 μm (974 cm^{-1}), which overlaps with molecular-selective absorption features (the CH_3–C bending vibration) of acetic anhydride. The emitted light from the QCL was pigtail-coupled into the GaAs planar waveguide, and radiation emitted at the distal end of the waveguide was detected via a liquid nitrogen–cooled MCT detector (Kolmar Technologies, Newburyport, MA, USA). Acetic anhydride was analyzed by depositing droplets with a volume of 0.5 μL onto the waveguide surface, with each droplet covering a diameter of approximately 3 mm. Sequential deposition of additional identical droplets along the light-propagating direction leads to further decrease of the light intensity propagating along the waveguide (I/I_0). According to Eq. (10.74), the absorption is linearly related to the absorption path length, i.e., here, the coverage length at the waveguide surface. The resulting response curve is shown in Figure 10.12, indicated by a solid line and with circular symbols. For comparison, the deposition of acetic anhydride was repeated at the surface of a 300-μm-thick multimode silver halide waveguide. The linear regression function obtained is shown as a dashed line and with triangular symbols. The slope of the response function obtained at the GaAs waveguide was 0.24, compared to 0.05 obtained for the silver halide waveguide. Based on the fact that the sensitivity of an analytical measurement is proportional to the slope of the response function, an enhancement factor of 5 is obtained for the GaAs waveguide compared to the multimode AgX waveguide. As predicted by computation, this enhancement is

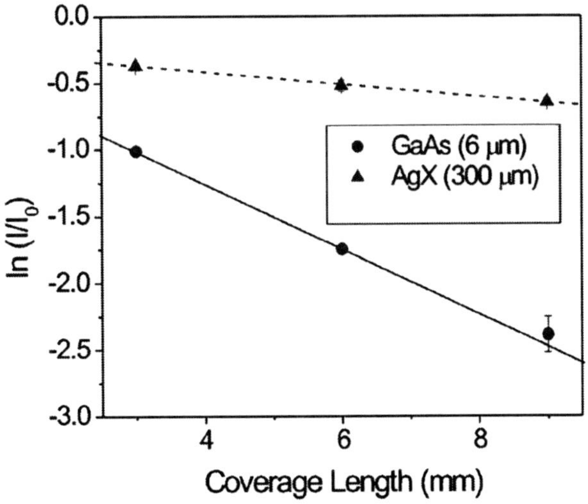

Figure 10.12. System response for multimode silver halide waveguide versus GaAs single-mode waveguide as a function of the waveguide surface coverage area along with linear regression fits. (Reprinted with permission from Charlton et al. 2006. Copyright 2006 American Chemical Society.)

attributed to an increase of r, i.e., the fraction of power residing outside the waveguide core within the evanescent field. It should be emphasized that the tested GaAs waveguide is a design optimized for low propagating losses rather than maximized evanescent field intensity. Consequently, the latter parameter is suspected to further improve the sensitivity of evanescent field absorption measurements by at least one order of magnitude, as discussed in Section 4.2.

In summary, guided by the motivation for more sensitive, robust, and potentially integrated MIR sensors, it is evident that advanced waveguide design and fabrication leading to propagating photons along single-mode frequency-matched GaAs/AlGaAs planar waveguides is a superior technology paving the way toward on-chip MIR sensing systems. Using unstructured waveguide slabs, this technique facilitates the one-dimensional (i.e., vertical to the direction of propagation) confinement of individual light modes, providing a well-defined evanescent field and enhanced sensitivity during quantitative analytical measurements of molecules present at or in the immediate vicinity of the waveguide surface. These next-generation MIR planar waveguides hold promising potential for on-chip infrared sensing platforms, which may integrate mode-matched QCLs and detectors to provide highly integrated and sensitive chem/bio sensing platforms.

A logical next step is to further enhance the intensity of the evanescent field, thereby ensuring more efficient interaction with the analyte molecules at the waveguide surface. Hence, three-dimensional structuring of the waveguide leading to further enhanced mode confinement and control appears to be a viable strategy.

5.2. STRIP/SLOT WAVEGUIDES

Most recently, ultrasensitive evanescent field chemical sensing was demonstrated by the research group of Mizaikoff based on microstructured GaAs/AlGaAs strip waveguides (Wang et al. 2012). The resulting two-dimensional confinement of the MIR beam was compared to the previously demonstrated one-dimensional confinement using a slab GaAs/AlGaAs waveguide. As shown schematically in Figure 10.13a, the same QCL MIR source and MCT detector were used to repeat the measurement of acetic anhydride at the surface of a 5 × 10 mm GaAs/AlGaAs planar waveguide. Instead of the aforementioned MBE technique, metal-organic vapor-phase epitaxy (MOVPE) was used for next-generation waveguide chips. Due to the much shorter deposition time, MOVPE is the method of choice for larger-scale MIR waveguide fabrication at low cost and with sufficient layer quality. For testing the sensitivity of the subsequently microstructured waveguide strips, a microcapillary was introduced to facilitate the liquid droplet deposition process, thereby generating droplets with a coverage length of approximately 50 μm at the waveguide surface. The transversal distribution of the optical intensity at the waveguide surface was mapped experimentally via evanescent field absorption

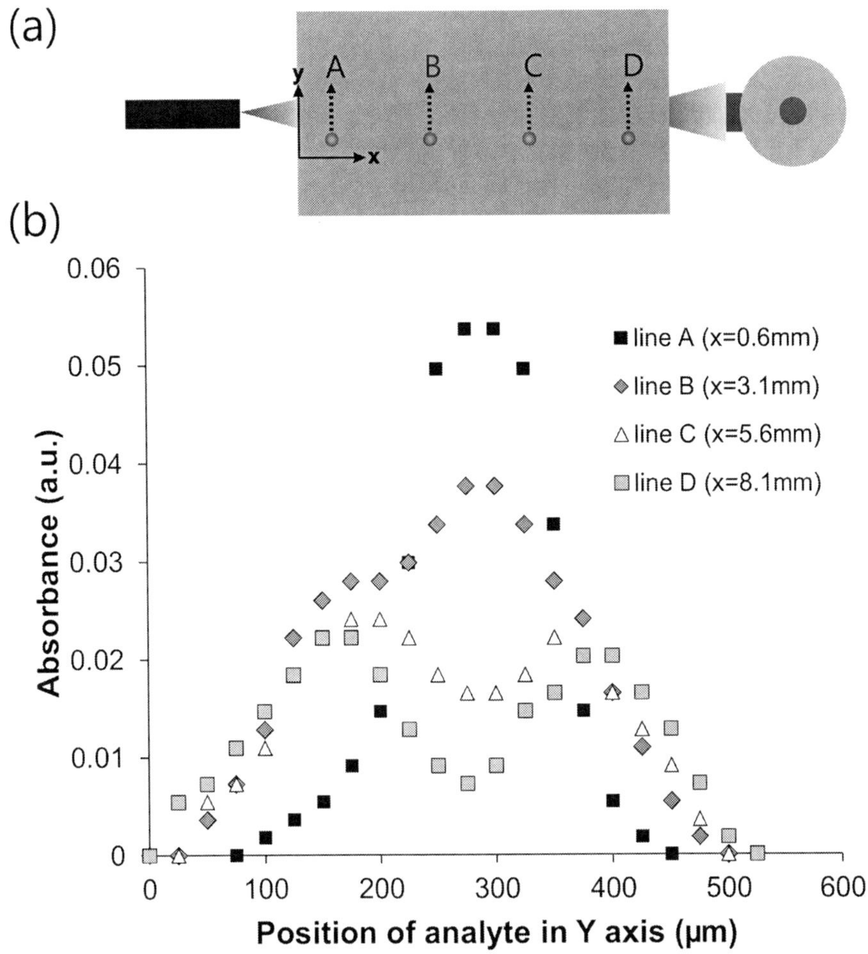

Figure 10.13. (a) Measurement scheme and (b) sensor response to acetic anhydride via absorbance within the evanescent field deposited perpendicular to the radiation propagation axis along the lines A–D at the surface of a planar $GaAs/Al_{0.2}Ga_{0.8}As$ slab waveguide (5 × 10 mm) without lateral confinement. (Reprinted with permission from Wang et al. 2012. Copyright 2012 Royal Society of Chemistry.)

measurements after depositing adjacent droplets of acetic anhydride in four consecutive regions of the waveguide across the waveguide (i.e., along lines A, B, C, and D shown in Figure 10.13). The absorbance profile along each line is shown in Figure 10.13b, and reveals the intensity distribution of the evanescent field, which reflects the propagation of the beam within the GaAs core layer. The absorbance profile on the y axis becomes broader along the beam propagation direction due to the absence of any lateral confinement of the EM field. Therefore, the sensing performance of a planar slab waveguide depends strongly on the analyte coverage length and position, thereby requiring two-dimensional mode confinement for enhanced mode control and sensitivity.

To realize 2-D mode control of MIR photons, strip waveguides were designed and fabricated via photolithography and reactive ion etching (RIE). The etching depth was controlled to approximately 12 μm. Strip waveguides with widths of 200, 50, and 10 μm were fabricated, and IR absorption measurements were performed at the strip waveguide surface using the same measurement setup already described for the analysis of slab waveguides. The system was aligned such that individual target waveguides could be addressed for optimizing light throughput of individual waveguide strips. In this study, acetic anhydride could be detected with a limit of detection (LOD) of 18 pL (19.4 ng) at the surface of a 200-μm-wide waveguide, which is a dramatic enhancement in sensitivity, by three orders of magnitude, compared to the sensitivity obtained using multimode silver halide planar waveguides.

Further measurements were performed at waveguides with a width of 50 μm, whereby each droplet provides a coverage length of 50 μm. The penetration depth of the evanescent field was calculated to be only 0.6 μm, which in turn indicates that the analytical interaction volume of the molecules is in fact not corresponding to the entire droplet volume, but to a 0.6-μm thin liquid film. Based on this argument, the LOD may be recalculated according to the effective interaction volume, thereby achieving an effective LOD of 0.05 pL for acetic anhydride, which is among the most sensitive direct evanescent field absorption measurements with mid-infrared sensor systems reported to date.

In the following, one may ask whether the intensity increase in the evanescent field might continue further if, e.g., the width of the strip waveguide is further decreased to a few micrometers or less. While experiments using such minute sample quantities require further refinement of the experimental setup, simulations on the governing parameters may assist in identifying potential sensitivity gains from tailoring and optimizing the waveguide structure for predicting achievable LODs.

In order to model the behavior of such strip waveguides, COMSOL Multiphysics 4.1 (COMSOL, Burlington, MA, USA) was used, taking advantage of the radio-frequency (RF) module, which simplifies modeling of electromagnetic wave propagation. For brevity, simulation details are omitted here so that we focus on the results obtained. The left side of Figure 10.14 shows a two-dimensional solution of the electric field strength of the propagating mode that can be expected in a 10-μm-wide GaAs strip waveguide with a core thickness of 6 μm. The refractive index for the GaAs core layer was set as n = 3.3. The 6-μm-thick cladding ($Al_{0.2}Ga_{0.8}As$) was assigned a refractive index of n = 3.2, while the surrounding environment was assumed to be air with n = 1; the structural support was considered to be a doped GaAs wafer, which was represented by n = 3.2 as material parameter. As the associated experimental studies are anticipated to be performed

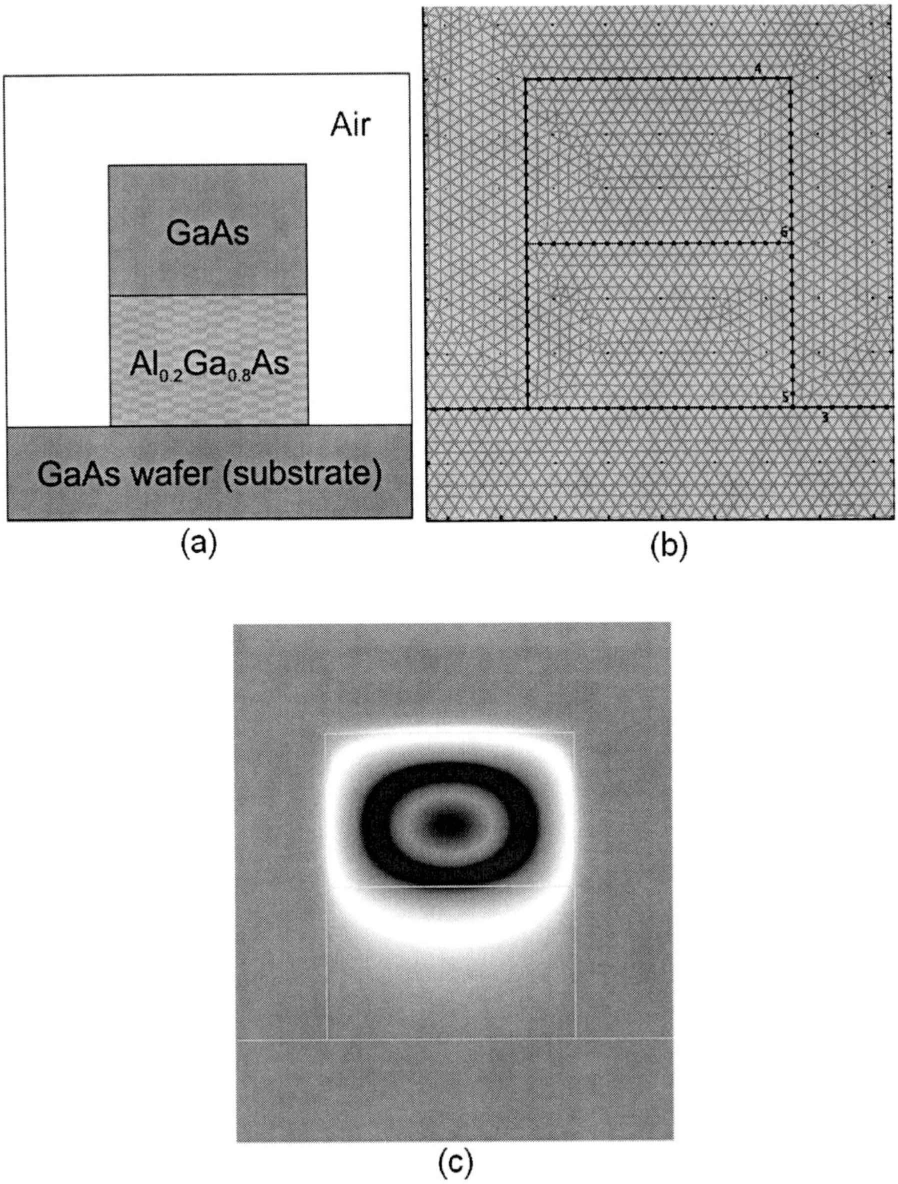

Figure 10.14. FEM simulation of a 10-μm-wide GaAs/AlGaAs strip waveguide: (a) strip waveguide schematic with ambient environment (air); (b) finite-element meshing process; (c) mode analysis of quasi-TM single-mode intensity profile.

with an external cavity coupled QCL (EC-QCL) (Daylight solutions, Poway, CA, USA) providing a tuning range of 1570–1735 cm^{-1} centered at 6.01 μm (1665 cm^{-1}), all calculations were performed in this wavelength regime. The element size parameters of the mesh were set as 0.015 μm (minimum) and 2 μm (maximum), as shown in Figure 10.14b. To complete the simulation, the minimum mesh size is required to be at least smaller than one-fifth of the set wavelength, such that a finer mesh results in significantly longer computing times, which may be critical in particular for 3-D simulation models. The simulation result of the E_y mode intensity profile is shown in Figure 10.14c, representing the fundamental TM_{00} mode. The evanescent field is evident within the GaAs core and within the ambient environment (air). Integration of electromagnetic field intensity was computed across the surrounding environment and was compared to the total intensity integrated across the entire cross-sectional area. The intensity ratio results in a value of approximately 0.01%. A similar simulation was performed for E_x polarized light generating an intensity ratio of 0.12%.

Based on the model above, the evanescent field ratio was then studied after tuning the width of the strip waveguide. As shown in Figure 10.15, the E_y mode shows a dramatic enhancement of evanescent field if the width of the waveguide approaches the cutoff width for the E_y single mode. Cutoff width in this context means that if the strip waveguide is narrowed beyond the cutoff value, propagation of a single mode is no longer supported by the strip waveguide, because of the asymmetric structure of the waveguide in the vertical direction. As an example, at 6.01 μm, a cutoff value of 3.5 μm is calculated with an intensity ratio of 1.1%.

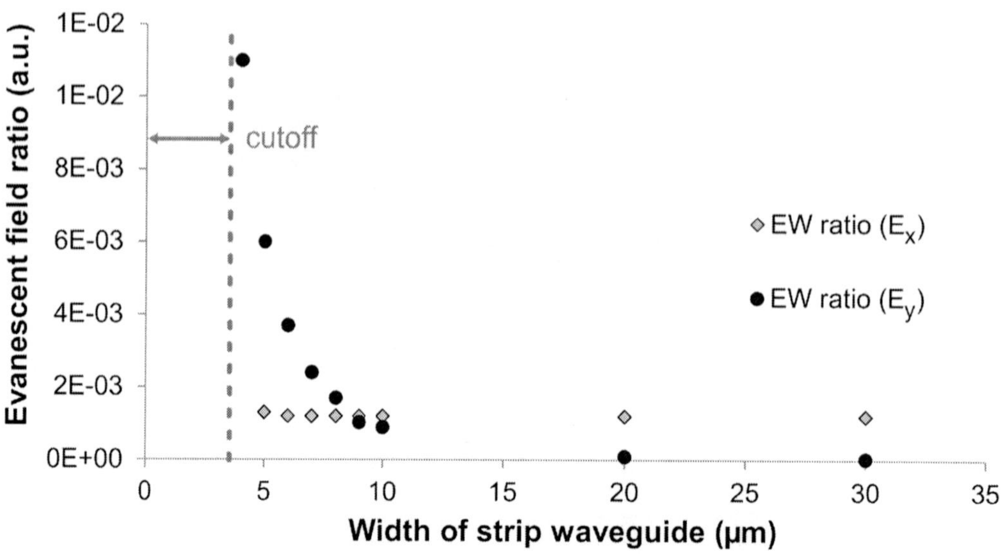

Figure 10.15. Computed evanescent field ratio versus width of the strip waveguide. E_x defines the in-plane horizontally polarized electric field, while E_y is polarized along the in-plane vertical direction. The wavelength in these simulations was set at 6.01 μm, resulting in a cutoff width of approximately 3.5 μm for associated strip waveguides.

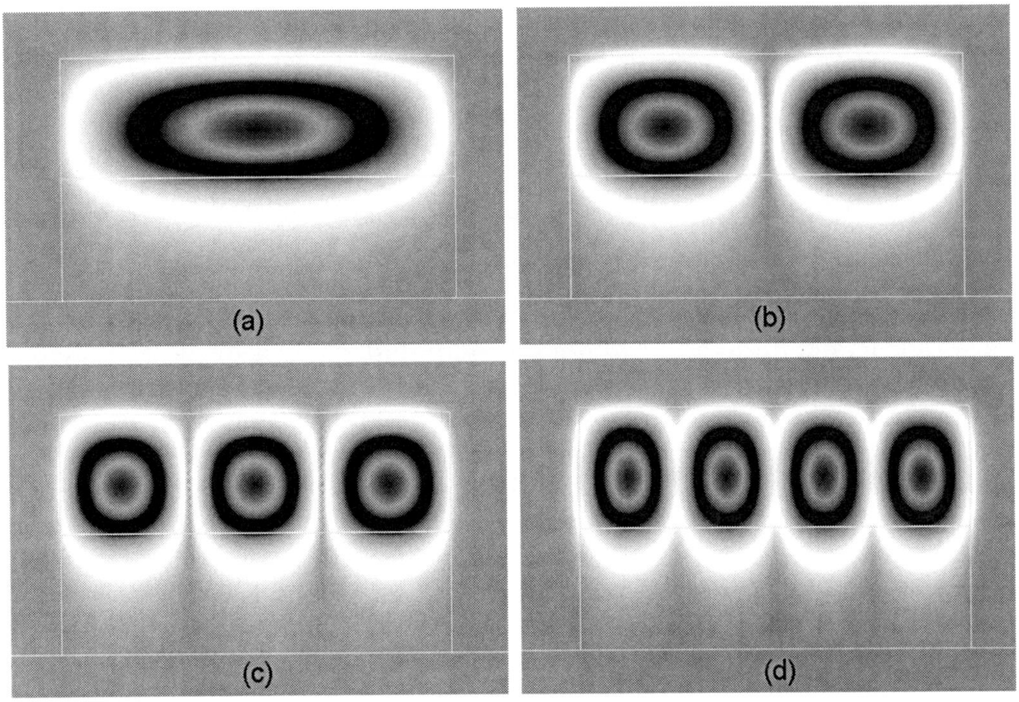

Figure 10.16. Mode analysis of a 10-μm-wide GaAs/AlGaAs strip waveguide. Possible modes supported for 6.01-μm MIR radiation are (a) $TM_{(0,0)}$ fundamental mode, (b) $TM_{(1,0)}$, (c) $TM_{(2,0)}$ and (d) $TM_{(3,0)}$.

In contrast, E_x polarized light maintains a stable value or an only slight increase of the intensity ratio when approaching the cutoff width. The different performance of strip waveguides with E_y and E_x polarized light is attributed to the structure of the strip waveguide. In conclusion, tight 2-D confinement of photons within the strip waveguide leads to a significant additional enhancement of evanescent field. By tuning the width of strip waveguides toward the cutoff width and maintaining the TM_{00} mode of E_y polarized MIR light, the evanescent field appears to be enhanced by at least one order of magnitude compared to wider strip waveguides.

While these examples show the behavior of single modes within GaAs/AlGaAs strip waveguides realizing 2-D confinement, one may ask whether higher-order modes are supported in such strip waveguides and whether their behavior is affected by the width of the strip waveguide in a similar fashion. As an example, the mode analysis of a 20-μm-wide strip waveguide is shown in Figure 10.16, which indicates four different mode profiles from the first-order fundamental mode (Figure 10.16a) to the fourth-order mode (Figure 10.16d).

To explain the existence of different modes within the waveguide structure, one needs to introduce the concept of the *effective mode index,* n_{eff}. In homogeneous media, the refractive index n is defined to quantify the change of wavelength as light propagates through the medium. The effective refractive index n_{eff} is analogously applied for light propagation in a waveguide, and is a function of the propagation constant β of the waveguide:

$$n_{\text{eff}} = \frac{\beta\lambda}{2\pi} \tag{10.75}$$

From Eq. (10.75) it is evident that the effective refractive index depends on both the wavelength and the mode within a certain multimode waveguide. Moreover, it is not just a property of the materials, but reflects the entire waveguide design.

Consequently, mode analysis was applied to GaAs/AlGaAs strip waveguides at a series of waveguide widths. The effective mode indices and corresponding guided modes are plotted in Figure 10.17. The E_y mode shows a cutoff width of 3.5 μm for the $TM_{(00)}$ fundamental mode, which fits well with the results shown in Figure 10.15. With an increase of the waveguide width, only the $TM_{(0,0)}$ mode is supported with rising effective mode index until the second-order mode becomes apparent at a width of approx. 9.5 μm. As the waveguide further increases in width, higher-order modes are evidently supported. Using the 20-μm-wide strip waveguide as an example, four low-order modes could be identified with their corresponding mode profiles as shown in Figure 10.16.

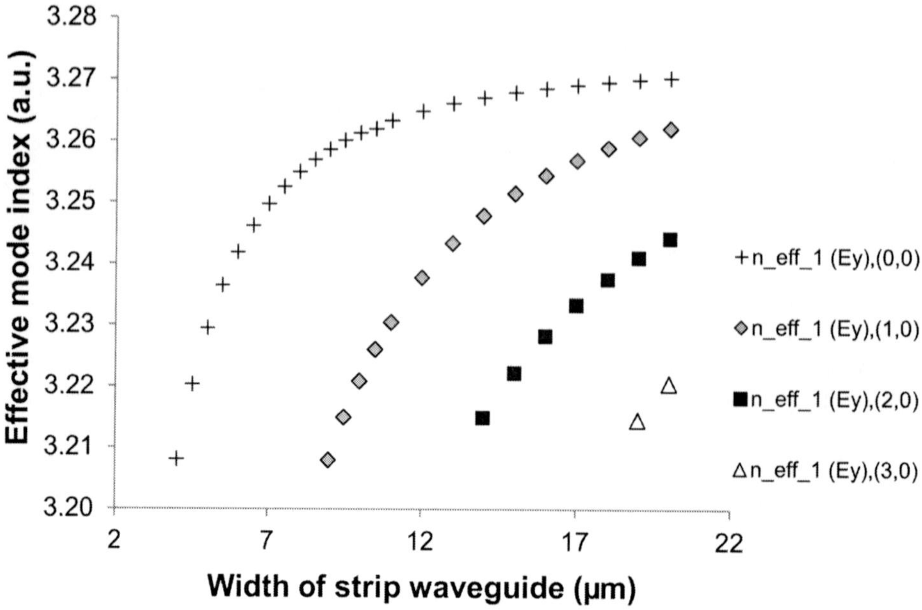

Figure 10.17. Mode study with effective mode index versus a series of waveguide widths.

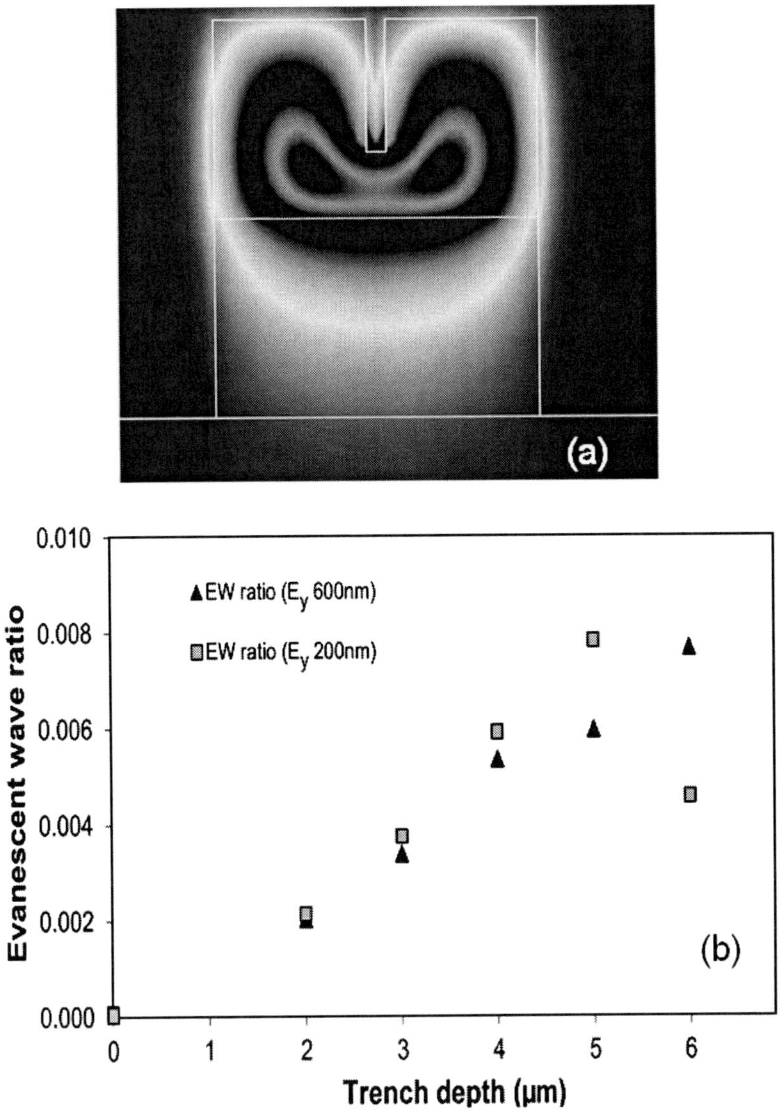

Figure 10.18. (a) 2-D mode propagation study showing the electric field amplitude of a confined mode in a slot-waveguide. (b) Evanescent field ration versus depth of the trench created within a 10-µm-wide strip waveguide. As an example, the results for trench widths of 200 nm and 600 nm are demonstrated in comparison.

Further expanding on the results obtained in the aforementioned study (Wang et al. 2012) providing previously unachieved sensitivity in MIR evanescent field absorption spectroscopy, further enhancements appear possible by, e.g., taking advantage of the evanescent field intensity present within slot-waveguide structures, as simulated in Figure 10.18.

Introducing a trench structure within a strip waveguide results in a so-called slot-waveguide. Mode analysis applied on the cross section of such slot-waveguides with comparable parameter settings from previously simulated strip waveguides reveals that the electric field protrudes from the GaAs–air boundaries with particular intensity inside the slot (Figure 10.18, top). With increased depth, both trenches (200 and 600 nm) show an overall trend toward an enhanced evanescent field, as shown in Figure 10.18 (bottom). These simulations predict significant further sensitivity enhancements by up to one or two additional orders of magnitude compared to strip waveguides without additional slot structure. Hence, another example is provided where simulations may guide future experiments prior to arbitrarily generating—frequently complex to fabricate—optical structures. Accordingly, such slot structures will be investigated in the near future for further enhancing the performance of real-world MIR sensing devices corroborating the central role of FEM simulations in accelerating the development of advanced waveguide designs.

Further studies aim at investigating resonant optical structures fabricated from thin-film MIR waveguides, as inspired by the progress in UV/Vis and NIR resonant cavities (Armani et al. 2007; Nitkowski et al. 2008). Consequently, molecules may be detected with so far unachieved sensitivity in the MIR by determining, e.g., a resonance peak shift, a change in Q factor, or a change in refractive index. This technology is considered particularly attractive for direct and label-free detection of biomolecular interactions within a portable MIR sensing system taking advantage of on-chip technology for establishing mid-infrared bioassays.

6. CONCLUSIONS AND OUTLOOK

This chapter has provided a brief introduction to finite-element modeling and has provided selected examples of its contribution to the development of modern infrared sensor technology. As a well-established simulation method, FEM fits well to complex optical domain structures and is particularly applicable to addressing situations where the electromagnetic fields are highly localized. Apart from the more common applications in MIR light source and detector design, current research in IR chem/bio sensor technology is focusing particularly on waveguide design because of the demand for more sensitive and highly integrated infrared sensing platforms applied in, e.g., increasingly extreme sensing environments, for increasingly complex sensing matrices, and where increasingly limited target analyte concentrations and/or volumes are present, such as in biodiagnostics and bioassays. Complementary to the development of micro- and nanotechnology along with appropriate micro- and nanofabrication techniques, recent developments in opto-electronics have been paving the way toward nanophotonic sensor technology inherently including the transition from conventional geometric optics to nano-optics. This development clearly demands appropriate modeling and simulation solutions, as provided by FEM, which is already an indispensable tool in UV/Vis device design. Given that NIR sensing techniques have significantly matured, and that MIR integrated photonics sensors are considered the most recent emerging optical platform, extending such modeling efforts to this spectral regime appears to be a logical consequence. While 3-D simulations of the comparatively large geometries in the MIR still do not necessarily facilitate a full-structure simulation for waveguide design, 2-D FEM modeling for, e.g., strip waveguides already fits well with the obtained experimental results for and assists in the design of next-generation evanescent field infrared sensing techniques.

With the introduction of the quantum cascade laser, a dramatic change in the research field of miniaturized MIR sensors toward wafer-scale technologies has been observed, which is now complemented by dedicated MIR waveguide research. It is anticipated that novel waveguide structures such as, e.g., slot waveguides, microresonators, etc., will meet the demand of label-free detection in chem/bio sensing applications within portable MIR sensing systems, taking full advantage of on-chip transducer technology. Additionally, multiphysics modeling based on FEM simulation will not only contribute to the analysis of the electromagnetic field distribution, it may combine optical simulations with, e.g., thermal or stress analysis of optical waveguides, which is essential for finally manufacturing advanced mid-infrared chem/bio sensing and assay platforms of practical relevance.

REFERENCES

Armani A.M., Kulkarni R.P., Fraser S.E., Flagan R.C., and Vahala K.J. (2007) Label-free, single-molecule detection with optical microcavities. *Science* **317**(5839), 783–787. DOI: 10.1126/science.1145002

Capasso F., Gmachl C., Sivco D.L., and Cho A.Y. (2002) Quantum cascade lasers. *Phys. Today* **55**(5), 34–40. DOI: 10.1063/1.1485582

Charlton C. (2005a) Quantum cascade laser for mid-infrared chemical sensing. Ph.D. thesis, Georgia Institute of Technology, Atlanta, GA..

Charlton C., De Melas F., Inberg A., Croitoru N., and Mizaikoff B. (2003). Hollow-waveguide gas sensing with room-termperature quantum cascade lasers. *IEE Proc.—Optoelectron.* **150**(4), 306–309. DOI: 10.1049/ip-opt:20030673

Charlton C., Giovannini M., Faist J., and Mizaikoff B. (2006). Fabrication and characterization of molecular beam epitaxy grown thin-film GaAs waveguides for mid-infrared evanescent field chemical sensing. *Anal. Chem.* **78**(12), 4224–4227. DOI: 10.1021/ac052214a

Charlton C., Katzir A., and Mizaikoff B. (2005b) Infrared evanescent field sensing with quantum cascade lasers and planar silver halide waveguides. *Anal. Chem.* **77**(14), 4398–4403. DOI: 10.1021/ac048189a

Charlton C., Temelkuran B., Dellemann G., and Mizaikoff B. (2005c) Mid-infrared sensors meet nanotechnology: Trace gas sensing with quantum cascade lasers inside photonic band-gap hollow waveguides. *Appl. Phys. Lett.* **86**(19), 194102. DOI: 10.1063/1.1925777

Chen J.Z., Liu Z., Gmachl C.F., and Sivco D.L. (2005) Silver halide fiber-based evanescent-wave liquid droplet sensing with room temperature mid-infrared quantum cascade lasers. *Opt. Express* **13**(16), 5953–5960. DOI: 10.1364/OPEX.13.005953

Faist J., Capasso F., Sivco D.L., Sirtori C., Hutchinson A.L., and Cho A.Y. (1994) Quantum cascade laser. *Science* **264**(5158), 553–556. DOI: 10.1126/science.264.5158.553

Gendron L., Koeniguer C., Berger V., and Marcadet X. (2005). High resistance narrow band quantum cascade photodetectors. *Appl. Phys. Lett.* **86**(12), 121116. DOI: 10.1063/1.1884257

Graf M., Hoyler N., Giovannini M., Faist J., and Hofstetter D. (2006). InP-based quantum cascade detectors in the mid-infrared. *Appl. Phys. Lett.* **88**(24), 241118. DOI: 10.1063/1.2210088

Hawkins A.R. and Schmidt H. (2008). Optofluidic waveguides: II. Fabrication and structures. *Microfluid Nanofluid.* **4**(1–2), 17–32. DOI: 10.1007/s10404-007-0194-z

Hrennikoff A. (1941). Solution of problems of elasticity by the frame-work method. *J. Appl. Mech.* **8**, A619–A715.

Huang M., Yang J.J., Jun S., Mu S.J., and Lan Y.Z. (2011) Simulation and analysis of a metamaterial sensor based on a microring resonator. *Sensors—Basel.* **11**(6), 5886–5899.

Jackson J.D. (1999). *Classical Electrodynamics.* John Wiley, New York.

Jenkins F.A. and White H.E. (1976) *Fundamentals of Optics.* McGraw-Hill, New York.

Kawano K. and Kitoh T. (2001) *Introduction to Optical Waveguide Analysis: Solving Maxwell's Equations and the Schrödinger Equation.* John Wiley, New York,

Kazarino R.F. and Suris R.A. (1971) Possibility of amplication of electromagnetic waves in a semiconductor with a superlattice. *Sov. Phys. Semicond.* (USSR) **5**(4), 707–709.

Kim S.S., Young C., and Mizaikoff B. (2008) Miniaturized mid-infrared sensor technologies. *Anal. Bioanal. Chem.* **390**(1), 231–237. DOI: 10.1007/s00216-007-1673-5

Kraft M., Jakusch M., Karlowatz M., Katzir A., and Mizaikoff B. (2003) New frontiers for mid-infrared sensors: Towards deep sea monitoring with a submarine FT-IR sensor system. *Appl. Spectrosc.* **57**(6), 591–599. DOI: 10.1366/000370203322005256

Messica A., Greenstein A., and Katzir A. (1996) Theory of fiber-optic, evanescent-wave spectroscopy and sensors. *Appl. Opt.* **35**(13), 2274–2284. DOI: 10.1364/AO.35.002274

Mizaikoff B. (1999) Mid-infrared evanescent wave sensors—A novel approach for subsea monitoring. *Meas. Sci. Technol.* **10**(12), 1185–1194. DOI: 10.1088/0957-0233/10/12/310

Mizaikoff B. (2003) Mid-IR fiber-optic sensors. *Anal Chem.* **75**(11), 258A–267A. DOI: 10.1021/ac031340g

Mortensen N.A., Xiao S.S., and Pedersen J. (2008) Liquid-infiltrated photonic crystals: Enhanced light-matter interactions for lab-on-a-chip applications. *Microfluid. Nanofluid.* **4**(1–2), 117–127. DOI: 10.1007/s10404-007-0203-2

Nitkowski A., Chen L., and Lipson M. (2008) Cavity-enhanced on-chip absorption spectroscopy using microring resonators. *Opt Express* **16**(16), 11930–11936. DOI: 10.1364/OE.16.011930

Pelosi G. (2007) The finite-element method, Part 1: R. L. Courant [Historical Corner]. *IEEE Antenn. Propag. Mag.* 49(2), 180–182. DOI: 10.1109/MAP.2007.376627

Rotermund F., Petrov V., and Noack F. (2000) Difference-frequency generation of intense femtosecond pulses in the Mid-IR (4-12 μm) using $HgGa_2S_4$ and $AgGaS_2$. *Opt. Commun.* **185**(1–3), 177–183. DOI: 10.1016/S0030-4018(00)00987-1

Simhony S., Schnitzer I., Katzir A., and Kosower E.M. (1988) Evanescent wave infrared-spectroscopy of liquids using silver-halide optical fibers. *J Appl. Phys.* **64**(7), 3732–3734. DOI: 10.1063/1.341369

Wang X., Kim S.S., Roßbach R., Jetter M., Michler P. and Mizaikoff B. (2012) Ultra-sensitive mid-infrared evanescent field sensors combining thin-film strip waveguides with quantum cascade lasers. *Analyst* **137**(10), 2322–2327. DOI: 10.1039/c1an15787f

Yang W.G., Conkey D.B., Wu B., Yin D.L., Hawkins A.R., and Schmidt H. (2007) Atomic spectroscopy on a chip. *Nat. Photonics* **1**(6), 331–335. DOI: 10.1038/nphoton.2007.74

Young C., Kim S.S., Luzinova Y., Weida M., Arnone D., Takeuchi E., Day T., and Mizaikoff B. (2009). External cavity widely tunable quantum cascade laser based hollow waveguide gas sensors for multianalyte detection. *Sens. Actuators B* **140**, 24–28. DOI: 10.1016/j.snb.2009.03.023

CHAPTER 11

Design and Optimization of Optical Gas Sensor Systems

I. Sieber

U. Gengenbach

1. INTRODUCTION

1.1. GAS SENSING BASED ON THE LAMBERT-BEER LAW

Identification of gaseous substances is a long-standing task for which numerous approaches exist. Besides analysis principles that rely on altering the reflective properties of functionalized surfaces, optical gas sensors frequently are based on the Lambert-Beer law:

$$I(z) = I_0 \cdot e^{-\alpha \cdot z} \tag{11.1}$$

where I_0 is the intensity of the ingoing light, $I(z)$ is the intensity of the outgoing light, α is the absorption coefficient, and z is the path length in the absorbing material.

The absorption coefficient α depends on the wavelength λ_0 of the light according to the following relation:

$$\alpha = 2k_0 \cdot n \cdot \kappa = \frac{4\pi \cdot n \cdot \kappa}{\lambda_0} \tag{11.2}$$

DOI: 10.5643/9781606503201/ch11

where n is the real part of the complex refraction index, κ is the attenuation coefficient of the refraction index. α can also be written as

$$\alpha = \varepsilon \cdot c \tag{11.3}$$

Here ε is the extinction coefficient and c is the molar concentration. From this relation the Lambert-Beer law yields for a measured α and a given substance-specific ε the concentration c of the substance.

The Lambert-Beer law holds under the following conditions:

- The molecular and atomic absorption centers of the substance do not interact.
- α is constant for λ_0.
- α is independent of the path length z.
- There is no saturation effect.

There are two main applications of the Lambert-Beer law:

1. The wavelength dependency of the extinction coefficient α leads to substance-specific relative absorption spectra which can be used to identify unknown substances. Figure 11.1 shows absorption spectra of atmospheric gases.
2. Measuring extinction (i.e., log I/I_0) of a known substance yields the concentration of the substance, provided the temperature and pressure dependence of the absorption coefficient α has been compensated either by calibration or by reference measurements.

Equations (11.1) and (11.2) can be translated into a schematic gas sensor design based on the Lambert Beer law. As shown in Figure 11.2, a schematic gas sensor cell based on the Lambert-Beer law is composed of a gas-filled cell of optical path length z, with a light source mounted on one end that sends out a light beam with intensity I_0 and a detector receiving the light with intensity I mounted on the opposite end. Light with wavelength λ passes through the gas and is attenuated by absorption, depending on the gas and on the wavelength-specific absorption coefficient α(λ), leading to $I < I_0$. The sensitivity of such a sensor obviously depends on three main factors:

1. The light source
2. The detector
3. The path length of the beam in the gas cell

The emission wavelength of the light source and the matching spectral sensitivity of the detector have to be chosen in such a way that the absorption coefficient α(λ)

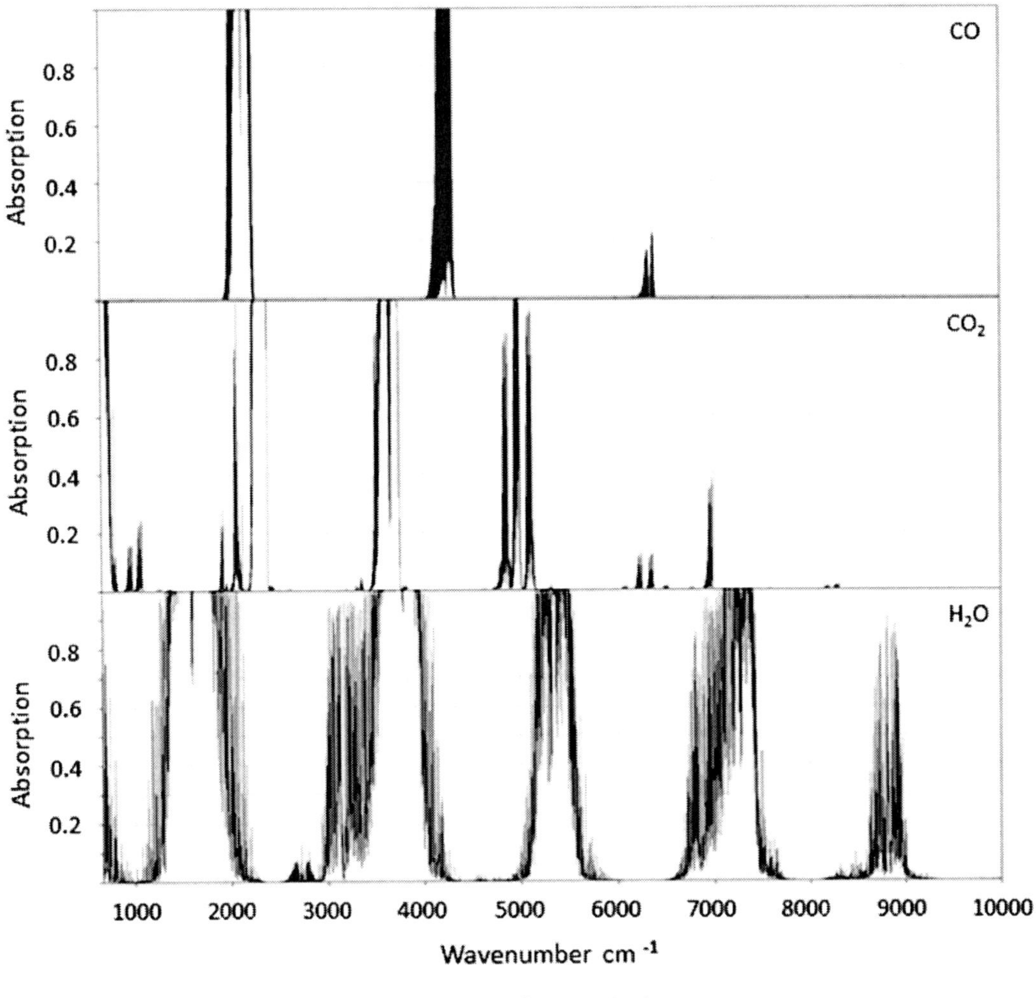

Figure 11.1. Low-resolution absorption spectra of atmospheric gases.

is as high as possible for the desired gas. In the above example of CO_2 preferably, the highest peak (wavenumber 2400, ~4.2 μm) would be selected. Figure 11.1 also illustrates that the target wavelength has to be selected carefully to obtain selectivity between gases. At wavenumbers around 2000, both water and CO_2 absorb, so in this wavelength range no discrimination between these gases is feasible without additional measures.

The path length of the light beam in the gas has to be dimensioned such that the lowest gas concentration still yields a detectable signal at the detector.

It is evident that these factors are not disjoint. They can be offset against each other in order to optimize system design, performance, and price.

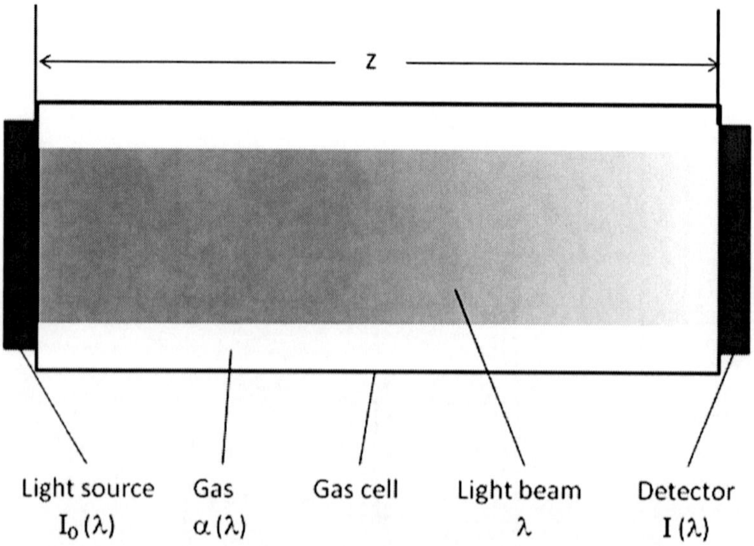

Figure 11.2. Schematic gas sensor design based on the Lambert-Beer law.

1.2. STATE OF THE ART OF FOLDED-OPTICAL-PATH GAS SENSOR CELLS

An early application of this measurement principle for gas sensing was developed by White (1942) for measurement of vapor-phase spectra in oil fabrication. It is customary to utilize spectral analysis for that purpose. White noted that measuring the spectra of substances with high boiling points requires either heating the absorption cell or making it very long. While the first approach decreases the accuracy of absorption measurement, long cells are not applicable for small sample volumes due to low interaction at low concentrations. On the contrary, for low concentrations the light has to pass through the sample as often as possible. Hence White developed a cell with a folded optical path based on three spherical concave mirrors.

The schematic design of a White cell is shown in Figure 11.3. The mirrors 1a and 1b, mounted close to each other, and the opposite mirror 2 have the same curvature. The centers of curvature of mirrors 1a and 1b are on the front surface of mirror 2. The center of curvature of mirror 2 is between mirror 1a and mirror 1b. This arrangement ensures that incoming light from a source (top right) is reflected from mirror 1a to mirror 2 to mirror 1b to mirror 2 and back to mirror 1a. If the centers of curvature of mirror 1a and mirror 1b coincided at one point, this would go on forever—without the light ever leaving the cell. However, since the two centers are offset from each other, the reflected image on mirror 2 moves slightly to one side until the final reflection on mirror 1b finally falls off mirror 2 and can

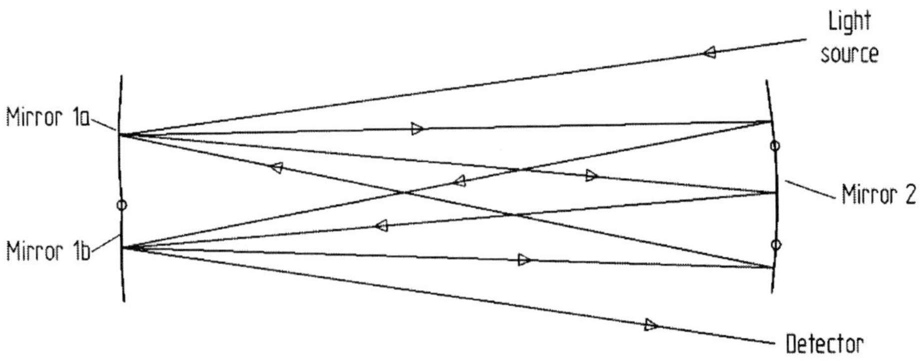

Figure 11.3. Schema of the original White cell. Idea from White (1942).

be picked up by a detector. By adjusting the curvature and the offset between the two centers of curvature of mirrors 1a and 1b, compact cells with very long optical paths can be designed.

In 1965 Herriott and Schulte also developed a cell with a folded optical path for optical data filtering and storage purposes (see Figure 11.4). This design has also been applied for optical gas sensing. The Herriott cell consists of two spherical or slightly aspherical mirrors positioned opposite each other. A laser beam is injected through either a hole or an opening in the reflective coating in one of the mirrors. In order to minimize losses in the chamber, the injected beam has to be matched by focusing optics to the natural modes of the mirror cavity. A small, so-

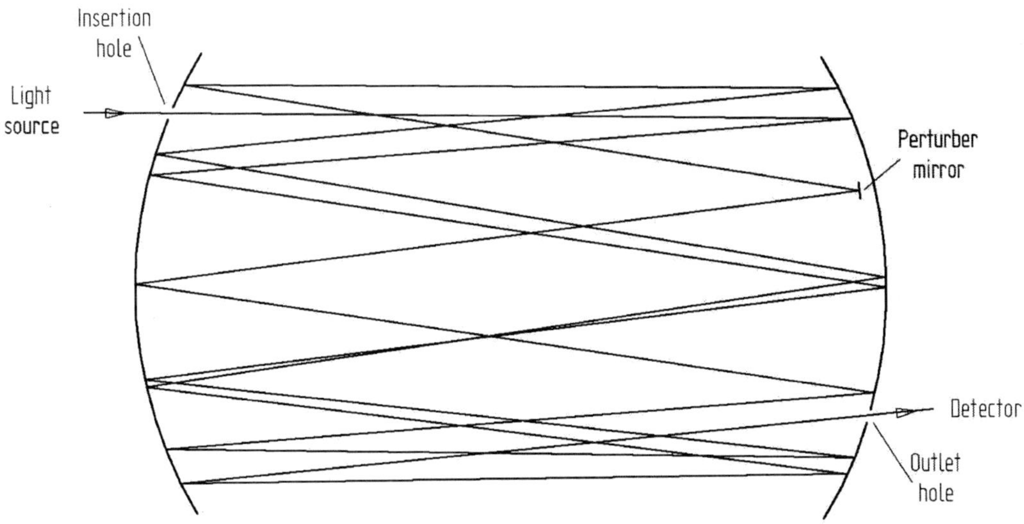

Figure 11.4. Schema of the original Herriott cell. Idea from Herriott and Schulte (1965).

called perturber mirror is positioned in front of the second mirror and is used to control the number of reflections in the cell as well as the exit of the beam. The beam can be coupled out of the cell again by a hole or an opening in the reflective coating in one of the main mirrors or by an optical fiber inserted in the chamber.

This arrangement can be used to build small and compact gas sensors. Steffen et al. (2007) introduced a modified Herriott cell design for an O_2 and CO gas sensor in which the cell volume is additionally reduced by a coaxial inner pipe. The reflection pattern is a circle about the mirror center.

By proper positioning and alignment of sources and detector, multiple beams can be simultaneously coupled into the cell without interference of the spot patterns on the mirror. This is the basis for measuring multiple gases in parallel with one cell. A compact, four-channel cell for a Mars science mission has been designed based on this principle (Tarsitano and Webster 2007). Four lasers are mounted on one mirror, and four detectors are mounted on the other mirror. The injection holes are positioned on a radius from the mirror center in order to obtain the same number of reflections. Altering the angle of the injecting beam turns the circular into an elliptical spot pattern and thus avoids spillover of beams at the output holes. Figure 11.5a shows the cell design. Figure 11.5b shows the resulting reflection pattern at the input mirror, with circles around the injection holes. This cell design permits the simultaneous detection of H_2O, CO_2, CH_4, H_2O_2, and N_2O with single-ppmv detection limits. In addition to the multipass cell, this instrument also contains a reference gas cell and a reference detector for each of the

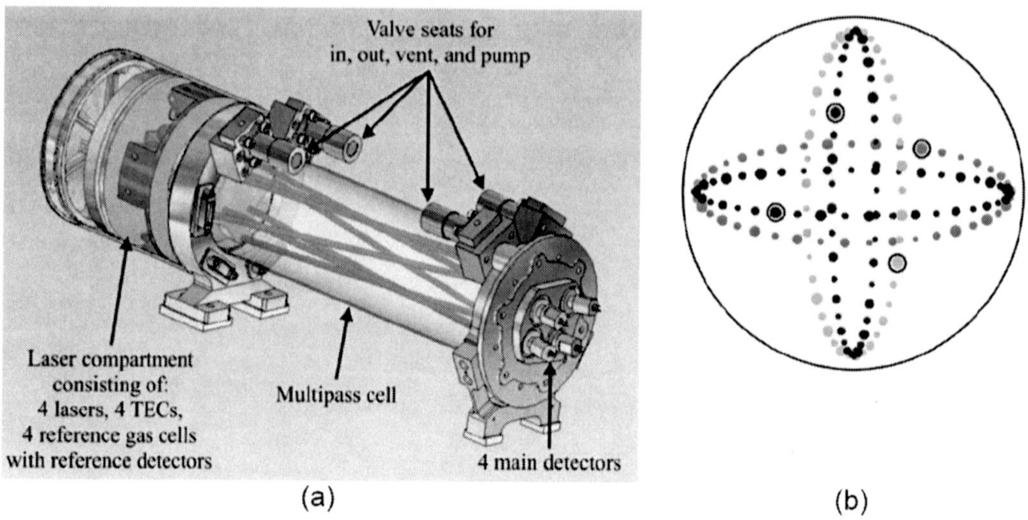

Figure 11.5. (a) Design of a four-beam multipass gas sensor cell. (b) Reflection pattern on the near mirror of the four-beam multipass cell. [(a) Reprinted with permission from Tarsitano and Webster 2007. Copyright 2007 The Optical Society of America.]

gases to be detected. This setup allows high-precision measurements without the need for separate calibration steps.

One criticism of both the White and Herriott cells is their poor space efficiency, which limits their potential for miniaturization. A modification proposed by Herriott alleviates this problem by replacing the spherical mirror with an astigmatic mirror. This creates elliptical reflection patterns, resulting in better usage of the mirror surface and allowing multichannel measurements. The trade-off is more complex mirror fabrication and alignment. Das and Wilson (2010) outlined a simpler multipass cell based on cylindrical mirrors. Their setup consists of two cylindrical mirrors of 40 mm diameter spaced at 90 mm. Twisting the cylinder mirror axes between 45° and 75° allows up to a 50-m path length. A sensitivity analysis, however, showed that this arrangement is very sensitive to mirror spacing and mirror-axis twist tolerances.

A rather new cell design is the quasi-chaotic cell (Narimanov et al. 2005). This design was developed to further improve on the main drawbacks of the Herriott and White cells:

- Necessity for precise alignment of the focusing elements
- Size

The need for precise alignment of the focusing elements is a problem in manufacturing, but it also a consideration with regard to long-term stability and freedom from maintenance. As to the second issue, Herriott, White, and the above cylindrical mirror cells are all usually at least several tens of centimeters in length. The setup with the focusing mirrors connected by a tube does not lend itself easily to miniaturization without running into manufacturing and assembly problems.

These issues can be alleviated by a monolithic optical cell design. Such a shape shows quasi-chaotic behavior in that rays injected into the cell can follow well-defined bowtie or chaotic orbits depending on the injection position and angle (see Figure 11.6). Ideally, in a stable orbit the ray has the same angle of incidence at all reflection points, allowing efficient use of ultrahigh-reflection interference coatings and thus minimizing the energy loss on every reflection (Narimanov et al. 2005).

Rays injected into such a stable bowtie orbit cannot cross into other orbits, hence the surface area of the cell that is used is well defined. This implies that the remaining cell surface does not have to meet such stringent requirements with respect to optical surface quality as the optically used surface. Moreover, the optically unused surface can be utilized for gas input and output. As the cell is an oval rotated about its minor z axis and thus can be conveniently manufactured by grinding or diamond-point turning, the stable orbit property can be exploited to lower fabrication costs by machining only these reflection areas with high precision.

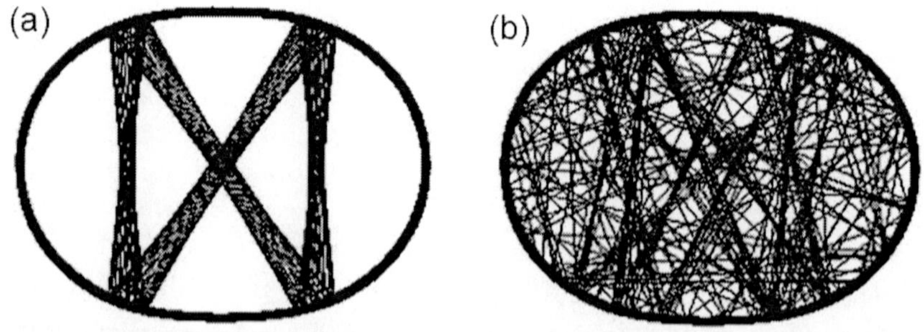

Figure 11.6. Bowtie orbit (a) and chaotic orbits (b) in a quasi-chaotic cell. (Reprinted with permission from Narimanov et al. 2005. Copyright 2005 The Optical Society of America.)

1.3. GENERAL SETUP OF MULTIPASS GAS SENSORS

The above discussion of the Lambert-Beer law and the state of the art leads to the following general constituents of optical multipass gas sensors for detection of a single gas:

- Gas feed
- Light source
- Detector
- Filter
- Optical multipass cell

The gas cell may be complemented by a reference cell to compensate the pressure and temperature dependence of the extinction coefficient (see Section 1.2, Figure 11.5). These components will be discussed in detail in the following paragraphs.

1.3.1. Gas Feed

The gas feed, i.e., the inlet and outlet of the gas to be measured, can be active or passive. "Active" means that the gas is being pumped through the multipass cell; "passive" means that the gas enters and leaves by means of diffusion. In both cases it has to be ensured that the gas reliably permeates the entire volume that is traversed by the light beams in an amount of time that is compatible with the application. Insufficient diffusion in the cell volume may lead to measuring errors. Active gas feeding has the advantage that the cell can be flushed for calibration or for measuring concentrations of different gases. This kind of feed is usually used for scientific instruments, while diffusion is used mainly for consumer applications.

1.3.2. Light Source

Important criteria for selection of a light source for a gas sensor are the wavelength and the modulation capabilities to suppress undesired effects in the signal such as noise.

Proper selection of the target wavelength to measure gas concentration is paramount. The absorption spectra illustrated in Figure 11.1 show that the peaks for CO_2 and H_2O are very close together. To achieve selectivity, i.e., to avoid the influence of adjacent peaks generated by other gases, the emission wavelength of the light source and/or the sensitivity of the detector have to be matched as closely as possible to the target wavelength.

Moreover, the absorption of many gases varies significantly between the mid- and near-infrared. In many cases, absorption in the mid-infrared is several magnitudes higher than in the near-infrared, which leads to significantly lower detection limits in the mid-infrared range. Table 11.1 gives an overview of the detection limits of a number of gases in the near- and mid-infrared range (Southwest Sciences 2011).

Table 11.1 also shows that several gases, such as HF, HI, and O_2, absorb only in the near-infrared; hence there is no choice between wavelength ranges. For the other gases, sensitivity in the mid-infrared range is so much higher that in recent years mid-IR sources have been increasingly applied for their detection.

The light sources can be broadly subdivided into narrow-band and broadband sources. The narrow-band sources have one main wavelength with a full-width half-maximum (FWHM) of up to a few hundreds of nanometers. They can be further subdivided into high-power (laser diodes) and low-power (light-emitting diode, LED) sources (Figure 11.7).

Since telecommunication wavelengths are in the near-IR range, a large variety of laser diodes (e.g., 1.3- to 2.3-μm GaSb, InAs material systems) are available. Due to the substantially higher absorption in the mid-IR range (see Table 11.1), laser sources for wavelengths beyond 5 μm are a continuous topic of research and development. Lead-salt lasers (PbSnSe, PbSSe material systems) with wavelengths up to 14 μm have been available for many years, with peak powers in the watt range in continuous-wave (cw) mode. The drawback of these high-power sources is the elaborate cooling that is needed, frequently even cryo-cooling. This makes this kind of sources less suitable for portable or miniaturized gas-sensing devices.

A recent development is quantum cascade lasers with Bragg feedback gratings based on GaAs/AlGaAs and GaInAs/AlInAs material systems. The material system determines the wavelength range, while the emission can be adjusted by the design of the Bragg grating. These laser diodes allow currently only cw powers in the milliwatt range. This drawback, however, is compensated by operation at room temperature with inexpensive cooling such as Peltier coolers.

Table 11.1. Detection limits of gases in the near and mid-infrared range

	Mid-infrared		Near-infrared	
Gas	Detection limit (ppb)	Lambda (nm)	Detection limit (ppb)	Lambda (nm)
H_2O	2.0	5,940	60	1,390
CO_2	0.13	4,230	700	2,040
			5,500	1,570
CO	0.75	4,600	30,000	1,570
			500	2,330
NO	5.8	5,250	60,000	1,800
	—	—	1,000	2,650
NO_2	3.0	6,140	340	680
N_2O	0.44	4,470	1,000	2,260
SO_2	14	7,280	—	—
CH_4	1.7	3,260	600	1,650
C_2H_2	3.5	7,400	80	1,520
HF	—	—	10	1,310
HCl	0.83	3,400	150	1,790
HBr	7.2	3,820	600	1,960
HI	—	—	2,100	1,540
HCN	12	6,910	290	1,540
H_2S	—	—	20,000	1,570
NH_3	0.80	10,300	800	1,500
H_2CO	8.4	3,550	50,000	1,930
PH_3	6.2	10,100	1,000	2,150
O_2	—	—	78,000	760
O_3	11	9,500	—	—

Source: Based on Southwest Sciences data.

In general, the far-field angles of laser sources are small (in the range of 50°), so simple collimation optics can be used.

Another narrow-band light source is infrared LEDs. The range from 1.6 to 2.3 μm is covered by GaSb-InAs and the range from 2.7 to 4.6 μm by InAs material systems. At room temperature they can be operated in cw or pulsed mode. For stable and increased optical power, thermoelectric cooling is recommended. The FWHM is usually in the range of several 100 nm up to 1 μm. Modulation ranges of up to several hundred megahertz can be achieved. For collimation purposes the LEDs are frequently sold in a housing with a parabolic reflector.

Broad-band light sources are black-body emitters and incandescent lamps. The black-body emitters are in the simplest case resistance wire wound around a cylindrical ceramic carrier. This type of emitter is frequently combined with a parabolic reflector for beam collimation (see Figure 11.8).

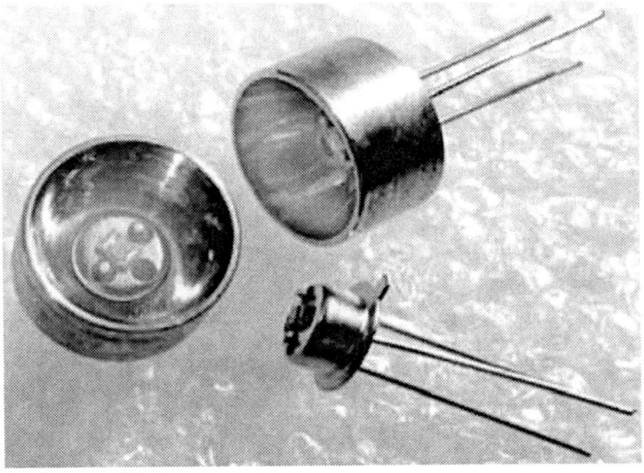

Figure 11.7. IR-LED. (Reprinted from www.laser2000.de/index.php?id=373850.)

Newer developments realize the emitter as foils or planar filaments in thin-film technology (Figure 11.9). This technology allows a reduction in thermal mass and an increase of modulation frequency to 100 Hz and beyond. Nevertheless, power consumption is still in the range of several hundred milliwatts, as the emitter is operated at 800–1000°C. This requires effort toward cooling the housing to keep the emitter in a stable operating range of 100–120°C. The spectral output ranges from below 1 μm to up to 20 μm.

Another broad-band emitter at low cost is a standard incandescent lamp, with an emission from the visible over the near-IR up to the lower mid-IR range. Modulation frequency is in the low 10-Hz range, power consumption in the watt range.

Figure 11.8. Filament black-body emitter with parabolic reflector. (Reprinted from www.laser2000.de/index.php?id=365586.)

Figure 11.9. Thin-ilm emitters in TO5 housing. (Reprinted from www.laser2000.de/index.php?id=365589.)

1.3.3. Detector

In the near IR-range, Ge photodiodes (0.8–1.8 μm) are the standard detector type. High sensitivities can be achieved with cryo-cooling, but these detectors can also be operated at room temperature with good sensitivity. An alternative in this wavelength range is a InGaAs detector with comparable sensitivity at room temperature and very low dark current. These frequently come in a robust TO housing with a quartz or sapphire window (see Figure 11.10).

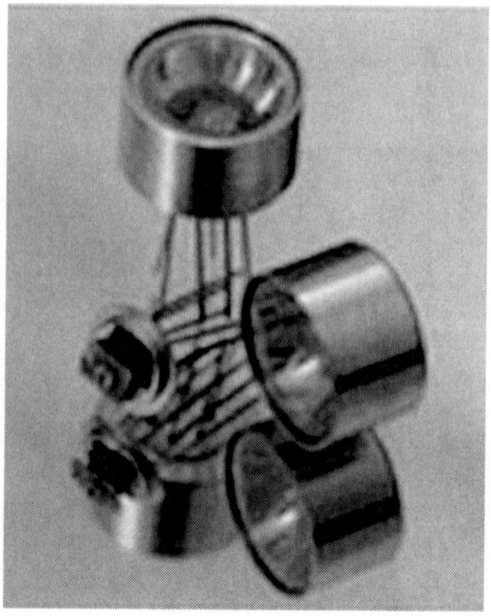

Figure 11.10. IR photodiodes in TO housing. (Reprinted from www.laser2000.de/index.php?id=365268.)

Figure 11.11. Thermopile in TO housing. (Reprinted from www.micro-hybrid.de.)

The so-called photovoltaic detectors are based on the InAs (1- to 3.6-μm) and InSb (2- to 5.5-μm) material systems. They are operated in DC mode and are particularly suitable for fast pulsed measurements. However, InSb requires cryo-cooling.

The detector of choice for high wavelengths, up to 26 μm, is the mercury cadmium telluride diode. In the mid-IR range it can be operated with a Peltier cooler; beyond that range, cryo-cooling is required.

An alternative measuring principle is the thermopile. Thermopile detectors can be used from the x-ray range to the far-IR. Because the measurement principle is based on the thermoelectric (Seebeck) effect, the detector is particularly sensitive to ambient temperature (conduction, convection). Hence housing temperature has to be closely monitored to achieve reliable measurements. In some devices this is realized with integrated thermistors. They usually come in TO housings with windows for cutoff of undesired wavelength ranges (see Figure 11.11).

1.3.4. *Window Materials*

In order to cut off undesired wavelengths emitted from a broad-band light source, filter materials are inserted as windows in front of a broad-band detector. Table 11.2 shows the wavelength ranges for a number of frequently used filter materials.

Table 11.2. Cutoff wavelengths of frequently used window materials*

MATERIAL	CUTOFF WAVELENGTHS	REMARKS
NaCl	0.25 μm/16 μm	Low cost, sensitive to water or moisture
KBr	0.23 μm/25 μm	Sensitive to water or moisture
CaF_2	0.13 μm/11 μm	Robust
BaF_2	0.15 μm/12 μm	Sensitive to water or moisture
CsI	0.23 μm/50 μm	Sensitive to water or moisture
ZnSe	0.50 μm/20 μm	Robust
Al_2O_3	0.15 μm/5 μm	Robust
Si	1.20 μm/8 μm	Low cost, brittle
Ge	2.00 μm/16 μm	Brittle
BK 7	0.35 μm/2 μm	Low cost, robust

*Data may differ due to different cutoff criteria, purity, or thermal treatment of material.

Source: Reprinted with permission from Sieber et al. 2001. Copyright 2001 VDE Verlag.

1.3.5. Optical Multipass Cell

The optical multipass cell is not standard component in the way the above-discussed subsystems are. As outlined in Section 1.2, it is usually tailored specifically to the application. It may be manufactured in metal and polished directly to obtain mirror surface quality. It may also be made of glass or polymer with surface coatings to obtain high reflectivity. The cell may be monolithic or be composed of several parts aligned to each other. In general, the optical multipass cell has to fulfill the following requirements:

- An optical path length sufficient to create enough light–gas interaction to detect the lowest gas concentration
- Well-defined light orbits with clear separation between reflection and non-reflection areas
- Mirror coating with sufficient reflectivity
- Well-defined openings for light inlet and outlet
- Gas inlet and outlet for homogeneous gas distribution in the chamber

2. DESIGN OF OPTICAL MULTIPASS GAS SENSOR CELLS

2.1. GENERAL INTRODUCTION TO DESIGN METHODOLOGY FOR OPTICAL MULTIPASS GAS SENSOR CELLS

In general, product design should follow the methodology laid out by Pahl and Beitz (1995) (see Figure 11.12). The first step is of paramount importance for the

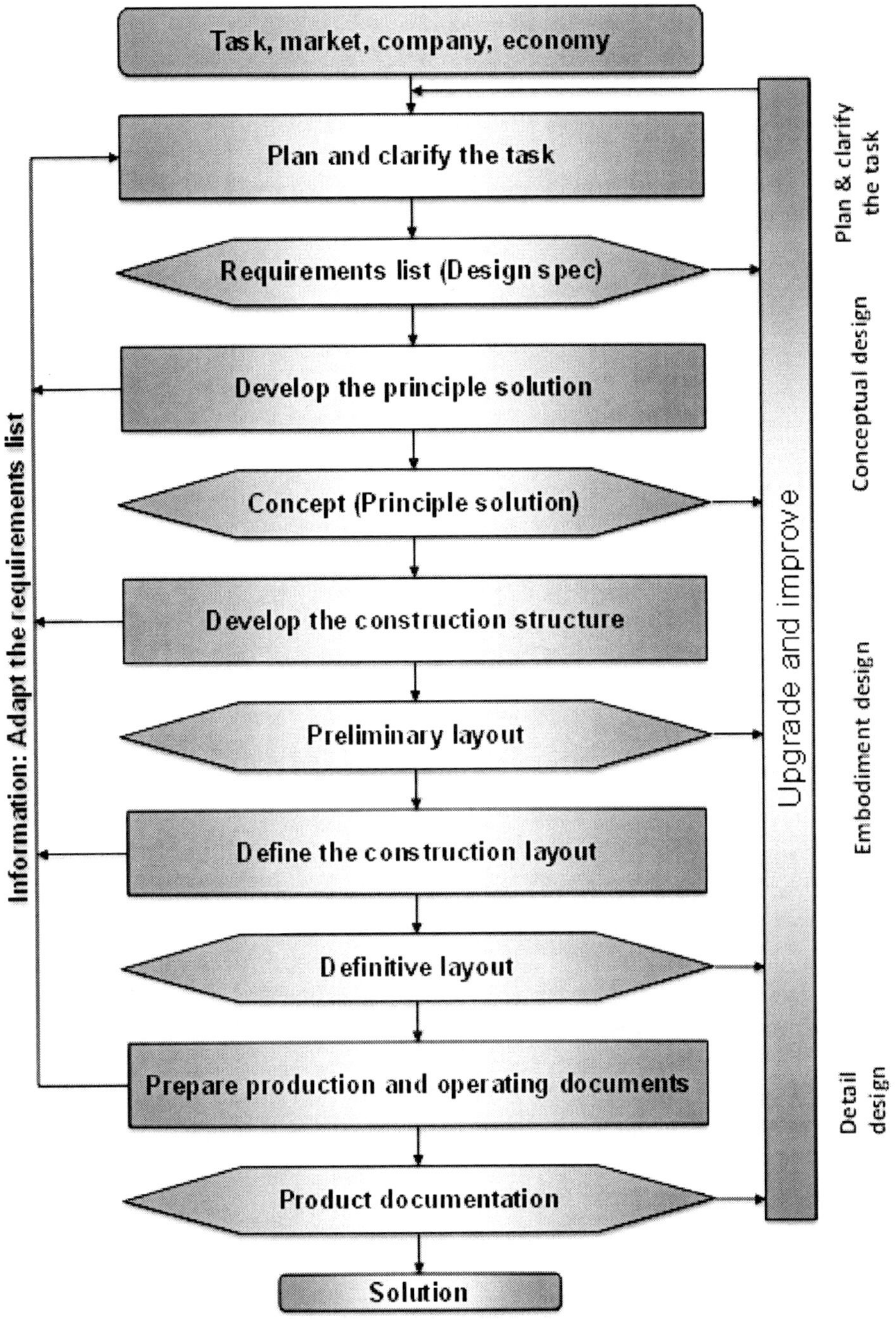

Figure 11.12. Design flow by Pahl and Beitz.

subsequent product development. Only if the market and task demands are properly collected can the requirements list be established.

This is the basis for product development. For gas sensors the requirements list encompasses a complex set of dimensions:

- Operating environment—e.g., a moderate environment in a household versus a harsh environment in industry or space applications.
- Precision—e.g., medium for consumer applications versus high for scientific instruments.
- Reliability and lifetime—e.g., short for consumer applications versus high for mission-critical applications.
- Quantity—e.g., unique or in small-quantity-produced scientific instruments versus large-series consumer devices. Production quantity is of course also correlated with price.

It is evident that there are relations between these dimensions so that compromises have to be found in the development process.

The next step in the design flow is selection of the principal solution. The principal solution for a gas sensor has already been established by deciding to use an optical multipass cell. Under this premise, the embodiment design can be started. The most elementary rules of embodiment design are that it be

- Clear
- Simple
- Safe

At first glance, it seems obvious that one should follow such an approach. Each of these rules, however, can be translated into tangible requirements on the design with respect to function, active principle, safety, ergonomics, fabrication and control, assembly and transport, usage, maintenance, and recycling. Violation of any of these fundamental rules may lead to serious consequences for product function and cost. Depending on the perspective in the design process, each fundamental rule may be applied in multiple ways for different aspects of the product. It is a good idea during product design to step back once in a while and reconsider the design with respect to these basic rules of embodiment design.

Apart from these basic rules, Pahl and Beitz postulate general design principles that can also be applied to the design of gas sensors:

2.1.1. Principle of the Division of Tasks

In macroscopic engineering it is good practice to subdivide tasks into subtasks and define specific subfunctions that fulfill these tasks. An example in optical gas sensor design is beam collimation. This task can be split among the multipass cell, the light source, and the detector. If light sources and detectors with integrated collimation structures—e.g., a parabolic reflector—are applied, the optical design of the multipass cell can be simplified, which in turn may lead to lower fabrication costs.

The principle of division of tasks is important for a clear design with parts optimally designed to their specific requirements. In optical gas sensor design it should be applied with good measure, since it may have to be brought in line with the objectives of function integration and miniaturization.

2.1.2. *Principles of Fault-Free Design*

The principle of fault-free design seems to be obvious, because any design is intended to be fault-free in the first place. The idea of this principle, however, is to consider explicitly in the design phase what faults might arise from either fabrication problems or proper/improper use of the component. Measures to achieve a fault-free design include:

- Simple component structure and piece parts
- Design to minimize the influence of errors
- Selection of active principles that are invariant to disturbances
- Mutual compensation of disturbances

It is evident that these measures revert to the basic rules and to some of the principles outlined above. This will be further detailed in the specific design rules below.

2.2. DESIGN RULES FOR OPTICAL MULTIPASS GAS SENSOR CELLS

It is a challenge to formulate rules for optical gas sensor design, since requirements for gas sensors differ widely. For scientific instruments for space applications, reliability, precision, and low payload are prime requirements which override other aspects such as cost and series manufacturability—which may be dominating for consumer application.

2.2.1. *Wavelength*

Proper selection of the target wavelength to measure gas concentration is of paramount importance. The absorption spectra in Figure 11.1 show that the peaks for

CO_2 and H_2O are very close together. In order to avoid measurement errors due to interference of absorption by another gas, the target wavelength has to be selected very carefully, which may put considerable requirements on the bandwidth of the light source and detector. To exclude the influence of water vapor, nonoptical measurement principles can also be applied, such as capacitive humidity sensors.

As outlined in Table 11.1, many technically relevant gases can be measured with lower detection limits at mid-IR wavelengths than at near-IR wavelengths, due to their higher absorption at the higher wavelengths. This may make decision for the mid-IR region attractive. It should be considered, however, whether these lower detection limits are really required by the application, since some light sources, e.g., laser diodes for the mid-IR, are substantially more expensive than near-IR laser diodes.

2.2.2. *Light Source*

The emission characteristics of the light source have a strong correlation with the beam path in the multipass cell. In order to obtain a well-defined path, the beam has to be collimated. Light sources with low divergence, such as laser diodes, can be used without collimation, though for long path lengths or to measure multiple gases with several beams in one multipass cell, collimation is indispensable. For these applications, laser diode sources with integrated slow and fast axis collimation optics are recommended, since they reduce alignment and integration effort.

For broad-band sources as LEDs, black-body emitters, or incandescent lamps, collimation can be realized in the light source by using integrated collimation elements such as parabolic reflectors or lenses, or in the cell by adjusting the mirror shape. Integrating the collimation function in the cell alone requires sophisticated modeling and optimization of the light source's emission characteristics, the cell shape, und the reflectance of the mirrors. If the emission characteristic varies from sample to sample (e.g., filament position and shape of black-body emitters or incandescent lamps), the influence of these tolerances has to be studied by means of sensitivity analysis to obtain a robust multipass cell shape. Cell design and fabrication can be considerably simplified if at least part of the beam collimation is already achieved on the level of the light source by selection of light sources with integrated collimation optics. This is an example of the principle of division of tasks (see Section 2.1.1).

In order to reduce fabrication costs and error sources, passive alignment of the light source should be preferred over active alignment. To achieve this, light sources with alignment aids on the housing should be preferred. For example, in the case of a TO-39 housing, the cylindrical face can be used for lateral, the rim for axial positioning, and the flap on the rim for orientation about the optical axis. Similar alignment aids are realized in laser modules. The optical axis and emitter position are specified in data sheets relative to these shapes. If mating features for such aids are designed into the multipass cell, passive alignment becomes feasible. It should be considered, however, that some housing types, e.g., TO housings, are fabricated by rather imprecise fabrication processes that in turn may lead to dimensional tolerances of tenths of millimeters. Hence it should be verified by simulation that the fabrication tolerances of the housing itself or the mounting tolerances of the source within the housing can still be accommodated by the optical design of the cell.

Thermal management is of paramount importance to achieve constant output power and wavelength stability. In particular, narrow-band sources such as laser diodes are very sensitive to temperature with respect to emission wavelength and lifetime. For applications such as tunable-diode laser spectroscopy (TDLAS), this effect is used to sweep a wavelength range by altering the source temperature. For single-gas sensing, a constant wavelength—and thus maintaining the source temperature in a well-defined range—is mandatory. Depending on the application and the source type, more or less sophisticated cooling solutions are available. As far as possible, passive cooling methods should be applied. As ambient temperature and ventilation can significantly impact thermal management, the source temperature should be closely monitored.

2.2.3. Detector

The spectral sensitivity of the detector has to be matched to the application requirements and to the emission characteristics of the light source. Usually the detectors accept a wide wavelength range. If the source is broad-band, the wavelength selection has to be done on the detector side by selection of proper window materials.

For most detector types, thermal management is of utmost importance to achieve high sensitivity and stable signals. To obtain reliable sensor signals, the detector has to be cooled, or at least its temperature has to be monitored and taken into account in evaluating the sensor signals.

The reasoning on passive versus active alignment outlined above for the light sources holds equally for detectors; i.e., alignment features on the detector housing should be applied as far as possible. Detector alignment is usually less critical than source alignment, since most detectors have rather large sensitive areas and acceptance angles.

2.2.4. Filter

As outlined in the description of the source and the detector, for broad-band sources the filter characteristics decide over the wavelength selection. It should be realized that the available filter materials do not allow sharp wavelength discrimination.

Besides the optical characteristics, mechanical and chemical filter properties may also be of importance for some applications. For harsh environments (thermal, mechanical, chemical stress), the remarks outlined in Table 11.1 should be considered.

2.2.5. Cell

2.2.5.1. GEOMETRY

The cell geometry should be designed as simple as possible (see Section 2.1.2 on the principles of fault-free design):

- As few parts as possible, to avoid tolerance problems and errors due to assembly and alignment. Ideally, the optical surfaces should be one monolithic part as far as manufacturability allows. If several parts are required, passive alignment structures should be provided to make assembly and alignment as simple and as precise as possible.
- With alignment structures for light source and detector. Alignment structures should be provided on the cell that mate with alignment structures on the source and detector housings (see above). This ensures that the light can be reproducibly coupled into and out of the cell without the need for expensive active alignment procedures.
- As few and as small optical surfaces as possible, since optical quality is expensive to manufacture. As outlined above in terms of the state of the art, some cell designs allow restricting the optical path to a few small optical surfaces. Only these should be manufactured with optical surface quality, with the rest only as precise as necessary.
- As simple optical surfaces as possible. Planar and spherical surfaces are most easy to manufacture with optical precision. These should be preferred over aspheric or free-form surfaces. Free-form surfaces, though expensive to manufacture, may still have merits in a design if their application allows a higher degree of integration (fewer parts) and more robustness.

Moreover the cell should be as robust as possible. The cell design should be robust against:

- Mechanical tolerances of the optical path geometry due to manufacturing and assembly.
- Source and detector tolerances. This includes provision for optical tolerances, i.e., emission field of the light source and acceptance angles of the detector. For standard source and detector housings such as TO cans, mechanical

tolerances are in the range of some tenths of millimeters. Depending on the fabrication technology, the interface to the cell may have tolerances in a similar range. The cell design should be insensitive to these tolerances to allow cost-effective passive alignment of light source and detector.
- Tolerances of the overall cell geometry due to thermal expansion or due to water absorption. Thermal expansion or water absorption in the case of polymer materials may alter the cell shape and thus the optical path. Depending on the thermal range of the application and the cell material, this may also be a factor to consider in cell design.

In order to develop a robust cell design, optical simulation and sensitivity analysis are recommended as illustrated in later sections.

2.2.5.2. MATERIAL

The choice of cell material depends on the application requirements and the target price of the instrument. For scientific instruments, metal and glass are the materials of choice, as they can be polished to high precision and surface quality. For consumer applications, polymers are the material for series fabrication by injection molding, hot embossing, etc. Frequently, the optical surfaces have to be coated to achieve proper reflection. The influence of thermal stress and water absorption on such composites has to be considered .

2.2.5.3. REFLECTIVE COATING

If the cell material itself is not reflective, a reflective coating has to be applied. The following aspects should be considered in selection and application of the coating layer.

- *Reflectivity.* The choice of reflective coating depends on the application. For large numbers of reflections, a multilayer dielectric coating with high reflectivity may be required; for fewer reflections, a gold or even an aluminum coating will be sufficient (see Figure 11.13).
- *Chemical stability.* An important aspect to consider besides the optical properties is the requirements imposed by the application or the ambiance of the sensor. Concentration measurement of aggressive gases (e.g., polluted air) requires a coating that can withstand the chemical attack over the intended lifetime of the sensor (e.g., gold).
- *Thermomechanical stability.* Thermocycling of the sensor may induce strain in the coating if the thermal expansion coefficients of cell material and

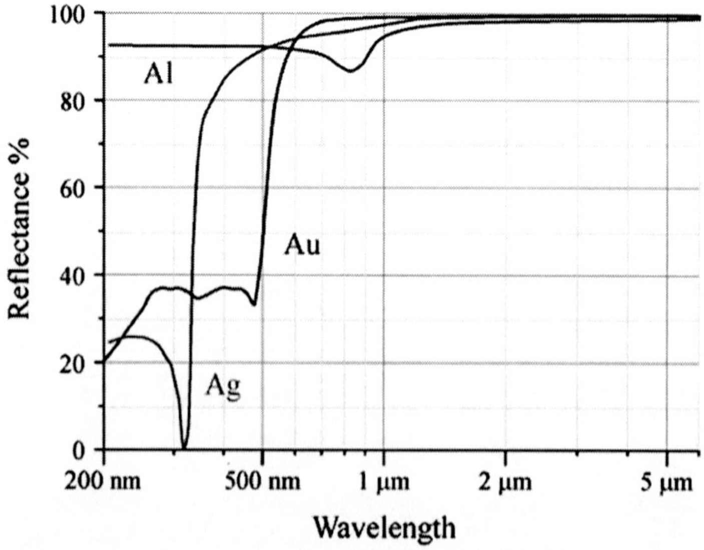

Figure 11.13. Reflectance over wavelength for the materials aluminum (AL), silver (Ag), and gold (Au). (Data from http://de.wikipedia.org/w/index.php?title=Datei:Image-Metal-reflectance.png&filetimestamp=20070301140531.)

coating do not match. This is an issue particularly with cells made from polymers that may potentially be aggravated by water absorption into the material. In the end this leads to delamination of the reflective coating and thus to failure of the device. This can be avoided by proper coating designs such as an additional adhesion layer (e.g., Cr) or multilayer systems which reduce strain and crack proliferation (see Section 2.1.2 on principles of fault-free design).

2.2.5.4. GAS INLET AND OUTLET

The gas flow in the cell can be either passive, by diffusion, or active, by pumping. In either case, two aspects have to be taken into account:

- To avoid unwanted optical paths, the gas inlet and outlet must not be in the optical region of the cell. It should be verified by optical simulation or by testing that this condition is fulfilled, even in the case of unfavorable tolerance situations (see above).
- The fluid flow in the cell must ensure complete and uniform gas distribution in the optical path in the timeframe specified by the application requirements. Incomplete and/or nonuniform gas distribution in the optical path will lead to incorrect und unreliable measurements. For complex cell

geometries and high sensitivity requirements, it is advisable to verify gas flow and distribution in the optical path by means of fluid simulation.

2.3. CASE STUDY: NDIR GAS SENSOR FOR CO_2 DETECTION: DETAILED DESIGN

The challenge of the case study at hand is the design of a low-cost device for infrared gas analysis for CO_2 detection that is robust with respect to mechanical and manufacturing tolerances. The principle of IR measurement used is a nondispersive approach. In contrast to systems employing dispersive spectral analysis, not a range of wavelengths is to be considered, but a separated, specific wavelength is to be analyzed. The nondispersive infrared (NDIR) measuring method allows for adaptation to the low-cost domain. Since the application at hand is for the consumer market, the requirements to be met are quite coarse (cf. requirements list, Section 2.1). The following considerations were identified as allowing a low-cost device:

- Adaptation of an original broad-band system for the gas to be analyzed by means of a narrow-band filter. This allows for the use of inexpensive standard incandescent lamps in the wavelength range of up to approximately 5 μm and follows the principle of the division of tasks (cf. Section 2.1.1).
- Design of the IR optical system as a single-cell device. Eliminating the reference cell, however, forces compromises with respect to resolution and stability of the overall system.

The simplest design would be geometric positioning of an IR source opposite a selective IR detector to a linear cell. The application of the gas sensor on gases with a low absorption coefficient and in spatially restrained areas requires a realization of the analysis cell as a multipass cell.

In this section the following components of the setup of the NDIR gas sensor are discussed with respect to the low-cost approach:

- Source
- Detector
- Gas analysis cell

2.3.1. Source

The IR source consists of a wide-band incandescent lamp. The spectral emission is shown in Figure 11.14. The radiation of the incandescent IR lamp was

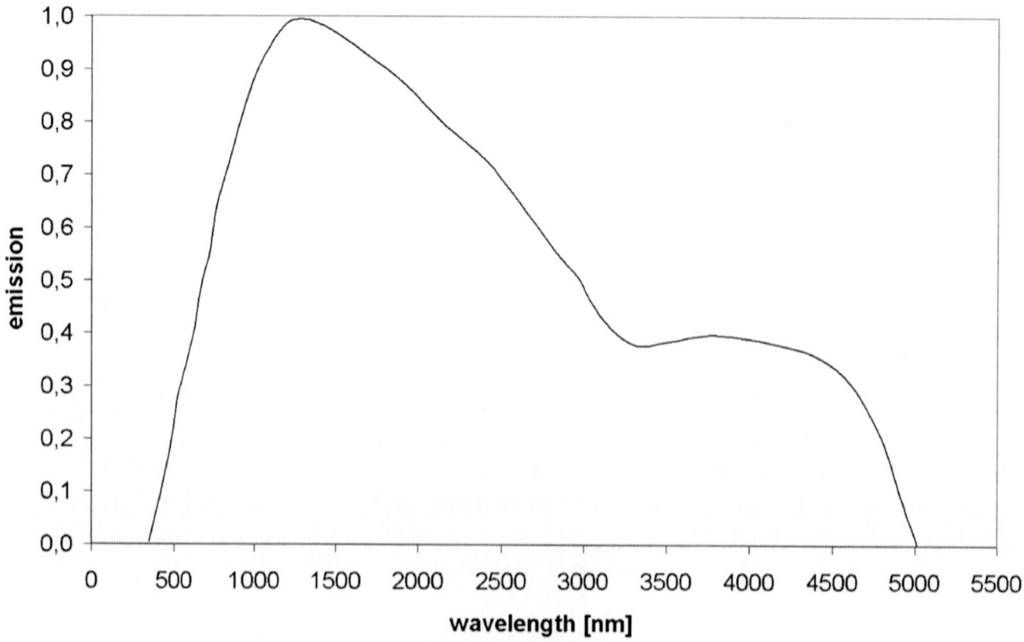

Figure 11.14. Spectral emission of the incandescent lamp.

measured in three sectional planes (Figure 11.15). The results of these emission measurements are shown in Figure 11.16 as dark areas for the three planes, respectively. These measurements lead to the typical "butterfly" characteristic of the

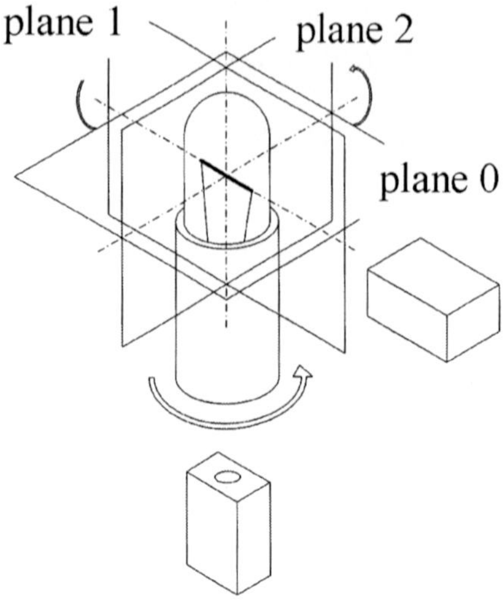

Figure 11.15. Setup of the measurement of the IR lamp's radiation.

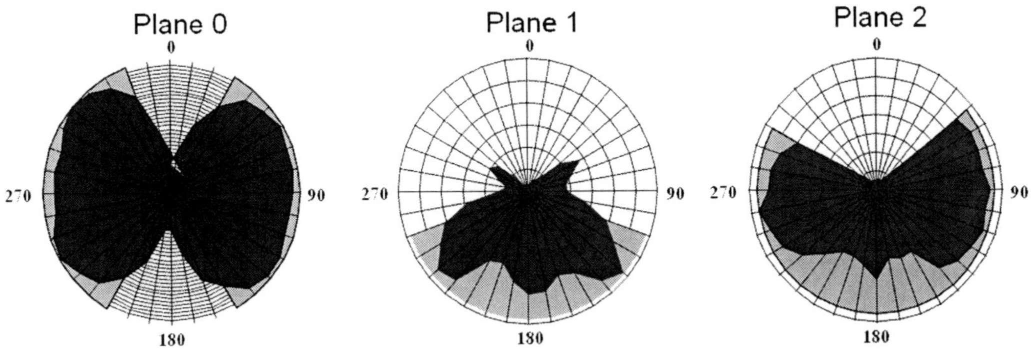

Figure 11.16. Results of the emission measurements of the incandescent IR lamp. (Reprinted with permission from Sieber et al. 2001. Copyright 2001 VDE Verlag.)

irradiance in plane 0 (the level plane). The zones of minimal irradiance are limited to an angular range of approximately 30° at the front of the helical filament of the lamp. In the perpendicular planes, angles of irradiance of at least 120° are determined. This minimum was measured along of the helical filament. In the cross direction, a value of 250° is obtained.

As a result of these measurements, the source model also incorporates the angles of radiation into the three space directions. These angular areas are shown bright in Figure 11.16.

Since radiation takes place approximately in the half-space, a source reflector was used for beam focusing to accomplish a high power density in the detector. A standard parabolic LED and IR reflector was chosen as the reflector.

Since the NDIR gas sensor is designed to analyze CO_2 concentration, the emission spectrum of the incandescent lamp has to be adapted. CO_2 has its principal absorption band centered at a wavelength of approximately 4.2 μm and a full width at half-maximum (FWHM) of 0.34 μm (see Figure 11.17).

The filter used for constraining the lamp's emission spectrum to the principal absorption band of CO_2 was the Jenofilt 711 (4.26/2%). The transmission characteristic of this filter is shown in Figure 11.18. The filter band is centered at 4.27 μm with a FWHM of 86 nm.

The incandescent lamp was operated in a pulsed mode. Three main advantages could be identified for using a pulsed mode for operation:

- Reduction in current need.
- Avoidance of warming of the probe gas. This would lead to a falsification of the measured values.
- Slowing of aging effects of the incandescent lamp. This point is of high importance in a single-cell system, since no correction of the system by means of a reference canal is possible.

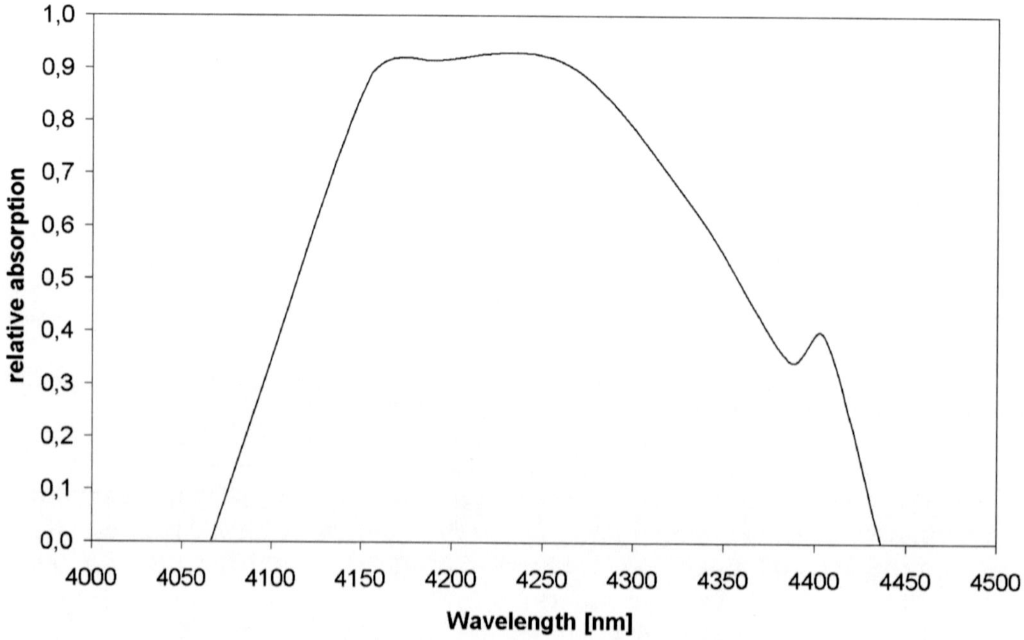

Figure 11.17. Principal absorption band of CO_2. (Reprinted with permission from Sieber et al. 2001. Copyright 2001 VDE Verlag.)

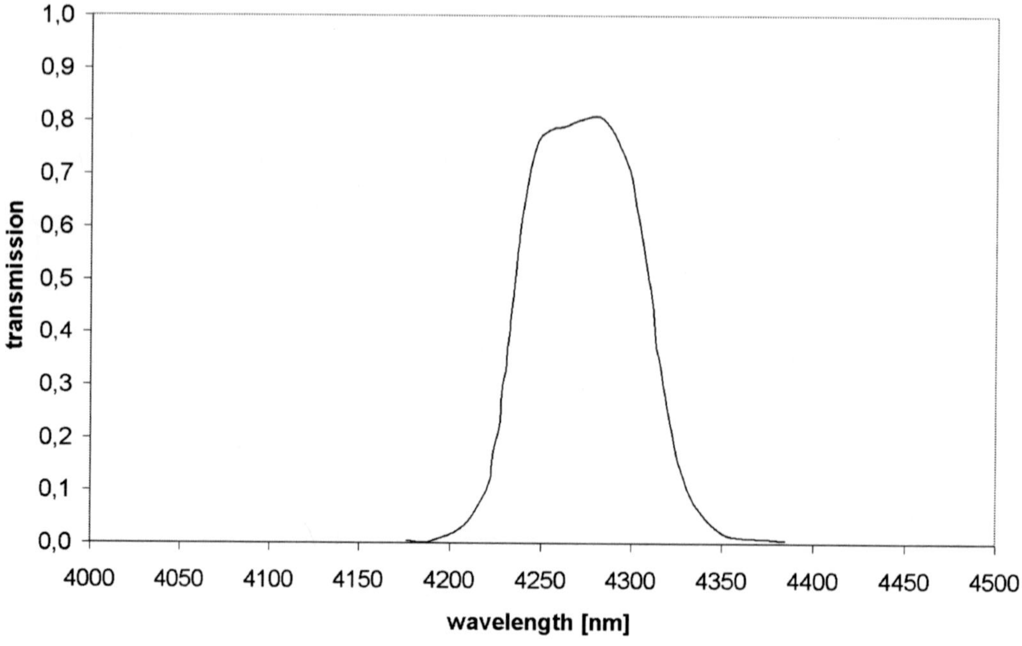

Figure 11.18. Transmission characteristic of Jenofilt 711 (4.26/2%).

For pulse keying, a relation of 1:5 to 1:10 between “lamp on” and “lamp off” with periods of 1–2 s were used.

2.3.2. Detector

A thermopile sensor serves as thermoelectric sensitive component. The functional principle of these thermo elements is based on the Seebeck effect. Each thermo element has two contacts. If there is a temperature difference between the two contacts, a measurable current flow is generated due to the thermo voltage.

To amplify the low thermo voltage, multiple thermo elements will be connected to form a thermopile, leading to addition of the individual voltages.

Since the thermo voltage does not depend on the dimensions of the thermo elements, miniaturization and integration of multiple thermopiles to a thermopile array is possible. The pixel voltage generated in the individual thermo elements is too low even after serial connection to thermopile arrays, so an additional amplification is necessary. For that purpose the thermopile array will be packaged together with a multiplexer and an operational amplifier (op-amp) to form a thermopile sensor element. The multiplexer is used to address the thermopiles to be read out and to switch the particular signal to the op-amp.

2.3.3. Gas Analysis Cell

The gas analysis cell was realized as a multipass cell (see Figure 11.19). The cell is designed to match the specific absorption properties of a single gas; i.e., the

Figure 11.19. Photograph of the multipass gas analysis cell.

cell is tuned to detect a single gas. To detect even weakly absorbing gases, the ray path through the analysis cell must be extended. As this is to be achieved without a significant increase in the dimensions of the analysis cell, the boundary surfaces of the cell must fulfill focusing and folding tasks. The openings for the source as well as the for the detector are arranged in one plane. Beam focusing and reflection are carried out by two paraboloid structures opposite to each other. In principle there are two possible ways of feeding the gas to the analysis cell: by means of pumping or by means of diffusion. Our example sensor system is designed to match both feeding principles: a pump inlet as well as holes in the side wall to allow for the diffusion of the gas into the analysis cell are possible. The pump inlet can be seen at the lower right of the photograph of the analysis cell. Reflection of the IR radiation at the focusing structures will be ensured by means of a gold coating. In principle, three different materials are suited for the reflection of IR radiation: aluminum (Al), silver (Ag), and gold (Au). Figure 11.13 shows the reflectance of these materials over the wavelength range of interest. In our case study, gold was chosen as the reflective coating.

3. MODELING AND SIMULATION

3.1. OPTICAL SIMULATION TECHNIQUES

The methods being used in simulation of optical phenomena can be divided into three main categories:

- Geometric optical methods
- Scalar wave optical methods
- Rigorous physical optical methods

Geometric optical methods include paraxial methods such as, e.g., the ray transfer matrix method (Hecht 2005) and nonparaxial methods such as ray tracing. The ray transfer matrix method, also known as the ABCD matrix method, is based on the so-called paraxial approximation, a small-angle approximation which holds as long as the approximation $\sin\theta \approx \theta$ is reasonable. Only rays making a small angle θ with the optical axis and having only a small distance to the optical axis are regarded in this approximation. Subject to these requirements, simple linear mathematical expressions of the paraxial meridional rays can be derived. Tracing a meridional ray through an optical system, the above-mentioned linear expressions are recurring, hence the use of matrices is possible. The refraction process at the surface of adjacent media as well as the transfer process of the ray passing through the medium can be described mathematically by matrix operations. The simple linear form of the expressions as well as the repetitive fashion of their application made the ABCD matrix method popular in the precomputer area. Today this method is still in use in the design of resonator optics.

The ray-tracing approach overcomes the limitation of the paraxial approximation. This technique is widely used and implemented in the majority of commercially available optical simulation tools. The ray-tracing technique is based on the approximation of the wave field propagation by means of idealized small beams, the so-called rays. These rays are perpendicular to the local wave front (in isotropic media) and point along the direction of energy flow (the direction of wave-front propagation). A wave front is an imaginary surface where the electric field of an optical beam has a uniform phase with respect to some reference point. It is numerically difficult to propagate wave fronts directly.

At any point on a wave front, some small area may be selected which can be thought of as a piece of a plane wave. The ray position and coordinates are defined by the position of the small area and the orientation of the normal vector to this small plane wave, respectively. The integrated energy or flux of the beam over this small area may be assumed to be entirely represented by the ray. Rays have position, direction, amplitude, phase, and polarization data associated with them. While propagating through a medium characterized by varying indices of refraction, absorption properties, and surface properties, such a ray may bend, be diminished, change its direction, or experience reflection. The course of the ray is discretized depending on the sequence of (approximated) layers of constant optical properties. Discrete straight portions of the ray belong to constant optical properties of a layer (see Figure 11.20).

Refraction of the ray at an intersection with a surface defining the boundary between media with different optical properties is according to Snell's law (see, e.g., Hecht 2005).

Ray tracing is a powerful tool in optical systems design, and a wide variety of optical problems can be simulated by means of this technique. However, ray tracing is limited to optical problems where diffraction effects can be neglected. Diffraction of a wave front at an aperture stop cannot be described by ray tracing; the ray approximation of the wave fronts breaks down. In such cases, other simulation techniques applying diffraction theory need to be employed.

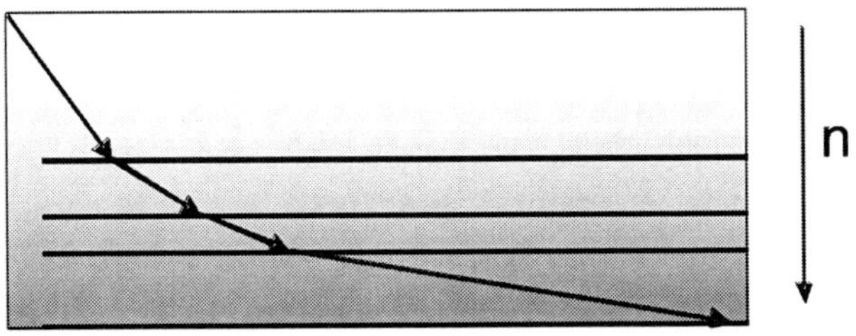

Figure 11.20. Ray tracing through a layered medium. Index of refraction n increases with depth.

Using scalar wave optical methods, the propagation of a light wave in a homogeneous isotropic medium can be simulated with consideration of diffraction effects on structures that are large compared with the wavelength of light (Goodman 1996). These wave optical methods are based on the scalar diffraction theory. Regarding the propagation of light as an electromagnetic wave in a homogeneous isotropic medium, the vectorial nature of Maxwell's equations can be disregarded. In this case the electric as well as the magnetic field can be described by a scalar function representing either the electric or the magnetic field amplitude. Two approximations to the general scalar diffraction theory are commonly made in calculations of diffraction patterns: the Fresnel approximation, often referred to as the near-field approximation, and the Fraunhofer approximation, the so-called far-field approximation. On the basis of these approximations the mathematical calculations can be reduced to comparatively simple manipulations. As a rule of thumb, the Fresnel approximation is valid when the curvature of the incoming and outgoing wave fronts at the position of the aperture is not negligible. The criterion of Fraunhofer diffraction is seen mathematically in the linearity between the aperture variables (Hecht 2005): As long as the incoming and outgoing waves are almost plane over the dimensions of the aperture, the Fraunhofer approximation is valid.

For simulation of photonic structures where the vectorial nature of Maxwell's equations cannot be disregarded, such as, e.g., nanophotonic devices such as photonic crystals, rigorous physical optical methods have to be employed. In general, two different approaches are used in computations of such structures: frequency-domain methods and time-domain methods. Joannopoulos et al. (2008) provided a detailed discussion of the advantages and disadvantages of these methods.

Following is a brief summary of the conclusions drawn in the manual of the MIT Photonic Bands software package for computing band structures of optical systems (Johnson 2011). Frequency-domain methods allow for direct computation of the eigenstates and eigenvalues of Maxwell's equations on the basis of a plane-wave approach. Each field computed has a definite frequency. In contrast, time-domain methods iterate Maxwell's equations in time. The fields have a definite time at each time step, but not a definite frequency. By far the most common technique for time-domain simulations is the finite-difference time-domain method (FDTD). In the FDTD method, time and space are divided into a grid of discrete points and approximate the derivatives of Maxwell's equations by finite differences. The main advantage of using the FDTD method is that for a linear system the response can be calculated at many frequencies with a single computation. However, there are also several disadvantages of the FDTD method, such as, e.g., that time-domain methods require a long time to resolve a sharp spectral feature.

In summary, every optical problem requires its own method to be solved. The modeling and simulation of the case study at hand requires geometric optical methods; in particular, the ray-tracing method employed by the commercial simulation software ZEMAX-EE was used for simulation and design optimization of the gas analysis cell of the IR gas sensor.

3.2. CASE STUDY: NDIR GAS SENSOR FOR CO_2 DETECTION: SIMULATION

Quantified IR gas analysis employs the Lambert-Beer law, by which absorption is described as a function of the thickness of a layer (and the ray path in the medium to be measured), the type of molecule, and the concentration of a dissolved substance as described in Section 1.1.

3.2.1. Study of Ray Paths in the Analysis Cell

3.2.1.1. CONTROL OF THE RAY PROPAGATION

The ray path in the absorbing medium is one of the crucial parameters in IR gas analysis. The Lambert-Beer relation (see Eq. (11.1) indicates the linear relationship between the ray path and absorption. The gas analysis cell is designed for the detection of a specific gas. This means that the ray path of radiation in the medium to be detected must be well defined. Rays whose paths differ will falsify the result and prevent the real gas concentration from being determined. As a consequence, the paths covered by the rays in the gas analysis cell must be studied. A detail important to know is what angle of emission gives rise to what ray path. The results of these studies are shown in Figure 11.21. The angular segment given by the angles −17° and +23° shown in the picture on the left gives rise to the "desired" rays. These rays have a mean ray path of 60.0 mm. A summary statistical evaluation of the paths of approximately 200 single rays, whose starting angles were randomly generated in the "desired" angular segment, can be seen in Table 11.3.

On the right side of Figure 11.21, the "undesired" rays can be seen, given by the angular segment confined by the angles 23° and 69°. The mean path of these rays is 41.2 mm, and the associated statistics are shown in Table 11.4.

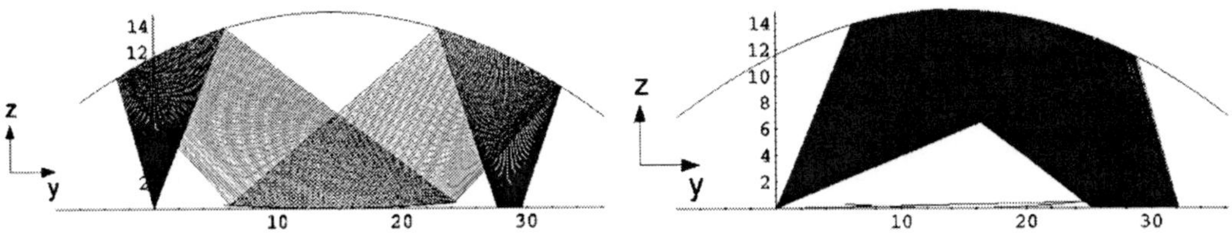

Figure 11.21. Possible ray paths: On the left the "desired" rays, and on the right the "undesired" ones. (Reprinted with permission from Sieber et al. 2001. Copyright 2001 VDE Verlag.)

Table 11.3. Statistical evaluation of approximately 200 randomly generated "desired" rays (all units in mm)

Mean ray path	Variance	Standard deviation	Max. deviation	Mean deviation
60	0.0203	0.1424	0.5405	0.1213

Source: Reprinted with permission from Sieber et al. 2001. Copyright 2001 VDE Verlag.

The distributions of points of impact of the rays at the cell outlet are shown in Figure 11.22 for these two cases. Focusing of the "desired" rays on the cell outlet is obvious from the narrow pattern. The points of impact of the "undesired" rays spread over the entire cell outlet. However, also in this case a concentration of spots can be seen in a similar area as for the "desired" rays. This can be seen by looking at the spot diagram of the "undesired" rays, where the contour line of the ray distribution of the "desired" rays is also shown. Elimination of the "undesired" rays by a diaphragm immediately upstream of the active detector layer thus is not possible. Figure 11.23 shows the ray path achieved by focusing the source emission by means of a standard parabolic mirror, such as can be ordered out of a catalog of optics suppliers. Especially the y–z section shown at the bottom clearly indicates that by far the largest part of the rays follow the "desired" ray path. Another major advantage of beam focusing over a diaphragm is the fact that no intensity is lost.

In the following treatment all rays are considered, including those that undergo side reflections. The path-length distribution versus the number of rays propagated is shown in Figure 11.24. The upper part of Figure 11.24 indicates a path-length division into discrete bands of distances. An assignment of the rays to a distance range enables quantification of how many rays travel which distances. This assignment is shown on the lower part of Figure 11.24. It is evident from the bar chart that by far the largest part of the rays follow the desired path.

Furthermore, it is of interest to know how many reflections the individual rays undergo until the detector is reached. Therefore, investigations of the distribution of the number of reflections were performed. Figure 11.25 shows the distribution

Table 11.4. Statistical evaluation of approximately 200 randomly generated "undesired" rays (all units in mm)

Mean ray path	Variance	Standard deviation	Max. deviation	Mean deviation
41.2	1.028	1.014	5.125	0.751

Source: Reprinted with permission from Sieber et al. 2001. Copyright 2001 VDE Verlag.

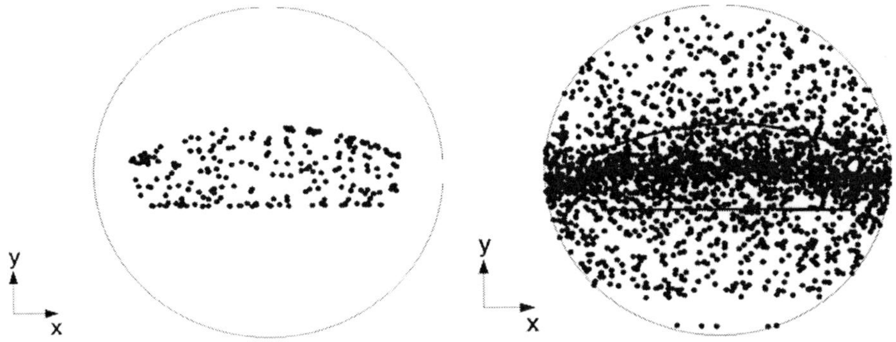

Figure 11.22. Points of impact of the "desired" rays (left) and the "undesired" rays (right) shown as spot diagrams. The contour line of the distribution of the desired rays is also shown in the right spot diagram. (Reprinted with permission from Sieber et al. 2001. Copyright 2001 VDE Verlag.)

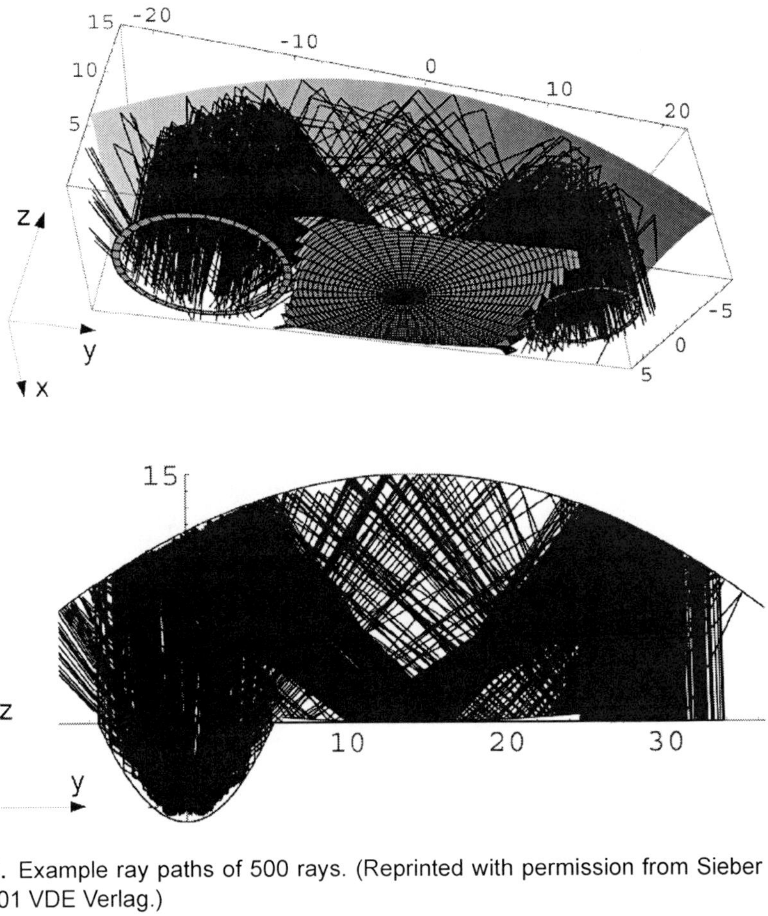

Figure 11.23. Example ray paths of 500 rays. (Reprinted with permission from Sieber et al. 2001. Copyright 2001 VDE Verlag.)

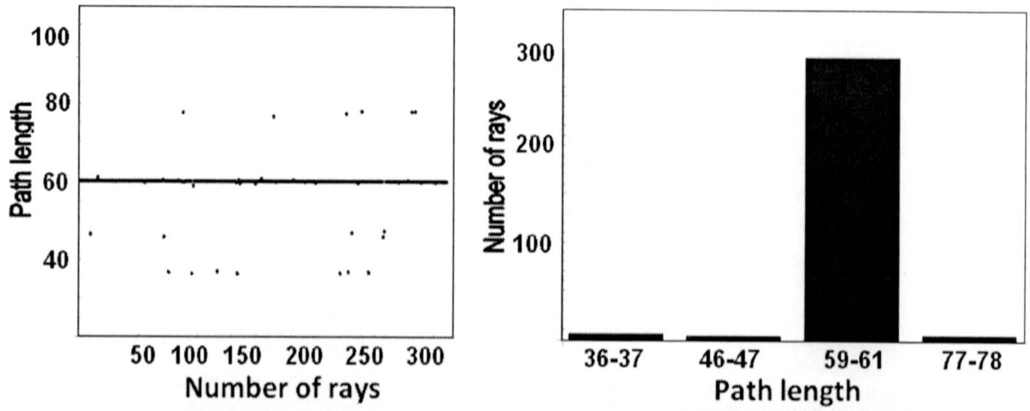

Figure 11.24. Path-length distribution versus number of rays propagated. [Reprinted with permission from Sieber et al. 2002. Copyright 2002 Society of Photo-Optical Instrumentation Engineers (SPIE).]

of the number of reflections versus the number of rays propagated. Here also the largest part of the rays undergo three reflections before hitting the detector.

The question of interest now is the correlation of the path length of the rays and the number of reflections. The assignment of the first bar of the two charts is obvious (Figure 11.24, bottom, and Figure 11.25, bottom), because it is evident that the rays with the shortest path length undergo the least reflections. Assignment of the other bars requires further investigations. The "straight" ray path shown in Figure 11.26 undergoes three reflections and has a mean path length of 60 mm. However, the rays with a path length in the range of [46–47 mm] also need three reflections to reach the detector (see Figure 11.27, left). Hence, the

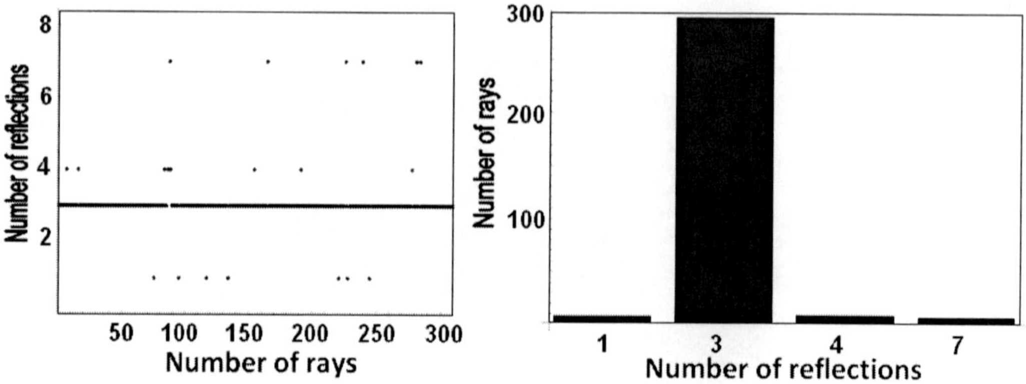

Figure 11.25. Distribution of the number of reflections versus the number of rays propagated. [Reprinted with permission from Sieber et al. 2002. Copyright 2002 Society of Photo-Optical Instrumentation Engineers (SPIE).]

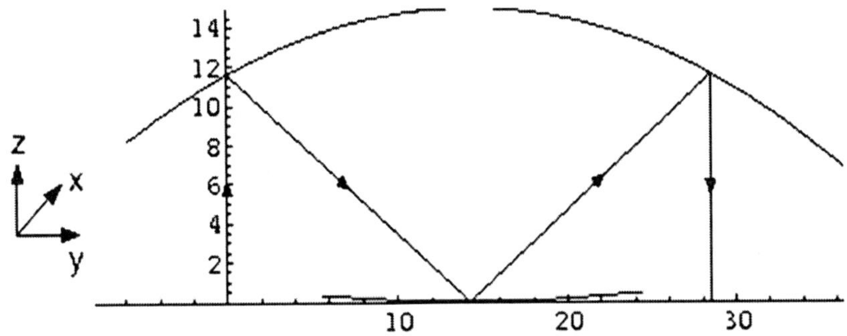

Figure 11.26. "Straight" ray path undergoing three reflections. [Reprinted with permission from Sieber et al. 2002. Copyright 2002 Society of Photo-Optical Instrumentation Engineers (SPIE).]

bar indicating three reflections must be subdivided into rays with a path length in the range of [46–47 mm] and rays with a path length in the range of [59–61 mm].

Furthermore, there are also rays in the path-length range of [59–61mm] undergoing four reflections, as shown in Figure 11.27, right. The "desired" rays traveling a path length in the range of [59–61mm] are composed of rays undergoing both three and four reflections.

These investigations show that by far most of the rays follow the "desired" ray path with a mean path length of 60 mm and three reflections. This leads to the conclusion that control of the ray propagation inside the analysis cell can be achieved by means of an adequate source reflector.

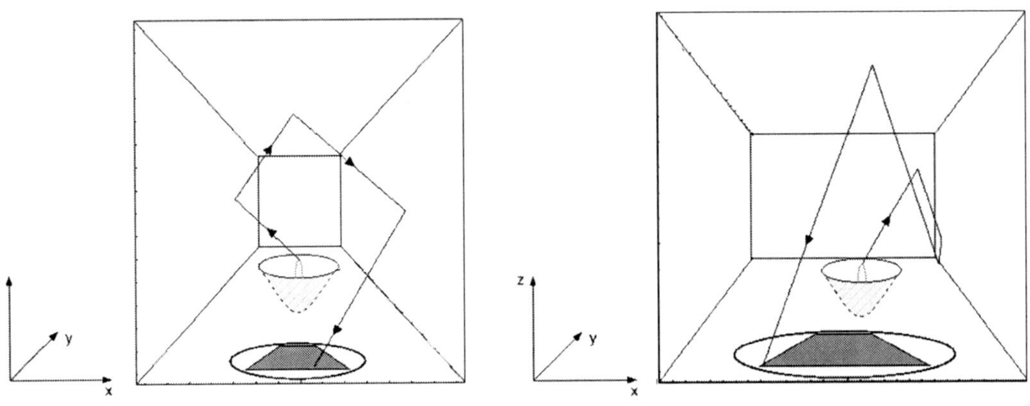

Figure 11.27. "Skewed" ray paths undergoing three reflections while traveling a path length in the range [46–47 mm] (left) and four reflections while traveling a path length in the range [51–61 mm] (right). [Reprinted with permission from Sieber et al. 2002. Copyright 2002 Society of Photo-Optical Instrumentation Engineers (SPIE).]

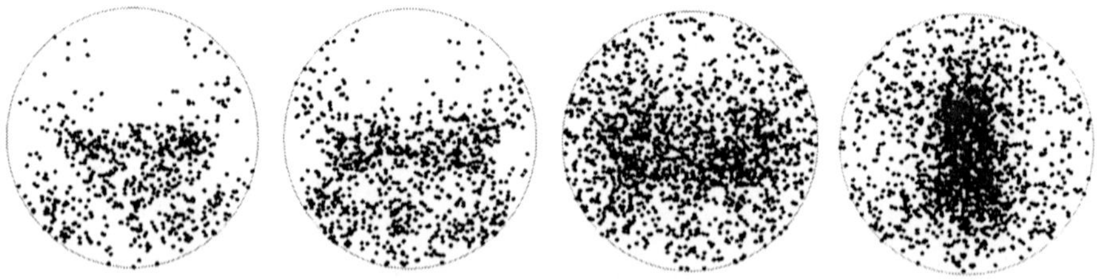

Figure 11.28. Spot diagrams for four different *z* positions of the helical filaments. (Reprinted with permission from Sieber et al. 2001. Copyright 2001 VDE Verlag.)

3.2.1.2. ANALYSIS OF SOURCE POSITION

Besides ray control, another crucial point is the ray distribution and the irradiance distribution, which depend on the source position relative to the source reflector—i.e., the sensitivity of the optics due to the position tolerance of the source.

Figure 11.28 shows the spot diagrams for four varying *z* positions of the source within the paraboloid source reflector. On the one hand, the different numbers of rays impinging upon the cell outlet and, on the other hand, the different distributions of the points of impact over the cell outlet are clearly evident. The degree of variation of the number of rays with the *z* position of the source is shown in Table 11.5.

A measurement of the irradiance distribution was carried out by means of a test setup. The results of these measurements show strong differences between the irradiance distribution of a simply assembled cell (see Figure 11.29, top) and the irradiance distribution for a optically maximum balanced cell (see Figure 11.29, bottom). The strong difference in the performance of the two setups results from the huge tolerance of the position of the helical filament in the source lamp. Moreover, Figure 11.29 shows that the performance and therefore the design of the IR sensor is extremely sensitive to correct positioning of the IR source.

The measurements of the irradiance distribution shown in Figure 11.29 were used for an evaluation of the model of the optical subsystem consisting of the analysis cell, the IR source, the source reflector, and the detector.

Table 11.5. Number of rays reaching the cell outlet with different source positions; in each case, 2800 rays with random starting angles were emitted

z POSITION	−3.0 mm	−3.6 mm	−4.2 mm	−5.0 mm
Number of impacts	631	1300	2000	2300

Source: Reprinted with permission from Sieber et al. 2001. Copyright 2001 VDE Verlag.

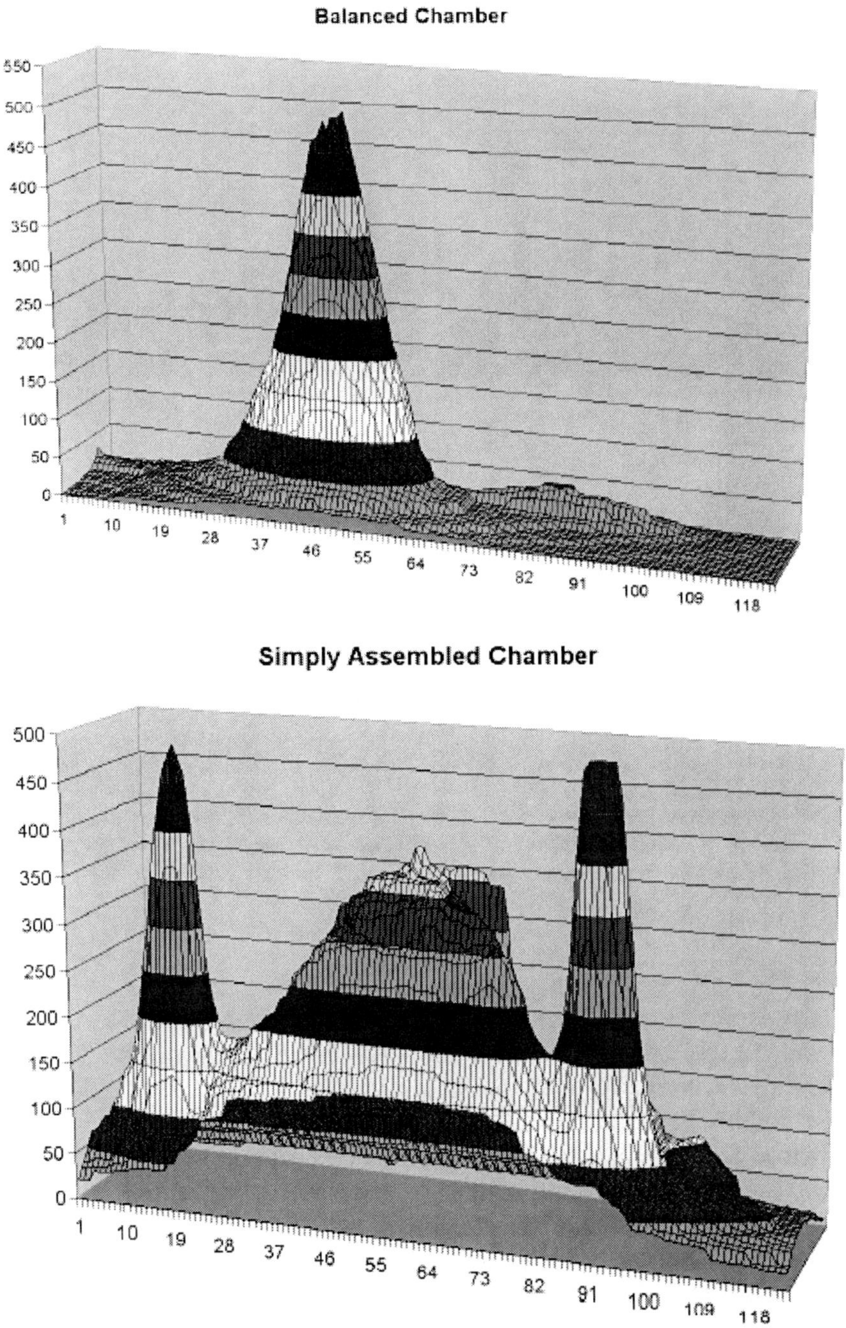

Figure 11.29. Irradiance distributions of the analysis cell. Top: The source was simply assembled without correction. Bottom: The cell was optically balanced. (Reprinted with permission from Sieber et al. 2002. Copyright 2002 VDE Verlag.)

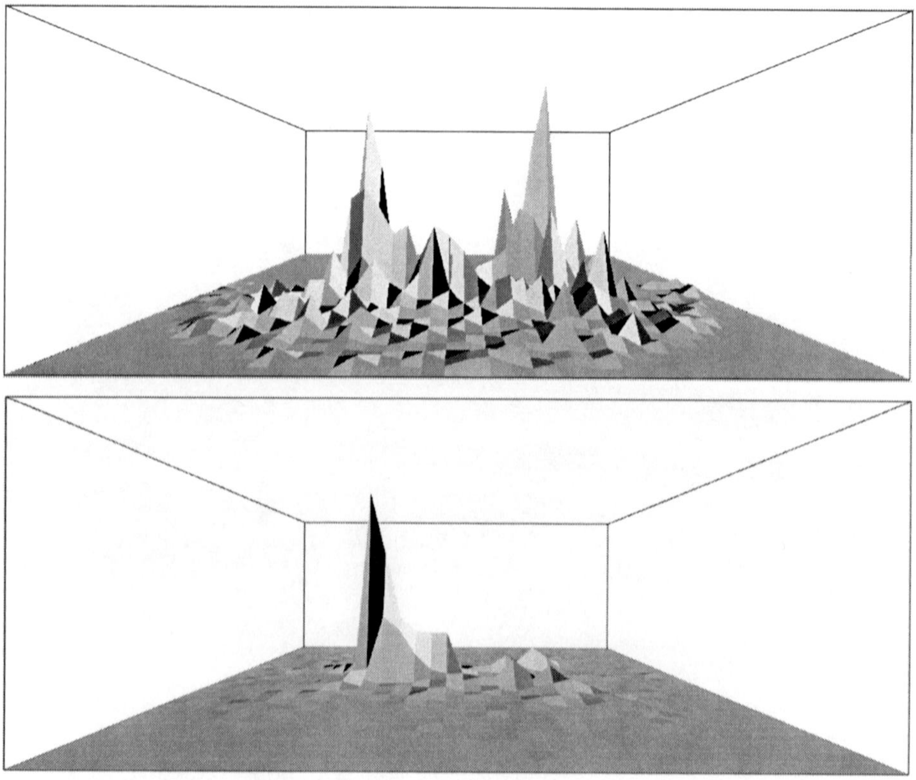

Figure 11.30. Calculated irradiance distributions of the analysis cell. Top: The source is misaligned. Bottom: The source is optimally positioned. [Reprinted with permission from Sieber et al. 2002. Copyright 2002 Society of Photo-Optical Instrumentation Engineers (SPIE).]

In a first step the measurements of the irradiance distribution were used for an evaluation of the model of the IR gas sensor. For this reason the irradiance distribution of the IR sensor in the case of a misaligned source was calculated using the system model. In Figure 11.30, top, the calculated irradiance distribution of a nonbalanced analysis cell is shown. The source position during the calculation was within the tolerance range of the position accuracy of the helical filament within the lamp. The results of the measurement (Figure 11.29, top) and calculation (Figure 11.30, left) were found to be in good agreement.

In a second step the system model was used to find the optimum source position for the given setup. For this position the irradiance distribution of the sensor was calculated (see Figure 11.30, bottom). A comparison of the measured and the calculated irradiance distributions in the case of an optimally positioned source (Figure 11.29, bottom, and Figure 11.30, right) leads to the same statement as above: The model and the real setup are found to be in good agreement. So the presented system model can be used for design optimization.

4. DESIGN OPTIMIZATION

4.1. OPTIMIZATION STRATEGY

4.1.1. Optimization Methods

This subsection covers a brief discussion of optimization strategies and methods, extracted from Jakob (2004). For a detailed treatise, consult the original reference (Jakob 2004).

Depending on the complexity of the search space of the optimization problem at hand, different optimization strategies and methods may be used. In terms of their search behavior, optimization methods can be divided into local search methods and global search methods.

Local search algorithms calculate solutions in the search space by applying local changes. They are deterministic by nature. One advantage of local search algorithms is the fast convergence, and the main drawback is that local search methods can only find local optima. So, for multimodal search spaces, it cannot be assured that the global optimum is found. Examples of local search methods are gradient methods, the Newton method, the Gauss-Seidel algorithm, the Complex-algorithm, and the least-squares algorithm.

In contrast to the local search methods, many of the *global search techniques* are heuristic, i.e., they make use of experience-based techniques. The main advantage of global search techniques is the ability to find global optima. This comes, however, at the cost of slow convergence. Examples of global search methods are evolutionary algorithms and simulated annealing.

The software used for optical simulation and optimization in this case study, ZEMAX-EE, uses a damped least-squares algorithm for the optimization.

4.1.2. Optimization Approach

The approach toward a robust design responds to the principles of fault-free design as described in Section 2.1.2. It can be described by the following items:

- Influences to be compensated
- Control parameters of the design
- Design parameters to be varied
- Setting up of a reasonable start design
- Development of the merit function

4.2. CASE STUDY: NDIR GAS SENSOR FOR CO_2 DETECTION: OPTIMIZATION OF THE ANALYSIS CELL

4.2.1. Starting Model for Optimization

The optimization approach was introduced in the previous section, so the first question to be answered is "What are the influences that have a severe impact on the system's performance and hence should be compensated?". This question can be answered by different approaches:

- Examination of the crucial functions of the system, as we did it in the section above by studying the ray paths in the analysis cell
- Accomplishing a sensitivity analysis to identify the worst offenders

The answer to the second question, with which design parameters these influences could be controlled, is more difficult to answer, since it depends on the respective application as well as on the experience of the designer. As a rule of thumb it can be said that all functional design parameters should be examined in this context. In the case study at hand, the optimization parameters should control the beam focusing structures to restrict the possible beam paths to the "desired" paths.

Directly associated with this aspect is the next question concerning the identification of the variable parameters. Not all parameters which influence the system's performance can be arbitrarily altered. The design rules of the individual manufacturing processes have to be considered as well as the manufacturability of the optimized design.

When all these questions have been answered, a starting model has to be defined in such a way that all parameters found to influence the control of disturbing influences can be optimized independent of one another.

Last but not least, the merit function, controlling the optimization procedure, has to be constructed. Here all the requirements and boundary conditions have to be considered and formulated.

Since the present design of the IR gas sensor is extremely sensitive to the exact positioning of the IR source, and the position of the helical filament within the lamp used for this setup is afflicted with a huge tolerance, an optimization strategy with respect to a robust design concerning the source position has to be chosen. The robust design has to be rugged enough to be used for mass fabrication. In mass fabrication, it is not the ideal arrangement of individual components which results in the most effective system. Instead, tolerances in positioning individual optical elements need to be taken into account already in modeling.

The absence of a standard material system for optical systems normally renders monolithic designs impossible. Consequently, complex optical systems are generally modular designs. The modular structure of these hybrid systems requires that single components be manufactured in isolation and later combined into one total system. So the system has to be simulated as a function of disturbing influences, and the system design has to be optimized with respect to minimizing the influences of any effects that affect performance. The result then represents a rugged, stable functional setup.

The design parameters which have the potential to control the tolerance of the source position are the geometry of the reflecting cell walls as well as the lamp reflector. Since the replication process chosen for production of the gas analysis cell is injection molding, the design of the cell geometry is free to a great extend. Of course, the requirements and design rules for injection molding have to be considered and mapped onto the merit function.

The starting model for optimization is shown in Figure 11.31. The most important change compared to the original design is splitting the top reflector into two structures. Thus the starting design consists of four reflectors: The first reflector has the task of collimating the source's radiation, the second reflector focuses the collimated beam on the detector module. Reflectors three and four are plane mirrors and serve only to fold the beam in this starting model.

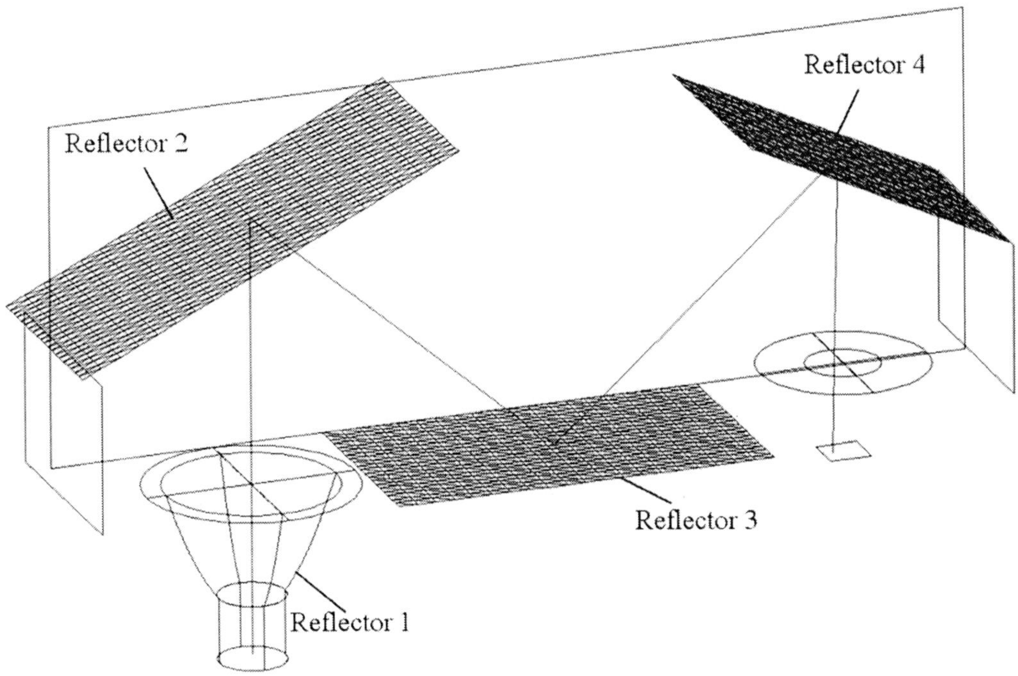

Figure 11.31. Starting model of the absorption cell.

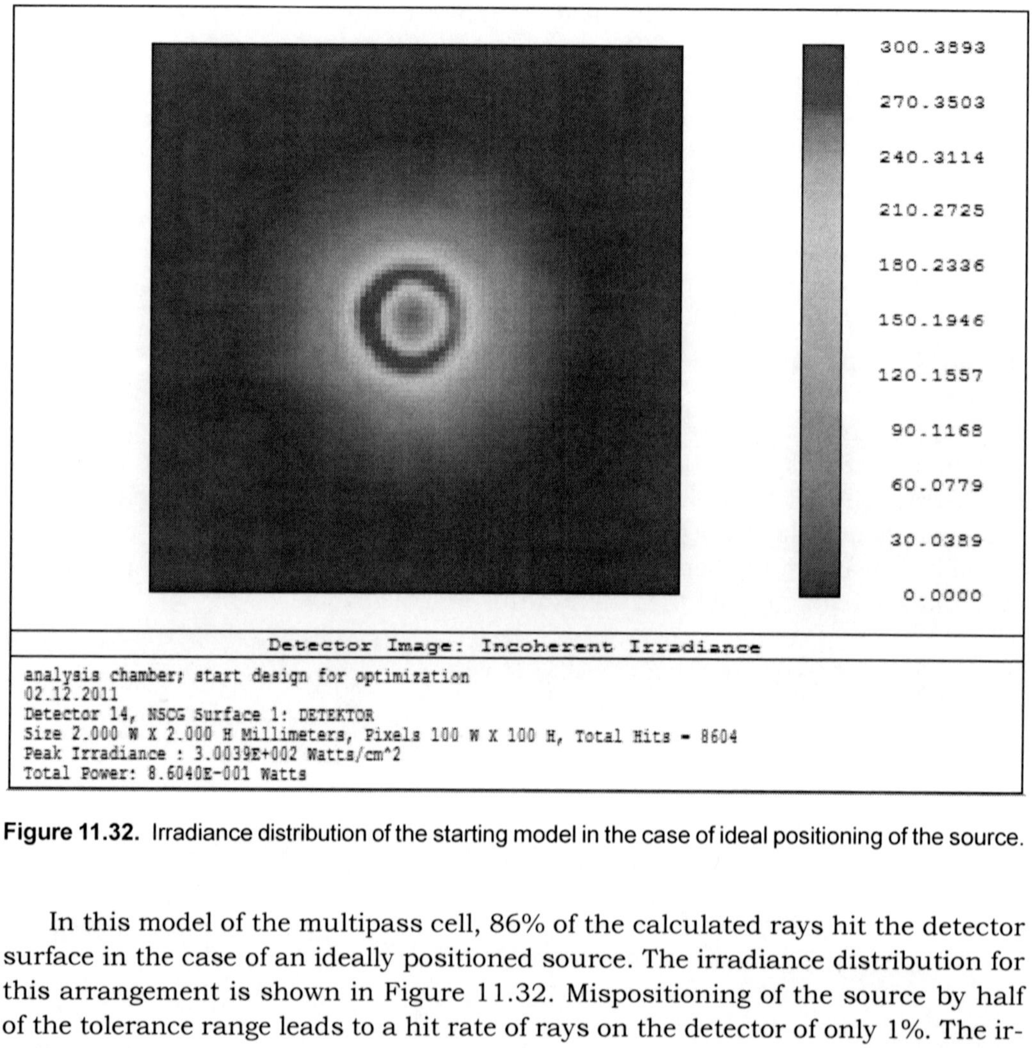

Figure 11.32. Irradiance distribution of the starting model in the case of ideal positioning of the source.

In this model of the multipass cell, 86% of the calculated rays hit the detector surface in the case of an ideally positioned source. The irradiance distribution for this arrangement is shown in Figure 11.32. Mispositioning of the source by half of the tolerance range leads to a hit rate of rays on the detector of only 1%. The irradiance distribution of this configuration is shown in Figure 11.33. On the basis of this starting model, optimization of the absorption cell's geometry was carried out. The goal of the optimization was to find a rugged design with consistent and satisfactory performance in the whole tolerance range of source positioning.

4.2.2. Optimization of the Analysis Cell

For optimization the following parameters were defined as variables:

- The position of the reflectors relative to each other
- The curvatures of the reflectors
- The source position inside of reflector 1 (see Figure 11.31)

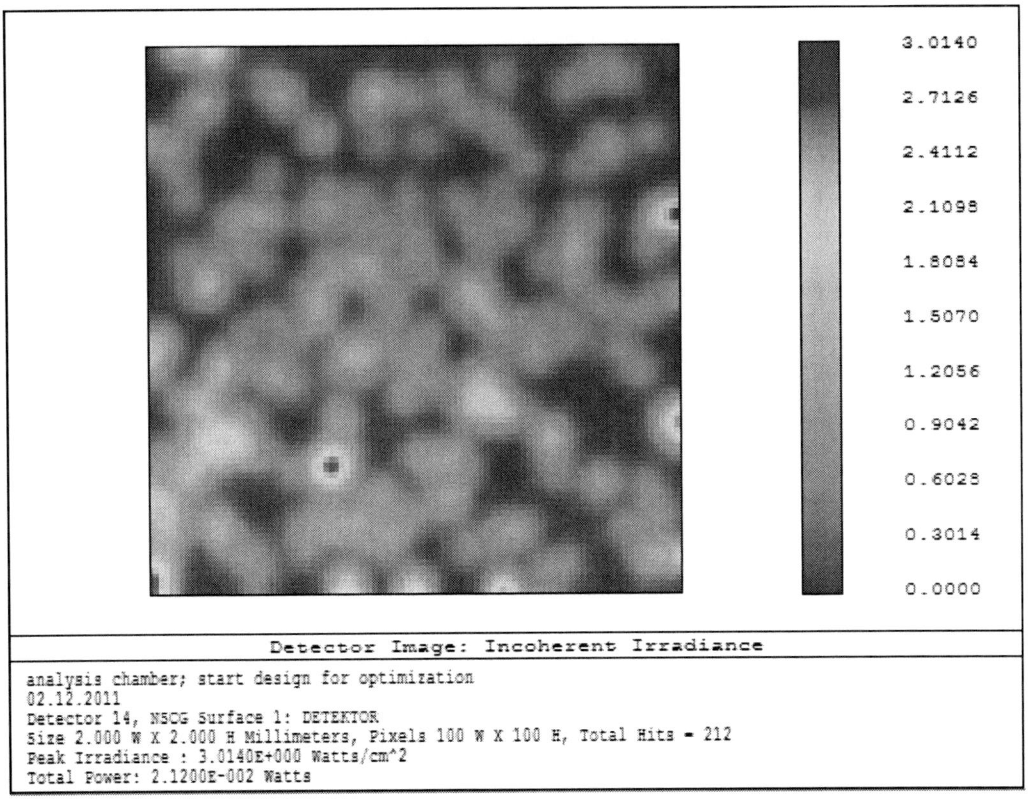

Figure 11.33. Irradiance distribution of the starting model in the case of mispositioning of the source by half the value of the tolerance range.

Boundary conditions of the optimization runs were the physical dimensions of the cell, the distance between source and detector, and the ray path inside the cell. For reasons of convergence of the optimization algorithm, the starting model has to be reduced in complexity.

One result of Section 3.2.1 was that control of the ray propagation inside the analysis cell can be achieved by means of an adequate source reflector. For this reason, the first step in optimization was the optimization of the shape of the source reflector. Therefore the model was reduced to the source reflector (reflector 1 in Figure 11.31) and the first top reflector (reflector 2 in Figure 11.31) (see Figure 11.34). The goal of this optimization step was to attain a robust performance of this subsystem over the entire tolerance range of the filament inside the light bulb.

The model of the optimized source reflector was inserted into the starting model for optimization of the cell geometry. Then, reflectors 3 and 4 (see Figure 11.31) were processed next. The result of this optimization step is an improved starting model. This improved starting model can be used in the next step for

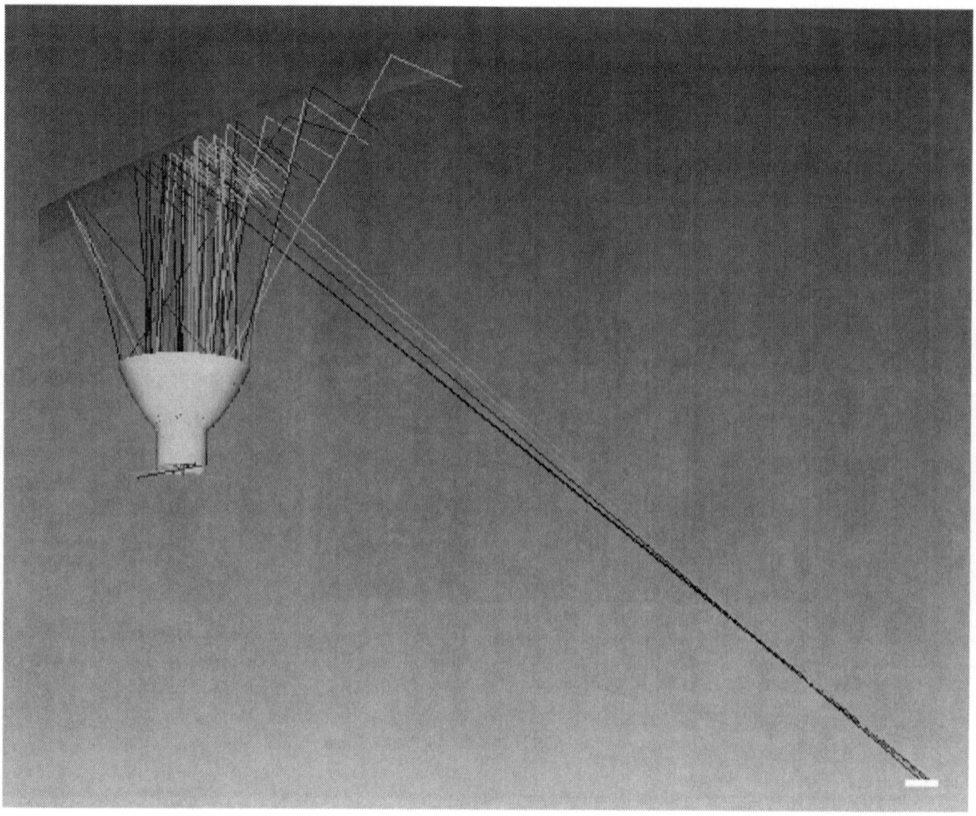

Figure 11.34. Optimization of the source reflector in terms of position tolerance of the source.

optimization of all four reflectors in one single optimization run without risking convergence of the algorithm. The resulting optimized model is shown in Figure 11.35. Remarkable is that reflector 3 conserves its plane shape; i.e., reflector 3 serves only as a folding element of the beam. The most significant change is obvious for reflector 4, which is strongly curved in both the x and y directions. Figure 11.36 shows the irradiance distributions for three positions of the filament inside the light bulb spanning the complete tolerance range. From left to right, the first and third distributions are calculated for the upper and lower boundaries of the tolerance range, respectively. The second distribution is determined for the center position of the tolerance range (the nominal position of the filament in the light bulb).

The optimized design of the analysis cell reaches a yield of 30% of the propagated rays on the detector in the whole tolerance range of the source position. Therefore, the optimized design ensures defined system performance in the complete tolerance range and enables automated assembly of the analysis cell.

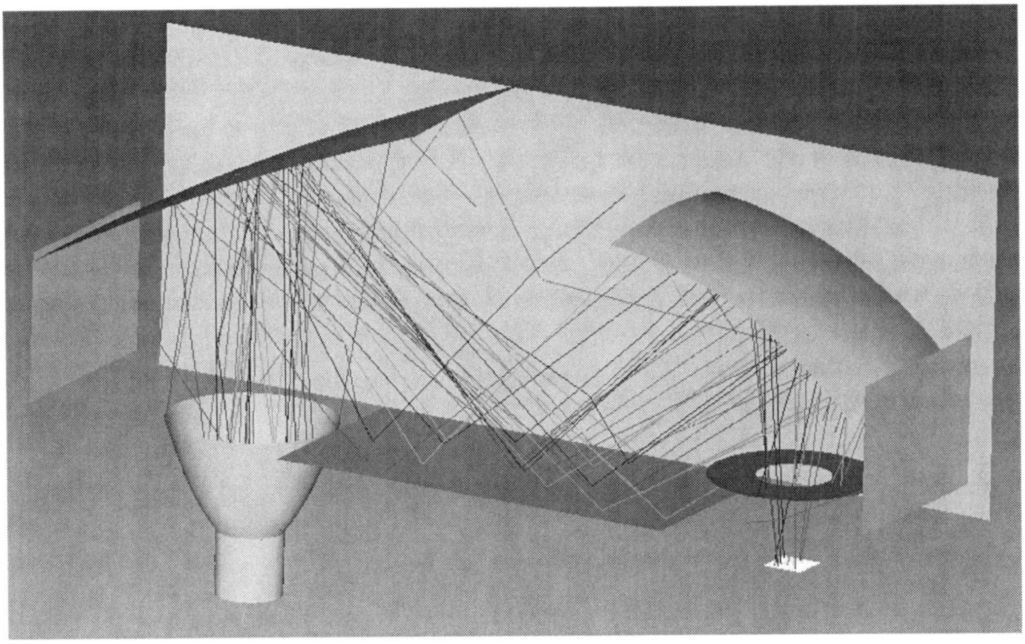

Figure 11.35. Optimized design of the absorption cell.

5. CONCLUSION

Starting with a brief overview of the historical development of gas sensors based on optical multipass cells, the main subsystems and technical components for their realization have been outlined. After a general introduction to design

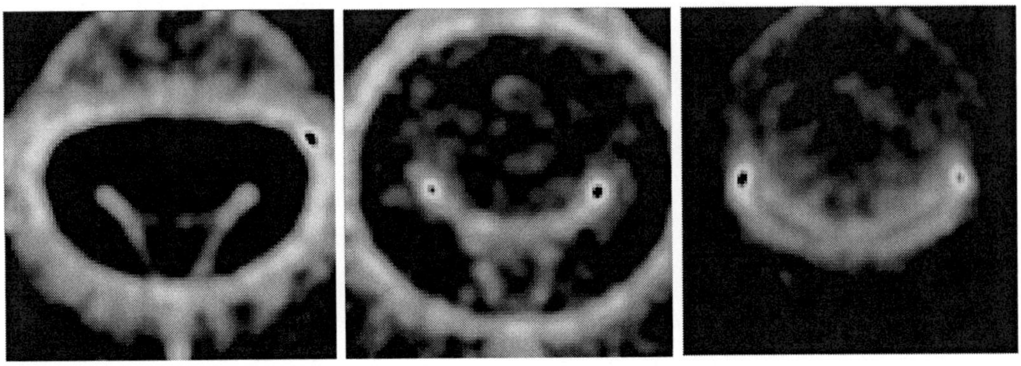

Figure 11.36. Irradiance distribution of the optimized design for three positions of the source inside the tolerance range.

methodology, design rules have been given for each of the subsystems. From these considerations it became evident that there are many interrelations between subsystems; hence they cannot be designed independently. This holds in particular if a trade-off between technical performance and cost has to be made. Thus a careful requirements analysis and development of principle solutions are of paramount importance. A very useful technique to developing the principle solution is optical modeling and simulation, which ideally result in a conceptual design. On the basis of the conceptual design, an enhancement of the model takes place as information of manufacturing processes as well as of the individual components by means of tolerance data are added. This comprehensive model can be used for a sensitivity analysis to show the performance of the system in terms of realistic manufacturing conditions. This analysis is the basis for a decision concerning the

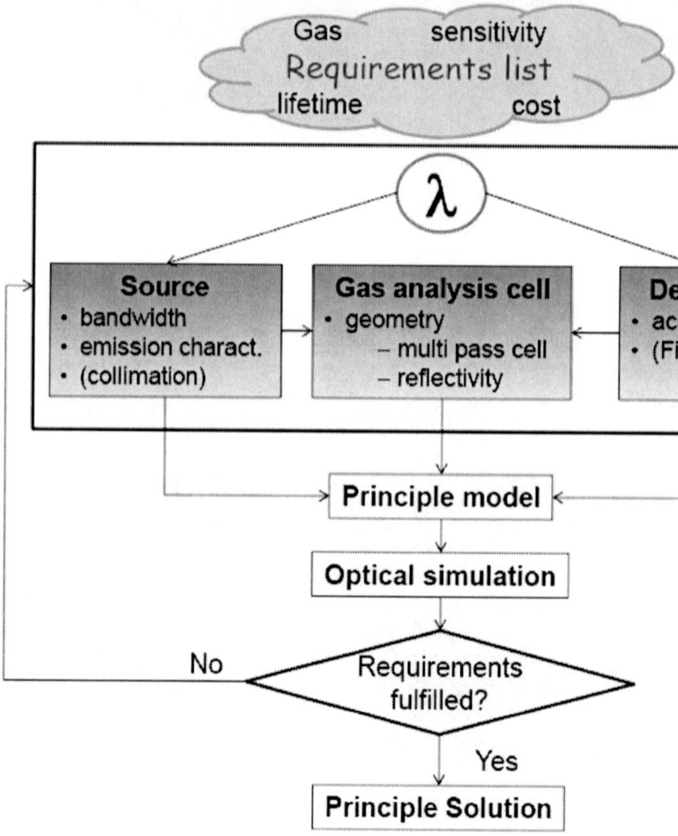

Figure 11.37. Optical modeling and simulation in conceptual design.

further approach: If the system's performance meets the specifications given by the requirements list, the embodiment design is found (at least the preliminary layout). If the system's performance fails, a decision has to be made whether a robust design approach could improve the system's performance, or whether other manufacturing techniques or components will have to be used, or even whether the conceptual design has to be changed.

This approach has been demonstrated in this chapter for modeling, simulation, and optimization of a low-cost gas sensor.

Already in conceptual design, optical modeling and simulation is an important tool. In early product development it is used for evaluating and comparing components such as light sources, detectors, and cell shapes (see Figure 11.37). In this phase, optical modeling and simulation yields first idealized principle solutions.

In particular, for turning the idealized conceptual design into a realistically producible embodiment design, not only the nominal component characteristics, but also

- Source and detector tolerances (both optical and dimensional)
- Cell fabrication tolerances (both optical and dimensional)
- Assembly tolerances
- Aging characteristics

have to be taken into account. Some of these can be derived from suppliers' data sheets; some tolerances may have to be gathered from one's own measurement data. The enhancement of the principle solution by means of these operational and manufacture-dependent parameters of both the gas sensor components and the assembled system leads to the comprehensive model. A sensitivity analysis is used to check whether the comprehensive model meets the system's requirements as defined in the requirements lists. If this holds true, the preliminary layout is identical to the principle solution. If the requirements are not met, an optimization of the comprehensive model with respect to the operational and manufacture-dependent parameters has to be carried out, leading to a more robust design. Again, it has to be checked whether the requirements are met. If not, a robust design based on the initial principle solution does not exist, and a redesign of the principle solution is mandatory. This entails reconsideration of the subsystems source, multipass cell, and detector. This design flow is summarized in Figure 11.38.

The merit of the above-outlined design rule and model-based approach to optical gas sensor design is that the majority of product development and optimization can be done in software. This allows shifting investments in hardware (part fabrication, tooling) to later stages in the product development process and thus saves considerable amounts of money.

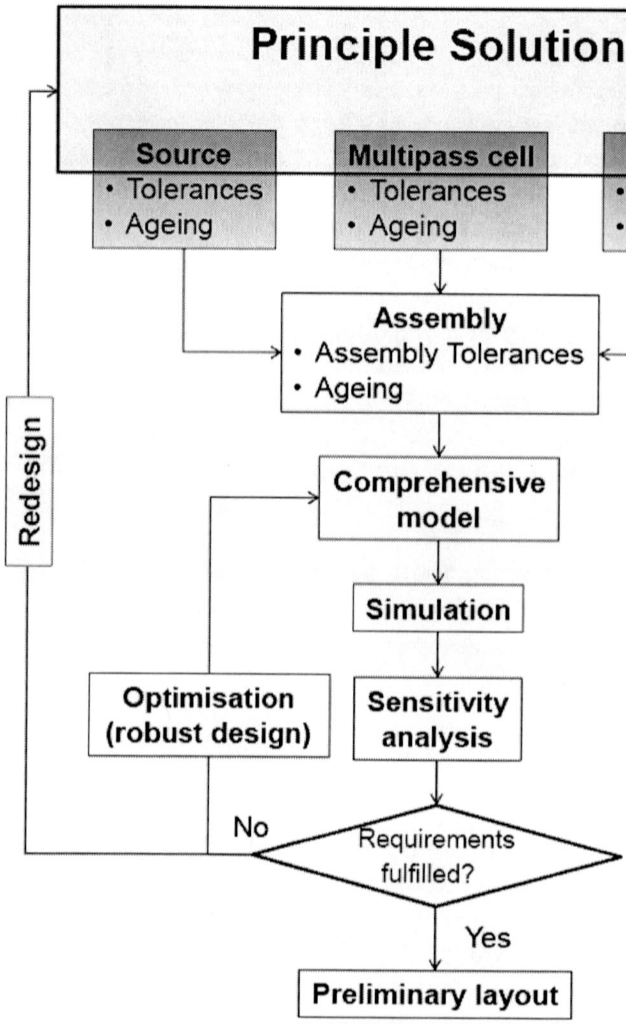

Figure 11.38. Optical modeling and simulation in embodiment design.

REFERENCES

Das D. and Wilson A.C. (2010) Very long optical path-length from a compact multi-pass cell. *Appl. Phys. B* **103**, 749–754. DOI: 10.1007/s00340-010-4337-7

Goodman J.W. (1996) *Introduction to Fourier Optics,* 2nd ed. McGraw-Hill, New York.

Günzler H. and Heise H. (1996) *MIR-Spektroskopie.* VCH, Weinheim, Germany.

Herriott D.R. and Schulte J.H. (1965) Folded optical delay lines. *Appl. Opt.* **4**, 883–889. DOI: 10.1364/AO.4.000883

HITRAN on the Web (2011) http://hitran.iao.ru.

Jakob W. (2004) Eine neue Methodik zur Erhöhung der Leistungsfähigkeit Evolutionärer Algorithmen durch die Integration lokaler Suchverfahren. Ph.D. thesis, Forschungszentrum Karlsruhe Wissenschaftliche Berichte FZKA 6965, Karlsruhe, Germany.

Joannopoulos J.D., Johnson S.G., Winn J.N., and Meade R.D. (2008) *Photonic Crystals: Molding the Flow of Light*, 2nd ed. Princeton University Press, Princeton, NJ.

Johnson S.G. (2011) *Manual of MIT Photonic-Bands*. http://ab-initio.mit.edu/wiki/index.php/MPB_Introduction.

Narimanov E., Fan J., and Gmachl C. (2005) Compact quasi-chaotic optical cavity. In: *Proceedings of Quantum Electronics and Laser Science Conference (QELS)*, IEEE, New York, May 22–27, Vol. 1, 421–423.

Pahl G. and Beitz W. (1995) *Engineering Design*. Springer-Verlag, Heidelberg, Germany.

Sieber I., Eggert H., Suphan K.-H., and Bechthold S. (2001) Simulation and modeling of a chamber of analysis of an IR gas sensor. In: *Proceedings of Intentaional Conference on Micro System Technologies*, March 27–29, Düsseldorf, Germany, 71–76.

Sieber I., Eggert H., Suphan K.-H., and Nüssen O. (2002) Optical modeling of the analysis chamber of an IR gas sensor. In: *Proceedings of Design, Test, Integration, and Packaging of MEMS/MOEMS 2002*, April 25–27, Cannes, France, *Proc. SPIE* **4408**, 272–282.

Southwest Sciences (2011) *Gas sensing tutorial*. www.swsciences.com/technology/sensors.html.

Steffen M., Herbst J., Woellenstein J., and Lambrecht A. (2007) Ultracompact Herriott cell for optical gas sensor applications. In: *Proceedings of 6th International Conference on Tunable Diode Laser Spectroscopy*, July 9–13, Reims, France, 102.

Tarsitano C. and Webster C. (2007) Multilaser Herriott cell for planetary tunable laser spectrometers. *Appl. Opt.* **46**, 6923–6935. DOI: 10.1364/AO.46.006923

White J.U. (1942) Long optical paths of large aperture. *J. Opt. Soc. Am.* **32**, 285–288. DOI: 10.1364/JOSA.32.000285

Wikipedia (2011) http://de.wikipedia.org/w/index.php?title=Datei:Image-Metal-reflectance.png&filetimestamp=20070301140531.

CHAPTER 12

SIMULATION AND MODELING FOR OPTICAL DESIGN OF LASER REMOTE SENSING FOR GAS MEASUREMENTS

Tatsuo Shiina

1. INTRODUCTION

Lidar (LIght Detection And Ranging) is an active optical sensor which transmits a pulsed beam to a target and detect its echo. Basically, by measuring the time of flight, one can know the distance between the apparatus and the target. The terms "ladar (LAser Detection And Ranging)" and "laser radar" have the same meaning.

Lidar development began just after the innovation of lasers. Before that, high-powered searchlights were used for trials; practical ones use the laser pulse as a light source. The difference between lidar and radar (RAdio Detection And Ranging) is the characteristics of the transmitting beam. Radar transmits the expanding beam, while lidar throws the collimated beam. This causes the different appearance of the receiving echoes. At first, lidar was used as a range finder for military purposes. It was noted that atmospheric echoes were overlapped on the observed signals, which became the start for monitoring the atmosphere (Fiocco 1963).

Light is scattered by tiny particles. Everyone has seen such scattering, as in clouds, fogs, smokes, and so on. Everything you see is caused by light scattering. Trees, buildings, mountains, sea, and even human bodies scatter lights. Thus they can become a lidar target: One only needs to select a certain signal from a

DOI: 10.5643/9781606503201/ch12

target. How does one accomplish this? Choose the light source and the detector, and set up the optics. An analysis method is also essential. Now lidar techniques are mature, and one need only decide on the target to monitor it.

Scattering—interaction between the light and the target material—may be understood in more than one way. The first understanding of "scattering" was due to Rayleigh in 1871. He explained the blue of the sky by using a dipole model of the atmospheric molecules (Rayleigh 1871). Later, in 1908, Mie solved the complicated scattering mechanism of a particle which has larger size than atmospheric molecules, by considering the boundary conditions of the Maxwell's equations (Mie 1908). His purpose was to study the phenomenon that tiny particles of gold scatter red-colored light. One can see this effect in old stained-glass windows in European churches. Now these effects are called Rayleigh scattering and Mie scattering, respectively.

Scientists today understand that the color of scattered light depends on the interaction between the characteristics of the target material and the wavelength of the incident light (Deirmendjian 1969; Born and Wolf 1980; Hulst 1981; Bohren and Huffman 1983). In 1928, Raman found that the wavelength of the scattered light is shifted according to certain rules (Raman 1928). The Raman scattering has several variations, such as vibrational and rotational or Stokes and anti-Stokes. The basic Raman scattering in lidar applications is Stokes light or vibrational Raman scattering. Other scattering phenomena were found during the same period by Bullurian, Thomson, Compton, and so on, but these are not utilized in lidar observations.

Not only scattering, but also other optical phenomena of the interaction between the target and the light are utilized in lidar target detection. Fluorescence is a kind of resonance which is pumped by an incident light and spreads its wavelength to the longer part. Fluorescence represents the characteristics and the conditions of the target material stimulated by the incident light. The first observation of fluorecence, from a quinine solution in sunlight, was reported by Sir John Frederik William Herschel in 1845 (Herschel 1845). The Doppler shift (the Doppler effect was found in 1842) is a well-known phenomenon evidenced by the fact that the siren of a fire engine changes its frequency as it passes just in front of you. A Doppler shift can also be found in an optical field. Moving targets shift the frequency (wavelength) of the scattering light. DIAL (DIfferental Absorption Lidar) is effective to monitor a certain gas and oil. It utilizes two light sources, whose wavelengths are tuned on the absorption line and nonabsorption line of the target molecules (Murray 1978).

Lidar techniques have been improved for various purposes. The type of scattering one wants to observe determines the lidar optics to be used. Mie lidar is the basic setup for lidar optics. Rayleigh lidar needs a narrow-spectrum band filter. Raman lidar should shift the monitoring light spectrum line due to the interaction between the incident beam line and the target. DIAL uses two light sources.

Lidar is used not only on the ground but also in vehicles, ships, aircraft, and satellites. The Space Shuttle was utilized to observe global water convection (Kovacs 2003). The project was called the LITE mission, and its data were made available to scientists. Lidar networks are also used for global atmospheric monitoring. EARLINET, MPLNET, SPALINET, and the ASIAN Dust Network are now activated. Monitoring of volcano smoke, yellow sand, and desert dust are the global monitoring targets (Welton 2002; Ansmann 2003; Chaikovsky 2006; Sugimoto 2006; Baars 2008; Hoff 2008; Sicard 2009).

In this chapter, Lidar techniques are explained. In Section 2 the structures and the devices to measure specified targets are described. Next the author shows how to estimate the scattering intensity in Section 3. It is important to select the light source power and the sensitivity of the detector. Standard analysis methods are also described. Not only classical lidar but also DIAL and HSRL (High Spectral Resolution Lidar) are mentioned. Typical lidar applications are introduced with practical observation data in Section 4. Mie and Raman lidar are explained. The LED mini lidar is also introduced as the current development. In the last section, the chapter is summarized.

2. LIDAR TECHNIQUES

2.1. LIDAR STRUCTURE

The lidar structure consists mainly of three parts: a light-transmitting part, an echo-receiving optics part, and a signal processing part. An illustration is shown in Figure 12.1. The light source is a pulsed laser beam. The pulse width τ decides the maximum resolution (shortest interval) of the distance, that is, $c\tau/2$, where c is the speed of light. Dividing by 2 means a round trip. In general, the pulse width is a few to a few tens of nanoseconds. It is equivalent to a distance of a few meters. The earliest laser was a ruby laser because of its giant pulse. Now usage has shifted to YAG (yttrium aluminum garnet) lasers and various other kinds of lasers such as excimer, YLF, DPSS, CO_2, and so on. High transmitting power is good, but nonessential power causes some problems such as large system size, high comsumed power, and risk to eye safety. The minimum peak power is only 100 mW (= 1 nJ/10 ns)! The power depends on the observation range and the density of the target atmosphere or gas. Laser wavelengths are utilized from the ultraviolet to near-infrared light. In practice, the wavelength usually ranges from 266 nm to 2 μm. A UV beam is used mainly in Raman lidar for the detection of certain gases such as O_3, H_2O (water vapor), NH_3, and so on. Visible light (355 nm to 1 μm) is utilized for Mie and Rayleigh lidar for the monitoring of aerosols, clouds, and high-altitude atmospheric characteristics such as temperature and pressure, because in this spectral range we have high-power YAG lasers and photomultipliers

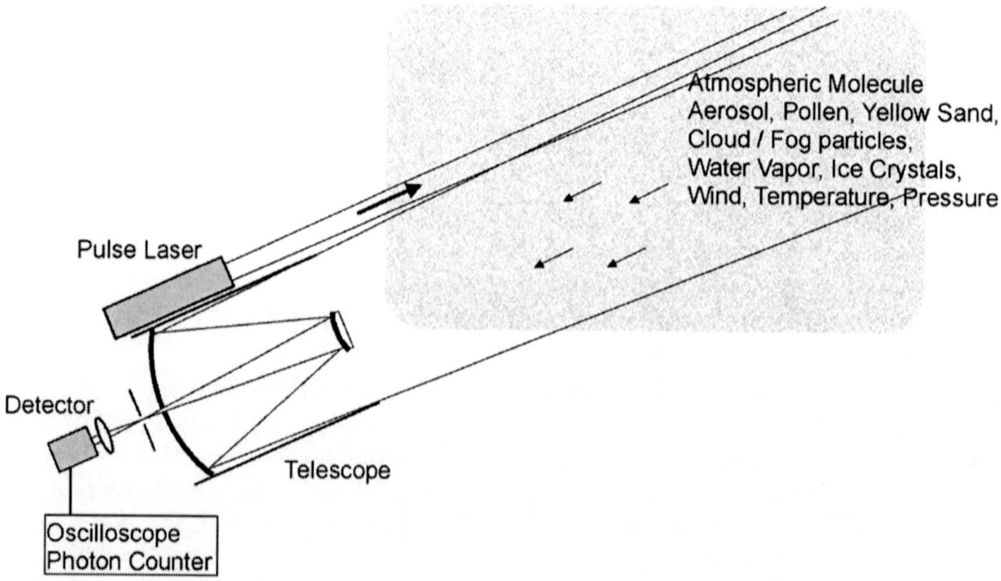

Figure 12.1. Lidar setup.

with high sensitivity. A near-infrared (NIR) beam is used in DIAL for quantitative investigation of certain molecules and gases such as NO_x, SO_x, CO_2, and so on. Lidar techniques are classified by the targets in Table 12.1.

To aim at a far target, the transmitting laser beam should be collimated. This is different from radar, which transmits a spreading electromagnetic wave. For lidar, beam collimation not only restricts the extinction of the power, it also make it possible to narrow the receiver's field of view. The scattered light from the target becomes diffused light, and goes generally in all directions. The scattered power is

Table 12.1. Lidar techniques and their main targets

TYPE	DESTINATION (MAIN TARGET)
Mie lidar	Aerosol, cloud, fog, atmosphere (troposphere)
Rayleigh lidar	Atmospheric molecules(stratosphere/mesosphere), temperature, pressure
Raman lidar	O_3, H_2O(water vapor), NH_3, N_2, H_2, . . .
DIAL	SO_x, NO_x, CO_2, . . .
Doppler	Wind speed and direction
HSRL	High-precision atmospheric parameters such as back-scatter cross section

inversely proportional to the square of the distance. To detect the scattered light, which is called an "echo," the optical receiver should have a wide aperture. The receiving intensity is proportional to the fourth power of the diameter. An astronomical telescope is often utilized. A reflecting telescope is better because of its large aperture and the independence of the wavelength of the receiving light.

The receiving optics consists of the telescope, field stop aperture (pinhole), eyepiece, optical filter, and detector. The field stop aperture (FSA) determines the field of view (FOV) of the telescope. It is written as

$$\theta_b = -\frac{A}{2f} \tag{12.1}$$

where θ_b is the spreading angle of the field of view, in units of radians. A is the diameter of the FSA. f is the focal length of the telescope. The geometric arrangement is shown in Figure 12.2. The receiver's FOV is an important parameter for lidar observation. When it is wide, the optical arrangement is easy, but a large amount of background light is incident into the lidar optics. When it is narrow, the background light is reduced, but the requirements for optical arrangement become stricter. A typical field of view of the telescope is 2–3 mrad for lidar applications. This is good for Mie lidar in both daytime and nighttime observation. For Raman lidar, however, this is not enough for daytime observation. Micro pulse lidar has a narrower field of view of 0.1 mrad, to observe the atmospheric echo with a low-powered laser beam. The narrow FOV also eliminates the depolarizaiton effect caused by multiple scattering of the atmospheric particles (see Section 4).

The detecter unit is shown in Figure 12.3. The eyepiece is used to control the size of the delivering echo light because the optical filter and detector have

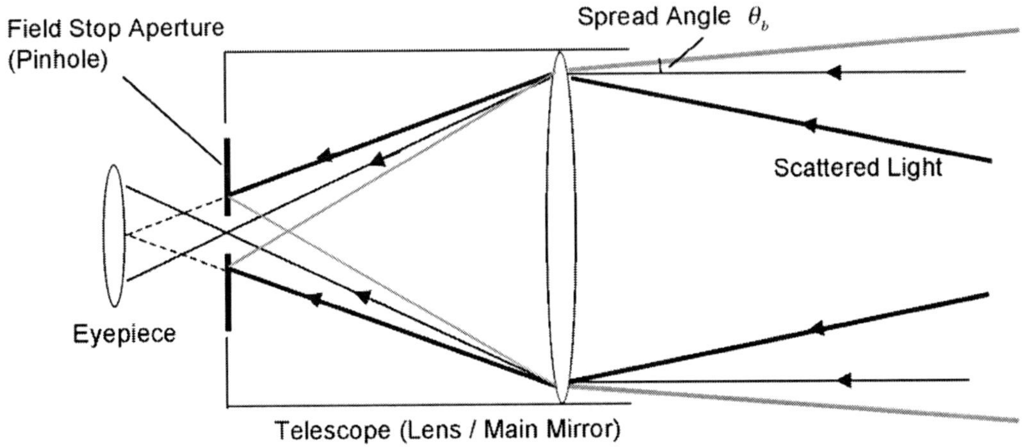

Figure 12.2. Geometric arrangement of field of view on receiver's optics.

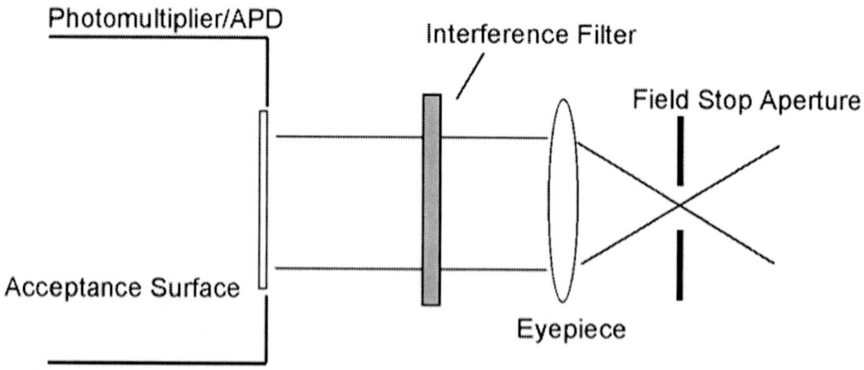

Figure 12.3. Optical receiver optics.

directional qualities. An optical filter is essential in lidar optics. The typical use is to select the same center wavelength for the transmitting laser beam for Mie and Rayleigh lidars. Shifting the center wavelength of the optical filter is adequate to monitor a given target for Raman and fluorescence lidars. The bandwidth of the filter is a few to 10 nm. An interference filter is easy to use because its optical characteristics are well controlled. Some other kinds of filters such as etalon and iodine filters have narrower spectrum bandwidth, and they may also be installed into the optics of Rayleigh lidar and HSRL. Those lidars should eliminate the background light drastically. The optical detectors used are photomultipliers (PMTs) and avalanche photo diodes (APDs). A photomultiplier is a quite sensitive photo detecting device. Its sensitivity is over the wavelength from 110 nm to 1 μm. To operate it, high voltage (a few hundred to a thousand volts) is needed. APDs are utilized mainly for near-infrared light detection of wavelengths of more than 1 μm. They are used with a Peltier cooling unit and a heat sink. The electrical noise of an APD is larger than that of a PMT, and it should be cooled. To keep the device temperature constant is necessary for stable measurement and is thus important in long-term observations. PMTs and APDs can be operated in analog mode and in photon counting mode. (For APDs the photon counting mode is called "Geiger mode.") Background light can be restricted severely by operating in photon counting mode.

The signal processing part consists of an amplifier, digitizer (A/D interface, oscilloscope), or scaler. In analog-mode detection, a digitizer is utilized. In that case, sampling speed determines the resolution of the distance. The memory depth represents the observation range. Typical Mie lidar has an observation range of less than 20 km with a spatial distance of a few tens of meters. Current commercial digitizers have enough sampling speed and memory to be used in a lidar system. The scaler, often called a "multichannel scaler," is a photon counter that counts scattered photons as a function of time. It segments photon count data into sequential time bins. The width of the bins represents the sampling resolution. The photon counter has a discriminater to distinguish the photon signal from the noise.

As the fundamental structure of the lidar optics, there are three types as shown in Figure 12.4. The biaxial type of optics shown in Figure 12.4a has separate optics for the optical transmitter and the optical echo receiver. The coaxial type of optics shown in Figure 12.4b transmits a pulsed laser beam from a portion of the receiver's aperture, along the optical axis of the receiver. The "in-line" type of optics shown in Figure 12.4c has common optics for the transmitter and the receiver. In the case of biaxial optics, an easy way to shorten the blind area at near range is to broaden the receiver's FOV at the expense of an increase of background light. The background light decreases the dynamic range of the lidar echo. As the path of the transmitted laser beam is set at a slant angle against the optical axis of the receiver, the blind area of the near range can be controlled with the fixed FOV. The slant path, however, will deviate from the receiver's field at a certain distance and beyond. To avoid this problem, the laser path is slanted at a small angle with respect to the optical axis of the receiver, which, however, results in a large blind area. With coaxial optics, the laser path is parallel to the optical axis of the receiver. The coaxial optics are set by reflecting the laser beam on a reflector located on the back side of the secondary mirror at the center of the receiver's aperture. This also includes the case where the laser beam is transmitted from a certain point in the telescope aperture by a small mirror or a prism. To narrow the receiver's FOV to eliminate background light, a field stop aperture

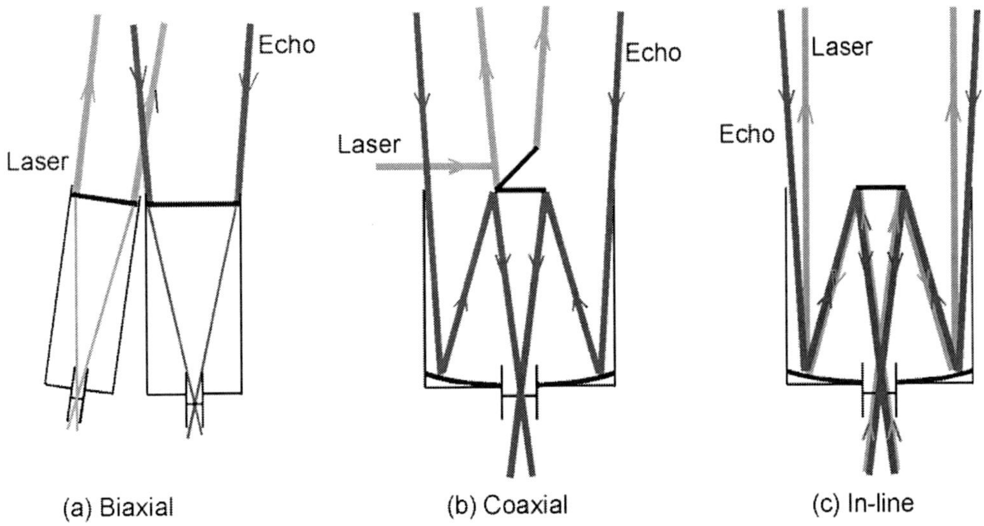

Figure 12.4. Optical setup for three types of transmitting and receiving lidar optics.

is installed at the focal point of the telescope. In this case, the near-range lidar echo is out of focus and cannot pass all of the echo light through the pinhole. Futhermore, the shadow of the secondary mirror will cover the pinhole in the near range. As a result, the coaxial lidar also has a blind area (Halldorsson 1978; Hams 1979; Sugimoto 1990; Chourdakis 2002). With in-line optics, the transmitting laser beam and the receivering echo have the same path in the lidar optics and in open air. That is, the laser path always overlaps with the receiver's FOV, even if the FOV is narrow, of the order of 0.1 mrad. As the intensity distribution of the transmitted laser beam can be enlarged up to the telescope aperture, there is no blind area in the near range (Shiina 2005a, 2005b). On the other hand, the lidar echoes should be separated from the transmitting path in the lidar optics. In general, this will make the optics complex.

2.2. LIDAR TECHNIQUES

Mie scattering is basic scattering for the lidar technique because the wavelength of the transmitting light is shorter than that of the scattering particles such as aerosols, cloud/fog particles, rain droplets, ice crystals, dusts, smokes, and so on. One can catch Mie scattering echoes easily using the ordinary receiving optics shown in Figure 12.3. Mie scattering has low dependency on the wavelength of the incident beam.

Rayleigh scattering is caused by atmospheric molecules. If one wants to capture Rayleigh scattering, Mie scattering should be removed, because the central wavelength of both types of scattering is the same. For observing the high-altitude atmosphere, the receiving optics are identical to that for Mie lidar, while the laser should be stable to lock the wavelength, and the optical filter should be narrowed in bandwidth by the monocrometer, etalon, or other filters. For observations at low-altitude, the receiving optics should have quite narrow bandwidth to eliminate Mie scattering. The Mie scattering has a narrow peak at the center of the wavelength, while the Rayleigh scattering has a broader band. The large particles which cause Mie scattering have a low speed of "wind" at low altitude, while the atmospheric molecules have high speed of "sound" at high altitude. The high speed produces a large Doppler shift in the scattering light, and the Rayleigh scattering echoes have broadened spectral bandwidth in comparison to the Mie scattering echoes, which have a narrow spectral peak, as shown in Figure 12.5. Etalon filters and iodine cells are utilized to separate the Mie scattering and Rayleigh scattering. Rayleigh lidar especially for low-altitude observation is called HSRL (High Spectral Resolution Lidar) (Shipley 1983; Weitkamp 2005).

Raman scattering shifts the wavelength of the scattering light from that of the incident light. In general, the intensity of Raman scattering is 3 digits smaller than that of the Mie scattering (Leonard 1967; Melfi 1972). When electrons are

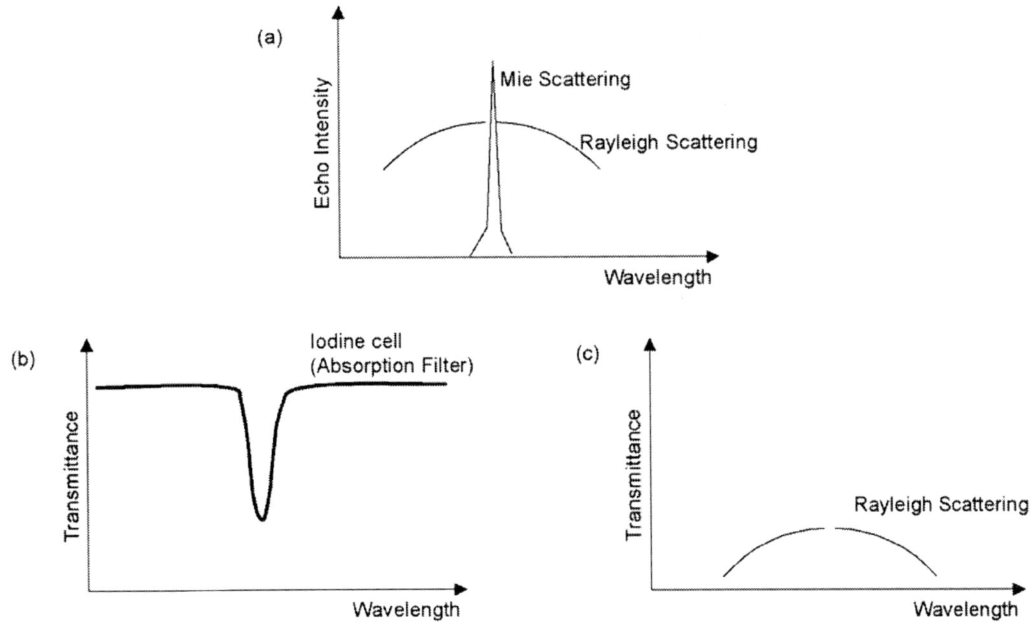

Figure 12.5. Raleigh scattering separated from Mie scattering.

scattered from the atmospheric molecule (or a particular gas, atom, and so on), most photons are elastically scattered (Rayleigh scattered), such that the scattered photons have the same frequency and wavelength as the incident light (photons). A small fraction of the scattered electrons are scattered by an excitation, with the scattered photons having a frequency different from the frequency of the incident photons. That is, there are two types of Raman scattering, Stokes scattering and anti-Stokes scattering. The interaction of light with matter in a linear regime allows the absorption and emission of a photon precisely matching the difference in energy levels of the interacting electron or electrons. Stokes scattering is caused by the energy absorption of the molecules, while anti-Stokes scattering is caused by the energy loss of the molecules. Stokes scattering shifts the wavelength of the incident light to the longer side, and anti-Stokes scattering shifts the wavelength to the shorter side, as shown in Figure 12.6.

The atmospheric molecules absorb light of certain wavelengths, while they do not absorb, that is, penetrate, lights of other wavelengths. When two lights of different wavelengths, that is, absorbed wavelength and nonabsorbed wavelength, for a certain kind of molecule, are thrown to the atmosphere, there is no difference in the echoes from the two lights if there are no molecules that absorb the alternate light. If such molecules are present, the echo of the absorbed light is decreased in comparison with the echo of the nonabsorbed light. The decrease will be large when the density of the molecules is high. If the absorption coefficient is

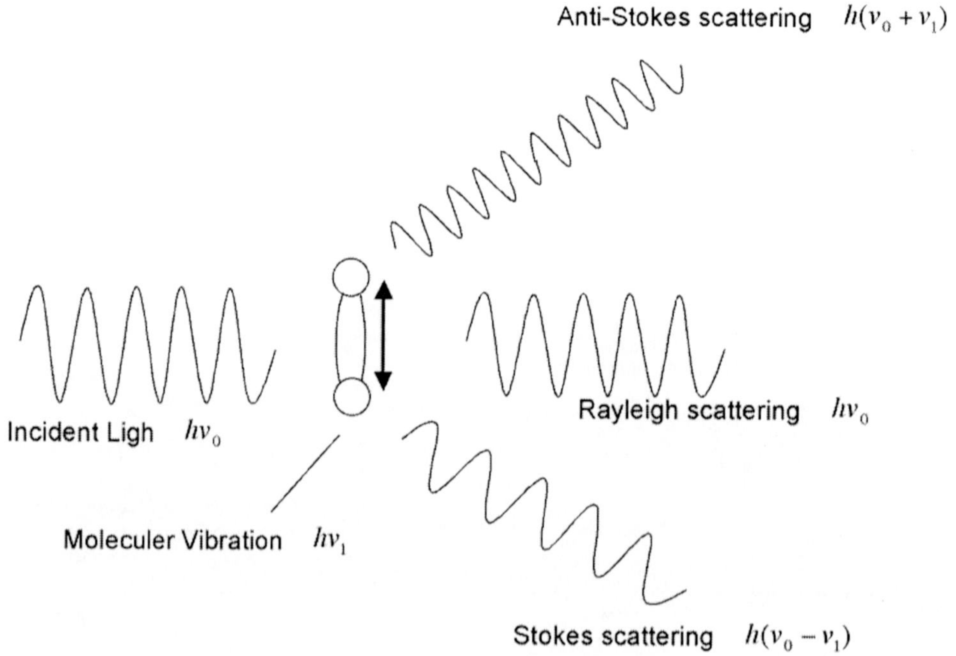

Figure 12.6. Raman scattering.

known, one can estimate the density of a target gas remotely, from the propagation distance (Measures 1984).

DIAL utilizes the above technique. In this case, two lights of different wavelength are arranged as the lidar light source. The absorbed light and the non-absorbed light are written as λ_{on} and λ_{off}, respectively (see Figure 12.7). The near wavelengths of these lights are utilized to ascribe the same optical characteristics to the wavelengths. To arrange the two wavelength-controlled laser lights is not easy. A wavelength-tunable laser such as a dye laser is utilized. CO_2 lasers and excimer lasers are also used because the wavelength of such lasers matches well with the absorption line of ozone (O_3). The wavelength range is from UV to IR.

3. CALCULATIONS

3.1. LIDAR EQUATION

When the collimated beam penetrates the atmosphere, it is attenuated by suspended particles in the atmosphere. As the atmospheric extinction coefficient is represented as α, the transmittance T obeys the Beer-Lambert law as follows (Fujii and Fukuchi 2005):

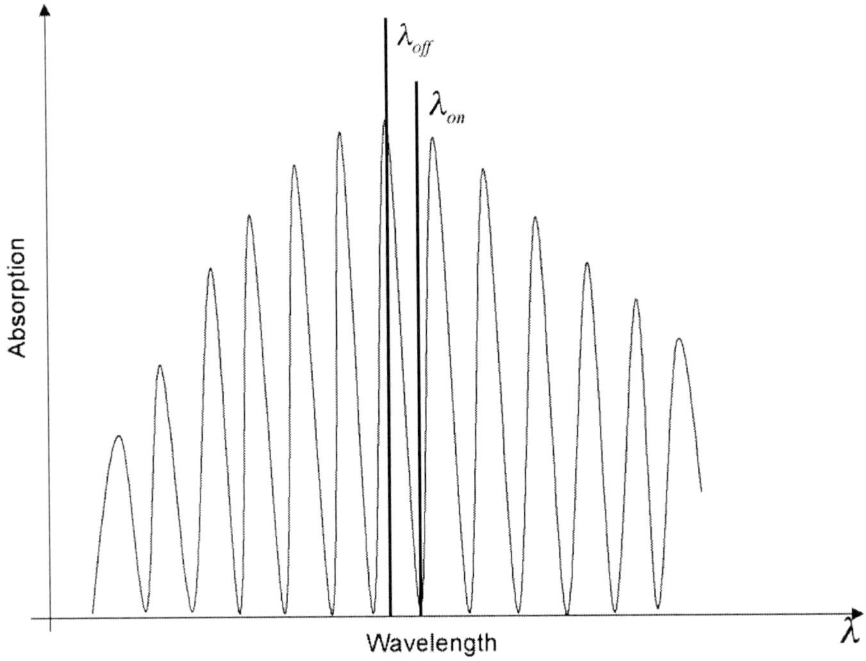

Figure 12.7. Differential absorption.

$$T(L) = \exp\left[-\int_0^L \alpha(x)\right]dx \tag{12.2}$$

where L is the propagation distance of the beam. When the beam hits the particles, the scattered light goes back to the same path. The echo, however, is also attenuated by the particle. Furthermore, the scattered echo spreads in all directions, and its intensity decreases in inverse proportion to the square of the distance. The lidar echo power depends on the atmospheric characteristics and also on the optical characteristics of the lidar apparatus, such as the aperture of the optical receiver, the pulse width of the transmitting beam, and the optical efficiency of the lidar system. The practical lidar echo power $P(L)$ is estimated as

$$P(L) = P_0 K Y(L) A_r \frac{c\tau}{2} \beta(L) T(L)^2 / L^2 \tag{12.3}$$

where P_0 is the transmitted power [W = J/s], K is the system optical efficiency, A_r is the aperture of the optical receiver [m^2], c is the speed of light [m/s], τ is the pulse width of the beam [tu]. $\beta(L)$ is the back-scattering cross section of the target [m^{-1}]. $Y(L)$ is the geometric form factor, which is the overlap function between the transmitting beam and the receiver's field of view. It is determined by the telescope specification, the beam divergence, and the size of the field stop aperture. The geometric form factor determines the initial receiving characteristics in the near range. When one observes a near-range target, its lidar echo can be corrected using this function. However, in the general lidar setup, there is never a detecting area in the near range. When the laser beam fires from the lidar system, it takes a certain distance to enter into the receiver's field of view (see Figure 12.8). This is called the "blind area." An echo cannot be obtained in this area. The biaxial type and the coaxial type lidars have a blind area, while the in-line type of lidar never has one. After the blind area, there is the another distance at which the beam partially crosses into the field of view. Here the lidar echo power rises until the whole of the beam enters into the field of view.

An estimate of the signal-to-noise ratio (SNR) is effective to evaluate practical lidar echoes under several noises. One should consider the influences of the background light and the noise factor of the detector. The signal-to-noise ratio is written as

$$\mathrm{SNR}(L) = \frac{\sqrt{M}\sqrt{\eta\,\Delta t / h\nu P(L)}}{\sqrt{\mu}\sqrt{P(L) + P_b + P_d}} \tag{12.4}$$

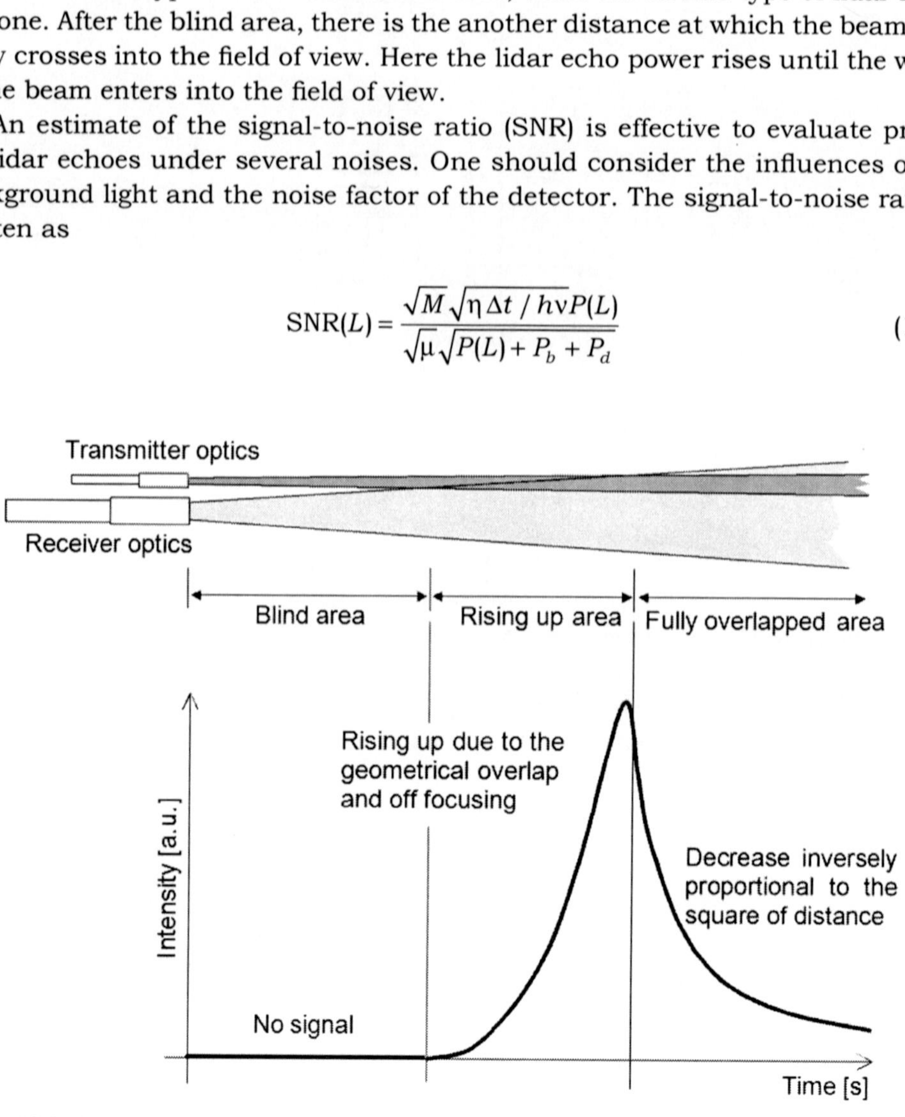

Figure 12.8. Blind area and rise of the echo power.

where η is the detector's quantum efficiency and Δt is the sampling interval [s]. μ is the detector's noise factor; it is 2–3 for a APD and 1 for a PMT. h is Planck's constant [J s], ν is the light frequency [Hz = s^{-1}]. P_b is the background light power [W]. P_d is the equivalent dark current power [W]. M is the number of signal summations. That is, the long-term summation will increase the SNR, but its effect does not increase linearly. The numerator of Eq. (12.4) improves SNR by the detector's sensitivity and the summation. The denominator weakens SNR by the noises of the atmosphere and the lidar system.

3.2. SIMULATION

The actual estimation will start from the design of the lidar optics. First, one must select the lidar types. Next. the designs of the transmitting and receiving optics and their arrangement should be decided. Figure 12.9 shows the geometric arrangement of the coaxial and in-line optics. In this illustration, the lens is a model of the primary mirror of the telescope, which is the same arrangement for a reflecting and a refracting telescope. The field stop aperture is arranged at the focal plane of the primary mirror. The laser beam is thrown from any position of the system. Its beam size and divergence are arbitrary. The illustration represents the coaxial-type lidar setup. If one selects a reflecting telescope, the shadow of the secondary mirror must also be considered (see Figure 12.10). Though the echo goes back to the telescope and is focused, its focal plane is after the FSA. Then the part of the echo is cut at the FSA, especially in the near range. That is, the

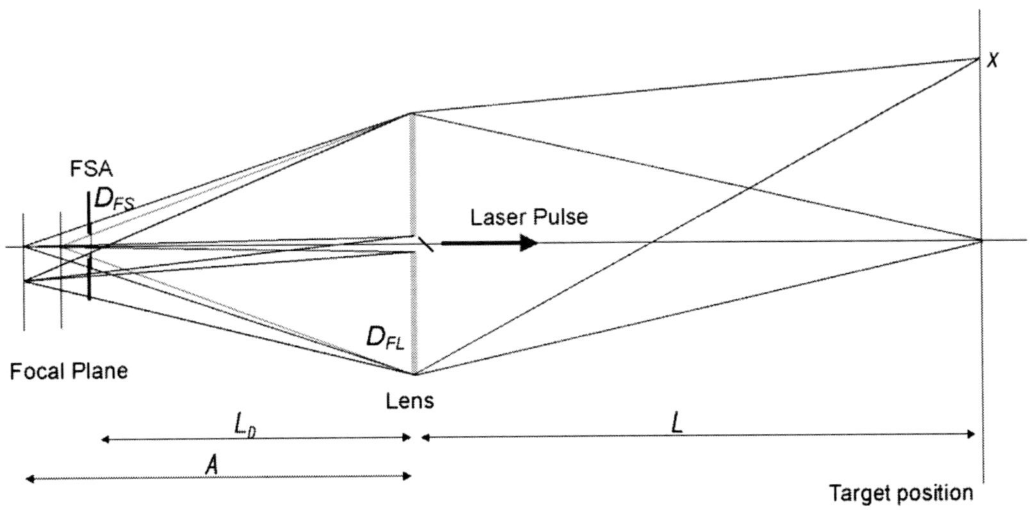

Figure 12.9. Geometric arrangement of lidar optics.

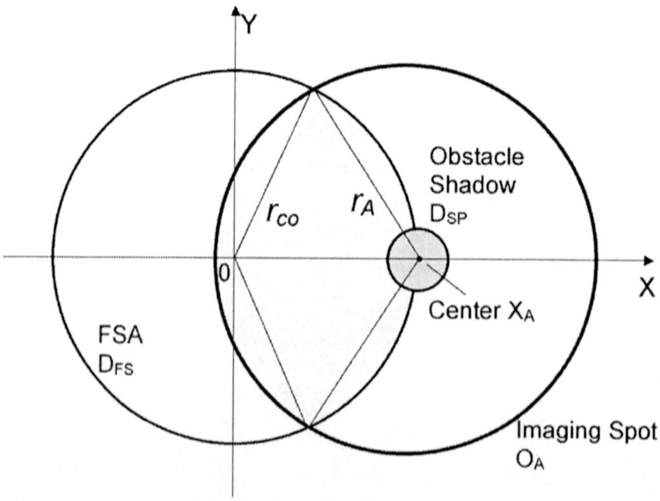

Figure 12.10. Boundary condition of FSA: on-axis view.

geometric form factor becomes the combination of the beam insertion into the FOV and the defocusing effect of the near-range echoes. The parameters in Figure 12.9 are summarized in Table 12.2.

The actual Lidar simulations were performed using the above optical designs (Shiina 2005a, 2005b). Three types of lidar were designed: a coaxial type, an improved coaxial type for near-range detection, and an in-line type. Their specifications are summarized in Table 12.3. The coaxial type and the improved coaxial type of lidar have been put to practical use. The transmitting beam was fired from the center of the telescope, generally on the back side of the secondary mirror. The improved coaxial lidar was designed for near-range echo measurement. The

Table 12.2. Design layout of lidar optics

Main lens	Aperture: D_{FL}
	Focal length: f_{FL}
Obstacle	Diameter: D_{ss}
FSA	Diameter: D_{FS}
Target	Distance: L
	Position : X
Focal distance	$A = L \cdot f_{\mathrm{FL}} / (L - f_{\mathrm{FL}})$
Image size at FSA	Diameter: $O_A = (A - L_D)D_{\mathrm{FL}} / A$ Center position: $X_A = L_D X / L$
Obstacle at FSA	Diameter: $D_{\mathrm{sp}} = (A - L_D)D_{\mathrm{ss}} / A$ Center position: $X_A = L_D X / L$

Table 12.3. Specifications for three types of lidar

(A) Coaxial type	(B) Improved coaxial type	(C) In-line type
Telescope parameters		
Aperture 304 mm^{ϕ}	Aperture 350 mm^{ϕ}	Aperture 304 mm^{ϕ}
Focal length 3048 mm	Focal length 1600 mm	Focal length 3048 mm
Field stop aperture 0.65 mm^{ϕ}	Field stop aperture 1.6 mm^{ϕ}	Field stop aperture 0.65 mm^{ϕ}
Field of view 0.1 mrad	Field of view 0.5 mrad	Field of view 0.1 mrad
Transmitting laser beam parameters		
Intensity distribution Gaussian	Intensity distribution Gaussian	Intensity distribution annular
Wavelength 1.047 μm	Wavelength 1.047 μm	Wavelength 1.047 μm
Power 16 kW	Power 20 kW	Power 16 kW
Pulse width 5 ns	Pulse width 10 ns	Pulse width 5 ns
Diameter 100 mm^{ϕ}	Diameter 100 mm^{ϕ}	Diameter 304 mm^{ϕ}
Spread angle 0.1 mrad	Spread angle 0.3 mrad	Spread angle <15 μrad
Detector parameters		
Responsivity 36 A/W	Responsivity 0.7 mA/W	Responsivity 36 A/W
Quantum efficiency 36%	Quantum efficiency 0.08%	Quantum efficiency 36%
Noise factor 3.2	Noise factor 1.0	Noise factor 3.2
Lidar parameters		
System optical efficiency 0.3	System optical efficiency 0.3	System optical efficiency 0.3
Data acquisition time 50 ns	Data acquisition time 50 ns	Data acquisition time 50 ns

divergence of the transmitting beam and the receiver's FOV were larger than those of the coaxial-type lidar. The detector of the improved coaxial lidar was a PMT, while that of the other lidars was an APD. The intensity of the transmitting beam was nearly equal in these models, while their intensity distributions and aperture sizes were different. First, the geometric form factor $Y(L)$ was examined. The result is shown in Figure 12.11. The $Y(L)$'s variation in the improved coaxial type of lidar increases fastest of the three. This is the reason that the beam divergence and the receiver's FOV are large. The $Y(L)$ of the in-line type lidar increases slowly and needs a longer distance to reach unity. The importance, however, is the receiver's efficiency at near distance for lidar echo estimation. The coaxial-type lidar cannot detect the echo within 0.2 km in the condition. The improved coaxial type also has a blind area.

The variation in $Y(L)$ of the in-line type of lidar never becomes 0. It is higher than that of the other two types of lidar at the nearest distance. This effect is due to the identification between the intensity distribution of the annular beam and the receiver's efficiency distribution in the near range. The lidar echo variations along the propagated distance are shown in Figure 12.12. The atmospheric extinction coefficient α is 1.70×10^{-4} m^{-1}, and the back-scattering coefficient β is 3.4×10^{-6} m^{-1}. That is, $\beta = \alpha / 50$. The usual lidar echo variation along the propagated distance has a peak like that of the coaxial type and the improved coaxial type. The peak is caused by the relation between the geometric form factor $Y(L)$ and the lidar echo intensity $P(L)$ which depend on the propagation distance. On the other hand, the echo variation of the in-line type of lidar has no peak, that is, the lidar

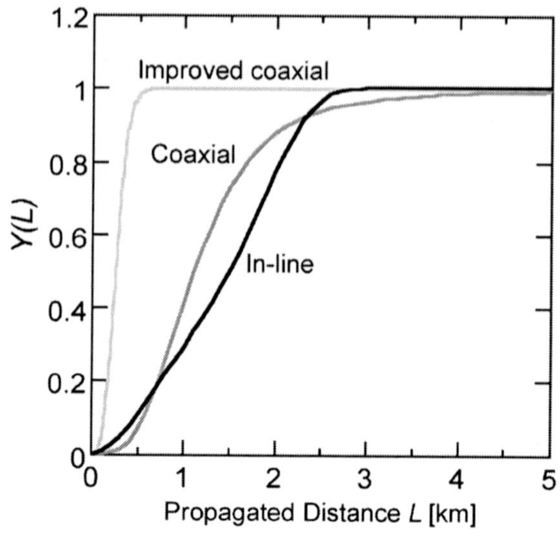

Figure 12.11. Variations of geometric form factors $Y(L)$ with three types of lidar.

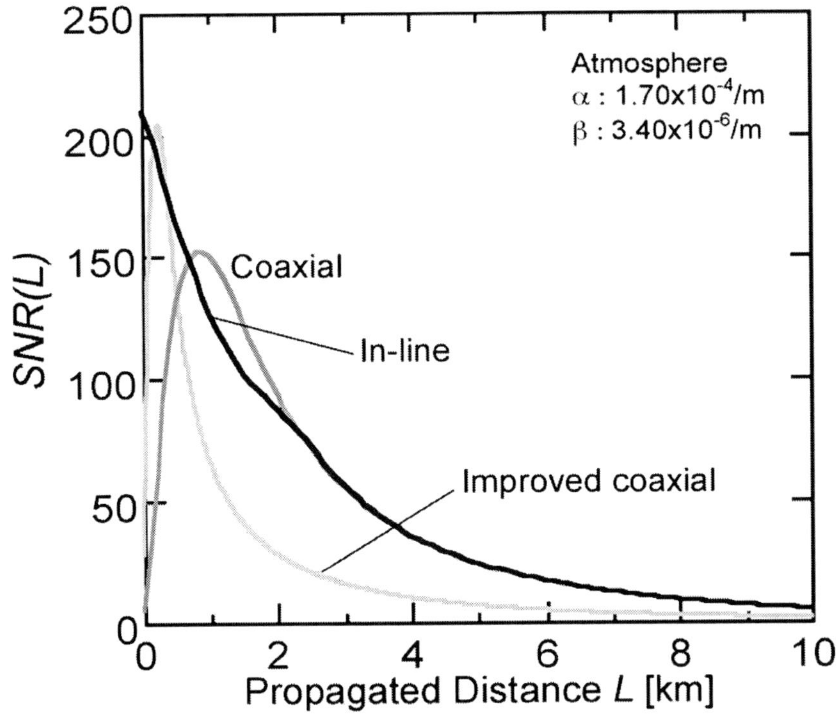

Figure 12.12. Estimates of signal-to-noise ratio with three types of lidar.

echo becomes the maximum at the nearest distance and decreases monotonically along the propagated distance, owing to the efficiency of the geometric form factor in the near range. It is becoming clear when it is indicated in the log-scale graph. Although the improved coaxial type of lidar has a high signal-to-noise ratio in the near range, the ratio decreases rapidly along the further propagated distance in comparison with those of the coaxial and the in-line types of lidar. The improved coaxial type of lidar needs a distance of more than a few kilometers to obtain the ideal echo variation, which is inversely proportional to the square of the propagated distance. The in-line type of lidar has approximately the ideal echo variation from the nearest distance. It is thus easy to analyze the near-range echo without any compensation for the variation of the factor $Y(L)$. The result also indicates that the in-line type of lidar can detect any obstructions (humans, buildings, etc.) without a blind area.

3.3. ANALYSIS

The lidar echo has characteristics of a decreasing function due to the Lambert-Beer law. The far distance echo is influenced by the closer distance atmosphere.

Lidar signal analysis solves the inversion problem. Here the author introduces the slope method, the Klett method, and the Fernald method.

By rewritting the lidar equation [Eq. (12.3)], we obtain

$$\ln\left[P(L)\cdot L^2\right] = \ln\left[P_0 K Y(L) A_r \frac{c\tau}{2}\beta(L)\right] - 2\int_0^L \alpha(x)\,dx \tag{12.5}$$

Here the first term on the right is replaced by C, and when the atmosphere is stable (calm), the integral of the second term is rewritten as

$$\ln\left[P(L)\cdot L^2\right] = C - 2\alpha L \tag{12.6}$$

Equation (12.6) shows that the slope is related to the atmospheric extinction coefficient α and the intercept indicates that the constant depends on the lidar apparatus and the target characteristics. In this case, the atmospheric back-scattering cross section β is also assumed to be constant. An illustration is shown in Figure 12.13. Then the echo corrected by the square of the distance in log scale yields the atmospheric extinction coefficient linearly. The slope is

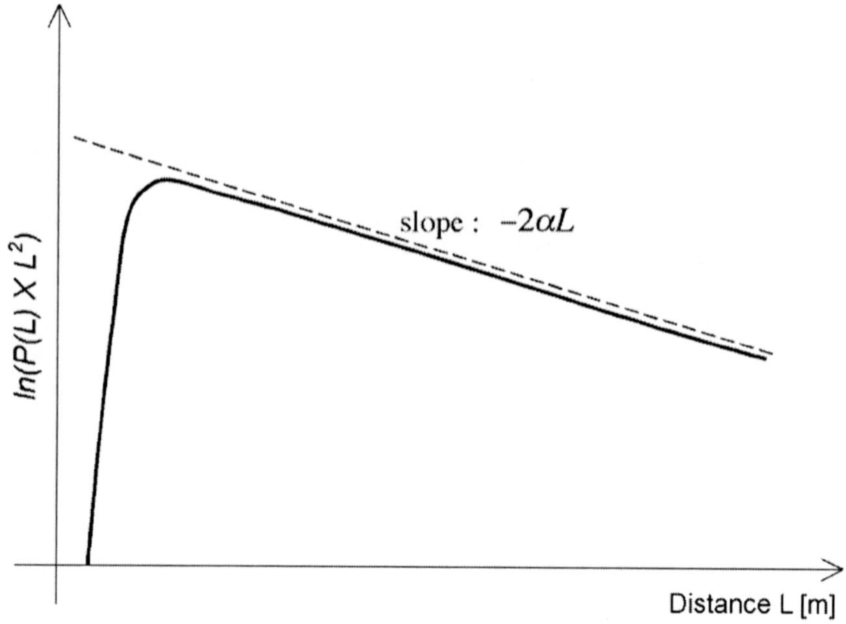

Figure 12.13. Slope method.

proportional to the atmospheric extinction coefficient α. This is known as the slope method. It is adaptable in calm atmosphere conditions, that is, clear days and nights. It is useful to calibrate the lidar system parameters (optical alignment, echo analysis, etc.).

When the atmosphere condition has changes due to the distance (point), the slope method cannot be adapted. There is no solution when the atmospheric conditions at each distance are obtained only in one directional observation (inversion problem). In such a case, the Fernald and Klett mothods are utilized (Davis 1969; Fernald 1972, 1984; Klett 1981). The Fernald method queries each atmospheric extinction coefficient step by step from the initial distance, while the Klett method solves each extinction coefficient step by step from the final distance.

As the distance-corrected echo is rewritten as

$$P(L)\times L^2 = X(L) \tag{12.7}$$

the lidar equation is represented as

$$X(L) = C\beta(L)\exp\left[-2\alpha(L)L\right] \tag{12.8}$$

Here $\beta(L)$ is different from the system constant. To derive the extinction coefficient, differentiation is performed.

$$\frac{dX(L)}{dL} = \frac{1}{\beta(L)}\frac{d\beta(L)}{dL} - 2\alpha(L)X(L) \tag{12.9}$$

If the back-scattering cross section is constant, Eq. (12.9) is equivalent to Eq. (12.6).

The ratio of the extinction cross section α to the back-scattering cross section β is represented as

$$s = \frac{\alpha}{\beta} \tag{12.10}$$

This ratio is called the lidar ratio, and it depends on the distribution of the particle size, features, and incident wavelength.

Then

$$\frac{dX(L)}{dL} = \frac{1}{\alpha(L)}\frac{d\alpha(L)}{dL} - 2\alpha(L)X(L) \tag{12.11}$$

Solving this ordinary differential equation, we obtain

$$\alpha(L) = \frac{X(L)}{[X(L_0)/\alpha(L_0)] - 2\int_{L_0}^{L} X(L')dL'} \tag{12.12}$$

L_0 is the starting point (initial distance) of the calculation. This equation explains that the extinction cross section α at the further distance can be calculated by the closer extinction. The first extinction $\sigma(L_0)$ is obtained by the slope method. This direction, which is away from the starting point, is the forward direction.

Fernald separated the back-scattering cross section of the Mie scattering part (aerosols, cloud particles, etc.) from the Rayleigh scattering part (atmospheric molecules). The ratio of the extinction to the back scattering can be represented as

$$S_1 = \frac{\alpha_1}{\beta_1} \qquad S_2 = \frac{\alpha_2}{\beta_2} \tag{12.13}$$

The subscript numbers identify the Mie scattering part as 1 and Rayleigh scattering part as 2. Fernald then derived the following equation:

$$\beta_1(L) = -\beta_2(L) + \frac{X(L)\exp[-2[(S_1/S_2)-1]\int_{L_0}^{L} \alpha_2(L')dL'}{\{X(L_0)/[\beta_1(L_0)+\beta_2(L_0)]\} - 2S_1\int_{L_0}^{L} X(L')\exp[-2[(S_1/S_2)-1]\int_{L_0}^{L'} \alpha_2(L'')dL'']dL'} \tag{12.14}$$

The parameter S_2 can be estimated by Rayleigh scattering targets as $S_2 = 8.38$. S_1 depends on the characteristics of the Mie scattering targets. It is in the range of 10–100 sr, and $S_1 = 50$ sr is often used as a rough estimate.

The Fernald method is unstable in thick optical depths such as clouds. Klett improved this defect by changing the inversion direction; that is, he put the starting point of the calculation at the farther distance, and obtained the closer extinction cross section.

$$\beta(L) = \frac{X(L)}{[X(L_m)/\beta(L_m)] + 2S\int_{L}^{L_m} X(L')dL'} \tag{12.15}$$

The subscripted characteristic m indicates the end point (final distance) of the calculation. The calculation is performed that from the end point of the calculation. This analysis is more stable for thick optical depths.

4. APPLICATIONS

4.1. MIE LIDAR

Mie lidar is a very basic setup to detect an echo of the same wavelength as the transmitted beam. The targets are aerosols, water vapors, droplets (rain, fog, cloud particles), and so on. By detecting the echoes of the orthogonal polarization components, the depolarization can be examined. Ice crystals, yellow sands, and pollens are distinguished by this depolarization. Multiple scattering of spherical particles also causes the depolarization effect.

The depolarization effect caused by ice crystals is illustrated in Figure 12.14. In general, ice crystals are hexagonal crystals. They flow horizontally in juxtaposition in the air. When the lidar beam is incident from the slant direction, the

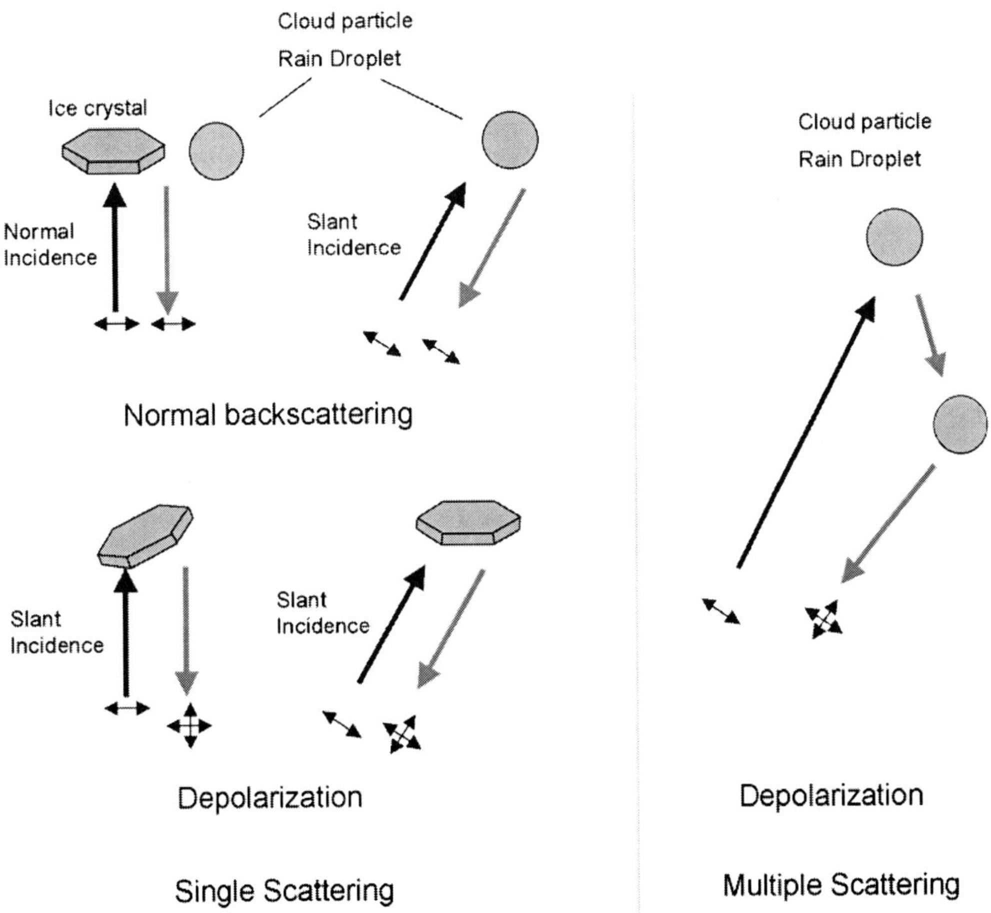

Figure 12.14. Depolarization effect caused by suspended particles.

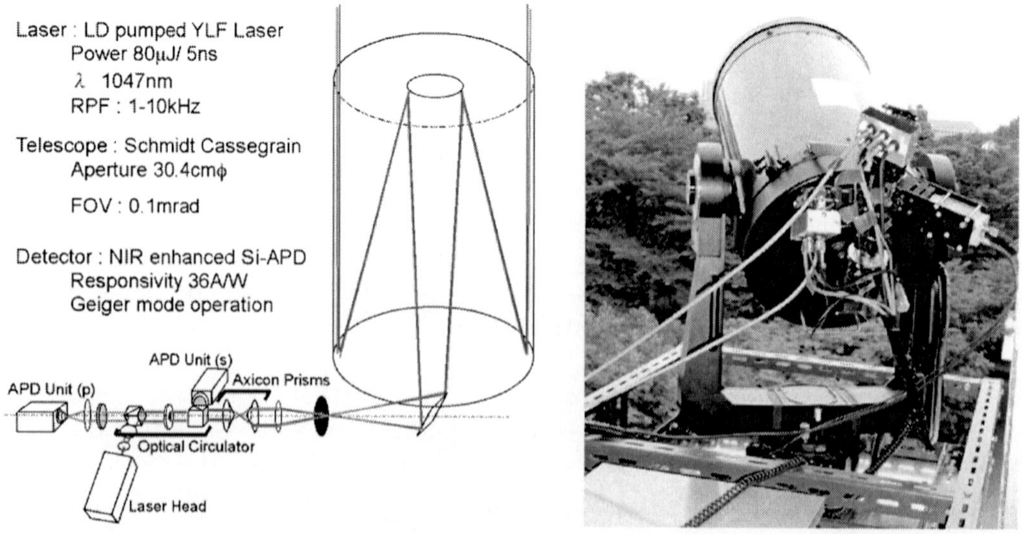

Figure 12.15. In-line type of micro-pulse lidar.

intensities of the back-scattered echoes depend on the Fresnel equations, that is, the echo intensities are different in polarization. The crystal size is small, of the order of a few to 10 μm. The scattering pattern of these nonspherical particles is complex. The depolarization effects of ice clouds and multiple scattering of spherical particles have been examined both theoretically and experimentally (Sasssen 1974, 1986; Allen 1977; Bissonnette 1990; Weigner 1993).

Figure 12.15 shows an in-line type of micro pulse lidar used for the prediction of heavy rain (Shiina 2005a, 2005b). To grasp the sign of the heavy rain, the system can detect near-range echoes. Furthermore, the receiver's field of view is quite narrow, so the depolarization is severely diminished. Nonspherical particles such as ice crystals cause depolarization, while spherical particles such as cloud particles do not cause depolarization under single scattering. The narrow FOV eliminates the multiple scattering echoes. The in-line optics realize near-range detection. The detection range of this device is 10 m to 15 km. The optical circulator and a couple of Axicon prisms were installed into the lidar optics (Shiina 2002). All of the optics, including the laser head and detectors (APDs), were mounted on the telescope tube. Observation is conducted in any direction. The laser power is micro joule energy, and its wave shape is expanded up to the telescope tube. The system is eye-safe. The beam is thrown as a p-component. When depolarization occurs, the echoes will appear at the orthogonal polarization detectors.

Long-term observation of ice clouds is shown in Figure 12.16. The observation was conducted from 22:40 to 4:20 on 24 January 2009. As the direction of the observation was nearly vertical, the cloud base appeared only in the graph of the p-component, because all the ice crystals were arranged in the horizontal direction in the cloud. When the ice crystals fell from the cloud base, they tilted and appeared under the cloud base in both the p- and s-components' graphs. The strong p- and s-echoes were captured at 2:30–3:00. One can clearly see that the ice crystals were falling from the cloud base and reached the ground, and that they melted on the ground because the s-component echo did not touch on the ground.

The optical system was improved by changing the interference filters to high-transparent narrow-bandpass ones and by installing a photon-counting board in place of the oscilloscope. Compensation by the overlap function was performed to evaluate the low-altitude atmosphere. The overlap function derived from the horizontal observation data was compared with that of the simulation. This is shown in Figure 12.17. The results are well matched. The interaction between the cloud and the atmosphere can be visualized by these improvements.

Figure 12.18 shows the observation example of rainfall on 17 October 2009. The observation direction was vertical. The summation was 2000 shots for each observation. The spatial resolution was 7.5 m. The cloud gradually lowered in altitude from 4 km to 3 km during the 2-h observation, while the dense atmosphere gradually gained altitude. The change of the atmosphere coincided with that of humidity. Droplets began to fall several times, at 16:15, 16:40, 17:00, and 17:30. This was captured as lidar echoes at the cloud base. The falling velocities at 17:00 and 17:30 were 5.8 and 8.2 m/s, respectively. The falling velocity was getting faster. This meant that the falling cloud droplets gradually grew and finally became rain droplets. The rise of the dense atmosphere led the rainfall. Rain started to fall on the ground at 17:33, and the observation was stopped at 17:38 because of the hard rain. The observed lidar echoes clearly indicate the interaction between the cloud and the low-altitude atmosphere.

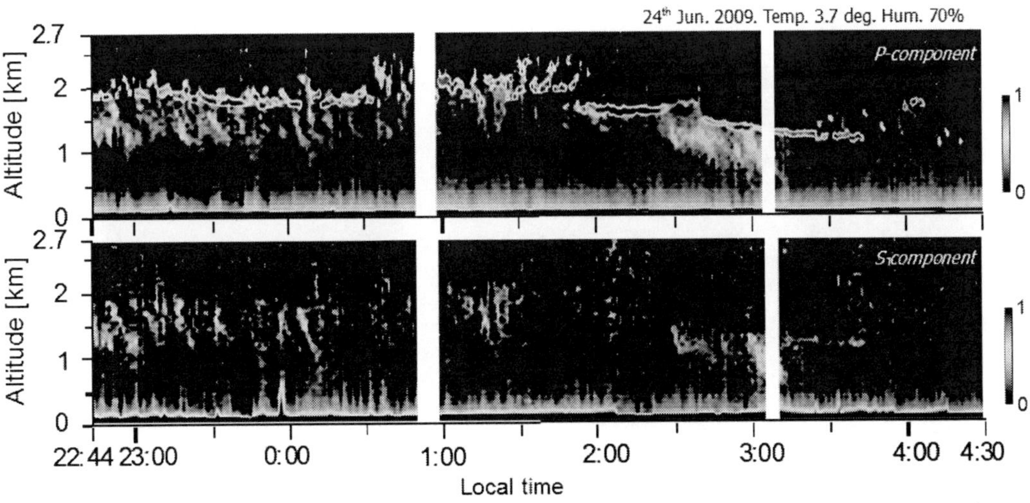

Figure 12.16. Long-term observation of ice clouds.

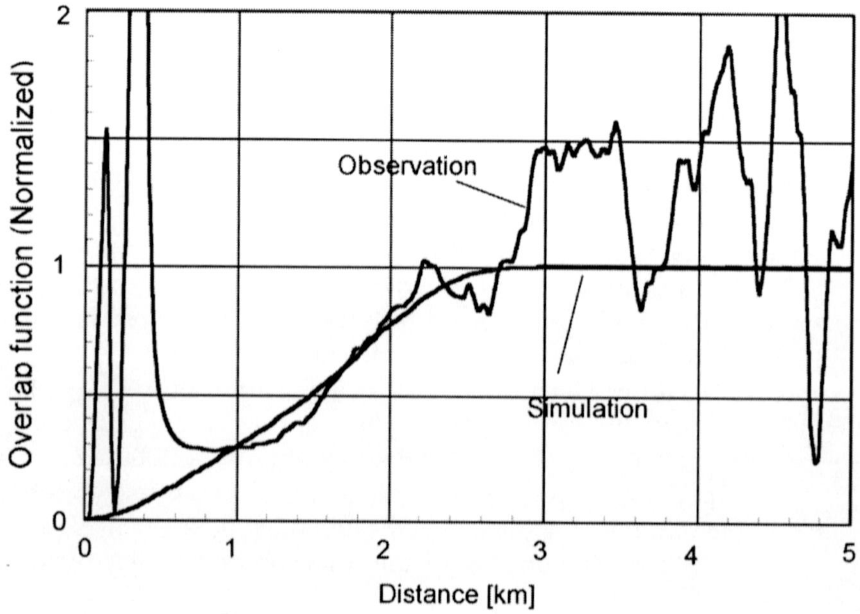

Figure 12.17. Overlap functions.

4.2. RAMAN LIDAR

A practical, compact Raman lidar system for the detection of leaked hydrogen gas has been developed for near-range detection (Ninomiya 2007). The aim is to use it on a mini-car. From a road, one can scan for leakage gas inside a hydrogen

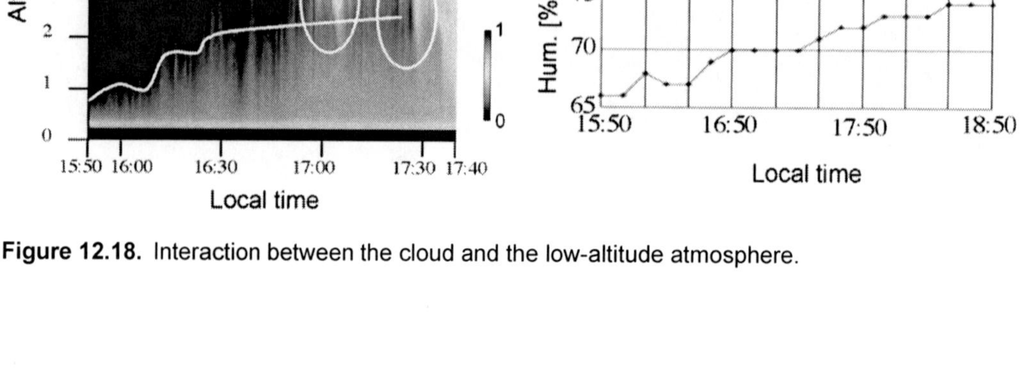

Figure 12.18. Interaction between the cloud and the low-altitude atmosphere.

station. The observation range is 0–50 m. The sensitivity should be higher than detection of a hydrogen concentration of 1%, because hydrogen gas of more than 4% will explode in air (Asahi 2010). To realize near-range detection with high precision as required, inline optics is adequate. Furthermore, the Raman scattering shifts the wavelength of the echo light, and it can be easily separated from the outgoing beam by a dichroic filter. In this study, the minimum requirement for beam power and the optical system size were examined (Noguchi 2010). The optical setup is shown in Figure 12.19. The outgoing laser power is 120 μJ, and it will become sufficiently eye-safe. The receiver's FOV is 2 mrad. A narrower FOV is selectable by using the FSA of the small aperture. The change of the echo characteristics in the difference of the receiver's FOV (FSA) was estimated as shown in Figure 12.20. One can see that the observation range is from 0 to 50 m with low-enough signal-to-noise ratio. When the receiver's FOV is 2 mrad (FSA: 400 μmϕ), the signal-to-noise ratio is different depending on whether the observation is made in daytime or at night. Gas leakage is caused in the gas station, and the observation will be conducted with sufficient ratio under the peaked roof of the station. When the smaller FSA of the diameter of <200 μm is inserted, that is, the receiver's FOV is 0.1 mrad, the echo characteristics become unity in the observation range, whether the observation is conducted during the day or at night. The optical stability, however, will be poor when the smaller FSA is used. In the actual setup, an FSA of diameter 400 μm was selected in a hands-on form because of the beam divergence of the laser source used. The system is also compact, as the size is 580 mm × 520 mm × 230 mm.

The overlap function derived from the nitrogen observation was compared with that of the simulation as shown in Figure 12.21. The actual optical setup of the lidar system had the receiver's FOV of 2.5 mrad., while the simulation set it to 2.0 mrad. The difference in the rise was caused by the difference in the receivers' FOV. Even so, the start of the observation range was less than 2 m.

Figure 12.22 shows one of the observation results (N_2 Raman signal). The N_2 echo rises from 0 m to 50 m, corresponding to the system design. At the near distance of <10 m, there is a large peak, which was caused by the direct reflection from the optics. Then a calibration was conducted by subtracting the direct reflection light (block) from the observation echo (outside). The peak of N_2 signal was caused by the calibration. It will be shortened by decreasing the direct reflected light. The decay curve of the observation echo is inversely proportional to the square of the distance. This is identical to the theoretical decay curve. It is confirmed that the H_2 Raman signal is also detected in the same observation range.

Hydrogen detection was performed using a quartz cell filled with hydrogen gas under concentration control. The result is shown in Figure 12.23. One can see that the Raman intensity is proportional to the concentration of the hydrogen gas. Now the limit of the sensitivity is improved, and it reaches a concentration of less than1%.

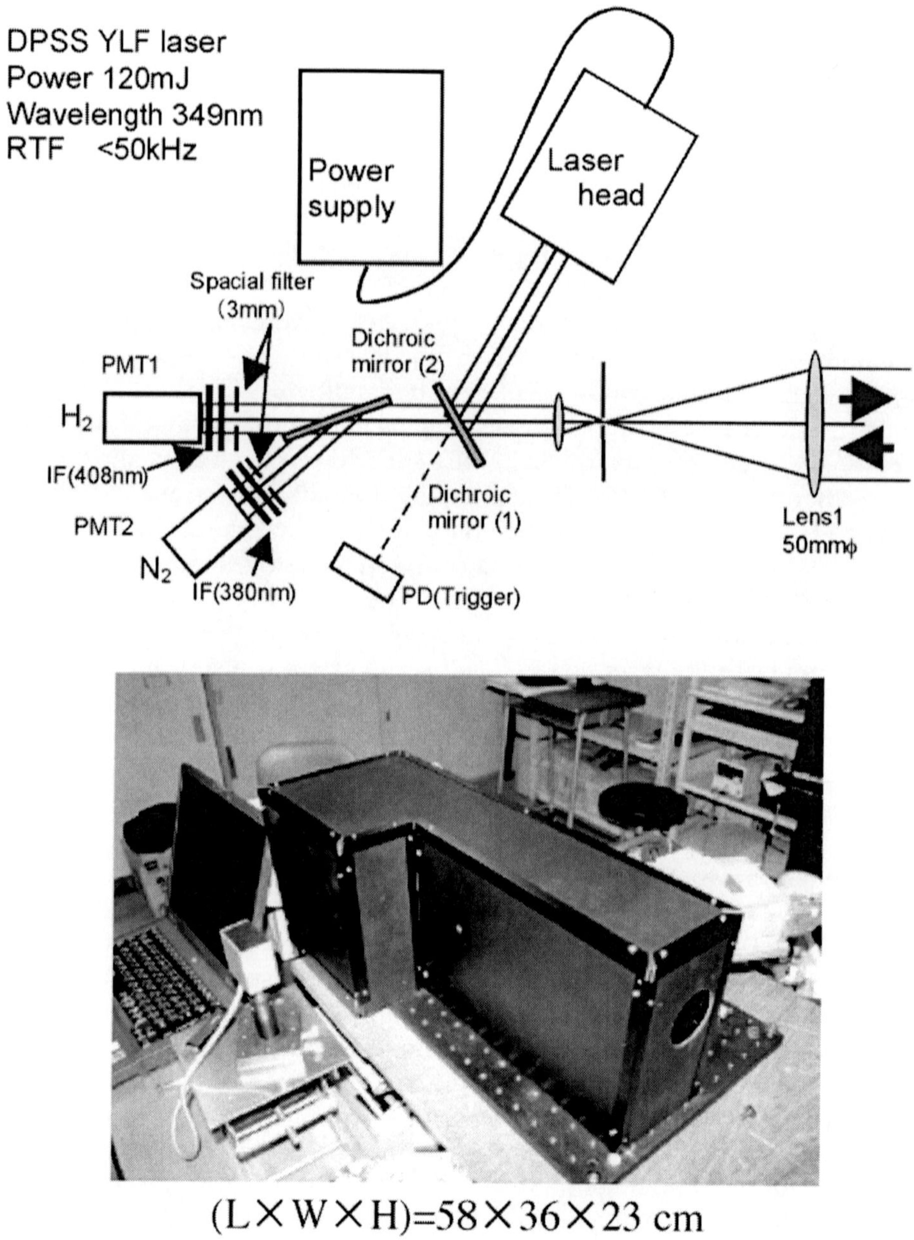

Figure 12.19. Compact Raman lidar for detection of leaked hydrogen.

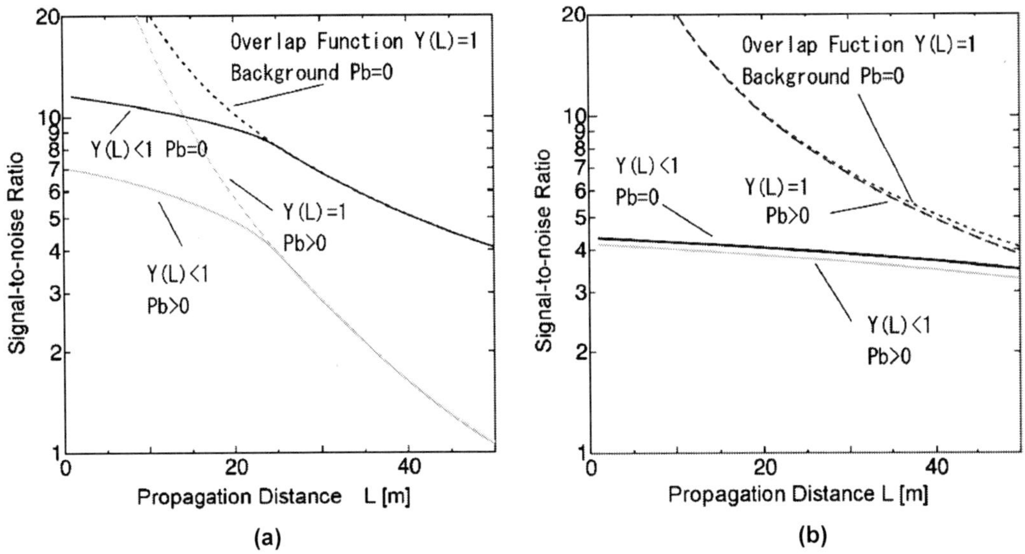

Figure 12.20. Estimate of signal-to-noise ratio for the compact Raman lidar: (a) receiver's FOV = 2 mrad. (FSA: 400 μm); (b) receiver's FOV = 0.1 mrad (FSA: 200 μm)

4.3. LED MINI-LIDAR

The problems limiting wider use of the lidar technique are cost, maintenance, and handling of the lidar system. The laser device is the primary cause. Giant pulse laser systems such as ruby, YAG, and YLF lasers need high electrical power and

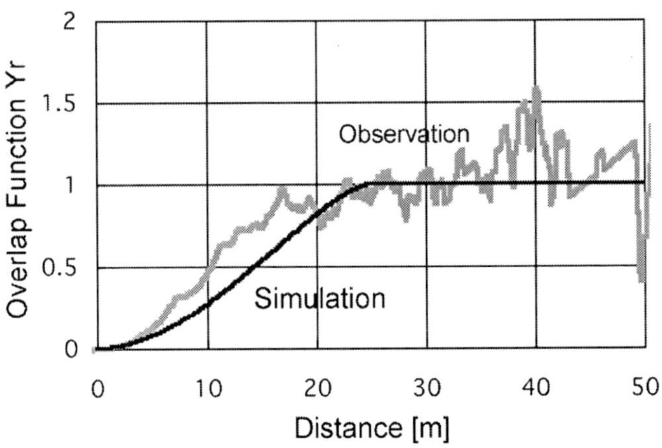

Figure 12.21. Overlap functions of the observation and the simulation.

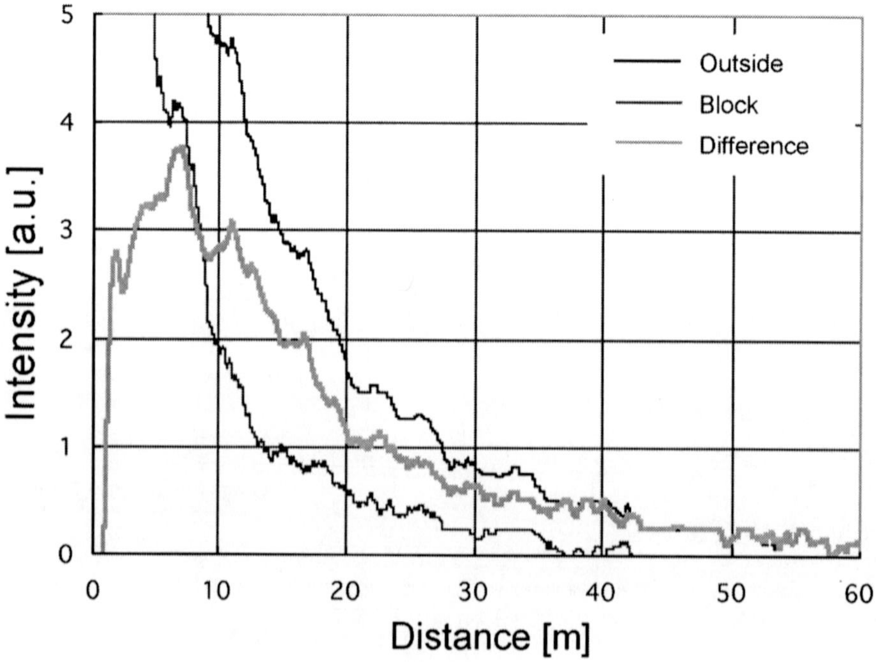

Figure 12.22. Observation of atmospheric nitrogen.

cooling systems. Maintenance is needed to change the flush lamp or to check the beam quality. The quality of the electrical power supply is also important for such a laser system. In developing countries, this often becomes a problem. The laser diode (LD) can also be utilized as a lidar light source. It is easy to use in

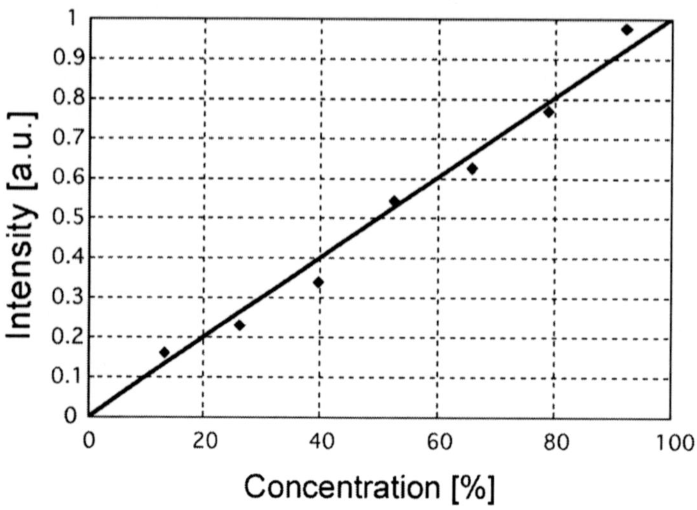

Figure 12.23. Relationship between H_2 gas concentration and lidar echo intensity.

comparison with classical solid-state lasers. However, the laser diode is semsitive to static electricity. It is hard to change by hand. It is also difficult to adjust the optical arrangement with the tiny LD device. Diode pumped solid-state lasers (DPSS lasers) are compact and easy to use, but temperature control is essential and there is no way to change the LD by oneself when the output power is reduced.

For increased usability, an LED mini-lidar has been proposed and developed for trial (Koyama 2011). LEDs (light-emitting diodes) are quite cheap, resistant to static electricity, and have wide selectivity of wavelength. Available optical power is improving year by year. As a concept lidar for near-range Mie lidar, a simulation was conducted wit;h an observation range of 0–300 m and an optical power of 100 mW of the pulse width of 10 ns. The estimated signal-to-noise ratio is shown in Figure 12.24. The result indicates that the tiny peak power (100 mW = 1 nJ/10 ns) will have enough margin on the ratio in the observation range. The optical system for an LED mini-lidar should be simple and compact. No particular optical arrangement is better. For this example, the LED mini-lidar has coaxial optics with a common main Fresnel lens as both transmitter and receiver. It is shown in Figure 12.25. The transmitting beam is collimated by a combination of the self-ramp lens and the center of the Fresnel lens. However, it is hard to collimate the LED beam perfectly. In this system, the divergence of the LED beam is 9.5 mrad, and the receiver's FOV is 10 mrad. This is wider than that of the usual lidar with a laser beam. To obtain the pulse characteristics for the LED lidar, a special driver was designed (Koyama 2011). The pulse repetition frequency is more than

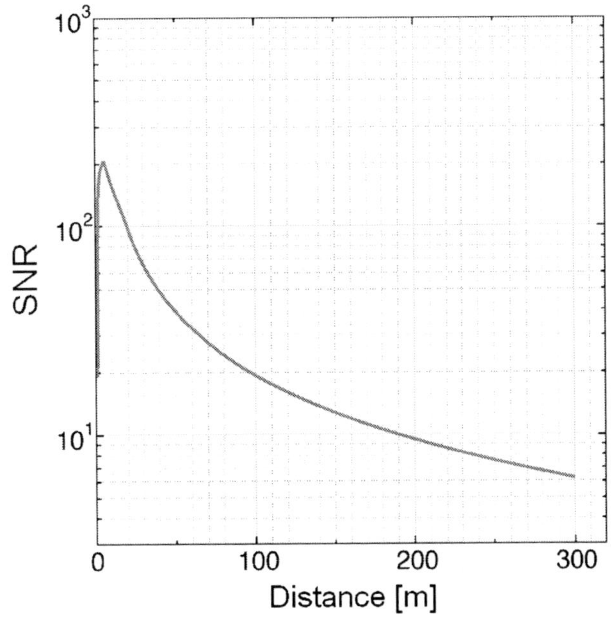

Figure 12.24. Simulation result for LED mini-lidar.

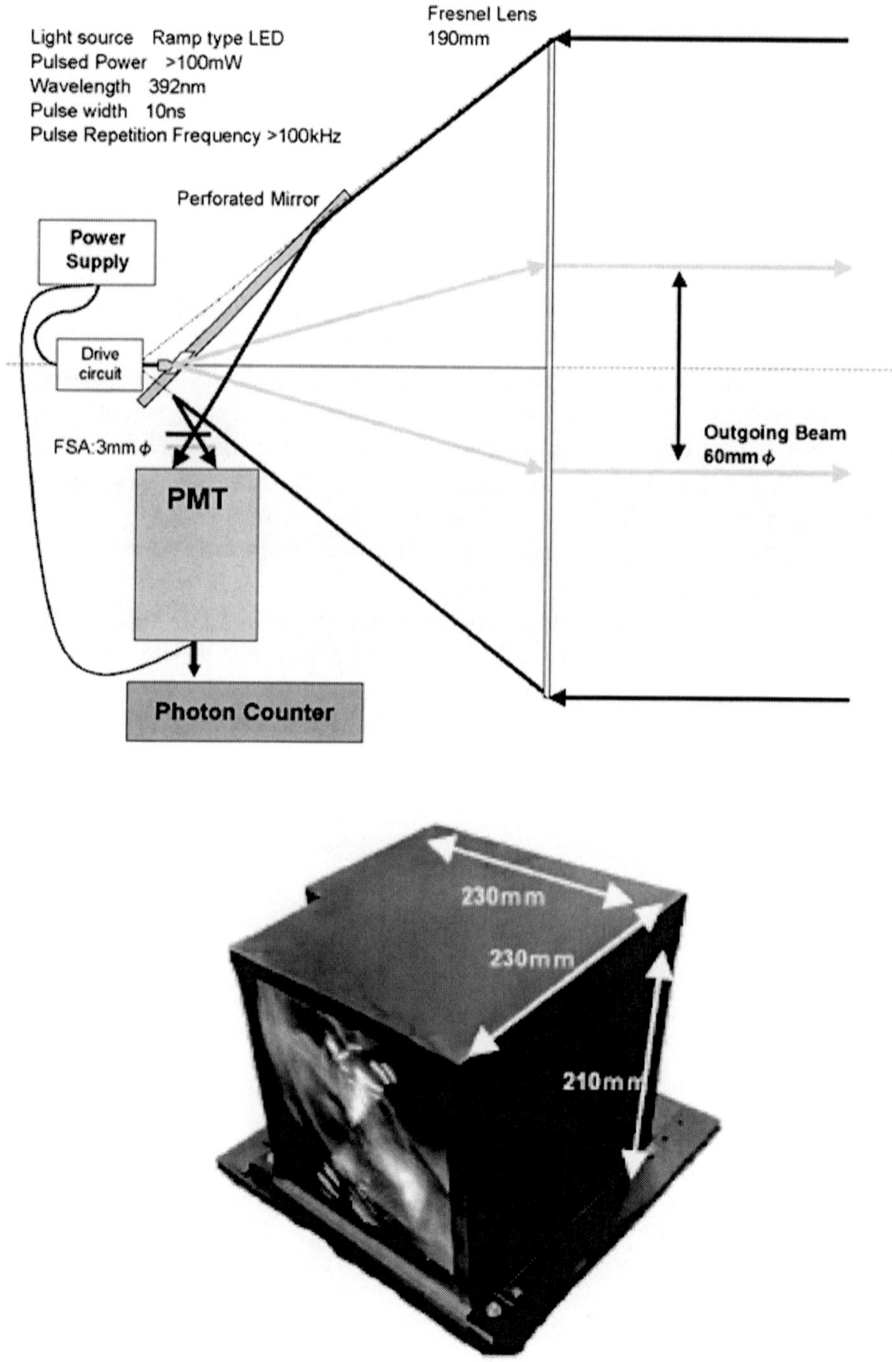

Figure 12.25. LED mini-lidar.

100 kHz. A blue color LED was used for the first trial. The echoes are reflected by a perforated mirror and detected by a photomultiplier. The main lens is common, and the transmitting beam is always within the receiver's field of view. The system size is a cube of 20 cm on a side. The LED and PMT drivers are placed inside this cube.

Observation results of the near-range atmosphere are shown in Figure 12.26. Thanks to the wide field of view, the observation range starts from almost 0 m. The atmospheric echoes were obtained on both clear and rainy days. At least, their range was more than 100 m in the horizontal direction. The peak echoes at 55 m and at 80–120 m come from trees. One can see the difference in the atmospheric echoes between the clear and rainy conditions. This is due to the suspended particles in the atmosphere, that is, aerosols or raindrops. The LED mini-lidar can be used for air monitoring in closed spaces such as mines, stadiums, and factories. It will be of benefit to monitor the flow of smoke from exhaust pipes or quantitative examination of air pollution, in particlar the concentration of black carbons, at roadway intersections.

With the system as originally designed, it was hard to make observations in the daytime because of the wide FOV. Therefore, a second system was developed to observe the atmosphere in both daytime and nighttime. It has been confirmed that it is possible to narrow the FOV with a large lens up to the same order with the usual lidar with a laser device. Now the high-speed photon counter is designed and fabricated to speed up the summation to follow the pulse repetition frequency of the LED lidar.

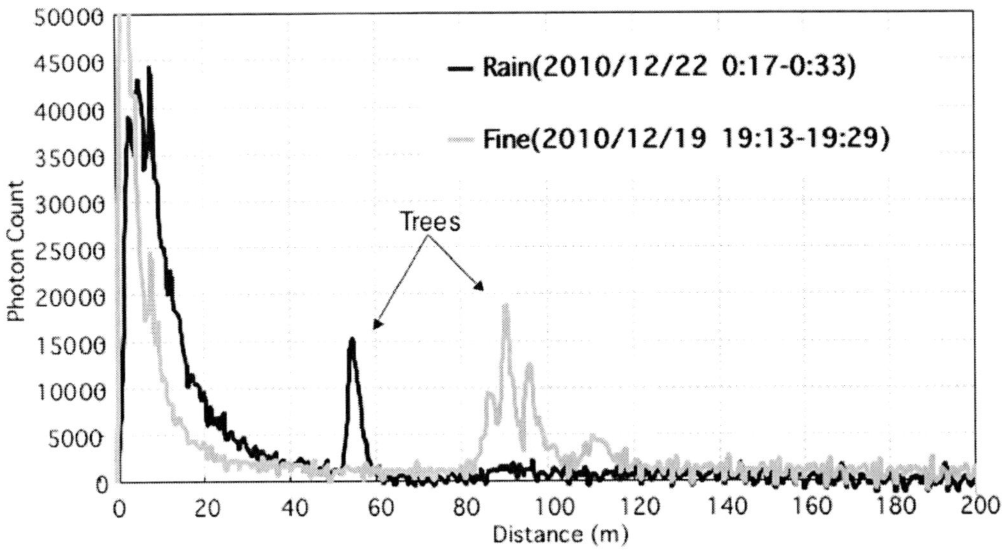

Figure 12.26. Atmosphere observations with LED mini-lidar.

5. SUMMARY

In this chapter, optical designs, simulations and practical applications of lidar systems have been explained. Lidar uses a giant pulse, and its optics require special care. The optical design and simulation are important to accomplish a foolproof strategy. In-line optics are adequate for near-range detection with a narrow field of view. Lidar systems have a history of 50 years. The techniques are various and these systems can be installed for use in a variety of fields, including not only meteorology, but also chemistry, biology, and antidisaster measures.

The future of lidar techniques will trend toward making systems more compact and easy to use. Laser devices are improving year by year, but they still have weaknesses in terms of use on-site. They need high electrical power, cooling, and maintenance. The device size is also large. To make lidar systems user-friendly, they need to be compact and robust to operate. One of the potential solutions is LED mini-lidar. These systems are quite compact, robust, and tough, and thus LED mini-lidars should become widely used in various fields in the near future.

REFERENCES

Allen R.J. and Platt C.M. (1977) Lidar for multiple backscattering and depolarization observations. *Appl. Opt.* **16**, 3193–3199. DOI: 10.1364/AO.16.003193

Ansmann A., Bösenberg J., Chaikovsky A.P., Comerón A., Eixmann R., Freudenthaler V., Ginoux P., Konguem L., Linné H., Márquez M.A.L., Manoj S., Matthias V., Mattis I., Mitev V., Müller D., Nickovic S., Pelon J., Sauvage L., Sobolewsky P., Stohl A., Torres O., Vaughan G., Wandihger U., and Wiegner M. (2003) Long-range transport of Saharan dust to northern Europe: 11–16 October 2001 outbreak with EARLINET. *J. Geophys. Res.* **108**, 4783. DOI: 10.1029/2003JD003757

Asahi I., Ninomiya H., and Sugimoto S. (2010) Remote sensing of hydrogen concentration by low power laser. *IEEJ Trans. Elec. Info. Sys.* **130**, 1145–1150 (in Japansese). DOI: 10.1541/ieejeiss.130.1145

Baars H., Ansmann A., Engelmann R., and Althausen D. (2008) Continuous monitoring of the boundary-layer top with lidar. *Atmos. Chem. Phys.* **8**, 7281–7296. DOI: 10.5194/acp-8-7281-2008

Bissonnette L.R. and Hutt D.L. (1990) Multiple scattering lidar. *Appl. Opt.* **29**, 5045–5046. DOI: 10.1364/AO.29.005045

Bohren C.F. and Huffman D.R. (1983) *Absorption and Scattering of Light by Small Particles.* John Wiley, New York.

Born M. and Wolf E. (1980) *Principles of Optics,* 6th ed. Pergamon Press, Oxford, UK.

Chaikovsky A., Ivanov A., Balin Y., Elnikov A., Tulinov G., Plusnin I., Bukin O., and Chen B. (2006) CIS-LINET—Lidar network for monitoring aerosol and ozone in CIS regions. In: *Proceedings of 23rd International Laser Radar Conference*, 24–28 June 2006, Nara, Japan, Part II, 671–672.

Chourdakis G., Papayannis A., and Porteneuve J. (2002) Analysis of the receiver response for a noncoaxial lidar system with fiber-optic output. *Appl. Opt.* **41**, 2715–2723. DOI: 10.1364/AO.41.002715

Davis P.A. (1969) The analysis of lidar signatures of cirrus clouds. *Appl. Opt.* **8**, 2099--2102. DOI: 10.1364/AO.8.002099

Deirmendjian D. (1969) *Electromagnetic Scattering on Spherical Polydispersions.* Elevier, New York.

Fernald F.G., Herman B.M., and Reagan J.A. (1972) Determination of aerosol height distributions by lidar. *J. Appl. Meteor.* **11**, 482–489. DOI: 10.1175/1520-0450(1972)011<0482:DOAHDB>2.0.CO;2

Fernald F.G. (1984) Analysis of atmospheric lidar observation. *Appl. Opt.* **23**, 652–653. DOI: 10.1364/AO.23.000652

Fiocco G. and Smullin L.D. (1963) Detection of scattering layers in the upper atmosphere (60-140km) by optical radar, *Nature* **199**, 1275–1276. DOI: 10.1038/1991275a0

Fujii T. and Fukuchi T. (eds) (2005) *Laser Remote Sensing.* CRC Press, Boca Raton, FL.

Halldorsson T. and Langerholc J. (1978) Geometrical form factors for the lidar function. *Appl. Opt.* **17**, 240–244. DOI: 10.1364/AO.17.000240

Harms J. (1979) Lidar return signals for coaxial and noncoaxial systems with central obstruction. *Appl. Opt.* **18**, 1559–1566. DOI: 10.1364/AO.18.001559

Herschel J.F.W. (1845) On a case of superficial colour presented by a homogeneous liquid internally colourless. *Phil. Trans. Roy. Soc. (Lond.)* **135**, 143–145. DOI: 10.1098/rstl.1845.0004

Hoff R.M., Bösenberg J., and Pappalardo G. (2008) The GAW aerosol lidar observation network (GALION). In: *Proceedings of IGARSS Meeting,* 6–11 July 2008, Boston, 1775.

Hulst H.C. (1981) *Light Scattering by Small Particles.* Dover, New York.

Klett J.D. (1981) Stable analytical inversion solution for processing lidar returns. *Appl. Opt.* **20**, 211–220. DOI: 10.1364/AO.20.000211

Kovacs T.A. and McCormick M.P. (2003) Observations of typhoon Melissa during the Lidar In-space Technology Experiment (LITE). *J. Appl. Meteor.* **42**, 1003–1013. DOI: 10.1175/1520-0450(2003)042<1003:OOTMDT>2.0.CO;2

Koyama M. and Shiina T. (2011) Development of LED mini-lidar. In: *Proceedings of CLEO-Pacific Rim,* 28 Aug.–1 Sept. 2011, Sydney, Australia, 544–545.

Koyama M. and Shiina T. (2011) Light source module for LED mini-lidar. *Rev. Laser Eng.* **39**, 617–621 (in Japanese).

Leonard D.A. (1967) Observation of Raman scattering from the atmosphere using a pulsed nitrogen ultraviolet laser. *Nature* **216**, 142–143. DOI: 10.1038/216142a0

Measures R.M. (1984) *Laser Remote Sensing—Fundamentals and Applications.* John Wiley, New York.

Melfi S.H. (1972) Remote measurement of the atmosphere using Raman scattering. *Appl. Opt.* **11**, 1605–1610. DOI: 10.1364/AO.11.001605

Mie G.A. (1908) Beiträge zur Optik trüber medien, speziell kolloidaler Metallösungen. *Annalen Phys.* **25**, 377–445. DOI: 10.1002/andp.19083300302

Murray E.R. and van der Laan J.E. (1978) Remote measurement of ethylene using a CO_2 differential-absorption lidar. *Appl. Opt.* **17**, 814–817. DOI: 10.1364/AO.17.000814

Ninomiya H. Yaeshima S., Ichikawa K., and Fukuchi T. (2007) Raman lidar system for hydrogen gas detection. *Opt. Eng.* **46**, 094301. DOI: 10.1117/1.2784757

Noguchi Y. Miya H. Shiina T., Noguchi K., Fukuchi T., Asahi I., Sugimoto S., Shimamoto Y., and Ninomiya H. (2010) Compact Raman lidar for hydrogen gas leak detection. In: *Proceedings of 25th ILRC,* 5–9 July 2010, Saint-Petersburg, Russia, 178–181.

Raman C.V. and Krishman K.S. (1928) A new type of secondary radiation. *Nature* **121**, 501–502. DOI: 10.1038/121501c0

Rayleigh L. (1871) On the light from the sky, its polarization and colour. *Phil. Mag.* **12**, 107–120, 274–279. DOI: 10.1080/14786447108640452

Sassen K. (1974) Depolarization of lidar light backscattered by artificial clouds. *J. Appl. Meteorol.* **13**, 921–929. DOI: 10.1175/1520-0450(1974)013<0923:DOLLBB>2.0.CO;2

Sassen K. and Petrilla R.L. (1986) Lidar depolarization from multiple scattering in marine stratus clouds. *Appl. Opt.* **25**, 1450–1459. DOI: 10.1364/AO.25.001450

Shiina T., Minami E., Ito M., and Okamura Y. (2002) Optical circulator for an in-line type compact lidar. *Appl. Opt.* **41**, 3900–3905. DOI: 10.1364/AO.41.003900

Shiina T., Yoshida K., Ito M., and Okamura Y. (2005a) In-line type micropulse lidar with an annular beam: Experiment. *Appl. Opt.* **44**, 7407–7413. DOI: 10.1364/AO.44.007407

Shiina T., Yoshida K., Ito M., and Okamura Y. (2005b) In-line type micropulse lidar with annular beam—Theoretical approach. *Appl. Opt.* **44**, 7467–7473. DOI: 10.1364/AO.44.007467

Shipley S.T., Tracy H., Eloranta E.W., Trauger J.T. Sroga J.T., Roesler F.L., and Weinman J.A. (1983) High spectral resolution lidar to measure optical scattering properties of atmospheric aerosols. 1: Theory and instrumentation. *Appl. Opt.* **22**, 3716–3724. DOI: 10.1364/AO.22.003716

Sicard M., Molero F., Guerrero-Rascado J.L., Pedrós R., Expósito F.J., Córdoba-Jabonero C., Bolarín J.M., Comerón A., Rocadenbosch F., Pujadas M., Alados-Arboledas L., Martínez-Lozano J.A., Díaz J.P., Gil M., Requena A., Navas-Guzmán F., and Moreno J.M. (2009) Aerosol lidar intercomparison in the framework of SPALINET—The Spanish lidar network: Methodology and results. *IEEE Trans. Geo Rem.* **47**, 3547–3559. DOI: 10.1109/TGRS.2009.2021525

Sugimoto N., Matsui I., and Sasano Y. (1990) Design of lidar transmitter-receiver optics for lower atmospheric observations: Geometrical form factor in lidar equation. *J. Jpn. Opt.* **19**, 687–693.

Sugimoto N., Shimizu A., Matsui I., Dong X., Zhou J., Bai X., Zhou J., Lee C.H., Yoon S.C., Okamoto H., and Uno I. (2006) Network observations of Asian dust and air pollution aerosols using two-wavelength polarization lidars, In: *Proceedings of 23rd International Laser Radar Conference,* 24–28 June 2006, Nara, Japan, 851–854.

Weitkamp C. (2005) *Lidar: Range-Resolved Optical Remote Sensing of the Atmosphere.* Springer-Verlag, New York, chap 5.

Welton E.J., Campbell J.R., Spinhirne J.D., and Scott V.S. (2002) Measurements of aerosol vertical profiles and optical properties during INDOEX 1999 using micro-pulse lidars. *J. Geophys. Res.* **107**, 8019. DOI: 10.1029/2000JD000038

Wiegner M. and Echle G. (1993) Lidar multiple scattering: Improvement of Bissonnette's paraxial approximation. *Appl. Opt.* **32**, 6789–6803. DOI: 10.1364/AO.32.006789

Index

G

H

P

S

W

Z

CPSIA information can be obtained at www.ICGtesting.com
Printed in the USA
BVOW021630250113

310843BV00003BA/3/P